I0047589

Marine Ecology: Current and Future Developments

(Volume 1)

Marine Pollution: Current Status, Impacts and Remedies

Edited by

De-Sheng Pei & Muhammad Junaid

Research Center for Environment and Health,
Chongqing Institute of Green and Intelligent Technology,
Chinese Academy of Sciences, Chongqing, China

Marine Ecology: Current and Future Developments

Volume # 1

Marine Pollution: Current Status, Impacts, and Remedies

Editors: De-Sheng Pei & Muhammad Junaid

ISBN (Online): 9789811412691

ISBN (Print): 9789811412684

© 2019, Bentham eBooks imprint.

Published by Bentham Science Publishers Pte. Ltd. Singapore. All Rights Reserved.

BENTHAM SCIENCE PUBLISHERS LTD.
End User License Agreement (for non-institutional, personal use)

This is an agreement between you and Bentham Science Publishers Ltd. Please read this License Agreement carefully before using the ebook/echapter/ejournal (**"Work"**). Your use of the Work constitutes your agreement to the terms and conditions set forth in this License Agreement. If you do not agree to these terms and conditions then you should not use the Work.

Bentham Science Publishers agrees to grant you a non-exclusive, non-transferable limited license to use the Work subject to and in accordance with the following terms and conditions. This License Agreement is for non-library, personal use only. For a library / institutional / multi user license in respect of the Work, please contact: permission@benthamscience.net.

Usage Rules:

1. All rights reserved: The Work is the subject of copyright and Bentham Science Publishers either owns the Work (and the copyright in it) or is licensed to distribute the Work. You shall not copy, reproduce, modify, remove, delete, augment, add to, publish, transmit, sell, resell, create derivative works from, or in any way exploit the Work or make the Work available for others to do any of the same, in any form or by any means, in whole or in part, in each case without the prior written permission of Bentham Science Publishers, unless stated otherwise in this License Agreement.
2. You may download a copy of the Work on one occasion to one personal computer (including tablet, laptop, desktop, or other such devices). You may make one back-up copy of the Work to avoid losing it.
3. The unauthorised use or distribution of copyrighted or other proprietary content is illegal and could subject you to liability for substantial money damages. You will be liable for any damage resulting from your misuse of the Work or any violation of this License Agreement, including any infringement by you of copyrights or proprietary rights.

Disclaimer:

Bentham Science Publishers does not guarantee that the information in the Work is error-free, or warrant that it will meet your requirements or that access to the Work will be uninterrupted or error-free. The Work is provided "as is" without warranty of any kind, either express or implied or statutory, including, without limitation, implied warranties of merchantability and fitness for a particular purpose. The entire risk as to the results and performance of the Work is assumed by you. No responsibility is assumed by Bentham Science Publishers, its staff, editors and/or authors for any injury and/or damage to persons or property as a matter of products liability, negligence or otherwise, or from any use or operation of any methods, products instruction, advertisements or ideas contained in the Work.

Limitation of Liability:

In no event will Bentham Science Publishers, its staff, editors and/or authors, be liable for any damages, including, without limitation, special, incidental and/or consequential damages and/or damages for lost data and/or profits arising out of (whether directly or indirectly) the use or inability to use the Work. The entire liability of Bentham Science Publishers shall be limited to the amount actually paid by you for the Work.

General:

1. Any dispute or claim arising out of or in connection with this License Agreement or the Work (including non-contractual disputes or claims) will be governed by and construed in accordance with the laws of Singapore. Each party agrees that the courts of the state of Singapore shall have exclusive jurisdiction to settle any dispute or claim arising out of or in connection with this License Agreement or the Work (including non-contractual disputes or claims).
2. Your rights under this License Agreement will automatically terminate without notice and without the

need for a court order if at any point you breach any terms of this License Agreement. In no event will any delay or failure by Bentham Science Publishers in enforcing your compliance with this License Agreement constitute a waiver of any of its rights.

3. You acknowledge that you have read this License Agreement, and agree to be bound by its terms and conditions. To the extent that any other terms and conditions presented on any website of Bentham Science Publishers conflict with, or are inconsistent with, the terms and conditions set out in this License Agreement, you acknowledge that the terms and conditions set out in this License Agreement shall prevail.

Bentham Science Publishers Pte. Ltd.
80 Robinson Road #02-00
Singapore 068898
Singapore
Email: subscriptions@benthamscience.net

BENTHAM SCIENCE

CONTENTS

Bhumi K. Sachaniya, Haren B. Gosai, Haresh Z. Panseriya, Anjana K. Vala
Bharti P. Dave

FOREWORD

This book, edited by De-Sheng Pei and Muhammad Junaid, emphasizes that the oceans are a vast but fragile resource that must be protected if we want to protect our livelihoods and our planet. Although marine pollution is a topic of concern for a long period of time, it has recently attracted the significant attention of scientific and non-scientific debate circles, including environmentalists, economists, and politicians. The chapters on methods to assess pollution provide important information for identifying, measuring, and remediating various pollutants, while the chapters on known pollutants and their management point out how widespread the problems are and how intense international effort is required to resolve the problems.

Besides providing food, transportation and lifestyle resources, the oceans serve as a vast sink to absorb increases in global heat, mitigating at least temporarily more extreme changes in global climate. But in doing so, oceans also present a threat to coastal communities by altering local weather patterns and disrupting local livelihoods with changes in acidity and temperature.

This book will prove to be a useful resource for students, researchers, and policymakers, who are working on the management and protection of the world's valuable marine resources and environment.

Phyllis R. Strauss, Ph.D.
Matthews Distinguished Professor
Northeastern University,
Boston, MA 02115,
USA

PREFACE

There are increasing environmental concerns about the current status of the world's oceans. The rapid development of industrial zones and growth of human population in coastal areas have led to exploitation of marine resources resulting in chemical pollution from industry, domestic wastewater intrusion, invasion of non-native species, toxic algal bloom, and microbial pathogens. On the other hand, earth's oceans offer abundant food resources, easy shipping, and coastal living. In this book, the experts from different countries in Asia, Europe, and America give their overviews and opinions about the current status of marine pollution, environmental impacts, and possible remedies.

Introductory Chapter 1 highlights the overall theme of this book: the importance of oceans in the 21st century. This chapter orderly presents an overview of pollution dynamics including inorganic pollutants (heavy metals, metalloids), organic pollutants (POPs-persistent organic pollutants, PAHs-polycyclic aromatic hydrocarbons, and PCBs-polychlorinated biphenyls), microplastics, and algal blooms in the marine environment. The second section specifically introduces the negative impacts of marine pollution and assessment methods to highlight the toxicity of marine pollutants. The last section of Chapter 1 is an overview of various remedial techniques, such as bioremediation, phytoremediation, and the challenges related to marine pollution. Chapter 2 describes common sampling procedures for the most diverse and abundant marine organisms that comprise ecosystem components under the Essential Ocean Variables (EOVs), such as phytoplankton, zooplankton, and fish. In this framework, biodiversity is assessed based on the status of ecosystem components, including phytoplankton biomass and diversity, zooplankton biomass and diversity, fish abundance and distribution, as well as marine turtle, bird and mammal abundance and distribution.

Chapters 3 & 4 highlight the important reactions of metals and non-metals with inorganic and organic constituents in marine water and sediments. In addition to these reactions, Chapter 3 also covers biokinetic aspects of two major marine environmental problems: eutrophication and the release of organotin compounds and copper from antifouling paints used on ships' hulls, as an example of the effects of uncontrolled introductions of metals and non-metals on marine ecosystems. Chapter 4 highlights natural and anthropogenic sources of metals and non-metals, as well as their toxicity and accumulation in different marine organisms. Chapter 5 discusses pollution dynamics of organic contaminants and associated impacts in marine ecosystems. These contaminants include persistent organic pollutants (POPs), such as pesticides, brominated flame retardants, perfluoroalkyl compounds, fluorotelomer alcohols, perfluoroalkyl sulfonic acids (PFSAs), perfluorocarboxylic acids (PFCAs), fluorotelomer carboxylic acids, fluorotelomer sulfonic acids, and fluorinated polymers. Apart from POPs, microplastics and accidental oil spills are also highlighted in terms of their growing concern in oceanic gyres. Chapter 6 explores monitoring of organic pollutants in the marine ecosystem, including fate, distribution, and behavior of PCBs, as well as uptake of organic contaminants/PCBs by marine organisms.

Chapter 7 describes pollution dynamics along the Pakistan coast with special reference of nutrient pollution. In this Chapter, the magnitude of pollution (organic and inorganic) in coastal environments of Pakistan is discussed including plastic pollution, and enrichment of macro-nutrients in coastal waters leading to the explosion in frequency of harmful algal blooms. Chapter 8 explores ecotoxicology of heavy metals in marine fish. The authors review the occurrence and chemistry of heavy metals in the marine environment, as well as the bioaccumulation and toxicity of heavy metals in marine fish. Chapter 8 also summarizes the public health risks due to the consumption of heavy metals' contaminated fish. Chapter 9 highlights the effects of microplastic on the marine ecosystem. Further, several aspects related to research gaps for the management of microplastic waste are proposed. Chapter 10 explores methods to measure toxicity in flora and fauna exposed to different categories of marine pollutants, their sources, various exposure routes, and associated toxicological impacts on marine organisms. Chapter 11 covers the topic of chemical toxicity screening by using marine medaka (*O. melastigma*) as a model system. This chapter provides the recent research progress in the toxicological impacts and responsive biomarker of *O. melastigma* caused by various marine pollutants, such as heavy metals, endocrine disruptors, and organic pollutants.

Chapter 12 reviews the problems of invasive species in Andaman and Nicobar Islands, Andaman Sea, India. Chapter 13 highlights the problems of dispersal of invasive species through marine ecosystems with a special focus on the case study of five invasive species and associated problems. Chapter 14 describes the effects of the disturbing unique island biodiversity of marine protected areas linked with the environmental changes influenced by anthropogenic activities, overexploitation of resources, and the habitat loss due to developmental activities and natural change in climate. Chapter 15 presents information on monitoring environmental indicators and bacterial pathogens in aquaculture practices impacted the Muthupettai Mangrove Ecosystem, Tamil Nadu, India. This chapter, a research article instead of a review, reports on the vulnerability of the mangrove ecosystems after continuous discharges of untreated aquaculture effluents have caused water quality to deteriorate so far that physiochemical parameters and bacterial pathogens highly exceed WHO, EU, and CPCB standard permissible limits.

Chapter 16 highlights the vast potential of marine microbes (bacteria and fungi) for their application in bioremediation of heavy metals. This chapter also discusses the specific factors influencing heavy metal bioremediation including biotic and abiotic factors. Chapter 17 focuses on bioremediation of low and high-molecular-weight polycyclic aromatic hydrocarbons (PAHs) in the marine environment through bacterial and fungal strains (lignolytic fungi and non-lignolytic fungi). Further, recent advancements in applications of genomics, proteomics and metabolomics technologies for in-depth investigation of microbial communities involved in PAHs remediation are summarized. Chapter 18 provides final thoughts and concluding remarks.

This book contains the latest progress in the theoretical background of marine pollutants, occurrence, distribution, risk assessment, and the bioremediation in the marine environment, which will be of specific interests for academic scientists, students, and government officials to develop background knowledge of marine pollution based multidisciplinary research.

De-Sheng Pei & Muhammad Junaid
Research Center for Environment and Health,
Chongqing Institute of Green and Intelligent Technology,
Chinese Academy of Sciences,
Chongqing,
China

DEDICATION

This book is in the memory of my mother, Ms. Jinhua Wang, who passed away on February 14, 2014. This book is dedicated to all the researchers around the globe, who are working effectively and finding novel ways to protect the environment.

De-Sheng Pei
Research Center for Environment and Health,
Chongqing Institute of Green and Intelligent
Technology, Chinese Academy of Sciences,
Chongqing, China

I would like to dedicate this book to my father, Mr. Muhammad Yahya (Late, January 30, 1992) and my beloved mother, Ms. Najma Jalil.

Muhammad Junaid
Research Center for Environment and Health,
Chongqing Institute of Green and Intelligent
Technology, Chinese Academy of Sciences,
Chongqing, China

ACKNOWLEDGEMENTS

The editors sincerely thankful to all contributing authors, who worked hard to improve the quality of this book. Further, we sincerely thankful to Ms. Naima Hamid for professional help, and her efforts immensely improved the quality of this book.

The editors are grateful for the support from the CAS Team Project of the Belt and Road (to D.S.P), the Three Hundred Leading Talents in Scientific and Technological Innovation Program of Chongqing (No. CSTCCXLJRC201714), and the Program of China–Sri Lanka Joint Research and Demonstration Center for Water Technology and China–Sri Lanka Joint Center for Education and Research by Chinese Academy of Sciences, China.

List of Contributors

Abida Farooqi
Department of Environmental Sciences, Faculty of Biological Sciences, Quaid-i- Azam University, Islamabad 45320, Pakistan

Akshai Raj
Department of Ocean Studies and Marine Biology, Pondicherry University Off Campus, Brookshabad, Port Blair – 744112, Andaman and Nicobar Islands, India

Altaf Hussain Narejo
Centre of Excellence in Marine Biology, University of Karachi, Karachi, Pakistan

Anjana K. Vala
Department of Life Sciences, Maharaja Krishnakumarsinhji Bhavnagar University, Bhavnagar-364 001, India

Araceli Verónica Flores Nardy Ribeiro
Coordenação de Licenciaturas – Instituto Federal de Educação, Ciência e Tecnologia do Espírito Santo, Avenida Vitória, 1729, CEP: 29.040-780, Vitória-ES, Brazil

Arnaud Victor dos Santos
Departamento de Ciências Exatas e da Terra – Universidade do Estado da Bahia, Rua Silveira Martins, 2555, CEP: 41.150-000, Cabula, Salvador-BA, Brazil

Asmat S. Siddiqui
Centre of Excellence in Marine Biology, University of Karachi, Karachi, Pakistan

Ayesha Gul
Department of Biosciences COMSATS, Institute of Information Technology Park Road, Islamabad, Pakistan

Bharti P. Dave
Department of Life Sciences, Maharaja Krishnakumarsinhji Bhavnagar University, Bhavnagar-364 001, India
Department of Biosciences, School of Sciences, Indrashil University, Rajpur - Kadi - 382740, India

Bhumi K. Sachaniya
Department of Life Sciences, Maharaja Krishnakumarsinhji Bhavnagar University, Bhavnagar-364 001, India

Barbilina Pam
Department of Ocean Studies and Marine Biology, Pondicherry University Off Campus, Brookshabad, Port Blair – 744112, Andaman and Nicobar Islands, India

Chengjun Sun
Marine Ecology Center, the First Institute of Oceanography, Ministry of Natural Resources, Qingdao 266061, China

Daniel Carneiro Freitas
Departamento de Ciências Exatas e da Terra – Universidade do Estado da Bahia, Rua Silveira Martins, 2555, CEP: 41.150-000, Cabula, Salvador-BA, Brazil

De-Sheng Pei
Research Center for Environment and Health, Chongqing Institute of Green and Intelligent Technology, Chinese Academy of Sciences, Chongqing, 400714, China

Deedar Nabi
Institute of Environmental Sciences and Engineering, School of Civil and Environmental Engineering, National University of Sciences and Technology, Sector H-12, Islamabad, 44000, Pakistan

Donat-P. Häder
Department of Biology, Friedrich-Alexander University Erlangen-Nürnberg, 91054 Erlangen, Germany

Dongdong Song	Guandong Provincial Key Laboratory of Applied Marine Biology, South China Sea Institute of Oceanology, Chinese Academy of Sciences, Guangzhou, China
D.B.K.K. Sabith	Department of Ocean Studies and Marine Biology, Pondicherry University Off Campus, Brookshabad, Port Blair – 744112, Andaman and Nicobar Islands, India
Fenghua Jiang	Marine Ecology Center, the First Institute of Oceanography, Ministry of Natural Resources, Qingdao 266061, China
Fakhar I. Abbass	Bioresource Research Centre Bazar road, Islamabad, Pakistan
Haresh Z. Panseriya	Department of Life Sciences, Maharaja Krishnakumarsinhji Bhavnagar University, Bhavnagar-364 001, India Department of Biosciences, School of Sciences, Indrashil Universit, Rajpur - Kadi - 382740, India
Haren B. Gosai	Department of Life Sciences, Maharaja Krishnakumarsinhji Bhavnagar University, Bhavnagar-364 001, India Department of Biosciences, School of Sciences, Indrashil Universit, Rajpur - Kadi - 382740, India
Joselito Nardy Ribeiro	Centro de Ciências da Saúde – Universidade Federal do Espírito Santo, Avenida Maruípe, S/N, CEP: 29.042-751, Vitória-ES, Brazil
Jingxi Li	Marine Ecology Center, the First Institute of Oceanography, Ministry of Natural Resources, Qingdao 266061, China
Kanza Naseer	Institute of Environmental Sciences and Engineering, School of Civil and Environmental Engineering, National University of Sciences and Technology, Sector H-12, Islamabad, 44000, Pakistan
Lizhao Chen	Guandong Provincial Key Laboratory of Applied Marine Biology, South China Sea Institute of Oceanology, Chinese Academy of Sciences, Guangzhou, China
Li Zhang	Guandong Provincial Key Laboratory of Applied Marine Biology, South China Sea Institute of Oceanology, Chinese Academy of Sciences, Guangzhou, China
Lourdes Cardoso de Souza Neta	Departamento de Ciências Exatas e da Terra – Universidade do Estado da Bahia, Rua Silveira Martins, 2555, CEP: 41.150-000, Cabula, Salvador-BA, Brazil
Madson de Godoi Pereira	Departamento de Ciências Exatas e da Terra – Universidade do Estado da Bahia, Rua Silveira Martins, 2555, CEP: 41.150-000, Cabula, Salvador-BA, Brazil
Mazhar Iqbal Zafar	Department of Environmental Sciences, Faculty of Biological Sciences, Quaid-i- Azam University, Islamabad 45320, Pakistan
Mehtabidah Ali	Department of Environmental Sciences, Faculty of Biological Sciences, Quaid-i- Azam University, Islamabad 45320, Pakistan
Muhammad Junaid	Research Center for Environment and Health, Chongqing Institute of Green and Intelligent Technology, Chinese Academy of Sciences, Chongqing, 400714, China University of Chinese Academy of Sciences, Beijing 100049, China

Muhammad Arshad　Institute of Environmental Sciences and Engineering, School of Civil and Environmental Engineering, National University of Sciences and Technology, Sector H-12, Islamabad, 44000, Pakistan

Muhammad Arif Ali　Department of Soil Science, Faculty of Agricultural Sciences and Technology, Bahauddin Zakariya University, Multan, Pakistan

Muhammed A. Gondal　Department of Biosciences COMSATS, Institute of Information Technology Park Road, Islamabad, Pakistan

M. Ramiro Pastorinho　CICS-UBI, Health Sciences Research Centre, University of Beira Interior, Covilhã, Portugal
Department of Biology, University of Evora, Evora, Portugal
Faculty of Health Sciences, University of Beira Interior, Covilhã, Portugal

Naima Hamid　ResearchCenterforEnvironmentandHealth, Chongqing Institute of Green and Intelligent Technology, Chinese Academy of Sciences, Chongqing, 400714, China
University of Chinese Academy of Sciences, Beijing 100049, China

Nayab Kanwal　Centre of Excellence in Marine Biology, University of Karachi, Karachi, Pakistan

Nirmaladevi D. Shrinithivihahshinia　Environmental Microbiology and Toxicology Laboratory, Department of Environmental Management, School of Environmental Sciences, Bharathidasan University, Tiruchirappalli – 620 024, Tamil Nadu, India

Noor U. Saher　Centre of Excellence in Marine Biology, University of Karachi, Karachi, Pakistan

Peng Zhang　Guandong Provincial Key Laboratory of Applied Marine Biology, South China Sea Institute of Oceanology, Chinese Academy of Sciences, Guangzhou, China

P.M. Mohan　Department of Ocean Studies and Marine Biology, Pondicherry University Off Campus, Brookshabad, Port Blair – 744112, Andaman and Nicobar Islands, India

Rajendran Viji　Environmental Microbiology and Toxicology Laboratory, Department of Environmental Management, School of Environmental Sciences, Bharathidasan University, Tiruchirappalli – 620 024, Tamil Nadu, India

Rabeea Zafar　Institute of Environmental Sciences and Engineering, School of Civil and Environmental Engineering, National University of Sciences and Technology, Sector H-12, Islamabad, 44000, Pakistan
Department of Home and Health Sciences, Faculty of Sciences, Allama Iqbal Open University, Sector H-8, Islamabad, 44000, Pakistan

Ricardo Teles Pais　CICS-UBI, Health Sciences Research Centre, University of Beira Interior, Covilhã, Portugal

Riffat Naseem Malik　Department of Environmental Sciences, Faculty of Biological Sciences, Quaid-i- Azam University, Islamabad 45320, Pakistan

Shahbaz Ahmad　Institute of Agricultural Sciences, University of the Punjab, Lahore 54590, Pakistan

Sen Du	Guandong Provincial Key Laboratory of Applied Marine Biology, South China Sea Institute of Oceanology, Chinese Academy of Sciences, Guangzhou, China

Thelma Jeyasingh	Department of Zoology, Thiagarajar College, Madurai-625009, Tamilnadu, India

Wei Cao	Marine Ecology Center, the First Institute of Oceanography, Ministry of Natural Resources, Qingdao 266061, China

Yan-Ling Chen	Research Center for Environment and Health, Chongqing Institute of Green and Intelligent Technology, Chinese Academy of Sciences, Chongqing 400714, China

Zaheer Ahmed	Department of Home and Health Sciences, Faculty of Sciences, Allama Iqbal Open University, Sector H-8, Islamabad, 44000, Pakistan

Zahid Iqbal	Department of Pharmacology, Al-Nafees Medical College & Hospital, Isra University, Islamabad Campus, Islamabad, Pakistan

CHAPTER 1

An Introduction to the Recent Perspectives of Marine Pollution

De-Sheng Pei[1,*], **Muhammad Junaid**[1,2] and **Naima Hamid**[1,2]

[1] *Research Center for Environment and Health, Chongqing Institute of Green and Intelligent Technology, Chinese Academy of Sciences, Chongqing 400714, China*

[2] *University of Chinese Academy of Sciences, Beijing 100049, China*

Abstract: Marine ecosystem covers two-thirds of the earth's surface, and is characterized by its rich biodiversity and endemism of marine life. However, like many other ecosystems, it has been subject to diverse anthropogenic pressures, such as climate change, pollution, and biodiversity losses. In the first part of the book, we discussed the pollution dynamics of the inorganic pollutants (heavy metals, metalloids) and organic pollutants including persistent organic pollutants (POPs), polycyclic aromatic hydrocarbons (PAHs), polychlorinated biphenyls (PCBs), microplastics, nutrients, and algal blooms in the marine environment. Marine pollutants can have a wide range of pollution sources that are able to cause deleterious effects on marine flora and fauna. The second section of the book specifically elucidates the toxicity assessment by using marine model organisms. It provides extensive new insight into screening biomarker genes combined with advanced gene editing applications. In the last section of the book, various remedial techniques, such as bioremediation and phytoremediation, were discussed whether it could be beneficial to deal with the challenges of marine pollution.

Keywords: Marine Ecosystem, Pollution Dynamics, Remedial Measures, Toxicity Assessment.

According to the United Nations Convention on the Law of the Sea (UNCLOS), the marine pollution is defined as "the introduction by man, directly or indirectly, of substances or energy into the marine environment, including estuaries, which results or is likely to result in such deleterious effects as harm to living resources and marine life, also hazardous to human health" (Williams, 1996). The driving factors for emissions of marine pollutants include infrastructure development, human settlements, anthropogenic interventions, resource utilization, agriculture

---

* **Corresponding author De-Sheng Pei**: Research Center for Environment and Health, Chongqing Institute of Green and Intelligent Technology, Chinese Academy of Sciences, Chongqing 400714, China; Tel/Fax: +86-23-65935812; E-mails: peids@cigit.ac.cn and deshengpei@gmail.com

De-Sheng Pei & Muhammad Junaid (Eds.)
All rights reserved-© 2019 Bentham Science Publishers

activities, industrialization, and tourism (Derraik, 2002). The prominent marine pollutants of major concern include inorganic elements, persistent organic pollutants, microplastics, radionuclides, and oil spills. Most of these pollutants are interlinked in terms of their sources, jeopardizing the marine environment, and ecological resources. However, the existing classification of marine pollutants needs to be redefined (Islam & Tanaka, 2004). Due to the marine fisheries and commercial exploitation of coasts, most of the coastal areas in the world have been severely affected by marine pollution. Therefore, control of marine pollution is critically important and immensely needed for the conservation of marine ecology and sustainable management of resources. In addition, there is a scientific knowledge gap about marine pollution, which is also a constraint for controlling marine pollution.

The problem of marine pollution is dated back to the history of human civilization due to the anthropogenic interventions (Islam & Tanaka, 2004). However, this issue failed to receive considerable attention until recently when the consequences of marine pollution reached a threshold level and resulted in adverse impacts on the ecosystem and climate change. Now, marine pollution and associated hazards have become major environmental concerns around the globe. Among marine pollutants, persistent organic pollutants (POPs) are carbon-based legacy organic pollutants, which exhibit a high environmental persistence and toxicity (Tieyu *et al.*, 2005). POPs have attained a considerable global attention due to their potentials for long-range transport, persistence behavior, lipophilic nature, bio-accumulation, and biomagnification in the ecosystems, as well as their pronounced adverse effects on the environment and human health (Harrad, 2009). POPs usually include polychlorinated biphenyls (PCB), organochlorine pesticides (OCPs), brominated flame retardants (FBRs), polyfluorinated sulfonamides (FSAs), and other industrial chemicals, such as unintentional by-products of many industrial processes, especially polychlorinated dibenzofurans (PCDF) and dibenzo-p-dioxins (PCDD), commonly known as 'dioxins' (Tieyu *et al.*, 2005). In 2001, the Stockholm Convention under the umbrella of United Nations Environment Programme (UNEP) enlisted the sources, behavior, fate, and effects of POPs. This Convention was enacted in 2004. In 2008, 180 parties had accredited the Stockholm Convention in order to cope with POPs mediated hazardous impacts on human health and the environment. Initially, the Convention had listed 12 POPs for eradication and named them as "dirty dozen" that included DDT, aldrin, dieldrin, chlordane, heptachlor, hexachlorobenzene, mirex, polychlorinated biphenyls, polychlorinated dibenzo-*p*-dioxins, polychlorinated dibenzofurans, and toxaphene (Xu *et al.*, 2013).

A comprehensive study reported the contamination of POPs (organochlorine compounds) in the coastal water samples collected from 30 beaches of 17 countries, and the highest concentration was found at the coasts of USA, followed by Western Europe and Japan; while the lowest levels of POPs were reported at the coasts of tropical Asia, Australia, and Southern Africa (Ogata *et al.*, 2009). POPs also include polycyclic aromatic hydrocarbons (PAHs) as the priority class of organic pollutants, which are primarily emitted from incomplete combustion of petroleum products in automobiles, industries and also through the pyrolysis of organic materials. In the marine environment, several processes, such as deposition through the atmosphere, industrial sewage, transport (marine ships), oil spills, and terrestrial runoff, are the potential sources of PAHs (Hamid *et al.*, 2016). POPs exhibit exceptionally long retention time in the living bodies, pass through different stages of the food chain, and result in biomagnification at higher trophic levels. Further, persistent compounds can be bio accumulated and bio concentrated at the low trophic levels (Hamid *et al.*, 2016).

PCBs, organochlorines, organometallics, polychlorinated dibenzodioxin (PCDDs), and polychlorinated dibenzofurans (PCDFs) are compounds, which are usually present in elevated concentrations in the tissues of the exposed animals at higher trophic levels (Pérez-Carrera *et al.*, 2007). Bioaccumulation and bioconcentration may be the consequences of biomagnification process along the food chain in the marine ecosystem. The vertebrates and invertebrates in the aquatic ecosystem absorb different pollutants that can cause acute and chronic toxicity after magnification (Islam & Tanaka, 2004). Although many studies are available on the levels of pollutants in the marine ecosystems and their consequences, the precise and conclusive review of those studies is still elusive, which has been summarized in this book. This issue of organic contamination in the marine pollution is alarming to the extent that the Scientific Committee of International Whaling Commission (IWC) devised and launched a comprehensive program "Pollution 2000+" to elucidate the cause-and-effect relationship in cetaceans (Helmerhorst *et al.*, 1999). The objective of this program was to develop a predictive model that can link the concentration of the pollutants in the tissues with its effects at the population level. Pollution 2000+ specifically focused on PCBs as model organic pollutants to determined effects for organochlorine pesticides (OCs) pollution (Helmerhorst *et al.*, 1999).

Inorganic components include inorganic nutritive ions such as phosphates and nitrates, sulfur, arsenic, aluminum, cadmium, lead, mercury, and nickel, gases like carbon dioxide and metals. All of these inorganic ions are essential for maintaining ecological balance (Islam & Tanaka, 2004). Nevertheless, when these ions occur in higher concentrations, they affect the natural ecological harmony also affect the aquatic organisms. For example, Nitrogen and Phosphorous act as a stimulus to increase the algal production. If the biomass production remained increased, then the algal layer becomes thick that prevent the sunlight and oxygen to reach the lower part (Almeida *et al.*, 2007). Hence, low level oxygen deteriorates the marine life mainly invertebrates such as mollusks, worms, crustaceans and fish. However, aluminum, cadmium, lead, and mercury represent the group of toxic metals that are hazardous for human health. It is worth mentioning that the concentrations of aluminum in marine organisms is of extreme importance because this element is neurotoxic to humans and it may be responsible for Alzheimer's disease (Xu *et al.*, 2013). Moreover, the increase in cadmium, sulphate, and nitrate concentrations with depth are due to the decomposition of organic matter and consequent release of both nutrients as well as cadmium (Fergusson, 1990). Elevated Cadmium ions are distributed and accumulated in marine organisms and caused kidney failure, bone diseases, infertility, and different types of tumors in humans (Craig & Jenkins, 2004).

Furthermore, some inorganic compounds are microelements such as organometallic compounds which are emitted from agriculture and industrial activities are exhibiting high toxic potentials for the marine ecosystem. The toxicity of such compounds depends on both the metal atom and the organic compound bound to the metal (Craig & Jenkins, 2004). The toxicity of metals (mercury, chromium, selenium, cobalt, molybdenum, vanadium, iron, rhodium, iridium, silicon, germanium, tin, tungsten, manganese, and platinum) bearing organic compounds is elucidated not only through the inductive and composition resonance with steric characteristics, but also by using polarizability (Almeida *et al.*, 2007; Mantoura, 1981).

Organometallic compounds, such as methylmercury, butyl tin, phenyl tin, and diethyl lead, are predominantly presented in the marine ecosystem (Wong *et al.*, 1982). These compounds enter the aquatic ecosystem from the shipyard cleaning activities and landfill leaching. Similarly, the organolead compounds are persistent in the marine environment, because they had been used as anti-knocking agents after 1920s for almost more than a half-century (Craig & Jenkins, 2004). Mercury is also abundantly distributed in the marine ecosystem due to the natural earth process and anthropogenic activities. Methylmercury (MeHg) is a highly toxic organometallic compound, which is produced by marine microbes through using inorganic mercury. MeHg exhibited an augmented potential for bioaccumulation and biomagnification in the food chain at higher trophic levels (Almeida *et al.*, 2007).

The 21st century has been termed as the plastic age because of the widespread use of plastics. Plastics are being abundantly consumed in the industries and households, which can be observed almost everywhere due to their specific characteristics, such as good malleability, low density, low cost, and durability. It was estimated that the global production of plastic reached 335 million tons in

2016 (Jambeck *et al.*, 2015). According to the annals of UNEP for 2014, marine plastic pollution was listed as one of the ten alarming environmental problems that need an urgent and sustainable solution (UNEP, 2014). The distribution of marine plastic greatly influenced by the water currents in the sea, whereas they are more evenly distributed in the oceans with high density in specific regions (Browne *et al.*, 2015). According to estimates, about 480-1279 tons of plastic debris entered the ocean annually (Jambeck *et al.*, 2015). Plastics are usually non-degradable and can last several hundred to thousand years in the environment. Alarming concerns of plastics are raised because of their persistent nature and their potential to transport POPs into the marine environment (Ng & Obbard, 2006).

Microplastics (MP) are known as the small particles of plastic size less than 5 mm defined by the National Oceanic and Atmospheric Administration (Jambeck *et al.*, 2015). They enter into the marine environment from the direct sources such as industrial accidental spillages and usage or the release of microbeads used in cosmetics through wastewaters (Browne *et al.*, 2015). MP is considered as a new emerging pollutant and concerned researchers have started to study their effects and risks in marine environment. MP pollution has been listed as the second major scientific problems in the field of environmental and ecological science in 2015. The plastic (including the MP) pollution of the marine environment was also considered as the major global environmental problems together with ocean acidification, de-oxygenation, ocean warming (Almeida *et al.*, 2007).

Due to the ever-increasing pollution in the marine environment, many of the aquatic organisms have been recommended as suitable environmentally relevant models, which are used as the indicators of ecotoxicity research. Among these model organism, fish species are of particular importance. Fish is more sensitive to many toxicants, compared to other invertebrates. The presence of pollutants in the marine environment can be monitored directly in the environmental matrices or through analyzing them in the fish, such as tissues, body fluids, and liver (Sures, 2001). In fact, the response of biomarkers is relatively quick and they prove as an indicator or initial warning system for biological effects to predict the toxicity of environmental pollutants. For instance, *cyp1a1* is a specific biomarker for organic aromatic chemical exposure, which can be estimated by determining the level of ethoxyresorufin-O-deethylase (EROD) activity in the liver of fish.

Among fish models, zebrafish (*Danio rerio*) and Japanese medaka (*Oryzias latipes*) are fresh water, while marine medaka (*Oryzias melastigm*) is an emerging marine model to study the ecotoxicology in the aquatic environment (Dodd *et al.*, 2000; Wittbrodt *et al.*, 2002). Further, the transgenic fish are also employed for toxicity screening, which offers more precise and advanced systems to unveil the mechanistic toxicity (Lele & Krone, 1996; Nebert *et al.*, 2002). Similarly, there are different fluorescent protein reporter systems (*e.g.*, GFP, RFP) that are capable of tracking pollutants by real-time visualization of fluorescence signals in living embryos and organisms (Sures, 2001). In addition, the toxicity of the organic chemical is also measured through quantifying the transcriptional levels of heat-shock proteins, which are activated *via* aryl hydrocarbon receptor pathway usually together with *cyp1a1*. Toxicity of estrogen-like compounds can be quantified through assessing the expression levels of vitellogenin (Vtg), choriogenin H (ChgH), and choriogenin L (ChgL) (Sures, 2001).

Regarding the marine flora, seagrass is well known for its potential to interact and bioaccumulate with pollutants. The hazardous pollutants severely affected its growth (Cabaço *et al.*, 2008; Gacia *et al.*, 2003). Sometimes the turbidity of marine water increase due to the high amount of suspended particles, which cause growth difficulties for photosynthetic micro aquatic plant species and other benthic organisms that need larval settlements (Gallegos, 2001). Moreover, turbidity also increases the temperature of water, because the suspended particles tend to absorb more heat (Glynn, 1993). It is worthy to mention here that some of the host plants are not merely affected by turbidity, but also leave elevated stress on their associated epiphytes, such as microalgae and microphytobenthos. (Glynn, 1993). Some of the coastal areas and estuarial environment are also observed with high eutrophication episodes. Eutrophication can be defined as "the process by which water becomes enriched in dissolved nutrients (such as phosphates), and the dissolved nutrients stimulate the growth of aquatic plant life". Various pollutants, such as domestic sewage, detergents, industrial effluents, and agriculture run-off, cause eutrophication when they enter the marine environment (Edinger *et al.*, 1998). Eutrophication can cause severe damage to the marine flora, *e.g.* the biodiversity of coral reefs reduced up to 60% due to the alarming levels of eutrophication caused by nutrients. (Edinger *et al.*, 1998).

For the bioremediation of marine contaminated media, such as water, sediments, and subsurface materials, many techniques have been employed based on microorganisms (Bouwer & Zehnder, 1993). Microorganisms can extract energy from all the organic and inorganic pollutants through various pathways, therefore 80% of the microbial bioremediation studies employed bacterial strain to achieve high treatment efficiency. Further, the microbial remediation of heavy metals' contamination has several benefits, such as environment-friendly process, cost-effective, self-reproducible, and bio-products reuse. In addition, the microorganisms are adaptable, therefore, they have intensively used for the treatment of inorganic and organic contamination (Biache *et al.*, 2017). Microbial remediation is usually a long-term approach to marine pollution (Alvarez *et al.* 2017). In fact, microbial bioremediation of heavy metals and other compounds lead to their immobilization and solubilization in the media, which is a critical step for treatment of pollutants (Kuppusamy *et al.*, 2017). Generally, oil or petroleum contaminated sites exhibit POPs (specifically PAHs) degrading microbial community to a large extent (Zafra *et al.*, 2017). Many of the previous studies have reported for isolating various types of bacteria from the contaminated sediments, which have been involved in the degradation of PAHs, especially the low molecular weight PAHs, such as naphthalene and phenanthrene, which usually present in high concentrations. (Kuppusamy *et al.*, 2017; Li *et al.*, 2017).

For the bioremediation, the bioavailability of the target pollutant is the most critical and important criterion. Under certain circumstances, the adverse effects caused by heavy metals and PAHs in the marine environment rehabilitate by surfactants up to a smaller extent (Ron & Rosenberg, 2002). Surfactants can act as metal complexing agent, increase the hydrophobicity of the cell surface, and promote the transmembrane transport (Zafra *et al.*, 2017). Microbial responses to pollutants through uptake, bioremediation, and tolerance vary with different organisms. Similarly, organisms use various strategies to cope with the stress caused by heavy metals and their responses may vary at genus as well as species level. For example, *A. sydowii* showed maximum tolerance to as among different species of marine-derived *Aspergillus* fungus, such as *A. sydowii, A. niger, A. flavus*, and *A. candidus* (K. *et al.*, 2011; Vala, 2010; Vala & Dave, 2017). The freshwater microalgal species, *C. reinhardtii, C. vulgaris*, and *C. miniata*, can be used to remove the divalent heavy metals (Cd, Pb, Hg, Cu, Ni, and Zn), whereas *S. platensis* and *C. vulgaris* can remove trivalent metals (Cr and Fe). *C. vulgaris* and *C. miniata* can remediate the hexavalent metals (Cr) (Suresh Kumar *et al.*, 2015).

CONCLUSION

This book comprehensively highlighted almost entire aspects of the marine pollution, initiating from the current status, distribution, sources of legacy and emerging pollutants in various marine environmental and biological matrices from diverse spatial and temporal scenarios, then advancing with the direct or indirect impacts of marine pollution on the marine flora and fauna in terms of mechanistic toxicity *via* various exposure routes and associated pathways. The second last portion of this book especially provided useful insights about the monitoring of marine pollution using advance, efficient, and environmental friendly biological methods. The last section of this book focused more about indirect effects of marine pollution or problems of invasive species and challenges to the marine protected area, and most importantly this section also provided biological solutions and remedies for marine pollution. Considering the vast nature of this book, editors are hopeful that it will prove as a useful resource for students, researchers, and policymakers, who are working on management and protection of marine resources and environment. Hence, the precise and filtered knowledge about the current dynamics of pollution in the marine ecosystem, associated ecological losses, and possible remediation strategies is still elusive. The implementation of existing regulations to abate and control marine pollution is urgently needed. It is essential to devise more international agreements to address the problems related to marine pollution to achieve the targets of sustainable development. A policy framework is also required that include marine environment protection laws, source-based monitoring, control of marine pollutants, waste disposal, and management strategies. Importantly, the rational uses of plastic and microplastic materials should be promoted to control plastic waste in the marine environment. Forums should be established for public awareness about the hazardous impacts of marine pollution and introduce an innovative solution to reduce the number of pollutants entering the marine environment. Lastly, the cutting edge scientific research should continue to understand the scope and scale of marine pollution. Future studies may focus on the most urgent topics, such as devising the standard methods for precisely analyzing the emerging pollutants in environmental. These studies ultimately can assist to understand the global distribution of marine pollution, and are also useful to assess the long-term effects of pollutants on marine biodiversity and trophic transfer of pollutants along the food chain.

CONSENT FOR PUBLICATION

Not applicable.

ACKNOWLEDGEMENTS

The editors are grateful for the support from the CAS Team Project of the Belt and Road (to D.S.P), the Three Hundred Leading Talents in Scientific and Technological Innovation Program of Chongqing (No. CSTCCXLJRC201714 to D.S.P), the Program of China–Sri Lanka Joint Research and Demonstration Center for Water Technology and China–Sri Lanka Joint Center for Education and Research by Chinese Academy of Sciences, China(to D.S.P), and the University of Chinese Academy of Sciences (UCAS) for CAS-TWAS Scholarship (No. 2017A8018537001 to N.H).

CONFLICT OF INTERESTS

The authors confirm that this chapter contents have no conflict of interest.

REFERENCES

Almeida, E, Diamantino, TC & de Sousa, O (2007) Marine paints: the particular case of antifouling paints.

Prog Org Coat, 59, 2-20.
[http://dx.doi.org/10.1016/j.porgcoat.2007.01.017]

Biache, C, Ouali, S, Cébron, A, Lorgeoux, C, Colombano, S & Faure, P (2017) Bioremediation of PAH-contamined soils: Consequences on formation and degradation of polar-polycyclic aromatic compounds and microbial community abundance. *J Hazard Mater,* 329, 1-10.
[http://dx.doi.org/10.1016/j.jhazmat.2017.01.026] [PMID: 28119192]

Bouwer, EJ & Zehnder, AJB (1993) Bioremediation of organic compounds--putting microbial metabolism to work. *Trends Biotechnol,* 11, 360-7.
[http://dx.doi.org/10.1016/0167-7799(93)90159-7] [PMID: 7764183]

Browne, MA, Chapman, MG, Thompson, RC, Amaral Zettler, LA, Jambeck, J & Mallos, NJ (2015) Spatial and temporal patterns of stranded intertidal marine debris: is there a picture of global change? *Environ Sci Technol,* 49, 7082-94.
[http://dx.doi.org/10.1021/es5060572] [PMID: 25938368]

Cabaço, S, Santos, R & Duarte, CM (2008) The impact of sediment burial and erosion on seagrasses: a review. *Estuar Coast Shelf Sci,* 79, 354-66.
[http://dx.doi.org/10.1016/j.ecss.2008.04.021]

Craig, PJ & Jenkins, R (2004) *Organometallic compounds in the environment: an overview Organic metal and metalloid species in the environment.* Springer 1-15.

Derraik, JG (2002) The pollution of the marine environment by plastic debris: a review. *Mar Pollut Bull,* 44, 842-52.
[http://dx.doi.org/10.1016/S0025-326X(02)00220-5] [PMID: 12405208]

Dodd, A, Curtis, PM, Williams, LC & Love, DR (2000) Zebrafish: bridging the gap between development and disease. *Hum Mol Genet,* 9, 2443-9.
[http://dx.doi.org/10.1093/hmg/9.16.2443] [PMID: 11005800]

Edinger, EN, Jompa, J, Limmon, GV, Widjatmoko, W & Risk, MJ (1998) Reef degradation and coral biodiversity in Indonesia: effects of land-based pollution, destructive fishing practices and changes over time. *Mar Pollut Bull,* 36, 617-30.
[http://dx.doi.org/10.1016/S0025-326X(98)00047-2]

Gacia, E, Duarte, C, Marba, N, Terrados, J, Kennedy, H, Fortes, M & Tri, N (2003) Sediment deposition and production in SE-Asia seagrass meadows. *Estuar Coast Shelf Sci,* 56, 909-19.
[http://dx.doi.org/10.1016/S0272-7714(02)00286-X]

Gallegos, CL (2001) Calculating optical water quality targets to restore and protect submersed aquatic vegetation: overcoming problems in partitioning the diffuse attenuation coefficient for photosynthetically active radiation. *Estuaries,* 24, 381-97.
[http://dx.doi.org/10.2307/1353240]

Glynn, P (1993) Coral reef bleaching: ecological perspectives. *Coral Reefs,* 12, 1-17.
[http://dx.doi.org/10.1007/BF00303779]

Hamid, N, Syed, JH, Kamal, A, Aziz, F, Tanveer, S, Ali, U, Cincinelli, A, Katsoyiannis, A, Yadav, IC, Li, J, Malik, RN & Zhang, G (2017) A review on the abundance, distribution and eco-biological risks of PAHs in the key environmental matrices of south Asia. *Rev Environ Contam Toxicol,* 240, 1-30. [Springer.].
[PMID: 26809717]

Harrad, S (2009) *Persistent organic pollutants.* John Wiley & Sons.
[http://dx.doi.org/10.1002/9780470684122]

Helmerhorst, EJ, Reijnders, IM, van't Hof, W, Simoons-Smit, I, Veerman, EC & Amerongen, AVN (1999) Amphotericin B- and fluconazole-resistant Candida spp., Aspergillus fumigatus, and other newly emerging pathogenic fungi are susceptible to basic antifungal peptides. *Antimicrob Agents Chemother,* 43, 702-4.
[http://dx.doi.org/10.1128/AAC.43.3.702] [PMID: 10049295]

Shahidul Islam, M & Tanaka, M (2004) Impacts of pollution on coastal and marine ecosystems including

coastal and marine fisheries and approach for management: a review and synthesis. *Mar Pollut Bull,* 48, 624-49.
[http://dx.doi.org/10.1016/j.marpolbul.2003.12.004] [PMID: 15041420]

Jambeck, JR, Geyer, R, Wilcox, C, Siegler, TR, Perryman, M, Andrady, A, Narayan, R & Law, KL (2015) Marine pollution. Plastic waste inputs from land into the ocean. *Science,* 347, 768-71.
[http://dx.doi.org/10.1126/science.1260352] [PMID: 25678662]

V A, K, Vishnu, S & R.V., U (2011) Investigations on trivalent arsenic tolerance and removal potential of a facultative marine Aspergillus niger. *Environmental Progress & Sustainable Energy,* 30, 586-8.

Kuppusamy, S, Thavamani, P, Venkateswarlu, K, Lee, YB, Naidu, R & Megharaj, M (2017) Remediation approaches for polycyclic aromatic hydrocarbons (PAHs) contaminated soils: Technological constraints, emerging trends and future directions. *Chemosphere,* 168, 944-68.
[http://dx.doi.org/10.1016/j.chemosphere.2016.10.115] [PMID: 27823779]

Lele, Z & Krone, PH (1996) The zebrafish as a model system in developmental, toxicological and transgenic research. *Biotechnol Adv,* 14, 57-72.
[http://dx.doi.org/10.1016/0734-9750(96)00004-3] [PMID: 14536924]

Li, Q, Gao, J, Zhang, Q, Liang, L & Tao, H (2017) Distribution and Risk Assessment of Antibiotics in a Typical River in North China Plain. *Bull Environ Contam Toxicol,* 98, 478-83.
[http://dx.doi.org/10.1007/s00128-016-2023-0] [PMID: 28084506]

Mantoura, R (1981) *Organo-Metallic Interactions in Natural Waters. 1. Elsevier Oceanography Series.* Elsevier 179-223.

Nebert, DW, Stuart, GW, Solis, WA & Carvan, MJ, III (2002) Use of reporter genes and vertebrate DNA motifs in transgenic zebrafish as sentinels for assessing aquatic pollution. *Environ Health Perspect,* 110, A15-5.
[http://dx.doi.org/10.1289/ehp.110-a15] [PMID: 11813700]

Ng, KL & Obbard, JP (2006) Prevalence of microplastics in Singapore's coastal marine environment. *Mar Pollut Bull,* 52, 761-7.
[http://dx.doi.org/10.1016/j.marpolbul.2005.11.017] [PMID: 16388828]

Ogata, Y, Takada, H, Mizukawa, K, Hirai, H, Iwasa, S, Endo, S, Mato, Y, Saha, M, Okuda, K, Nakashima, A, Murakami, M, Zurcher, N, Booyatumanondo, R, Zakaria, MP, Dung, Q, Gordon, M, Miguez, C, Suzuki, S, Moore, C, Karapanagioti, HK, Weerts, S, McClurg, T, Burres, E, Smith, W, Van Velkenburg, M, Lang, JS, Lang, RC, Laursen, D, Danner, B, Stewardson, N & Thompson, RC (2009) International Pellet Watch: global monitoring of persistent organic pollutants (POPs) in coastal waters. 1. Initial phase data on PCBs, DDTs, and HCHs. *Mar Pollut Bull,* 58, 1437-46.
[http://dx.doi.org/10.1016/j.marpolbul.2009.06.014] [PMID: 19635625]

Pérez-Carrera, E, León, VML, Parra, AG & González-Mazo, E (2007) Simultaneous determination of pesticides, polycyclic aromatic hydrocarbons and polychlorinated biphenyls in seawater and interstitial marine water samples, using stir bar sorptive extraction-thermal desorption-gas chromatography-mass spectrometry. *J Chromatogr A,* 1170, 82-90.
[http://dx.doi.org/10.1016/j.chroma.2007.09.013] [PMID: 17915232]

Ron, EZ & Rosenberg, E (2002) Biosurfactants and oil bioremediation. *Curr Opin Biotechnol,* 13, 249-52.
[http://dx.doi.org/10.1016/S0958-1669(02)00316-6] [PMID: 12180101]

Sures, B (2001) The use of fish parasites as bioindicators of heavy metals in aquatic ecosystems: a review. *Aquat Ecol,* 35, 245-55.
[http://dx.doi.org/10.1023/A:1011422310314]

Suresh Kumar, K, Dahms, H-U, Won, E-J, Lee, J-S & Shin, K-H (2015) Microalgae - A promising tool for heavy metal remediation. *Ecotoxicol Environ Saf,* 113, 329-52.
[http://dx.doi.org/10.1016/j.ecoenv.2014.12.019] [PMID: 25528489]

Tieyu, W, Yonglong, L, Hong, Z & Yajuan, S (2005) Contamination of persistent organic pollutants (POPs)

and relevant management in China. *Environ Int,* 31, 813-21.
[http://dx.doi.org/10.1016/j.envint.2005.05.043] [PMID: 15982740]

Vala, AK (2010) Tolerance and removal of arsenic by a facultative marine fungus *Aspergillus candidus*. *Bioresour Technol,* 101, 2565-7.
[http://dx.doi.org/10.1016/j.biortech.2009.11.084] [PMID: 20022490]

Vala, AK & Dave, BP (2017) Marine-Derived Fungi: Prospective Candidates for Bioremediation. *Mycoremediation and environmental sustainability* Springer International Publishing, Cham 17-37.
[http://dx.doi.org/10.1007/978-3-319-68957-9_2]

Williams, C (1996) Combatting marine pollution from land-based activities: Australian initiatives. *Ocean Coast Manage,* 33, 87-112.
[http://dx.doi.org/10.1016/S0964-5691(96)00046-4]

Wittbrodt, J, Shima, A & Schartl, M (2002) Medaka-a model organism from the far East. *Nat Rev Genet,* 3, 53-64.
[http://dx.doi.org/10.1038/nrg704] [PMID: 11823791]

Wong, P, Chau, Y, Kramar, O & Bengert, G (1982) Structure–toxicity relationship of tin compounds on algae. *Can J Fish Aquat Sci,* 39, 483-8.
[http://dx.doi.org/10.1139/f82-066]

Xu, W, Wang, X & Cai, Z (2013) Analytical chemistry of the persistent organic pollutants identified in the Stockholm Convention: A review. *Anal Chim Acta,* 790, 1-13.
[http://dx.doi.org/10.1016/j.aca.2013.04.026] [PMID: 23870403]

Zafra, G, Absalón, ÁE, Anducho-Reyes, MÁ, Fernandez, FJ & Cortés-Espinosa, DV (2017) Construction of PAH-degrading mixed microbial consortia by induced selection in soil. *Chemosphere,* 172, 120-6.
[http://dx.doi.org/10.1016/j.chemosphere.2016.12.038] [PMID: 28063314]

CHAPTER 2

Sampling Pelagic Marine Organisms

Ricardo Teles Pais[1] and **M. Ramiro Pastorinho**[1,2,3,*]

[1] *CICS-UBI, Health Sciences Research Centre, University of Beira Interior, Covilhã, Portugal*

[2] *Department of Biology, University of Evora, Evora, Portugal*

[3] *Faculty of Health Sciences, University of Beira Interior, Covilhã, Portugal*

Abstract: Marine life remains far less well documented than terrestrial biodiversity. The main reason resides in the vastness of the ocean. Ocean waters, with an average depth of ≈3,800 m, cover 71% of the world's surface. The difficult access, the complexity of the logistics (any study below the top few meters of the ocean requires large means, specialized personnel, and equipment), and the high cost of research have determined the majority of studies being performed in the terrestrial environment. However, in recent times, this severe imbalance has started to reverse. This is mainly due to the implementation of supra-governmental cooperation programs. Due to human-driven ecosystems alteration, over-fishing, ocean acidification, and chemical pollution (together with other threats), multiple marine species are endangered, so this effort is more than ever relevant and eminently urgent. Recently, the Global Ocean Observing System (GOOS) has proposed, the development of an integrated framework for continued and systematic ocean observation. This framework is based on Essential Ocean Variables (EOVs) aiming to provide a credible response to scientific and societal issues, a high feasibility for sustained observation, and cost-effectiveness. Ecosystem EOVs have been developed. In this framework, biodiversity will be assessed based on the status of ecosystem components, nominate phytoplankton biomass and diversity, zooplankton biomass and diversity, fish abundance and distribution (as well as marine turtle, bird and mammal abundance and distribution). Recommendations for each EOV, including what measurements are to be made, but up to this point those recommendations do not exist. This chapter will try to identify common sampling procedures for the most diverse and abundant marine organisms considered as ecosystem components under the EOVs, *i.e.*, phytoplankton, zooplankton, and fish.

Keywords: Marine Environment, Essential Ocean Variables (EOVs), Phytoplankton, Zooplankton, Fish.

---

* **Corresponding author M. Ramiro Pastorinho:** Department of Biology, University of Evora, Evora, Portugal; Tel: +351 234370350/768; Fax: +351 234372587; E-mail: rpastorinho@uevora.pt

De-Sheng Pei & Muhammad Junaid (Eds.)
All rights reserved-© 2019 Bentham Science Publishers

INTRODUCTION

The most consensual agreed definition of biodiversity, "the variability among living organisms from all sources including, inter alia, terrestrial, marine and other aquatic ecosystems and the ecological complexes of which they are part; this includes diversity within species, between species and of ecosystems", can be found in article 2 of the Rio de Janeiro Convention on Biological Diversity (GBO, 2014). This binding agreement had the conservation of biodiversity at its core, and it makes clear that already by 1992 (when realities like global warming and climate change were just the concern of a few), biodiversity was recognizably facing accentuated alteration under the pressure of growing anthropogenic impact. Two and a half decades later, protection measures, either at species or ecosystems levels, are still infrequent. Moreover, a broad understanding of all of the components and functions of marine ecosystems as well as a thorough registry of marine biodiversity are lacking. Biological diversity has to be documented and understood before it can be totally preserved (Zampoukas *et al.*, 2014).

Marine life remains far less well documented than terrestrial biodiversity. Considering the major taxa, current knowledge indicates that diversity is much greater in the sea as compared to freshwater or land. Thirty-two of the currently recognized 34 animal phyla occur in oceanic waters, being 16 exclusively marine. Other major animal phyla, including the cnidarians, sponges, as well as the non-metazoan brown (Phaeophyta) and red algae (Rhodophyta) are largely marine (Chapman, 2009). This reflects the ocean as the cradle of life. However, species diversity is far lower in the sea (≈250,000 species registered) than on land (1.4 - 1.7 million). The main reason possibly resides in the vastness of the ocean. Ocean waters, with an average depth of ≈3,800 m, cover 71% of the world's surface. As a result, the marine environment is physically much less variable in space and time than the terrestrial environment, lowering genetic connectivity and speciation rates (Paulay & Meyer, 2002). Additionally, the most diverse group within the animal kingdom, the insects, together with that in the plant kingdom, the angiosperms, is largely restricted to terrestrial and freshwater environments. The higher species richness of the terrestrial and freshwater habitats together with a comparatively higher easiness of access (any study below the top few meters of the ocean requires large means, specialized personnel and equipment being, thus, highly expensive) have determined the majority of studies being performed in the terrestrial environment. According to Hendriks and Duarte (2008), of the 13336 articles concerning biodiversity published between 1987 and 2005, 72% addressed terrestrial ecosystems. However, in recent times, this severe imbalance has started to reverse. This is mainly due to the implementation of supra-governmental international cooperation programs, such as the United Nations' The World Ocean Assessment, the Oslo and Paris Commissions (OSPAR and cooperating entities on data collection, *e.g.*, ICES), the HELCOM Monitoring and Assessment Strategy, the Convention on Protection of the Black Sea Against Pollution, and multiple EU funded projects.

Human-driven ecosystems alteration, over-fishing, ocean acidification and chemical pollution (together with other threats) endanger marine species. Many mammals, birds, reptiles, and fish are currently in danger of extinction (Mark J. Costello, 2015; Mark J. Costello & Scott Baker, 2011; Webb & Mindel, 2015). Global, regional, and local scale assessments need data collected by similar methods and procedures in order to produce variables that can be integrated for analyses (Pereira *et al.*, 2013). The EU Marine Strategy Framework Directive (2008/56/EC) requires that European marine waters achieve a Good Environmental Status (GES) by 2020 (Boero, Dupont, & Thorndyke, 2015). It links ecosystem components, anthropogenic pressures and impacts on the marine environment and it contains the explicit regulatory objective that "biodiversity is maintained by 2020", as the cornerstone for achieving GES. For this, an extensive system of measures of biodiversity and ecosystem functioning was determined. The European Commission produced a set of detailed criteria and methodological standards (Commission Decision 2017/848 of 17 May 2017) to obtain and report those measures in order to help Member States implement the Marine Directive at local and regional scales. Similarly, UN's World Ocean Assessment emphasizes the need for more standardized reporting of information (Inniss *et al.*, 2016).

The ocean environment is vast, the marine biosphere difficult to access. The remoteness, harshness, and depth of the ocean make them challenging to study and dramatically raise the cost involved in its observation. Duplication of efforts should be avoided. Cutting across observing platforms and networks, and the adoption of common standards for data collection and dissemination to maximize the utility of data are imperative. Recently, the Global Ocean Observing System (GOOS) has proposed, under the umbrella of the Intergovernmental Oceanographic Commission (IOC) of UNESCO, to develop an integrated framework for continued and systematic ocean observation. This framework is based on what was defined as Essential Ocean Variables (EOVs). By definition, an EOV should provide i) a credible response to scientific and societal issues; ii) a high feasibility for sustained observation, and; iii) cost-effectiveness. Among other domains, ecosystem EOVs have been developed in collaboration with the Group on Earth Observations (GEO BON) (Pereira *et al.*, 2013). Up to this point, the defined ecosystem component EOVs directly dealing with biodiversity consist of those related to the status of ecosystem components and those related to the extent and health of ecosystems. The former is phytoplankton biomass and diversity, zooplankton biomass and diversity, fish abundance and distribution, marine turtle, bird and mammal abundance and distribution; the latter, cover and composition of hard coral, seagrass, mangrove, and macroalgal canopy. Additionally, "emerging" EOVs "benthic invertebrate abundance and distribution" and "microbe biomass and diversity" are being put forward "to be developed based on emerging requirements and new technologies" (Miloslavich *et al.*, 2018). In order to maintain common standards, GOOS will put forward a series of recommendations for each EOV including what measurements are to be made, various observing options, and data management practices. Up to this point, those recommendations do not exist.

Miloslavich *et al.* (2018) recently applied a driver-pressure-state-impact-response (DPSIR) model testing the EOVs on such principles such as relevant for science, society information, and technologically feasibility. They concluded upon the relevance of the different priority variables (based on the measurement of undergoing or past monitoring programs) that microorganisms, birds, and mammals were the ones with lower Relevance Index (RI) that based on the SCOPUS database estimates how each of the variables addresses the convention's drivers and pressures.

For the above reasons, this chapter will try to identify common sampling procedures for the organisms considered as ecosystem components under the EOVs, *i.e.*, phytoplankton, zooplankton, and fish.

The structure and organization of aquatic communities are molded in each environment by combinations of abiotic factors, recruitment, and productivity rates, and rely upon complex interactions among organisms that are both pairwise and transitive (Piraino, Fanelli, & Boero, 2002). As such, biological diversity can solely be maintained when the quality and occurrence of habitats and the distribution and abundance of species are in line with favorable physiographic, geographic and climatic conditions that need to be measured and correctly interpreted. Spatial and temporal variation in biodiversity (and its differential loss) from local to global scales drive the need for measurement. So in order to be able to know and understand the role and patterns of marine biodiversity, marine ecological research should take resource on experts from all scientific (Guerra-Garcia, Espinosa, & Garcia-Gomez, 2008).

Along these lines of thought, the measurements being proposed by GOOS will have to reflect the idea that only monitoring biodiversity with a long-term approach at a large scale can fulfill the objectives underlying the formulation of EOVs (Andréfouët, Costello, Rast, & Sathyendranath, 2008; Bianchi *et al.*, 2000; Mark J. Costello *et al.*, 2017; Krug *et al.*, 2017). Moreover, despite their formulation in the context of a big consortium, consideration on field observations and sampling operability and robustness should be granted in order to allow their scaling down and adaptation to local or regional surveys.

Since, as yet, how to measure and manage these variables and analyze the resulting data have not been object of detailed technical reports (only generic spec sheets are available), as already noted, here we review methods used for field observations and sampling marine biodiversity that seem to us as being good candidates for inclusion in those recommendations. However, a complete review of methods for the study of all marine biodiversity is outside the scope of this chapter. We will include those conventional methods that have been established as suitable, valid and cost-effective for monitoring biodiversity as well as less prominent, new approaches being currently implemented in hopes that this will provide a window into future marine biodiversity biomonitoring and assessment.

SAMPLING THE MARINE ENVIRONMENT

Sampling is the process of picking out single objects, items, or organisms on which to take the measurement(s). The collection of measurements (the sample) should reflect or represent the population under study. In practical terms, this means that the sample should be a small version of the studied population and should contain all the characteristics (and their variability) inherent to the population in general (Bar-On, Phillips, & Milo, 2018). For marine species, a wide array of methods has been used in order to achieve this objective. In general terms, they allow capture, detection and/or observation of marine organisms and include observation instruments (*e.g.*, cameras), nets, hooks, traps, grabs, sound, chemicals and electricity (Eleftheriou, 2013; Elliott & Hemingway, 2002; Hiscock, 2014; Kingsford & Battershill, 1998; Santhanam & Srinivasan, 1994; Tait & Dipper, 1998). Since any of these methods is selective (a good example are nets that according to mesh size, select organism size by exclusion/inclusion), a comprehensive sampling of marine biodiversity ideally needs to apply complementary methods. The overlapping of methods can produce an inventory of species present that reflect the environment, habitats, and ecology of an area.

In classic ecology, biodiversity measurement has two components: species richness and evenness. Of the two, species richness is by far the most common measure of 'diversity' used in science and conservation management and it is usually reached by the establishment of species inventories than can estimate that parameter (M. Costello, Pohle, & Martin, 2004; Gotelli & Colwell, 2001). Only by knowing which species are present can we determine those that are pests, those who are introduced, endemic, of ecological importance, socio-economically relevant, or most importantly, threatened with extinction (McGeoch *et al.*, 2016).

On the other hand, there is not a priority method to evenness, and several metrics have been applied including numbers of individuals, areal cover, and/or biomass within samples (Hiscock, 2014). Measurements in marine biodiversity thus need, as already established, to acknowledge that due to intrinsic bias, various methods have to be utilized, but also that different methods assess different components of biodiversity. For instance, the measurement of particular species populations dynamics is better achieved through quantitative sampling, the same not being true for measuring biodiversity across species. For the latter, the comparison of relative abundance of species on semi-quantitative (*e.g.*, log 10) abundance scales delivers much better results (Haegeman *et al.*, 2013; Hiscock, 2014).

Communities and biotopes are defined by the more abundant species and provide an indication on the functioning of the ecosystem functions in parameters like habitat, productivity, and food webs. A dominant species change can be interpreted as a space-time change in the communities present, and consequently ecosystem change. This, however, can be proven to not hold true as often abundance or body size do not relate to their effects upon the ecosystem, being top predators a fine example. This leads to the evidence that in order to monitor ecosystems, species from different guilds and body sizes should be simultaneously sampled (Hiscock, 2014). Additionally, makes evident that the time and place of sampling are also determinant.

If the objective is to maximize the potential biodiversity encountered, then the selected site should have high habitat heterogeneity. However, considerations of the area to sample and accessibility to a laboratory should weigh in on this decision. The past has proven that for the results of a survey, it is most effective to choose accessibility over diversity. Covering remote, extensive areas from a shore-based field lab is logistically difficult and ultimately inefficient. When using small vessels in pelagic and benthic sampling, the nearby presence of a mooring station equipped to provide logistic support (ideally, a lab equipped for the immediate sorting and processing of samples), has proven well advised. Sampling stations should be selected in order to span the range of target habitats, at each of which a broad range of sampling techniques should be utilized. Collecting background information from the literature, and the inclusion, if possible, of preliminary sampling of the area, are very useful in acquiring a scope of a region, identifying a balanced survey site, and on the choice of sampling stations.

During the field surveys design phase the stratification of sampling has to be made very clear: what habitats, body sizes, and taxa are the focus on, and what has to be excluded. These decisions should be as well informed as possible and the knowledge of environmental variables such as depth, salinity, temperature, substratum, topography variation can provide a powerful aid. The geographical mapping of these variables has been effectively performed by remote sensing (from satellites, aircraft and ships) (Andréfouët *et al.*, 2008) constituting a powerful aid in the determination of essential variables modelling ecosystems providing information on seabed depth, topography, and roughness, temperature, salinity, estimations of phytoplankton biomass and dominance, and acoustic signatures of zooplankton and pelagic megafauna. More than environmental variables, remote sensing provides information on distribution and extent of intertidal and shallow-water habitats such as coral reefs, kelp and seagrass beds, mangrove forests, and salt marshes (M. Costello *et al.*, 2004; Mark J. Costello, 1992; M. J. Costello, 2009; Hiscock, 2014). The use of new technologies does not end at remote sensing. The proliferation of remotely operated (ROVs) and autonomous vehicles (AUVs) has made them affordable. Together with the development of highly sensitive, miniaturized, detection devices that can be mounted into these ROVs and AUVs, have created a new era of underwater and aerial monitoring capability. As a result, undergoing research is evaluating the potential of sound signatures as indicators of biodiversity in the marine environment (Harris, Shears, & Radford, 2016). These advices not only act as effective data gatherers but also can effectively complement and fill in gaps in larger-scaled datasets obtained by more complex and sophisticated means (*e.g.*, satellites, airplanes) thus complementing *in situ* observations and enabling habitat and biotope mapping (Hiscock, 2014; Leleu, Remy-Zephir, Grace, & Costello, 2012; Remy-Zephir, Leleu, Grace, & Costello, 2012). Other technological contributions come in the form of data processing. Examples are the automated recognition of species from videos or photos aiding in population census or the growing computing power used in bioinformatics.

BIODIVERSITY OF OCEAN ZONES

The different ocean zones and major ecosystem are shown in Fig. (1). The pelagic realm is characterized by marked vertical gradients of light, temperature, pressure, nutrient availability and salinity that contribute to creating a vertical stratification of pelagic species assemblages that normally fluctuate in time and space. The surface, euphotic waters are dominated in numerical terms by plankton. These are passively floating, drifting or weakly swimming organisms, and include a wide range of bacteria, protists, tiny algae, small animals, and developmental stages (eggs, larvae, *etc.*) of larger organisms. Those that possess the ability to swim freely are collectively termed nekton. Both plankton and nekton usually concentrate along gyres (major circulation currents), upwelling areas and contact zones, resulting in wide variations in abundance and diversity. Some elements of the epipelagic and mesopelagic plankton and nekton (in its majority) execute diel migrations. They will rise to surface waters during the night only to dive during the day (Groombridge & Jenkins, 2002). These migrations aim at feeding and predator avoidance, respectively. For the pelagic inhabitants of the aphotic zone, the main sources of foods are the molted exoskeletons, feces, and corpses from the organisms inhabiting the euphotic zone. The biomass of the mesopelagic area has been proved to be high, with around 160 fish genera being recognized as important components of the mesopelagic fauna (Groombridge & Jenkins, 2002).

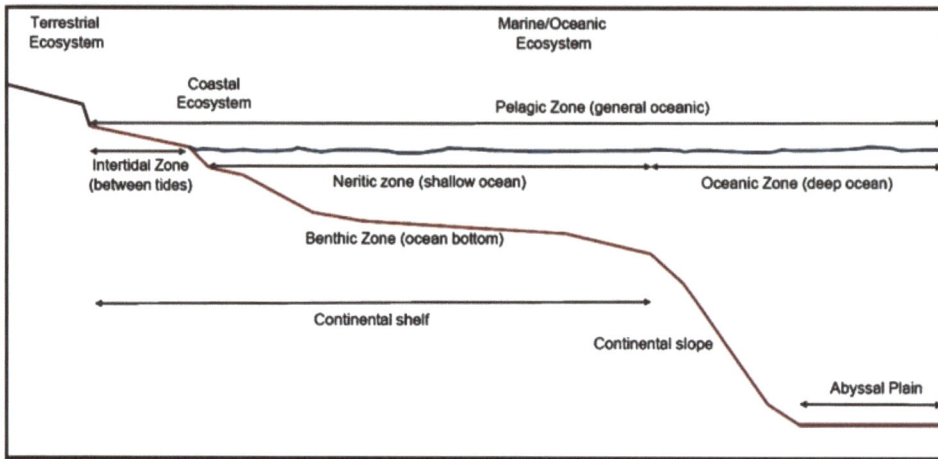

Fig. (1). Ocean zones and major ecosystems.

The sampling methods used in the study of pelagic fauna are mainly those employed in fisheries and oceanography, needing the use of sea-worthy vessels. There is an entire branch of science dedicated to nektonic species ("fisheries oceanography") with extensive bibliography published in dedicated journals and well-established techniques, among which are tagging and real-time tracking, which can provide estimates of the global distribution and abundance of the largest animals in this realm. Only consortia with large financial capacity can perform this expensive research. This assertion is even truer regarding deep-sea. There is not one overall accepted definition for "deep-sea" (for a review see M. J. Costello (2009). However, the more frequent operative definition is usually the one that considers it as the part of marine realm from zero light compensation depth (usually considered 200m) up to the bottom of an ocean, which in the case of approximately 50% of the Earth's surface has a depth of over 3,000 m. This makes deep-sea sampling costly and time-consuming. For instance, collecting a sample from a depth of 8,000m with towed gear, requires a very powerful winch with and at least 11 km of cable, taking up to 24 hours to let out the wire, obtain a sample, and retrieve it, costing on the excess of 20,000$ in ship time. Trawls, bottom sledges, dredges, grabs, box samplers and corers, as well as several acoustic and optical approaches have been classically used in these surveys. During the 1970s, the introduction of submersibles in deep-sea research became well established during the FAMOUS project (French-American Mid-Ocean Underwater Study) which used the manned submersibles Cyana, Archimède, and Alvin (Heirtzler & Grassle, 1976). More recently, several types of remotely operated vehicles (ROVs) from surface vessels or autonomous underwater vehicles (AUVs) pre-programmed to perform their tasks independently of direct human control, have been using recording equipment to document deep-sea organisms, with some being equipped to collect samples (Chave, 2004; Clarke, 2003).

The discovery, in 1978, of new and abundant sea life around deep-sea hydrothermal vents near the Galapagos Islands subverted the idea that the deep oceans to be relatively simple ecosystems that made little contribution to global species diversity, and proving that this diversity could rival that of coral reefs (Grassle & Maciolek, 1992). It has been determined that the distribution is patchy, which makes sampling more difficult, but the increase on gear sophistication has contributed to the knowledge of deep-sea biodiversity. The growing knowledge as indicated that areas such as seamounts and rock outcrops, submarine canyons, beds of manganese nodules, deep-water reefs of ahermatypic corals, hydrothermal vents, cold seeps, and other chemosynthetic ecosystems such whale skeletons or sunken wood are areas of unusual diversity, particularly that of benthic organisms (Dybas, 2004).

It is easy to acknowledge that deep-sea and deep-water samplings demand a set of procedures, techniques, and equipment that are different from those used in near-shore surveys, and that only great joint, usually international efforts can proceed these kinds of studies. For that reason, this review will not undergo their description. Several authors have extensively reviewed deep-water (Arkhipkin *et al.*, 2015; Gabriel, Lange, Dahm, & Wendt, 2007; Siebert & Nielsen, 2001) and deep-sea (Gage & Tyler, 1991; Wenneck, Falkenhaug, & Bergstad, 2008) and can be consulted by for those searching for more information on these disciplines.

Coastal waters are richer than open ocean areas due to the greater range of habitats. This is particularly true for benthic ecosystems whose biodiversity is much higher than that of pelagic ecosystems (Grassle & Maciolek, 1992). Despite representing less than 10% of ocean area, the continental shelve sustains most of the documented marine biodiversity, with estimations pointing to more than 75% of known marine species (Appeltans *et al.*, 2012). Because of the economic benefits obtainable from access to coastal fisheries, ocean navigation, tourism, and recreation, human settlements are concentrated in the coastal zone. According to the UN Atlas of the Oceans (http://www.oceansatlas.org/subtopic/en/c/114/), 44% of the world's population (more people than inhabited the entire globe in 1950) live within 150 kilometers of the coast. Simultaneously, the coastal population is increasing disproportionately to the global population increase. As population density and economic activity in the coastal zone increases, pressures on coastal ecosystems increase, with key coastal habitats being lost globally at rates 2 to 10 times faster than those in tropical forests (M Reaka-Kudla, 1997). This awards enormous importance and urgency to the study of coastal areas biodiversity.

Fig. (2). illustrates the sampling techniques used in biodiversity surveys for each of the EOVs defined by GOOS regarding ecosystem component, more specifically those related to the status of ecosystem components, nominated phytoplankton, zooplankton and fish. The details are discussed in the following section.

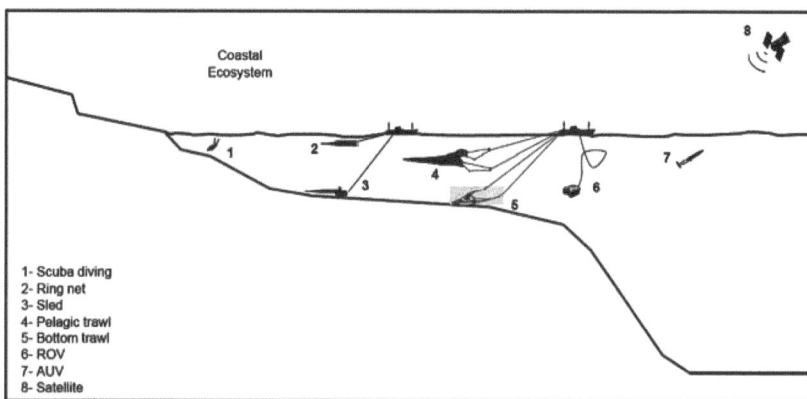

Fig. (2). Generic illustration of the several sampling methods described. ROV – Remotely Operated Vehicle; AUV- Autonomous Underwater Vehicle.

Phytoplankton

Phytoplankton, also known as micro-algae, are single-celled, photoautotrophic microorganisms. Phytoplankton are present in all surface ocean waters although densities can vary greatly between localities and with seasons. Photosynthesis by marine phytoplankton contributes roughly half of the total global primary production, being oceanic phytoplankton the major contributor, with a smaller contribution (approximately an order of magnitude lower) belonging to coastal phytoplankton (Cermeño, Teixeira, Branco, Figueiras, & Marañón, 2014). Phytoplankters dependence on sunlight for photosynthesis means they are restricted to the euphotic zone in the upper 50–100 m of the pelagic area.

Towed nets have been and are still the primary means of collecting many plankters, being this true for both phyto and zooplankton. Nevertheless, even when only nets are considered, both groups cover a large number of taxa whose members vary widely in size, meaning that there is not an "ideal" net for the entire plankton community. The choice is conditioned by which organisms are to study and by the available resources. In general, nets vary in length, shape and mesh size but all are designed to extract from the water column drifting or relatively slow-moving organisms, which will be retained by the mesh. The simplest model for these nets ("ring net") consists of a cone-shaped mesh with a wide open end supported by a metallic ring, whereas the opposite side (the "cod end") is closed by a collecting jar. This kind of net can be towed vertically, horizontally, or obliquely through the desired sampling depths. More sophisticated nets can be opened and closed at selected depths in order to obtain more detailed information on the vertical distribution of plankton. The additional possibility of attaching several of such nets to the same frame exists (*e.g.*, the computer controlled MOCNESS - Multiple Opening/Closing Net and Environmental Sensing System(Wiebe *et al.*, 1985) and the, either manual or automated MultiNet - Multi Plankton Sampler (Weikert & John, 1981) allowing sampling of different discrete depths during a single tow. Many authors have described plankton with different, complementary degrees of detail and their reading can cater to different needs in terms of sampling (Goswami, 2004; Harris *et al.*, 2016; Suthers & Rissik, 2009; Tranter and Fraser, 1968).

A phytoplankton net is a generic term for a sampling net with a mesh size of 150 μm or less. The phytoplankton version of the ring net has 10μm mesh size, a 36 cm diameter mouth (0.1 m²), and a length of 90 cm. The collection bottle has side windows covered in the same mesh as the body of the net allowing water to pass while retaining organisms. The net can be pulled vertically from depths of 50 m to the surface at 0.1 m.sec⁻¹.

An alternative in terms of quantitative phytoplankton studies in the open sea, is the collection of samples using a hose. An integrated sample from between 0-10 m depth is obtained by pooling equal amounts of water from the depths 0 – 1 m, 2.5 m, 5 m, 7.5 m and 10 m with the several fractions collected to the same bucket and thoroughly mixed (Lindahl, 1986). Quantitative phytoplankton counts can be obtained from an aliquot of 200 cm³ of this composite, well-mixed sample. If logistics allow, an additional sample should be obtained from 10 – 20 m by a similar process at the same depth intervals. A similar volume (200 cm³) of the same integrated sample should be used for chlorophyll a determination and, if desired, primary production. A ring net sample from the 0 – 20 m water column should be collected. This will render a concentrated plankton sample that will support species identification. In cases of high phytoplankton densities, it is advisable to use a net with 25 μm mesh-size instead of the regular 10μm. It is important to make here a note regarding the referenced volume. Cermeño *et al.* (2014) examined large volumes of seawater to explore the limits of phytoplankton diversity and concluded that estimates of species richness depend critically on sampling effort. Necessary volumes of sample needed to detect 90% of the species varied between 250 and 1000 cm^3 depending on the concentration of phytoplankton biomass.

Abundance is strongly influenced by the highly abundant picoplankton and small nanoplankton, which can be analyzed only with limited certainty. Biomass data are a much better descriptor of phytoplankton, being preferred for characterizing spatial and temporal phytoplankton patterns and modelling. Phytoplankton biomass can be expressed as cell volume (or weight), and Olenina *et al.* (2006) thoroughly describes the necessary steps to perform the necessary calculations. However, carbon content should also be calculated, since organic carbon is the universal energy currency transported along the food chain.

Additionally, to direct sampling methods such as fixed survey stations, and ships-of-opportunity transects, indirect observation methods such as satellite images and aerial surveillance can add to the identification of temporal and spatial variability in phytoplankton. By using sensors, aircraft or satellites can detect reflected (external) or emitted (internal) energy from the earth. Electromagnetic radiation (EMR) from the sun, reflected back from the ocean is one example of an external force detectable by satellites. EMR received contains information, being this information influenced by particulates in the atmosphere or in the ocean water. This knowledge allows the interpretation of phytoplankton variation, since photosynthetic pigments contribute to the spectral change in radiance, particularly during blooms. Pigment composition varies with the type of microalgae, absorbing a specific wavelength of the visible spectrum. This, together with the specific light scattering pattern promoted by differing structural components (*e.g.*, size, shape, type of cell covering) creates an optical signature for each microalgae species, permitting recognition by the satellite. A number of tools, including computer algorithms, makes this recognition. The evolution of algorithms has been very important in the detection and monitoring of algal blooms (Blondeau-Patissier, Gower, Dekker, Phinn, & Brando, 2014).

Planktonic assemblages are strongly affected by physical and chemical characteristics of water masses on scales ranging up to entire ocean circulations. The vertical structure of the water column is also important, especially the depth of the mixed layers, as this influences nutrient and light levels that control phytoplankton growth and assemblage composition. For this reason, phytoplankton shows a substantial seasonal variation, and obliging sampling to cover the entire growth season, which in parts of the ocean extend over the entire year.

Pronounced nutrient enrichment may give rise to dramatic shifts in phytoplankton species composition and biomass, with these blooms becoming dominated by harmful species. This proliferation of microalgae in marine or brackish waters is known as Harmful Algal Blooms (HAB) and can cause massive fish kills, contaminate seafood with toxins, and alter ecosystems in ways that humans perceive as harmful. When such events take place, sampling methods include plankton net hauls, water samples collection and detection of species-specific algal toxins in fish and shellfish using molecular techniques (Reguera, Alonso, Moreira, & Méndez, 2011). To this day, even by using a light microscope, the identification of certain HAB species is difficult or impossible. There has been a great effort from the scientific community on developing a consistent taxonomy and make observation and quantitative methods uniform (Babin, Cullen, & Roesler, 2008; Hallegraeff, Anderson, & Cembella, 2003; Karlson, Cusack, & Bresnan, 2010). Despite these efforts, synoptic surveys are necessary for the study of the extension of phytoplankton blooms.

Due to its uniqueness, a survey has to be mentioned here: The Continuous Plankton Recorder (CPR). Being held since 1931, it is the longest sustained and geographically most extensive marine biological survey in the world regarding phyto and zooplankton. The survey is operated by the Sir Alister Hardy Foundation for Ocean Science (SAHFOS), which by cooperating with 30 different shipping companies, collects samples at monthly intervals on 50 trans-ocean routes. These ships of opportunity cover a total of approximately 20,000 km/month and collect plankton on a band of silk that subsequently undergoes identification by experts, with 698 taxa being routinely surveyed (Martin Edwards, Beaugrand, Hays, Koslow, & Richardson, 2010). In 2011 SAHFOS, joined forces with 12 other research organizations using a CPR type of approach on plankton survey (low cost, application of new technologies) formed a Global Alliance of CPR surveys (GACs). The objectives are the harmonization and development of new surveys and the establishment of a centralized global database, and producing a global ocean status report (M Edwards *et al.*, 2012). Working together, centralizing the database and working in close partnership with the maritime shipping industry, this global network of CPR surveys with its low costs and new technologies makes the CPR an ideal tool for an expanded and comprehensive marine biological sampling program.

Zooplankton

Zooplankton are consumers that eat other plankton and thus provide an important link between primary producers (phytoplankton) and higher trophic levels. Some are herbivorous, whilst others are carnivorous predating on smaller zooplankters. Zooplankters are found in all oceans, from the surface to the deepest trenches. However, despite the extremely high biomass, their diversity is low. This reality, extensible to phytoplankton, finds its foundations in the dynamic mixing of oceans that heavily contributes to limiting geographic differentiation. Of the animals that spend their entire lives as plankton, the holoplanktonic zooplankton, which dominate both numerically and in total mass, only ≈3,700 species were described (Groombridge & Jenkins, 2002). However, since many benthic species have a planktonic larval or egg phase (meroplankton), most animal phyla are represented in the zooplankton. Of the 34 marine animal phyla, only 13 have representatives in the holoplankton.

Similar to phytoplankton, zooplankton pumps can also be used, operating at selected depths; these will pull water that will be filtered through a mesh. These can be portable or moored with some of the latter being automated (*e.g.*, the Moored Automated Serial Zooplanktic Pump - MASZP). In this case, discrete plankton samples are immediately preserved within two interwoven meshes and stored for posterior microscopic identification. Since this process generates a large number of samples, species-specific immunofluorescent markers have been developed in order to accelerate processing (Garland & Buntman, 1996).

Small dredges, grabs, traps, plankton nets, and other gear have traditionally been deployed from small boats in zooplankton research, bypassing the necessity of large, expensive research vessels. Small boats can accommodate small cranes and/or pulley systems, aiding on the retrieval of deployed gear.

It should be noted that, evidently, large surveys using great means are also undertaken in the context of institutional and international initiatives and that zooplankton is one of the focal points of the already described Continuous Plankton Recorder (CPR) survey.

The standard zooplankton net is a simple, conical shaped ring net. It is usually employed to sample planktonic organisms of sizes 0.2–10 mm from the upper 200 m of the water column. Among many varying designs, maybe the most famous model is the WP-2 net. This net was designed in an effort to standardize field equipment and best practices by a group of scientists that named themselves "the Working Party no. 2", hence the name WP-2 (Tranter & Fraser, 1968). The WP-2 has a diameter of 0.57 m (0.25 m^2) mouth opening and mesh size of 200 μm. The cod end is removable, being fitted with filtering side panels of similar mesh to the body of the net, so that the sample can be effectively concentrated. The end of the net is weighted down (15–20 kg) to aid sinking. A bottle type-closing device (*e.g.*, Van Dorn, Nissen) can be fitted to the net in order to obtain samples from distinct depths.

To obtain vertical samples, the net is hauled from the desired depth at a steady speed of 0.5m sec^{-1}, after having been lowered at no more than 1 m sec^{-1}. A wire angle of less than 30° has to be maintained during ascension. The calculation of filtered water volume is made by multiplying mouth area by distance. However, the generalization of flowmeters has permitted more accurate calculations, since the net can be clogged by, *e.g.*, neston and not filtering as much volume of water as originally calculated. The sample is retrieved from the cod end with the help of a hose or squirt bottle with seawater, and taken into the lab for processing. It is of paramount importance that the samples are transported under stable, adequate temperature conditions. Other very frequently used net is the WP-3. Similar to the WP-2, but with an opening diameter of 113 cm (1 m²) and a bag with a length of 200 cm, this net was designed to sample large sized plankton (>1 mm). Due to the increased size and weight, towing the WP-3 has to be performed at speeds below 0.3 m sec⁻¹. The cod end in this net does not possess side windows in order to be able to capture delicate zooplankton. Captures of delicate, sensitive zooplankton can also be attained by direct sampling by scuba diving or from submersibles. This enables the collection of important components in pelagic environments, such as large-bodied jelly-plankton (colonial radiolarians, medusae, ctenophores, salps, *etc.*) that were either undersampled or destroyed by traditional methods.

Near-bottom zooplankton (usually meso- and macro-zooplankton) can be sampled by sledges (or sleds) towed from a boat. There are many designs of sledges, varying from exceedingly simple (*e.g.*, a metal frame with a collecting net) to complex, *e.g.*, multiple nets to allow sampling various distances to the bottom within one haul. Since damages to the net can occur, due to contact with the bottom (especially when operating above hard substrata), simple designs present an advantage by being simple to repair, which can be performed on board. These sleds typically are built in steel with some variations in size. A popular layout is a frame made with 2.4 mm thick, 20.6 mm diameter tubing and 4.8 thick stainless-steel sheet metal (for the skis), with a circular mouth 50 cm in diameter (0.16 m² area). A conical macrozooplankton net, 2m long with a 500 μm mesh is secured to the frame, and the cod end receives a 1000 mL polyethylene jar screened with 400-500 μm wire mesh. An inner mesozooplankton housing is added, consisting of a 150 μm mesh conical net, with a 250 mL polyethylene jar with 140-150 μm wire mesh at the cod end. Since the overall weight is around 15 kg, a buoy is secured to the top of the frame, and differentially inflated to maintain neutral buoyancy at the desired depth.

Multiple holoplanktonic species are identifiable by modern acoustic or optical imagery. These resources belong to the realm of Remote Sensing. Despite not being, by definition a sampling method, Remote Sensing is a powerful approach to the marine environment enabling better quantitative estimates of biomass and distribution of zooplankton. In the broadest sense, remote sensing is the measurement or acquisition of information of an object or phenomenon, by a recording device that is not in physical or intimate contact with the object. This is particularly important when dealing with delicate plankton forms that usually are destroyed when direct sampling techniques. In practical terms, it consists in the utilization from a distance (*e.g.*, from aircraft, spacecraft, satellite, or ship) of one or various devices that allow gathering information about the marine environment.

The instruments/techniques used include acoustics, optics, and Satellites. The deployment of the equipment used to measure or acquire information is frequently performed using autonomous underwater vehicles (AUV, essentially oceanographic robotic systems) and remotely operated vehicles (ROVs). The most commonly used AUVs are gliders: they use wings and small changes in buoyancy to move in an ascending and descending oscillating pattern. This will produce a profile of biological (and physical) measurements obtained by the sensors on-board that are transmitted each time the glider surfaces. Since typically these gliders need large horizontal pathways, they are more suitable for open ocean operation.

An example of the use of optics in zooplankton science is the Video Plankton Recorder (VPR). It is essentially an underwater microscope that records images of plankton (from 100 μm up to a few centimeters in size). The VPR can be towed behind a research vessel at speeds ≈4-5 knots or mounted in ROVs. A much older technique, acoustics have been used to locate and visualize distribution, abundance, and behavior of living organisms for almost a century (Horne, 2000; Simmonds & MacLennan, 2005). In marine terms, acoustics is the sending of waves of sound energy through the water from an acoustic transducer, which is also capable of receiving sound. When hitting an object with a density contrast (positive or negative) to the surrounding seawater, the transmitted sound is reflected, and the transducer will receive part of the reflection (Simmonds & MacLennan, 2005). Originally, acoustics were used to locate dense aggregations by commercial fishers, but advances and developments resulted in the ability to make quantitative measurements of abundance and studies of fish behavior. The constant development and enhancement of equipment allowed for more extensive estimates, a growth in accuracy, affordability, and more importantly the possibility of transposition into plankton studies.

Fish

Fish make up the largest fraction of the nekton (large, pelagic, marine animals able to move independently of water currents). Fish species richness is much higher in coastal areas (≈13,000 species) than in the open sea (≈1,200 species) (Angel, 1993).

In an Editorial covering the 'Fish Sampling with Active Methods' (FSAM) conference, Kubečka *et al.* (2012) indicated the relative frequencies of different gear used in marine sampling by the conference contributors. The most frequent was trawling, followed by acoustic surveys, visual surveys and gillnets, and the last with a significant use was seining. Since we already approached acoustic and visual surveys in the previous sections, we will focus on the remaining gears.

Trawls are used to sample throughout the water column, from seabed to surface, providing samples and allowing estimation of species distribution and abundance, and biological parameters (*e.g.*, age, sex, diet). A golden rule is that the trawl has to be towed at higher speed than the one that the target species can attain.

Pelagic or midwater trawls are used to capture pelagic or benthopelagic species. The trawl is composed of pelagic trawl doors, bridles, and a cone-shaped net. The opening of the net has a frame with a headrope (on top) that receives floats and a footrope (the bottom) that is weighted down. The function of the trawl doors is to open the trawl horizontally by exerting an outward force while being dragged through the water column. The vertical opening is assured by the equilibrium between headrope, footrope and the pull form the vessel. Openings can measure tens of meters in height and width. These prevent fish from escaping an approaching trawl. It is frequent for these trawls to present large mesh in the forward section to herd fish to the bottom section (cod end) where mesh sizes are sufficiently small to capture species of interest by sieving. Towing speeds are usually superior to the ones used in other sampling gears (*e.g.*, 2.5 m sec^{-1} when targeting fast swimming species).

Several strategies of operation can be used with a trawl. If the aim is to determine species distributions or define community composition, then the used approach is fishing at fixed depth intervals. Diurnal changes in abundance or composition require sampling throughout the day and night. Investigating how different layers/patches/schools differ with respect to species, size composition or diurnal variations requires detection using acoustics and following these aggregations of fish and their regular sampling. This technique is commonly referred to as target registration towing.

Originally designed for fish stock assessment, bottom (or demersal) trawl surveys are now increasingly being used to analyze trends in the abundance, distribution and diversity of both commercial and non-commercial species of fish and epibenthos (Atkinson, Leslie, Field, & Jarre, 2011). The design is similar to a pelagic trawl, but the trawl doors, lower bridles, and groundline are designed to be in constant contact with the seabed and adapted to operating over smooth seabed or rocky habitats. Demersal trawls used in research surveys are usually smaller than pelagic trawls (net opening of 2-4 m in height). Bottom trawls are customized for catching fish on different types of seabed. A horizontal metal beam is used in Beam Trawls to keep the mouth open while targeting flatfish and other near-bottom species. In otter trawls, the mouth of the net is kept horizontally open by attaching the doors to the trawl using wires. Besides sprawling the trawl horizontally, they create noise and sand clouds (especially, when being operated over fine sediments) as they travel across the seabed, which in combination with the bridles, herds fish into the path of the trawl opening. Mesh size in the front portion of demersal trawls rarely exceeds 200 mm. The cod end mesh size is, as in the pelagic versions, chosen to retain the target size individuals. A Typical trawling speed would be in the range of 1-2 m sec^{-1}.

MultiSampler equipment can replace the standard single cod end on the same logic as that used for zooplankton (Engås, Skeide, & West, 1997; Madsen, Hansen, Frandsen, & Krag, 2012) obtaining samples from distinct vertical layers, both when using a midwater trawl or from specific areas of seabed when using a bottom trawl.

Some standardized bottom trawl surveys have been conducted for the last 50 years. For example, ICES created trawl survey data database the north eastern Atlantic (www.ices. dk/ marine-data/ data-portals/Pages/ DATRAS.aspx). In addition, NOOA possesses an online database for the Eastern Bering Sea and Gulf of Alaska (www.afsc.noaa.gov/RACE/ groundfish/survey_data). These datasets constitute a powerful window into marine fish distribution, abundance, and diversity.

Nearshore areas with a low degree of inclination of the sublittoral slope can effectively be sampled by beach seine fishing. This method utilizes a seine or drag net, which consists of a bag with long wings that end in long ropes for towing the gear to the beach. The net is operated from the shore, usually with help of a small boat. The headrope is fitted with floats staying on the surface whereas the footrope is weighted down being in permanent contact with the bottom. The net is deployed in a wide arch, being one of the towing lines either fastened at shore or held by at least two people. The deployment is performed by wading, swimming or by a small boat being brought back to shore where other people are waiting to secure it. The drag lines are then hauled to shore simultaneously having care so that the footrope arrives first so that a barrier from which fish cannot be escaped is maintained it has become evident that the gear needs at least 4 different operators. This gear targets small fish and nurseries of larger species, being a prime instrument for monitoring annual variations in abundance. However, since disturbance of breeding or capture of commercial species' juveniles can occur, the beach seine is regulated or restricted in a high number of countries. There are extensive data series obtained by beach seine, such as the "Flødevigen-series", that represents results from 100 fixed stations since 1919 (Smith, Gjosater & Stenseth, 2002; Stenseth, 1999).

There is a set of "passive gears" (*e.g.*, gill, entangling and fyke nets; traps) that can be deployed either on the bottom or in the water column, and have the capacity of catching fish as they come in contact with these gears. In fact, this is one of their major criticisms: that the efficiency of sampling with passive gear depends on the activity of the fish to encounter the gear and the retention probability once a gear has been encountered.

In gillnets and entangling nets, panels of single, double, or triple netting, are set at the surface, midwater, or at the bottom. Driftnets (those deployed near the surface), as the name implies, drift freely within the bound of their connection with a vessel or a buoy. They are used to catch schooling species (*e.g.*, herring, tuna), but because of indiscriminate captures of non-target species, they have been forbidden, for instance in the EU. Bottom nets (attached to the seabed) target species living in the dependence of the bottom (*e.g.*, monkfish).

The fish are captured in by several methods. They are either gilled, *i.e.*, the mesh becomes stuck behind the operculum; become wedged when attempting to swim through the net, as this becomes tight around the body; become entangled in a pocket made of net; or, by being caught by their own projections (*e.g.*, spines, maxillaries). By using specific mesh size these are selective nets, targeting particular species and being able to capture certain length classes (Engås *et al.*, 1997).

Fyke nets are fish traps made to a cylindrical or cone shape by rings or other rigid structure forming a net bag. One or two leading nets will lead fish through a funnel structure to the bag from where they cannot exit.

Hooks are probably one of the most ancient form of catching fish, appearing nowadays in a large range of hook-and-line gears. Some of these gears can be passive, such as the longlines that can be set vertically or horizontally, being one end of the line (at least) attached to a buoy for retrieval, and being frequently secured by an anchor to avoid drifting. Other gears are used actively, such as when they are fixed to a boat. However, recreational fishing is the best and better-known example. Angling can be useful for the capture of large predators, as these tend to evade capture by trawling. Hook-and-line gear includes the terminal tackle (*e.g.*, floats, sinks, hooks, lure, and bait), the fishing line, and the rod and/or reel. Target species mostly determines tackle. Smaller lures and baits should be used for smaller individuals, the type of hook will determine the location of the hooking injury and potentially if the captured specimen will live, so this should be considered, as well as the weights since they regulate depth at which the lure or bait are going to be positioned. Other factors to be considered should be the area where captures are to be attempted and weather conditions.

A FINAL WORD ON GENETIC SAMPLING

The capacity to identify all living organisms from a specific sequence of their genome ("Molecular barcodes" - molecular sequence data, typically consisting of one or several small stretches of DNA) has become an important character set for delineating species in all eukaryotic kingdoms of life. In the case of animals, the proposed standardized DNA region was Cytochrome Oxidase 1 or COI (Hebert, Cywinska, Ball, & deWaard, 2003). Since high-throughput sequencing (HTS) platforms became widely available, the simultaneous detection of tens to hundreds of species in a matter of days became a reality. Reference libraries of DNA barcodes are used for comparisons of the results and permitting richness and community composition estimates (Hajibabaei, 2012). "DNA Barcoding", "metabarcoding" or "metagenetics" is currently organized through CBOL (Consortium for the Barcode of Life: www.barcoding.si.edu) with membership from across the globe. More than the identification of organisms presents in the field, this capacity represents a window into (an uncertain) future. The fieldwork resulting from the sampling techniques described above provides continuous opportunity for the creation of DNA collections. Marine species biodiversity is under severe threat from pollution, climate change, and ocean acidification. The lack of a global effort for the creation of these libraries envisaging future genetic work (while is still possible), would be (literally) an irretrievable misstep.

CONSENT FOR PUBLICATION

Not applicable.

CONFLICT OF INTEREST

None to Declare

ACKNOWLEDGEMENTS

We are grateful to project ICON (Reference CENTRO010145FEDER000013) for logistic support. We also appreciate the review and suggestions made by Dr. Ana Sousa.

REFERENCES

Andréfouët, S, Costello, MJ, Rast, M & Sathyendranath, S (2008) Earth observations for marine and coastal biodiversity and ecosystems. *Remote Sens Environ,* 112, 3297-9.
[http://dx.doi.org/10.1016/j.rse.2008.04.006]

Angel, MV (1993) Biodiversity of the Pelagic Ocean. *Conserv Biol,* 7, 760-72.

[http://dx.doi.org/10.1046/j.1523-1739.1993.740760.x]

Appeltans, W, Ahyong, ST, Anderson, G, Angel, MV, Artois, T, Bailly, N, Bamber, R, Barber, A, Bartsch, I, Berta, A, Błażewicz-Paszkowycz, M, Bock, P, Boxshall, G, Boyko, CB, Brandão, SN, Bray, RA, Bruce, NL, Cairns, SD, Chan, TY, Cheng, L, Collins, AG, Cribb, T, Curini-Galletti, M, Dahdouh-Guebas, F, Davie, PJ, Dawson, MN, De Clerck, O, Decock, W, De Grave, S, de Voogd, NJ, Domning, DP, Emig, CC, Erséus, C, Eschmeyer, W, Fauchald, K, Fautin, DG, Feist, SW, Fransen, CH, Furuya, H, Garcia-Alvarez, O, Gerken, S, Gibson, D, Gittenberger, A, Gofas, S, Gómez-Daglio, L, Gordon, DP, Guiry, MD, Hernandez, F, Hoeksema, BW, Hopcroft, RR, Jaume, D, Kirk, P, Koedam, N, Koenemann, S, Kolb, JB, Kristensen, RM, Kroh, A, Lambert, G, Lazarus, DB, Lemaitre, R, Longshaw, M, Lowry, J, Macpherson, E, Madin, LP, Mah, C, Mapstone, G, McLaughlin, PA, Mees, J, Meland, K, Messing, CG, Mills, CE, Molodtsova, TN, Mooi, R, Neuhaus, B, Ng, PK, Nielsen, C, Norenburg, J, Opresko, DM, Osawa, M, Paulay, G, Perrin, W, Pilger, JF, Poore, GC, Pugh, P, Read, GB, Reimer, JD, Rius, M, Rocha, RM, Saiz-Salinas, JI, Scarabino, V, Schierwater, B, Schmidt-Rhaesa, A, Schnabel, KE, Schotte, M, Schuchert, P, Schwabe, E, Segers, H, Self-Sullivan, C, Shenkar, N, Siegel, V, Sterrer, W, Stöhr, S, Swalla, B, Tasker, ML, Thuesen, EV, Timm, T, Todaro, MA, Turon, X, Tyler, S, Uetz, P, van der Land, J, Vanhoorne, B, van Ofwegen, LP, van Soest, RW, Vanaverbeke, J, Walker-Smith, G, Walter, TC, Warren, A, Williams, GC, Wilson, SP & Costello, MJ (2012) The magnitude of global marine species diversity. *Curr Biol,* 22, 2189-202.
[http://dx.doi.org/10.1016/j.cub.2012.09.036] [PMID: 23159596]

Arkhipkin, AI, Rodhouse, PGK, Pierce, GJ, Sauer, W, Sakai, M, Allcock, L, Arguelles, J, Bower, JR, Castillo, G, Ceriola, L, Chen, C-S, Chen, X, Diaz-Santana, M, Downey, N, González, AF, Granados Amores, J, Green, CP, Guerra, A, Hendrickson, LC, Ibáñez, C, Ito, K, Jereb, P, Kato, Y, Katugin, ON, Kawano, M, Kidokoro, H, Kulik, VV, Laptikhovsky, VV, Lipinski, MR, Liu, B, Mariátegui, L, Marin, W, Medina, A, Miki, K, Miyahara, K, Moltschaniwskyj, N, Moustahfid, H, Nabhitabhata, J, Nanjo, N, Nigmatullin, CM, Ohtani, T, Pecl, G, Perez, J A A, Piatkowski, U, Saikliang, P, Salinas-Zavala, CA, Steer, M, Tian, Y, Ueta, Y, Vijai, D, Wakabayashi, T, Yamaguchi, T, Yamashiro, C, Yamashita, N & Zeidberg, LD (2015) World Squid Fisheries. *Rev Fish Sci Aquacult,* 23, 92-252.
[http://dx.doi.org/10.1080/23308249.2015.1026226]

Atkinson, LJ, Leslie, RW, Field, JG & Jarre, A (2011) Changes in demersal fish assemblages on the west coast of South Africa, 1986–2009. *Afr J Mar Sci,* 33, 157-70.
[http://dx.doi.org/10.2989/1814232X.2011.572378]

Babin, M, Cullen, J & Roesler, C (2008) *Real time observations systems for marine ecosystem dynamics and harmful algal blooms: Theory, instrumentation and modelling.* UNESCO Publishing, Paris.

Bar-On, YM, Phillips, R & Milo, R (2018) The biomass distribution on Earth. *Proceedings of the National Academy of Sciences.*
[http://dx.doi.org/10.1073/pnas.1711842115]

Bianchi, G, Gislason, H, Graham, K, Hill, L, Jin, X, Koranteng, K, Manickchand-Heileman, S, Payá, I, Sainsbury, K, Sanchez, F & Zwanenburg, K (2000) Impact of fishing on size composition and diversity of demersal fish communities. *ICES J Mar Sci,* 57, 558-71.
[http://dx.doi.org/10.1006/jmsc.2000.0727]

Blondeau-Patissier, D, Gower, J F R, Dekker, AG, Phinn, SR & Brando, VE (2014) A review of ocean color remote sensing methods and statistical techniques for the detection, mapping and analysis of phytoplankton blooms in coastal and open oceans. *Prog Oceanogr,* 123, 123-44.
[http://dx.doi.org/10.1016/j.pocean.2013.12.008]

Boero, F, Dupont, S & Thorndyke, M (2015) Make new friends, but keep the old: towards a transdisciplinary and balanced strategy to evaluate Good Environmental Status. *J Mar Biol Assoc U K,* 95, 1069-70.
[http://dx.doi.org/10.1017/S0025315415000557]

Cermeño, P, Teixeira, IG, Branco, M, Figueiras, FG & Marañón, E (2014) Sampling the limits of species richness in marine phytoplankton communities. *J Plankton Res,* 36, 1135-9.
[http://dx.doi.org/10.1093/plankt/fbu033]

Chapman, AD (2009) Numbers of Living Species in Australia and the World. Australian Biological

Resources (ABRS), Camberra 84.

Chave, A (2004) Seeding the seafloor with observatories. *Oceanus,* 42, 28-31.

Clarke, T (2003) Oceanography: Robots in the deep. *Nature,* 421, 468-70.
[http://dx.doi.org/10.1038/421468a] [PMID: 12556859]

Costello, M, Pohle, G & Martin, A (2004) *Evaluating biodiversity in marine environmental assessments Research and Development Monograph Series.* Canadian Environmental Assessment Agency, Ottawa, Canada.

Costello, MJ (1992) Abundanet alce and spatial overlap of gobies (Gobiidae) in Lough Hyne, Ireland. *Environ Biol Fishes,* 33, 239-48.
[http://dx.doi.org/10.1007/BF00005868]

Costello, MJ (2009) Distinguishing marine habitat classification concepts for ecological data management. *Mar Ecol Prog Ser,* 397, 253-68.
[http://dx.doi.org/10.3354/meps08317]

Costello, MJ (2015) Biodiversity: the known, unknown, and rates of extinction. *Curr Biol,* 25, R368-71.
[http://dx.doi.org/10.1016/j.cub.2015.03.051] [PMID: 25942550]

Costello, MJ, Basher, Z, McLeod, L, Asaad, I, Claus, S, Vandepitte, L, Yasuhara, M, Gislason, H, Edwards, M, Appeltans, W, Enevoldsen, H, Edgar, GJ, Miloslavich, P, De Monte, S, Pinto, IS, Obura, D & Bates, AE (2017) Methods for the Study of Marine Biodiversity. *The GEO Handbook on Biodiversity Observation Networks* Springer International Publishing, Cham 129-63.
[http://dx.doi.org/10.1007/978-3-319-27288-7_6]

Costello, MJ & Scott Baker, C (2011) Who eats sea meat? Expanding human consumption of marine mammals. *Biol Conserv,* 144, 2745-6.
[http://dx.doi.org/10.1016/j.biocon.2011.10.015]

Dybas, CL (2004) Close Encounters of the Deep-Sea Kind. *Bioscience,* 54, 888-91.
[http://dx.doi.org/10.1641/0006-3568(2004)054[0888:CEOTDK]2.0.CO;2]

Edwards, M, Beaugrand, G, Hays, GC, Koslow, JA & Richardson, AJ (2010) Multi-decadal oceanic ecological datasets and their application in marine policy and management. *Trends Ecol Evol (Amst),* 25, 602-10.
[http://dx.doi.org/10.1016/j.tree.2010.07.007] [PMID: 20813425]

Edwards, M, Helaouet, P, Johns, DG, Batten, S, Beaugrand, G & Chiba, S (2012) Global marine ecological status report: Results from the global CPR survey 2010/2011 *SAHFOS Technical Report,* 9, 1-40.

Eleftheriou, A (2013) *Methods for the study of marine benthos.* Wiley-Blackwell, Chichester.
[http://dx.doi.org/10.1002/9781118542392]

Elliott, M & Hemingway, K (2002) *Fishes in estuaries.* Blackwell Science, Oxford.
[http://dx.doi.org/10.1002/9780470995228]

Engås, A, Skeide, R & West, CW (1997) The 'MultiSampler': a system for remotely opening and closing multiple codends on a sampling trawl. *Fish Res,* 29, 295-8.
[http://dx.doi.org/10.1016/S0165-7836(96)00545-0]

Gabriel, O, Lange, K, Dahm, E & Wendt, T (2007) *Fish catching methods of the World.* Blackwell Publishing, London.

Gage, JD & Tyler, PA (1991) *Deep-sea biology A natural history of organisms at the deep-sea floor.* Cambridge University Press, Cambridge.
[http://dx.doi.org/10.1017/CBO9781139163637]

Garland, ED & Buntman, CA (1996) Measuring diversity of planktonic larvae. *Oceanus,* 39, 12.

GBO (2014) Global biodiversity outlook 4. *Secretariat of the convention on biological diversity* 155p.

Goswami, SC (2004) *Zooplankton methodology, collection & identification – a field manual.* National

Institute of Oceanography, Goa.

Gotelli, N & Colwell, R (2001) Quantifying biodiversity: procedures and pitfalls in the measurement and comparison of species richness. *Ecol Lett,* 4, 379-91.
[http://dx.doi.org/10.1046/j.1461-0248.2001.00230.x]

Grassle, JF & Maciolek, NJ (1992) Deep-Sea Species Richness: Regional and Local Diversity Estimates from Quantitative Bottom Samples. *Am Nat,* 139, 313-41.
[http://dx.doi.org/10.1086/285329]

Groombridge, B & Jenkins, MD (2002) *World Atlas of Biodiversity.* UNEP World Conservation Centre, Berkeley.

Guerra-Garcia, J M, Espinosa, F & Garcia-Gomez, J C (2008) Trends in Taxonomy today: an overview about the main topics in Taxonomy. *Zoologica baetica,* 19, 15-50.

Haegeman, B, Hamelin, J, Moriarty, J, Neal, P, Dushoff, J & Weitz, JS (2013) Robust estimation of microbial diversity in theory and in practice. *ISME J,* 7, 1092-101.
[http://dx.doi.org/10.1038/ismej.2013.10] [PMID: 23407313]

Hajibabaei, M (2012) The golden age of DNA metasystematics. *Trends Genet,* 28, 535-7.
[http://dx.doi.org/10.1016/j.tig.2012.08.001] [PMID: 22951138]

Hallegraeff, GM, Anderson, DM & Cembella, AD (2003) *Manual on harmful marine microalgae.* UNESCO Publishing, Paris.

Harris, S, Shears, N & Radford, C (2016) Ecoacoustic indices as proxies for biodiversity on temperate reefs. *Methods Ecol Evol,* 7, 713-24.
[http://dx.doi.org/10.1111/2041-210X.12527]

Hebert, PDN, Cywinska, A, Ball, SL & deWaard, JR (2003) Biological identifications through DNA barcodes. *Proc Biol Sci,* 270, 313-21.
[http://dx.doi.org/10.1098/rspb.2002.2218] [PMID: 12614582]

Heirtzler, JR & Grassle, JF (1976) Deep-sea research by manned submersibles. *Science,* 194, 294-9.
[http://dx.doi.org/10.1126/science.194.4262.294] [PMID: 17738036]

Hendriks, ie & Duarte, CM (2008) Allocation of effort and imbalances in biodiversity research. *J Exp Mar Biol Ecol,* 360, 15-20.
[http://dx.doi.org/10.1016/j.jembe.2008.03.004]

Hiscock, K (2014) *Marine biodiversity conservation: A practical approach.* Routledge, Abingdon.
[http://dx.doi.org/10.4324/9781315857640]

Horne, JK (2000) Acoustic approaches to remote species identification: a review. *Fish Oceanogr,* 9, 356-71.
[http://dx.doi.org/10.1046/j.1365-2419.2000.00143.x]

Inniss, L, Simcock, A, Ajawin, A, Alcala, A, Bernal, A & Calumpong, H (2016) The First Global Integrated Marine Assessment.

Karlson, B, Cusack, C & Bresnan, E (2010) *Microscopic and molecular methods for quantitative phytoplankton analysis.* UNESCO, Paris.

Kingsford, M & Battershill, C (1998) *Studying temperate marine environments A handbook for ecologists.* Canterbury University Press, New Zealand.

Krug, CB, Schaepman, ME, Shannon, LJ, Cavender-Bares, J, Cheung, W, McIntyre, PB, Metzger, JP, Niinemets, Ü, Obura, DO, Schmid, B, Strassburg, BBN, Van Teeffelen, AJA, Weyl, OLF, Yasuhara, M & Leadley, PW (2017) Observations, indicators and scenarios of biodiversity and ecosystem services change — a framework to support policy and decision-making. *Curr Opin Environ Sustain,* 29, 198-206.
[http://dx.doi.org/10.1016/j.cosust.2018.04.001]

Kubečka, J, Godø, OR, Hickley, P, Prchalová, M, Říha, M, Rudstam, L & Welcomme, R (2012) Fish sampling with active methods. *Fish Res,* 123-124, 1-3.

[http://dx.doi.org/10.1016/j.fishres.2011.11.013]

Leleu, K, Remy-Zephir, B, Grace, R & Costello, MJ (2012) Mapping habitats in a marine reserve showed how a 30-year trophic cascade altered ecosystem structure. *Biol Conserv,* 155, 193-201.
[http://dx.doi.org/10.1016/j.biocon.2012.05.009]

Lindahl, O (1986) *A dividable hose for phytoplankton sampling Report of the ICES Working Group on Exceptional Algal Blooms.* ICES.

Reaka-Kudla, M (1997) The global biodiversity of coral reefs: a comparison with rain forests.*Biodiversity II: Understanding and Protecting our Biological Resources.* In: Kudla, M.R.-, Wilson, D. E., Wilson, E. O., (Eds.), Joseph Henry Press, Washington, D.C. 83-108.

Madsen, N, Hansen, KE, Frandsen, RP & Krag, LA (2012) Development and test of a remotely operated Minisampler for discrete trawl sampling. *Fish Res,* 123-124, 16-20.
[http://dx.doi.org/10.1016/j.fishres.2011.11.016]

McGeoch, MA, Genovesi, P, Bellingham, PJ, Costello, MJ, McGrannachan, C & Sheppard, A (2016) Prioritizing species, pathways, and sites to achieve conservation targets for biological invasion. *Biol Invasions,* 18, 299-314.
[http://dx.doi.org/10.1007/s10530-015-1013-1]

Miloslavich, P, Bax, NJ, Simmons, SE, Klein, E, Appeltans, W, Aburto-Oropeza, O, Andersen Garcia, M, Batten, SD, Benedetti-Cecchi, L, Checkley, DM, Jr, Chiba, S, Duffy, JE, Dunn, DC, Fischer, A, Gunn, J, Kudela, R, Marsac, F, Muller-Karger, FE, Obura, D & Shin, YJ (2018) Essential ocean variables for global sustained observations of biodiversity and ecosystem changes. *Glob Change Biol,* 24, 2416-33.
[http://dx.doi.org/10.1111/gcb.14108] [PMID: 29623683]

Olenina, I, Hajdu, S, Andersson, A, Edler, L, Wasmund, N, Busch, S, Göbel, J & Gromisz, S (2006) Biovolumes and size-classes of phytoplankton in the Baltic Sea. *Baltic Sea Environment Proceedings No106*

Paulay, G & Meyer, C (2002) Diversification in the tropical pacific: comparisons between marine and terrestrial systems and the importance of founder speciation. *Integr Comp Biol,* 42, 922-34.
[http://dx.doi.org/10.1093/icb/42.5.922] [PMID: 21680372]

Pereira, HM, Ferrier, S, Walters, M, Geller, GN, Jongman, RHG, Scholes, RJ, Bruford, MW, Brummitt, N, Butchart, SHM, Cardoso, AC, Coops, NC, Dulloo, E, Faith, DP, Freyhof, J, Gregory, RD, Heip, C, Höft, R, Hurtt, G, Jetz, W, Karp, DS, McGeoch, MA, Obura, D, Onoda, Y, Pettorelli, N, Reyers, B, Sayre, R, Scharlemann, JPW, Stuart, SN, Turak, E, Walpole, M & Wegmann, M (2013) Ecology. Essential biodiversity variables. *Science,* 339, 277-8.
[http://dx.doi.org/10.1126/science.1229931] [PMID: 23329036]

Piraino, S, Fanelli, G & Boero, F (2002) Variability of species' roles in marine communities: change of paradigms for conservation priorities. *Mar Biol,* 140, 1067-74.
[http://dx.doi.org/10.1007/s00227-001-0769-2]

Reguera, B, Alonso, R, Moreira, A & Méndez, S (2011) *Guía para el diseño y puesta en marcha de un plan de seguimiento de microalgas productoras de toxinas.* COI de UNESCO y OIEA, Paris.

Remy-Zephir, B, Leleu, K, Grace, R & Costello, M J (2012) *Geographical information system (GIS) files of seabed habitats and biotope maps from Leigh marine reserve in 1977 and 2006.*

Santhanam, R & Srinivasan, A (1994) *A manual of marine zooplankton.* Oxford and IBH Pub. Co..

Siebert, JR & Nielsen, JL (2001) *Electronic Tagging and Tracking in Marine Fisheries Reviews: Methods and Technologies in Fish Biology and Fisheries.* Kluwer Academic Publishers, Dordrecht.
[http://dx.doi.org/10.1007/978-94-017-1402-0]

Simmonds, J & MacLennan, DN (2005) *Fisheries Acoustics - Theory and Practice.* Blackwell Publishing, London.
[http://dx.doi.org/10.1002/9780470995303]

Smith, TD, Gjosater, J & Stenseth, NC (2002) A century of manipulating recruitment in coastal cod populations: the Flodevigen experience. *ICES Marine Science Symposia.*

Stenseth, NC, Bjornstad, ON, Falck, W, Fromentin, JM, Gjosaeter, J & Gray, JS (1999) Dynamics of Coastal Cod Populations: Intra- and Intercohort Density Dependence and Stochastic Processes. *Proc Biol Sci,* 266, 1645-54.
[http://dx.doi.org/10.1098/rspb.1999.0827]

Suthers, IM & Rissik, D (2009) *Plankton A guide to their ecology and monitoring for water quality.* CSIRO Publishing, Sydney.
[http://dx.doi.org/10.1071/9780643097131]

Tait, R & Dipper, F (1998) *Elements of marine ecology.* Butterworth-Heinemann, Oxford.

Tranter, D & Fraser, J (1968) *Zooplankton Sampling.* UNESCO, Paris 174p.

Webb, TJ & Mindel, BL (2015) Global patterns of extinction risk in marine and non-marine systems. *Curr Biol,* 25, 506-11.
[http://dx.doi.org/10.1016/j.cub.2014.12.023] [PMID: 25639240]

Weikert, H & John, HC (1981) Experiences with a modified Bé multiple opening-closing plankton net. *J Plankton Res,* 3, 167-76.
[http://dx.doi.org/10.1093/plankt/3.2.167]

Wenneck, TL, Falkenhaug, T & Bergstad, OA (2008) Strategies, methods, and technologies adopted on the R.V. G.O. Sars MAR-ECO expedition to the Mid-Atlantic Ridge in 2004. *Deep Sea Res Part II Top Stud Oceanogr,* 55, 6-28.
[http://dx.doi.org/10.1016/j.dsr2.2007.09.017]

Wiebe, PH, Morton, AW, Bradley, AM, Backus, RH, Craddock, JE, Barber, V, Cowles, TJ & Flierl, GR (1985) New development in the MOCNESS, an apparatus for sampling zooplankton and micronekton. *Mar Biol,* 87, 313-23.
[http://dx.doi.org/10.1007/BF00397811]

Zampoukas, N, Palialexis, A, Duffek, A, Graveland, J, Giorgi, G, Hagebro, C, Hanke, G, Korpinen, S, Tasker, M, Tornero, A, Abaza, V, Battaglia, P, Caparis, M, Dekeling, R, Frias, V, Haarich, M, Katsanevakis, S, Klein, H, Krzyminski, W, Laamanen, M, Gac, J L, Leppanen, J, Lips, U, Maes, T, Magaletti, E, Malcolm, S, Marques, J, Mihail, O, Moxon, R, O'brien, C, Panagiotidis, P, Penna, M, Piroddi, C, Probst, W, Raicevich, S, Trabucco, B, Tunesi, L, Graaf, S V d, Weiss, A, Wernersson, A & Zevenboom, W (2014) Technical guidance on monitoring for the Marine Stategy Framework Directive.

CHAPTER 3

Macroelements and Microelements in Marine Ecosystems: An Overview

Madson de Godoi Pereira[1,*], **Daniel Carneiro Freitas**[1], **Lourdes Cardoso de Souza Neta**[1], **Arnaud Victor dos Santos**[1], **Joselito Nardy Ribeiro**[2] and **Araceli Verónica Flores Nardy Ribeiro**[3]

[1] *Departamento de Ciências Exatas e da Terra – Universidade do Estado da Bahia, Rua Silveira Martins, 2555, CEP: 41.150-000, Cabula, Salvador-BA, Brazil*

[2] *Centro de Ciências da Saúde – Universidade Federal do Espírito Santo, Avenida Maruípe, S/N, CEP: 29.042-751, Vitória-ES, Brazil*

[3] *Coordenação de Licenciaturas – Instituto Federal de Educação, Ciência e Tecnologia do Espírito Santo, Avenida Vitória, 1729, CEP: 29.040-780, Vitória-ES, Brazil*

Abstract: In this chapter, aspects concerning the complexity of marine chemistry were discussed. In this scope, important reactions of metals and non-metals with inorganic and organic constituents of water and sediments were considered. In addition to these reactions, this chapter considers biokinetic aspects, which are responsible for very important regulations concerning the assimilation and biotransformation of many chemical elements. Finally, two major environmental problems (eutrophication due to the excessive supply of nitrogen and phosphorus, and release of organotin compounds and copper from antifouling paints used on ships' hulls) were presented with the intention of discussing some forms in which uncontrolled introductions of metals and non-metals can change negatively the quality of marine ecosystems.

Keywords: Bioaccumulation, Essentiality, pE X pH Diagrams, Toxicity.

INTRODUCTION

The oceans cover approximately 71% of the Earth's surface, thereby playing an important role in human activities, including the transportation of millions of tons of cargo, as well as fishing activities. Additionally, the oceans are an important source of commercial extraction of sodium, magnesium, chlorine, and bromine. A large part of the world's population conglomerates in coastal environments, and this situation is worrying because of the bulky discharges of domestic and industrial wastes. Approximately 40% of the world's population lives near

* **Corresponding author Madson de Godoi Pereira:** Departamento de Ciências Exatas e da Terra – Universidade do Estado da Bahia, Rua Silveira Martins, 2555, CEP: 41.150-000, Cabula, Salvador-BA, Brazil; Tel/Fax: +55-71-31-7-2372; E-mail: madson.pereira444@gmail.com

De-Sheng Pei & Muhammad Junaid (Eds.)
All rights reserved-© 2019 Bentham Science Publishers

coastlines (up to 100 km away), thereby offering many opportunities for pollutant loading, including nutrients (nitrogen and phosphorus) that are able to promote eutrophication, as discussed later in the text (Wallace *et al.*, 2014). As early as 1977, serious environmental problems had already been reported in coastal environments, when wastewater discharges, in the vicinity of five domestic outfalls in southern California, were responsible for remarkable decreases in the biodiversity of benthic organisms (Reish *et al.*, 1977). According to the authors, the population of benthic organisms decreased because of the presence of high concentrations of some elements (nitrogen, for example) in the marine water, which were able to cause severe biological damages upon pelagic larvae of benthic organisms. Nowadays, many underdeveloped and developing countries still have serious problems related to basic sanitation and, consequently, coastal pollution.

Besides the domestic wastewater discharges, several agrochemical and industrial pollutants are brought by the rivers, and this pollution has been responsible for severe degradation of estuarine ecosystems, as well as worrying decreases in the population of many species of fish and other marine organisms. Obviously, because of this estuarine pollution, fishing industries have been suffering economic losses for decades.

The oceans are also hugely important from a biogeochemical point of view, since marine environments play important roles in the planetary distribution of several chemical elements. This fact is very important, for example, for regulating the atmospheric levels of CO_2 and O_2 with imperative consequences for the maintenance of life on our planet.

Regardless of whether for economic purposes or for environmental regulation and preservation, the importance of the oceans is based on their biodiversity. In this regard, it is necessary to maintain satisfactory physical and chemical conditions in the oceans, including adequate concentrations of trace elements, such as cobalt, copper, iron, manganese, and zinc, in order to promote the development of life. At the same time, the contents of toxic elements, such as cadmium, lead, and mercury, should be very low. In this scenario, human activities are potential sources of harmful wastes capable of causing serious disequilibria in marine life and in its ability to control biogeochemical cycles over long periods of time. The next section deals with the capacity of marine life to regulate two of the most important of these cycles, namely the biogeochemical cycles of carbon and oxygen.

The Importance of Marine Life for Carbon and Oxygen Biogeochemical Cycles

The marine life has extreme importance concerning the regulation of the global climate. This control is realized by means of thermohaline currents, but the fluxes of CO_2 between marine water and the atmosphere also effectively contribute to the regulation of atmospheric levels of this gas (vanLoon & Duffy, 2005), which is accomplished mainly by means of photosynthesis and burial of $CaCO_3$ (in marine sediments). As discussed below, both situations need the participation of marine life.

Photosynthesis is one the most important mechanisms for keeping concentrations of CO_2 in Earth's atmosphere well below those found in the atmosphere of Venus, for example (Rothschild & Lister, 2003; Allègre & Dars, 2009). In marine environments, a large proportion of the photosynthesis is carried out by phytoplankton, so that this wide group of living organisms is considered the fuel that moves marine ecosystems (Boyce *et al.*, 2017). Photosynthesis uses atmospheric CO_2 that is dissolved in marine water with the consequent production of carbohydrate, whose minimum formula is $[CH_2O]$, and oxygen gas. As indicated in Equation 1, this vital biochemical process also needs water and sunlight, whose energy is given by hv, where h is the Planck constant (6.63×10^{-34} J s) and v is the electromagnetic radiation frequency (Hz).

$$CO_{2(g)} + H_2O_{(l)} + hv \rightarrow [CH_2O]_{(s)} + O_{2(g)} \qquad (1)$$

The marine photic zone, with an average depth of approximately 250 m, is considered a soup of living organisms, and many of them belong to phytoplankton and zooplankton. In this zone, zooplankton eats phytoplankton and both classes of organisms are consumed by bigger animals. All these organisms breathe aerobically, thus releasing CO_2 to marine water, the pH of which favors its conversion to bicarbonate (HCO_3^-). After the death of marine organisms, bacteria decompose their soft tissues, releasing organic molecules that are also dissolved in marine water. These bacteria continue decomposing the dissolved organic matter and more CO_2 is returned to the photic zone, where, as discussed above, the formation of HCO_3^- is favored. In this sense, marine life recycles both organic and inorganic carbon in the photic zone. A large part of the atmospheric CO_2 assimilated by the oceans tends to be returned to the atmosphere within days or months. However, carbon recycling is not 100% efficient, and approximately 4-5 $\times 10^{12}$ kg C year^{-1} escapes from the superficial waters, a small part of which is deposited, buried, and accumulated in deep marine sediments. The buried carbon, which is not recycled by benthic and pelagic organisms, does not return to the water column, and consequently, more atmospheric CO_2 should be assimilated to support phytoplankton growth. This constant marine assimilation of CO_2 to sustain the biological processes is called the biological pump of carbon dioxide (Cockell, 2008).

The burial of $CaCO_3$ in marine environments is a biogeochemical process, beginning with the weathering of silicate minerals (Equation 2) that removes atmospheric carbon dioxide (Cockell, 2008).

$$2NaAlSi_3O_{8(s)} + 2CO_{2(g)} + 3H_2O_{(l)} \rightleftharpoons Al_2Si_2O_5(OH)_{4(s)} + 2Na^+_{(aq)} + 2HCO_3^-{}_{(aq)} + 4SiO_{2(aq)} \tag{2}$$

In Equation 2, the mineral albite ($NaAlSi_3O_8$) was used as an example, but other minerals containing silicon could be considered. The bicarbonate ions are carried by rivers to the oceans, where they are combined with Ca^{2+} to form $CaCO_3$ and release CO_2 (Equation 3). $CaCO_3$ is biochemically formed by several aquatic species, including those belonging to phytoplankton (Cockell, 2008).

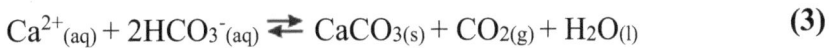

$$Ca^{2+}{}_{(aq)} + 2HCO_3^-{}_{(aq)} \rightleftharpoons CaCO_{3(s)} + CO_{2(g)} + H_2O_{(l)} \tag{3}$$

It is important to note that albite decomposition consumes 2 mol of atmospheric CO_2, while the formation of $CaCO_3$ by the marine organisms releases only 1 mol of CO_2, which can return to the atmosphere. After dying, carbonate-containing marine organisms sink into deep waters, where their carbonate shells can be dissolved due to the greater acidity. As the acidic dissolution of carbonate has a stoichiometry of 1:1 in relation to the production of CO_2, it is expected a gradual balance of the total amount of CO_2 withdrawn from the atmosphere along the weathering of silicate minerals. Nevertheless, if part of the carbonate-containing marine organisms is buried in the deep sediments, $CaCO_3$ dissolution does not occur, thus allowing an effective removal of carbon dioxide from our atmosphere, thus regulating the global climate (Cockell 2008).

Photosynthesis (Equation 1) is one of the most important planetary sources of atmospheric oxygen (Rotschild & Lister 2003, Gargaud *et al.* 2012), thereby highlighting how marine life also contributes to the oxygen biogeochemical cycle. Thus, changes in the marine phytoplanktonic population can promote disturbances in the oxygen biogeochemical cycle.

The participation of marine life in the control of atmospheric CO_2 and O_2 is enough to highlight the importance of marine life, and its preservation. Moreover, marine preservation has an imperative ecological relevance since the oceans house thousands of vegetal and animal species, thus representing a huge source of natural products that still are largely unknown.

In the next section, we discuss how chemical elements can have toxic or beneficial effects in relation to all forms of life, including marine life.

ESSENTIALITY AND TOXICITY OF THE ELEMENTS

All forms of life need carbon, nitrogen, hydrogen, oxygen, phosphorus, and sulfur to produce vital biomolecules, such as nucleic acids, amino acids, proteins, lipids, and carbohydrates (de Duve 1995). Calcium, magnesium, potassium, and sodium are also needed in high amounts to sustain life, and they are called primary macronutrients. In turn, phosphorus and sulfur are classified as secondary macronutrients because they are required in small amounts. On the other hand, other metals and non-metals (trivalent Cr, Cu, Co, Mn, Ni, Se, V, Zn, among others) are required at trace levels (microelements) to assure a healthy life cycle for organisms. The microelements, which are essential elements at adequate doses, can cause several types of biological damage if they are excessively assimilated. In this case, they cease to be essential and become toxic (Tchounwou *et al.*, 2014). In turn, there are elements (Cd, hexavalent Cr, Hg, Pb, Tl, among others) that will always be toxic, even when considering very small concentrations. The boundary between being essential or toxic is very subtle because it depends on aspects concerning the general health status of an organism and its age.

The toxic effects of microelements also depend largely on their bioavailability, which is the capacity of a chemical species for being assimilated by cells of living organisms. For this, an element needs to be in a form in which it is possible to establish interactions with chemical groups belonging to the cell membrane or wall. It is necessary to predict how an element is found in aquatic ecosystems, and thus knowledge is required about the stability of chemical species.

In this sense, it is necessary to know the oxidation state of the chemical element as well as whether it exists as an organic compound. This is the case for mercury, which is much more toxic as alkyl mercury than mercuric cation. In turn, organic forms of arsenic are less toxic than its inorganic compounds (Fergusson 1990, Emsley 2005).

The effects of toxic metals and non-metals comprise a huge set of biological damage, but many of these elements replace enzyme cofactors, thus partially or totally inactivating these vital macromolecules (Nordberg *et al.* 2007). This is the typical case of cadmium in relation to zinc. Cadmium, a very toxic metal, has a strong chemical similarity to zinc, which is an essential element largely involved in several metabolic processes including those concerning the activity of enzymes.

In the next section, we focus our attention on the processes that are able to change the chemical forms and bioavailability of the elements in marine waters.

AQUATIC CHEMISTRY

In marine ecosystems, several chemical and physical processes can modify the bioavailability of elements. In this section, we consider the combination of redox and proton-transfer reactions (pE x pH diagrams), as well as complexation reactions.

Chemical equilibria in marine environments, where there is a high ionic strength, should be described in terms of chemical activity. However, many sources in the literature approximate the chemical activities to molar concentrations. Here, we follow the notation of vanLoon & Duffy, (2005).

Although conditions away from equilibrium may occur in marine environments, we disregard these situations to simplify the explanations. All constants reported in the following sections were obtained considering the temperature to be 25° C.

pE X pH Diagrams

Of the forms that can be used to predict the chemical form of an element in the oceans, pE X pH diagrams are especially useful. Here, pE is defined as the anti-logarithm of electron activity, so that high values of pE indicate low electron activity (oxidating environment), while small values of pE correspond to high electron activity (reducing environment). In such diagrams, it is possible to identify the conditions responsible for changes in the oxidation number as well as the formation of insoluble compounds.

In aquatic systems, pE X pH diagrams are limited by water stability. Thus, the inferior limit is defined in terms of water reduction (Equation 4), while the upper limit is demarcated by water oxidation (Equation 5).

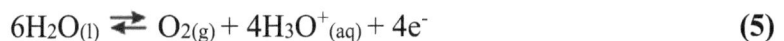

$$2H_2O_{(aq)} + 2e^- \rightleftharpoons H_{2(g)} + 2OH^-_{(aq)} \qquad \qquad \textbf{(4)}$$

$$6H_2O_{(l)} \rightleftharpoons O_{2(g)} + 4H_3O^+_{(aq)} + 4e^- \qquad \qquad \textbf{(5)}$$

Both reactions listed above have corresponding equations that relate pE and pH. These equations can be deduced using a general equilibrium condition (Equation 6), where A and B are the oxidized and reduced chemical species, respectively, while a and b are the respective stoichiometric coefficients of A and B. In turn, n is the number of electrons (e^-) involved in the redox process.

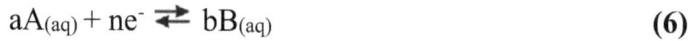

$$aA_{(aq)} + ne^- \rightleftharpoons bB_{(aq)} \tag{6}$$

The constant of equilibrium concerning the general reaction listed above (Equation 6) is given by Equation 7, where a is the chemical activity:

$$K_{eq} = a^b B/[(a^a A).(a^n e^-)] \tag{7}$$

Equation 7 can be rearranged to give rise to Equation 8, which can be rewritten as Equation 9.

$$1/a^n e^- = K_{eq}.[(a^a A)/(a^b B)] \tag{8}$$

$$-\log a^n e^- = \log K_{eq} + \log [(a^a A)/(a^b B)] \tag{9}$$

Equation 10 arises from Equation 9.

$$-n\log ae^- = \log K_{eq} + \log [(a^a A)/(a^b B)] \tag{10}$$

K_{eq} is related to ΔG° (standard Gibbs free energy variation) according to Equation 11, where R is the universal constant of gases (8.314 J K^{-1} mol^{-1}) and T is the absolute temperature (K).

$$2.303.\log K_{eq} = -\Delta G^\circ/RT \tag{11}$$

In turn, Equation 12 defines ΔG° as the electrical work performed by the evaluated electrochemical system. In this Equation, F is the Faraday constant (96,485 C mol^{-1}), and n is the number of electrons (in mol) involved in the electrochemical process, where ε° is the standard potential concerning the electrochemical system.

$$\Delta G^\circ = -nF\varepsilon^\circ \tag{12}$$

Substituting Equation 12 into Equation 11, it is possible to obtain Equation 13.

$$\log K_{eq} = nF\varepsilon^\circ/2.303RT \tag{13}$$

Given the values of F and R and assuming that T is 298.15 K (25°C), Equation 13 can be rewritten as Equation 14.

$$\log K_{eq} = n\varepsilon°/0.0591 \tag{14}$$

When Equation 14 is inserted into Equation 10, Equation 15 is obtained.

$$-n\log ae^- = n\varepsilon°/0.0591 + \log [(a^aA)/(a^bB)] \tag{15}$$

Equation 16 is obtained by dividing Equation 15 by n.

$$-\log ae^- = \varepsilon°/0.0591 + 1/n.\log [(a^aA)/(a^bB)] \tag{16}$$

Rewriting Equation 16, we obtain Equation 17, where the term pE means -log ae⁻.

$$pE = \varepsilon°/0.0591 + 1/n.\log [(a^aA)/(a^bB)] \tag{17}$$

The mass action law concerning the equilibrium indicated in Equation 6 is written as $Q = [(a^bB)/(a^aA)]$, so that Equation 18 is defined as indicated below.

$$pE = \varepsilon°/0.0591 - 1/n.\log Q \tag{18}$$

According to an IUPAC (International Union of Pure and Applied Chemistry) convention, water oxidation (Equation 5) must be inverted and, thus, represented by a reduction reaction (Equation 19). After applying Equation 18 to the reaction indicated in Equation 19, we obtain Equation 20. Here, we considered $\varepsilon° = 1.229$ V, and a partial oxygen pressure of 101,325 Pa, or 1 atm (minimum pressure required for gas evolution from an aqueous ecosystem).

$$O_{2(g)} + 4H_3O^+_{(aq)} + 4e^- \rightleftarrows 6H_2O_{(l)} \tag{19}$$

$$pE = 20.80 - pH \tag{20}$$

To deduce Equation 21, which relates -log ae⁻ (or pE) with pH for the water reduction (Equation 4), we also used Equation 18 in which we considered the value of $\varepsilon°$ (-0.828 V) and a partial hydrogen pressure of 101,325 Pa, or 1 atm (minimum pressure required for gas evolution from an aqueous ecosystem).

$$pE = -14.0 - \log (aOH^-) \Rightarrow pE = -14.0 + pOH \Rightarrow pE = -pH \qquad (21)$$

The borders of water stability (Equations 20 and 21) in all pE X pH diagrams for aquatic ecosystems are illustrated in Fig. (1).

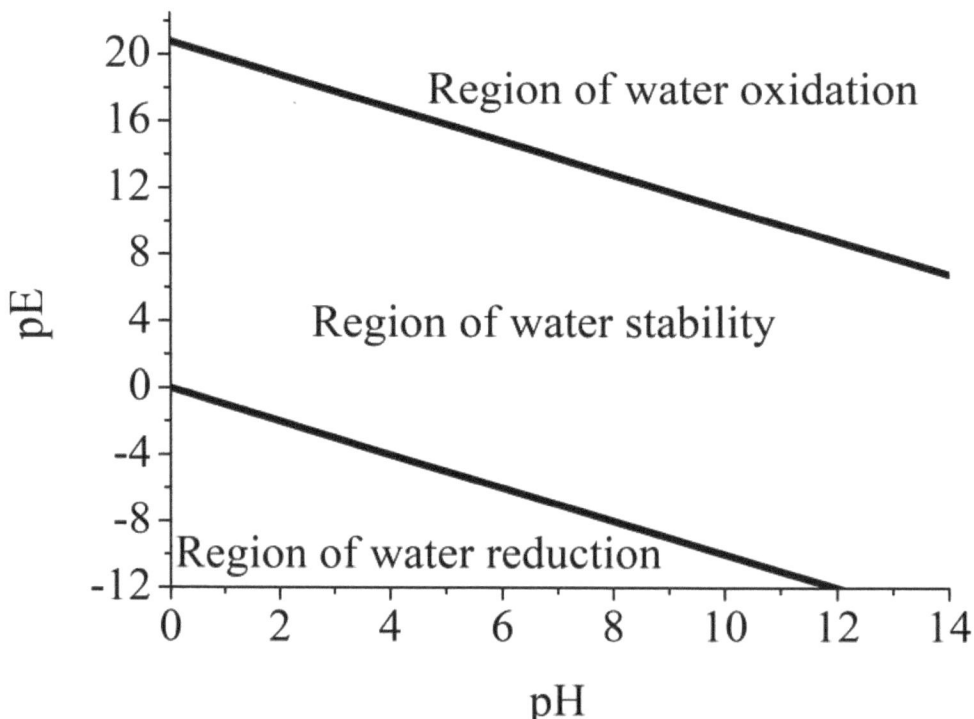

Fig. (1). pE X pH diagram showing the boundaries of water stability. Lines 1 and 2 correspond to Equations 20 and 21, respectively.

To illustrate that pE X pH diagrams can be used for a specific chemical element, we consider sulfur as an example, but the reasoning described below can be extended to any other element. In aquatic ecosystems, sulfur is involved in several equilibria, some of which are listed in Equations 22 to 28.

$$SO_4^{2-}{}_{(aq)} + H_3O^+{}_{(aq)} \rightleftarrows HSO_4^-{}_{(aq)} + H_2O_{(l)}, \text{ whose } K_{eq} = 10^{1.995} \qquad (22)$$

$$HSO_4^-{}_{(aq)} + 7H_3O^+{}_{(aq)} + 6e^- \rightleftarrows S_{(s)} + 11H_2O_{(l)}, \text{ whose } \varepsilon^° = +0.33 \text{ V} \qquad (23)$$

$$SO_4^{2-}{}_{(aq)} + 8H_3O^+{}_{(aq)} + 6e^- \rightleftarrows S_{(s)} + 12H_2O_{(l)}, \text{ whose } \varepsilon^° = +0.35 \text{ V} \qquad (24)$$

$$S_{(s)} + 2H_3O^+{}_{(aq)} + 2e^- \rightleftarrows H_2S_{(aq)} + 2H_2O_{(l)}, \text{ whose } \varepsilon^\circ = +0.14 \text{ V} \qquad \textbf{(25)}$$

$$SO_4{}^{2-}{}_{(aq)} + 10H_3O^+{}_{(aq)} + 8e^- \rightleftarrows H_2S_{(aq)} + 14H_2O_{(l)}, \text{ whose } \varepsilon^\circ = +0.30 \text{ V} \qquad \textbf{(26)}$$

$$H_2S_{(aq)} + H_2O_{(l)} \rightleftarrows HS^-{}_{(aq)} + H_3O^+{}_{(aq)}, \text{ whose } K_{eq} = 10^{-6.995} \qquad \textbf{(27)}$$

$$SO_4{}^{2-}{}_{(aq)} + 9H_3O^+{}_{(aq)} + 8e^- \rightleftarrows HS^-{}_{(aq)} + 13H_2O_{(l)}, \text{ whose } \varepsilon^\circ = +0.25 \text{ V} \qquad \textbf{(28)}$$

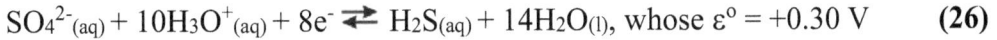

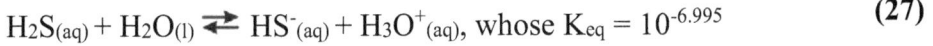

After applying Equation 18 to each reaction indicated in Equations 22-28, we obtained, respectively, Equations 29-35. Equations 29-35 represent the stability borders in a sulfur pE X pH diagram. To deduce Equations 29-35, we replaced the molar concentration of $SO_4{}^{2-}$, $HSO_4{}^-$, HS^-, and/or H_2S in Equation 18, which were arbitrarily equal to 10^{-2} mol L^{-1}. Moreover, we inserted in Equation 18 each value of ε°, respectively indicated in Equations 23, 24, 25, 26, and 28 in order to achieve the Equations 30, 31, 32, 33, and 35, respectively.

As the reactions indicated in Equations 22 and 27 are not redox reactions, values of ε° were not considered. In this case, the stability borders related to the reactions indicated in Equations 22 and 27 do not contemplate pE, and Equation 18 cannot be applied. Thus, these stability borders are constant values of pH, as indicated in Equations 29 and 34.

$$pH = 1.995 \qquad \textbf{(29)}$$

$$pE = 5.293 - 1.167pH \qquad \textbf{(30)}$$

$$pE = 5.625 - 1.333pH \qquad \textbf{(31)}$$

$$pE = 3.440 - pH \qquad \textbf{(32)}$$

$$pE = 5.079 - 1.25pH \qquad \textbf{(33)}$$

$$pH = 6.995 \qquad \textbf{(34)}$$

$$pE = 4.204 - 1.125pH \qquad \textbf{(35)}$$

To illustrate how Equations 29-35 were deduced, we selected Equations 29 and 30 as examples.

For Equation 29, we started by considering the equilibrium constant for Equation 22, whose value is $10^{1.995}$. The corresponding mass action law is defined by Equation 36, where $[HSO_4{}^-]$ and $[SO_4{}^{2-}]$ mean, respectively, the molar concentrations of $HSO_4{}^-$ and $SO_4{}^{2-}$.

$$K_{eq} = [HSO_4^-]/([SO_4^{2-}].aH_3O^+) \tag{36}$$

As $[HSO_4^-]$ and $[SO_4^{2-}]$ are equal to 10^{-2} mol L^{-1} (arbitrary value), Equation 36 can be simplified to Equation 29.

$$K_{eq} = 1/aH_3O^+ \Rightarrow \log K_{eq} = \log(1/aH_3O^+) \Rightarrow \log K_{eq} = -\log aH_3O^+ \Rightarrow$$
$$1.995 = pH \tag{29}$$

Equation 34 was deduced in a similar way.

In turn, all steps concerning the deduction of Equation 30 from the reaction indicated in Equation 23 are listed below. In this case, we applied Equation 18 using $\varepsilon^o = +0.33$ V and $[HSO_4^-] = 10^{-2}$ mol L^{-1}.

$$pE = +0.33/0.0591 - 1/6.\log(1/[HSO_4^-].a^7H_3O^+), \text{ or}$$
$$pE = 5.58 + 1/6.\log(10^{-2}) + 7/6.\log(aH_3O^+) \Rightarrow \tag{30}$$
$$pE = 5.25 - 1.17pH$$

Equations 31, 32, 33, and 35 were similarly deduced.

When Equations 20, 21, 29, 30, 31, 32, 33, 34, and 35 are plotted, the pE X pH diagram for sulfur in a typical marine ecosystem is built (Fig. **2**).

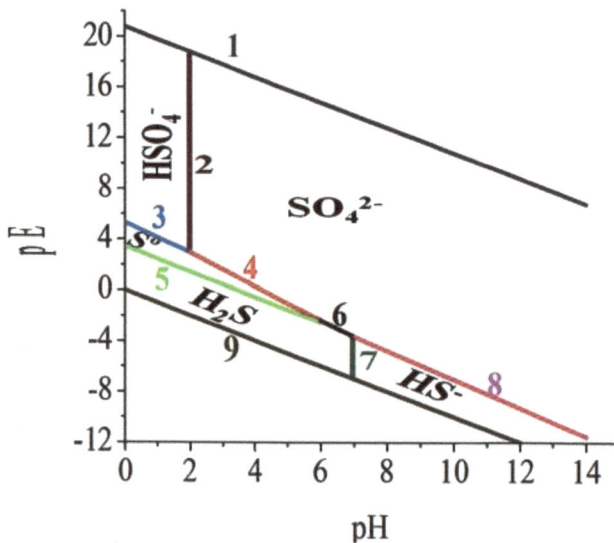

Fig. (2). pE X pH diagram showing boundaries of stability for different species containing sulfur. Lines 1, 2, 3, 4, 5, 6, 7, 8, and 9 correspond to Equations 20, 29, 30, 31, 32, 33, 34, 35, and 21, respectively.

From the pE X pH diagrams, it is possible to predict situations in which an element can exist in a soluble form in aquatic ecosystems. These structural features are important prerequisites for its bioavailability, since its assimilation by a living organism requires high solubility, since these elements should firstly be adsorbed on chemical groups of the cell membrane. Nevertheless, high solubility is not enough to assure high bioavailability, because metallic cations dissolved in marine water can be complexed by several inorganic and organic ligands, thus decreasing their capacity of being retained on cells, and consequently their bioavailability (vanLoon & Duffy, 2005).

To understand how metallic cations interact with other chemical species able to form complexes, in marine environments, the properties of different types of metals should be discussed. The next section deals with this subject.

Types of Metallic Cations and their Tendencies to form Complexes

Chemically, metallic cations can be divided into three classes: type A, type B, and borderline. Type A metallic cations have the inert gas electronic configuration, thus presenting small polarizability, and low capacity for establishing covalent interactions with ligands (chemical species with free electron pairs). In the type A metallic cations class, we have Ca^{2+}, Mg^{2+}, and Al^{3+}, which interact predominantly with ligands by means of electrostatic interactions, and the stability of these interactions is directly related to Z^2/r ratios, where Z and r are the charge and radius of the ion, respectively (vanLoon & Duffy, 2005).

Type B metallic cations form complexes by means of covalent bonds, since these cations have electronic configurations of nd^{10}or $nd^{10}(n + 1)s^2$. For this type of metallic cation, there is a direct relationship between the electronegativity (in relative terms) of the metal and the ligand. In this context, stable complexes are formed by type B metallic cations with higher electronegativity (or smaller electropositivity) and ligands with low electronegativity (vanLoon & Duffy, 2005). Hg^{2+}, Cd^{2+}, Pb^{2+} and Ag^+ are examples of type B metallic cations.

Finally, there are those metallic cations with nd^x ($0 < x < 10$) electronic configuration, which are classified as borderline metallic cations. These metallic cations exhibit intermediate chemical properties between type A and type B metallic cations, and they form complexes with all ligands able to donate electrons. The stability of a complex containing a borderline metallic cation depends on different aspects, including the atomic radius which varies across a period of the periodic table. For example, according to vanLoon & Duffy, (2005), for divalent cations belonging to the fourth period of the periodic table (Sc^{2+}, Ti^{2+}, V^{2+}, Cr^{2+}, Mn^{2+}, Fe^{2+}, Co^{2+}, Ni^{2+}, Cu^{2+}, and Zn^{2+}), Z^2/r ratios increase because the atomic radius decreases across the period (from left to right). Thus, from Sc^{2+} to Zn^{2+}, electrostatic forces assume an important role for the stability of complexes.

Besides the classifications discussed above, metals can have another classification, which exhibits great environmental importance. This classification is based on the covalent index, defined as $X_m^2 r$ where X_m is the metal electronegativity and r is the metallic ion radius (Nieboer and Richardson, 1980). Ultimately, the covalent index is related to the ability of the metal to accept electrons from ligands. Type A metallic cations have the low covalent index, while type B metallic cations exhibit high values of this index. In turn, the ionic index (Z^2/r, where Z is the ion charge and r is the metallic cation radius) reveals the degree to which an ionic bond tends to be formed.

Complexation of Metallic Ions in Marine Environments

Marine water has several dissolved compounds able to form stable and soluble complexes with some metallic ions, thus also controlling the bioavailability of several elements. Equation 37 represents a reaction in which a metallic ion (M) is complexed by four molecules of a ligand (L). With the intention of simplifying the reaction description, the charges of the metallic ion and ligand (considering that it has a charge) were omitted.

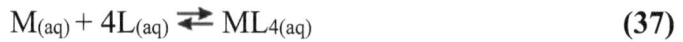

$$M_{(aq)} + 4L_{(aq)} \rightleftarrows ML_{4(aq)} \tag{37}$$

Equation 37 represents the general formation of ML_4, but four steps (reactions) occur to form this complex, as indicated in Equations 38-41, the sum of which leads to Equation 37.

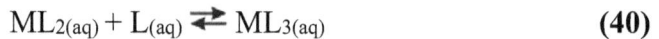

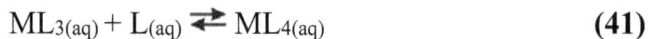

$$M_{(aq)} + L_{(aq)} \rightleftarrows ML_{(aq)} \tag{38}$$

$$ML_{(aq)} + L_{(aq)} \rightleftarrows ML_{2(aq)} \tag{39}$$

$$ML_{2(aq)} + L_{(aq)} \rightleftarrows ML_{3(aq)} \tag{40}$$

$$ML_{3(aq)} + L_{(aq)} \rightleftarrows ML_{4(aq)} \tag{41}$$

Equations 42-45 represent the mass action law concerning the reactions indicated in Equations 38-41, respectively, where K_f is the absolute constant of formation.

$$K_{f1} = [ML]/[M].[L] \tag{42}$$

$$K_{f2} = [ML_2]/[ML].[L] \tag{43}$$

$$K_{f3} = [ML_3]/[ML_2].[L] \qquad\qquad (44)$$

$$K_{f4} = [ML_4]/[ML_3].[L] \qquad\qquad (45)$$

When we make the product $K_{f1}.K_{f2}.K_{f3}.K_{f4,}$ we achieve the general absolute constant, or K_f (Equation 46), related to the general complexation reaction indicated in Equation 37.

$$K_f = [ML_4]/[M].[L]^4 \qquad\qquad (46)$$

It is important to note that the absolute constants do not consider aspects that are responsible for decreasing the tendency of ligands to form complexes with metallic ions. Thus, a single analysis of the absolute constants is not sufficient to predict the real tendency of a metallic ion to be complexed. Because of this, it is necessary to evaluate the conditional constants, which take into account aspects that limit the formation of complexes such as the pH of the aqueous medium.

In marine environments, there are several chemical species that can complex metallic ions. For example, we can consider ammonia (NH_3), which is a very effective ligand for Cu^{2+} and whose partial equilibria are indicated in Equations 47 to 50. In turn, Equation 51 indicates the global formation of $Cu(NH_3)_4^{2+}$, which is the sum of Equations 47 to 50. In marine ecosystems, ammonia is derived from anaerobic and aerobic organic matter decomposition as well as excretion of merozoites (Guimarães & de Mello, 2006).

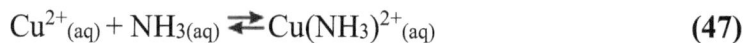

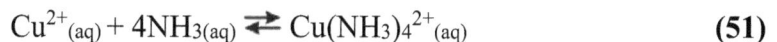

$$Cu^{2+}_{(aq)} + NH_{3(aq)} \rightleftarrows Cu(NH_3)^{2+}_{(aq)} \qquad\qquad (47)$$

$$Cu(NH_3)^{2+}_{(aq)} + NH_{3(aq)} \rightleftarrows Cu(NH_3)_2^{2+}_{(aq)} \qquad\qquad (48)$$

$$Cu(NH_3)_2^{2+}_{(aq)} + NH_{3(aq)} \rightleftarrows Cu(NH_3)_3^{2+}_{(aq)} \qquad\qquad (49)$$

$$Cu(NH_3)_3^{2+}_{(aq)} + NH_{3(aq)} \rightleftarrows Cu(NH_3)_4^{2+}_{(aq)} \qquad\qquad (50)$$

$$Cu^{2+}_{(aq)} + 4NH_{3(aq)} \rightleftarrows Cu(NH_3)_4^{2+}_{(aq)} \qquad\qquad (51)$$

Not every inorganic ligand forms complexes of great stability. This is the case, for example, for cadmium complexes with chloride, as indicated in Equations 52 to 55. As can be seen, except for the first reaction, the other constants are relatively small when compared with the complexes of ammonia and Cu^{2+} as well as complexes with the dissolved humic organic matter. The partial formation constants for the chemical species indicated in Equations 52 to 55 are, respectively: 790, 4.0, 2.0, and 0.6 (vanLoon & Duffy, 2005). In this case, the general absolute formation constant is 3,792. In comparison, the complexation of Cu^{2+} with ammonia has a general absolute formation constant of $2x10^{12}$ (Russell, 1982).

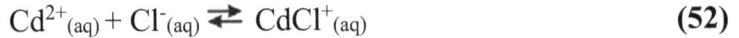

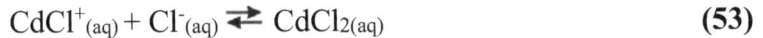

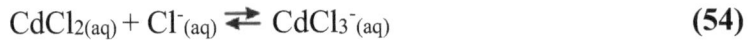

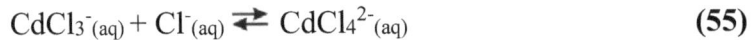

$$Cd^{2+}_{(aq)} + Cl^-_{(aq)} \rightleftharpoons CdCl^+_{(aq)} \tag{52}$$

$$CdCl^+_{(aq)} + Cl^-_{(aq)} \rightleftharpoons CdCl_{2(aq)} \tag{53}$$

$$CdCl_{2(aq)} + Cl^-_{(aq)} \rightleftharpoons CdCl_3^-_{(aq)} \tag{54}$$

$$CdCl_3^-_{(aq)} + Cl^-_{(aq)} \rightleftharpoons CdCl_4^{2-}_{(aq)} \tag{55}$$

Other kinds of ligands that can be present in marine water comprise humic organic matter, which is classified as polydentate ligands. Because of this, humic ligands are more effective ligands than inorganic species (NH_3, for example) that have only one electron pair to be used to form covalent bonds with metallic ions.

Humic organic substances are formed by means of chemical and/or microbial decomposition of known organic precursors, including proteins, carbohydrates, lipids, and nucleic acids. The fragments of these precursors are randomly recombined to form very intricate molecular structures that are very different from their organic precursors. Humic organic matter can be divided into three classes according to its solubility in relation to pH. In this sense, humines are very insoluble, independent of the pH, while fulvic acids are completely soluble within a wide range of pH. In turn, humic acids have the intermediate solubility, so that they are soluble in moderately acidic, neutral, or alkaline aqueous media (Pereira, *et al.*, 2014). From humines to fulvic acids, the carbon content decreases while the amounts of hydrophilic chemical groups increase.

Fulvic acids are especially efficient at forming soluble and stable complexes with different metallic cations, thus the presence of these acids in marine water can decrease the bioavailability of several metallic elements. Fig. (**3**) illustrates some humic functional groups that are able to interact covalently with metallic cations.

Fig. (**4**) represents the complexation of Pb^{2+} by humic functional groups. The constants listed in Table **1** are related to controlled conditions which are favorable to complexation reactions, including the pH value. In marine environments, the pH tends to be higher than seven, thus favoring the deprotonation of chemical groups. However, marine water has high salinity (high ionic strength) which diminishes the interactions among dissolved metallic ions and chemical groups of humic molecules such as those belonging to the fulvic acids class. Thus, marine environments offer intricate chemical conditions that can control the bioavailability of metallic ions, so it is very important to know the predominant conditions to predict the real toxicological effects of a particular chemical species.

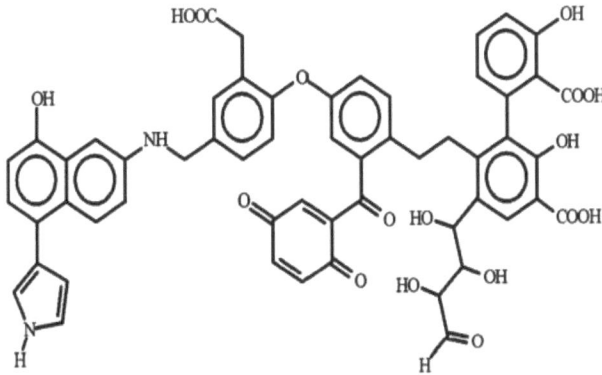

Fig. (3). Some functional groups of humic molecules.

Table 1 lists the conditional stability constants for complexes of fulvic acids and metallic ions (vanLoon & Duffy, 2005).

Table 1. Conditional stability constants (K'f) at pH 5.0.

Metallic Cation							
Ca^{2+}	Co^{2+}	Cu^{2+}	Mg^{2+}	Mn^{2+}	Ni^{2+}	Pb^{2+}	Zn^{2+}
1.2×10^3	1.4×10^4	1.0×10^4	1.4×10^2	5.0×10^3	1.6×10^4	1.1×10^4	4.0×10^3

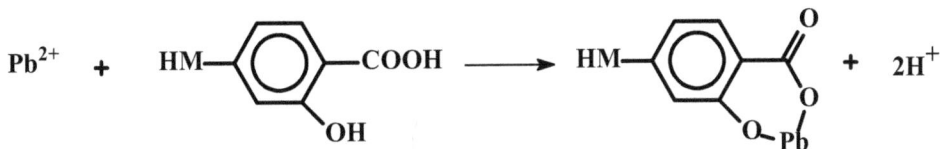

Fig. (4). Complexation of Pb^{2+} by humic functional groups, where HM means humic matter.

In addition to the ligands that are naturally found in all marine ecosystems, marine environments can be polluted with ligands derived from industrial, urban, and agricultural practices, including ammonia, sulfides, sulfates, phosphate, and ethylenediaminetetracetic acid (EDTA), among others. These anthropogenic ligands can form stable complexes with metallic cations, thus presenting a high potential for altering the chemistry of macro- and microelements in marine water.

Ammonia is derived from the degradation of organic matter which comprises domestic discharge, for example. Phosphates are present in many formulations of detergents, thus being introduced into the oceans from domestic and/or industrial discharge. Sulfides and sulfates make up the effluents of pulp and paper mills. In turn, EDTA can be found in some formulations of detergents and shampoos as well as in the discharge from photographic, textile, and paper industries (vanLoon & Duffy, 2005). Phosphates deserve special attention because they can also promote the eutrophication of marine environments, which is responsible for the severe consumption of dissolved oxygen, generation of reducing conditions, and changes in the stability of several chemical species.

Dissolved elements, either in ionic or complexed forms, can be adsorbed onto sediment particles. This process can immobilize elements and withdraw them from the liquid phase for long periods of time, thus decreasing the ability of these elements to be assimilated by pelagic organisms. However, the elements initially retained on sediments can be ingested by benthic forms of life, thus permitting a new and slow dissemination through the marine food chain.

The next section discusses the mechanisms by which sediments act on the mobility of elements in marine ecosystems. It is important to emphasize that not metallic elements (arsenic and sulfur, for example) tend to form anions in water, so these anions can only be electrostatically adsorbed on sediment particles if these particles have an excess of positive charges. This condition is not, however, easily achieved in marine environments because it would be necessary to have excess H_3O^+ in marine water. Nevertheless, elements such as arsenic and sulfur can settle if they are precipitated as their elementary forms. In turn, metallic cations are easily adsorbed on sediments, since their particles are generally negatively charged.

THE IMPORTANCE OF SEDIMENTS ON THE MOBILITY OF ELEMENTS IN MARINE ECOSYSTEMS

Geologically, sediments are fragments that are collected on the bottom of aquatic ecosystems. In marine environments, sediments can be deposited over several centuries so that layers of sediments with several hundred meters of thickness are observed. The chemical composition of sediments is extremely varied, ranging from inorganic particles such as metal oxides, aluminosilicates, and silicates to particles containing high levels of humified organic matter.

The degree of organic matter humification in sediments, as well as their granulometry are determining factors for controlling the mobility of elements between the water column and settled particles. Furthermore, the presence of clays, which have particle diameters ≤ 0.002 mm and high surface area, favors adsorption processes of ionic chemical species. Considering that marine water pH tends to be alkaline, it is possible to expect considerable deprotonation of inorganic/organic acidic chemical groups and the prevalence of negative charges on the sediment particles.

Tetrahedra of silicon and oxygen, as well as octahedra of aluminum and oxygen, are the structural unities of clays. In 1:1 clays, one layer of silica tetrahedra is chemically bound to one layer of aluminum octahedra, and this bond is made through sharing of oxygen atoms among the tetrahedra and octahedra. In turn, 2:1 clays have one layer of octahedra bound to two layers of tetrahedra, and again, tetrahedra and octahedra share oxygen atoms (de Aguiar *et al.* 2002). 2:1 clays are expansible, thus offering large surface area and high cationic exchange capacity. Kaolinite [$Al_2Si_2O_5(OH)_4$] and vermiculite [$(MgFe, Al)_3(Al,Si)_4O_{10}(OH)_2.4H_2O$] are examples of 1:1 and 2:1 clays, respectively.

These clays exhibit terminal hydroxyls, which can be deprotonated at values of pH higher than 5.0 (Essington, 2004), thus generating superficial negative charges. Furthermore, the ends of the crystals present oxygen atoms with unsatisfied valence, also generating negative charges. As previously discussed, these charges, which are placed on oxygen atoms, are especially capable of interacting electrostatically with type A metallic cations. However, type B metallic cations have lower affinity in relation to these negative charges on oxygen atoms. For borderline metals, the negative charges on clay particles have a significant effect on those elements with high Z^2/r.

This same interpretation is expected for humic organic matter present in the sediments in an associated form with 1:1 and 2:1 clays (Fig. **5**). These humic molecules adsorbed on clay particles can form complexes as stable as those formed in the aqueous phase.

Adsorptive processes concerning the metallic cations or other chemical species can be thermodynamically and quantitatively described by different mathematical models, but two of them, the Langmuir and the Freundlich models, are especially used. The Langmuir model is based on the following assumptions: i) adsorptive sites are energetically homogeneous, ii) adsorption occurs in monolayers, iii) adsorbed species do not interact, iv) temperature does not influence the monolayer volume and adsorption energy, and v) chemical equilibrium is attained (Essington, 2004). Assumption v) requires, at least, a long duration of contact between sediment particles and marine water as well as a relatively constant chemical composition of the water. In marine environments as well as in other aquatic ecosystems, these conditions are not ideally achieved. Therefore, all results concerning the capacity of marine sediments for adsorbing chemical species (adsorbate) are derived from controlled experimental conditions, and they offer only an estimate of the real sediment adsorptive potential. In these experiments, fixed masses of sediments are mechanically stirred with fixed volumes of solutions containing increasing adsorbate concentrations, thus tending to a saturation condition.

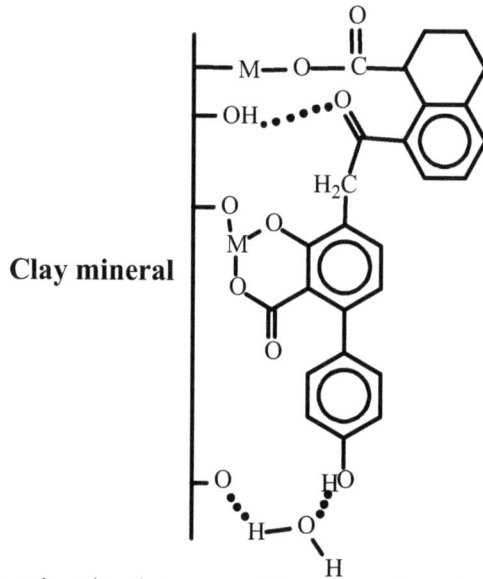

Fig. (5). Association between clay mineral structure and humic organic matter.

The Langmuir model is described by Equation 56, where q (mg kg^{-1}) is the amount of adsorbate that is adsorbed in a fixed mass of adsorbent (sediment particles, in this case), b (mg kg^{-1}) is the maximum adsorptive capacity of sediments for an adsorbate, C_{eq} is the equilibrium adsorbate concentration (mg L^{-1}) in the supernatants (or marine water, in this case), and K_L is the Langmuir constant (L kg^{-1}) which is related to energy adsorption (Essington, 2004).

$$q = b.K_L.C_{eq}/(1 + K_L.C_{eq}) \qquad (56)$$

Equation 56 can be linearized to generate Equation 57.

$$C_{eq}/q = (1/b).K_L + (1/b).C_e \qquad (57)$$

As can be seen, the maximum adsorptive capacity (b) can be estimated by the inverse of the angular coefficient of Equation 57.

In turn, the Freundlich model, which is mathematically represented by Equation 58, considers the heterogeneity of the adsorptive chemical groups. In this equation, q (mg kg^{-1}) is the amount of adsorbate that is adsorbed in a fixed mass of adsorbent (sediment particles, in this case), K_F (mg kg^{-1}) is the maximum adsorptive capacity of sediments for an adsorbate, C_{eq} is the equilibrium adsorbate concentration (mg L^{-1}) in the supernatants (or marine water, in this case), and N is a measure of the heterogeneity of adsorption sites on the sediment particles. The parameter N measures the heterogeneity of adsorption sites, so that as N approaches zero, this heterogeneity increases (Essington, 2004).

$$q = K_F.C^N_{eq} \qquad (58)$$

The parameters K_F and N can be calculated after linearization of Equation 58, which gives Equation 59.

$$\log q = \log K_F + N.\log C_{eq} \qquad (59)$$

Besides the great importance of sediments for the mobility of chemical species in marine environments, biokinetic aspects should also be considered in order to evaluate the bioaccumulation and biomagnification of metals and non-metals through the marine food chains. Therefore, it is important to emphasize that all biotic interaction involves assimilation, biotransformation, and elimination processes, as discussed in the next section.

BIOKINETIC ASPECTS INVOLVED IN THE BIOLOGICAL IN-TERACTION BETWEEN ELEMENTS AND LIVING ORGANISMS

According to Reinfelder, *et al.* (1998), the kinetic model of bioaccumulation of elements is based on linear expressions for uptake and elimination from dissolved and dietary sources. According to this model, the transfer of an element within a food chain is described for four parameters: ingestion rate (IR), assimilation efficiency (AE), physiological loss rate constant (k_e), and growth rate (g). The combination of these parameters defines the trace element trophic transfer potential (TTP) with the formula: TTP = IR.AE/[k_e + g]. The value of TTP can be interpreted as the ratio of the steady-state trace element concentration in a consumer (as a function of trophic accumulation) to that in its prey. Trace element uptake considers contributions from food (AE.IR.C_f) and water (α_w.FR.C_w), where AE and IR assume the definitions previously given while C_f and C_w are, respectively, the concentrations of an element in food (µg g^{-1}) and water (µg L^{-1}), α_w is the dissolved uptake efficiency (%), and FR is the filtration rate (L g^{-1} day^{-1}).

Reinfelder, *et al.* (1998) affirm that overall element dilution in a consumer organism is derived from growth, g (day^{-1}), elimination that includes physiological loss, k$_e$ (day^{-1}), and chemical transformation, k$_R$ (day^{-1}). This last parameter is very important because elements such as mercury and tin can be inserted in organic structures while iron and cobalt can be involved in redox reactions. In turn, these chemical and biochemical transformations are responsible for changes in the profiles concerning the elimination of elements. However, this aspect is not explicitly considered in models such as those described by Equations 60 to 62. These equations relate parameters involved in the assimilation and transfer of elements among abiotic and biotic compartments (Reinfelder, *et al.* 1998).

Equation 60 is used to calculate the time-dependent concentration (C$_t$) of an element in a consumer, where k$_u$ (L^{-1} g^{-1} day^{-1}) is defined as α_w.FR.

$$C_t = [(k_u.C_w) + (AE.IR.C_f)]/(k_e + g).(1 - e^{-(k_e + g)t}) \qquad (60)$$

In turn, Equation 61 indicates that at steady-state, uptake is balanced by elimination and growth, thus resulting in a constant concentration (C$_{ss}$) in the consumer.

$$C_{ss} = [(k_u.C_w) + (AE.IR.C_f)]/(k_e + g) \qquad (61)$$

Finally, Equation 62 describes a consumer accumulating an element only due to the ingestion of food (C$_{ss, f}$).

$$C_{ss, f} = (AE.IR.C_f)/(k_e + g) \qquad (62)$$

It is important to note that Equations 60 to 62 consider that physiological loss rates of an element accumulated from food and the dissolved phase are comparable, thus allowing the use of a single rate constant (k$_e$).

Nascimento, *et al.* (2017) evaluated the bioaccumulation rate of Cr, Cu, and Zn in shrimp (*Litopenaeus schmitii*) that were collected in Guanabara Bay, Rio de Janeiro, Brazil. The authors performed a very interesting approach to evaluate the pollutant absorption kinetics throughout the animals' life cycle. For this purpose, a study based on the dose-response relationship for the cited metals was carried out in which biological variations of the individuals were considered.

Total concentrations (mean ± standard deviation) of Cr, Cu, and Zn in 112 muscle samples of *Litopenaeus schmitii* were, respectively: 0.44 ± 0.27 µg g^{-1}, 23.46 ±

9.65 µg g^{-1}, and 34.55 ± 3.02 µg g^{-1}. However, a total metal concentration is not enough to evaluate the real capacity of a marine organism for assimilating pollutants from the environment. To address this, the authors estimated the time required for *L. schmitti* to attain its maximum (asymptotic) length (230 mm) to be 950 days (2.6 years). In turn, the time required by *L. schmitti* to achieve half of this asymptotic length is 140 days, and this time is important for providing information concerning the potential exposure time for pollutants (Nascimento, *et al.*, 2017), as discussed below.

Nascimento, *et al.*, (2017) observed that the metal uptake kinetics decreased as concentrations of Cr, Cu, and Zn in the shrimp tissues increased, and the incorporation rates (in µg metal per gram of animal tissue per day) fell from 0.009 to 0.0001 µg Cr g^{-1} day^{-1}, from 0.62 to 0.06 µg Cu g^{-1} day^{-1}, and from 0.83 to 0.18 µg Zn g^{-1} day^{-1}. For all evaluated metals, these reductions occurred from the age of 71 days to 240 days. According to the authors, the more accelerated metabolism of the younger individuals was responsible for the decreases that were observed in the incorporation rates as the shrimp age increased. This type of investigation is very important because it highlights that intrinsic biological factors, not just the external concentration of a pollutant, can effectively control its assimilation by a given organism.

Pavlaki, *et al.*, (2017) conducted an interesting investigation concerning the influence of environmental conditions (pH, temperature, and salinity) on the toxicokinetics of cadmium in a marine copepod (*Arcatia tonsa*). The authors concluded that cadmium assimilation in *Arcatia tonsa* increased when salinity decreased, but different salinities did not change cadmium uptake rates, thus indicating a simultaneous combined effect of Cd^{2+} speciation and metabolic rates for osmoregulation. Moreover, it was observed that cadmium uptake rates increased with increasing pH values (maximum uptake at pH 7.5), while depuration rates fluctuated at different pH values, thus suggesting that both processes are ruled by metabolic mechanisms and ion competition at membrane chemical groups.

EFFECTS OF ANTHROPOGENIC SOURCES OF NITROGEN, PHOSPHORUS, TIN, AND COPPER ON MARINE ENVIRONMENTS

The excessive anthropic supply of some elements can promote serious environmental disturbs in marine ecosystems. Due to a large number of marine pollutants, we focus our attention on some of them: nitrogen, phosphorus, tin, and copper.

Nitrogen and Phosphorus

One of the most important environmental disturbances that can occur in marine environments is eutrophication, in which excessive amounts of nitrogen and phosphorus are offered as their different anions, which are carried out from the continents to the oceans. Fig. (**6**) shows the Caspian Sea from space, where a huge green spot can be seen, which indicates severe eutrophication. This eutrophication is attributed to diffuse discharges containing nitrogen (as NO_3^-) and phosphorus (mainly as $H_2PO_4^-$ and HPO_4^{2-}), which are brought by the rivers that form the basin of the Volga River, whose waters carry pollutants generated by a large part of the Russian population. High concentrations of nitrogen and phosphorus come from the surface run-off of fertilized soils, untreated domestic wastes, and industrial wastes.

Fig. (6). Eutrophication of part of the Caspian Sea (Tsado/Alamy Stock Photo).

In eutrophication, several species of plants and algae grow quickly, thereby creating an environment of strong competition concerning nutrients. In turn, when this biomass dies, aerobic microorganisms decompose it, and the concentrations of dissolved oxygen are largely decreased with a consequent generation of hypoxia and reduction of the biodiversity (Wallace, *et al.*, 2014).

The biodiversity can also be decreased because of toxins that are released when plants and algae die. Moreover, the accumulation of suspended organic particles is responsible for blocking sunlight and decreasing the depth of the photic zone, with serious consequences for marine water oxygenation and photosynthesis (Miller & Spoolman, 2013).

According to Wallace, *et al.*, (2014), besides the increase of marine hypoxic areas around the world, aerobic decomposition of dead biomass derived from eutrophication releases large amounts of CO_2 into the marine water. The accumulation of carbon dioxide in water with eutrophication is also performed by the fact that the decreases in photosynthesis (due to the blocking of sun radiation) are responsible for the decrease in the assimilation of carbon dioxide dissolved in water, which can be derived, for example, from the decomposition of organic matter (Aparicio, *et al.*, 2016).

In coastal zones, this excessive production of CO_2 raises the concentration of dissolved carbonic acid (H_2CO_3), which dissociates into bicarbonate (HCO_3^-) and carbonate (CO_3^{2-}) with the consequent releasing of H_3O^+, and acidification of marine water. From a chemical point of view, this acidification consumes carbonate that is used by several organisms, both pelagic and benthic, to form their CaCO3 shells.

Table **2** lists some important works concerning the environmental consequences of eutrophication in marine ecosystems.

Table 2. Some important works concerning the eutrophication of marine environments.

Place	Environmental Consequences	Reference
Estuarine systems across Northeast US	Increases in CO_2 concentration, decrease in O_2 levels, and slight water acidification	Wallace *et al.* (2014)
Coastal water from different European areas	Modification of chemical composition of edible algae	Gao *et al.* (2018)
Australian Great Barrier Reef	Ecological disequilibrium in reefs, including the excessive growth of macro and filamentous algae	Bell (1992)
Mediterranean Sea	Modification of dissolved organic matter into more complex material	Aparicio *et al.* (2016)

Tin and Copper

In this section, we consider an important anthropic source of both elements, which is navigation routes, since ships' hulls are covered with antifouling paints (source of tin or copper) to prevent the adhesion and growth of many organisms, which reduce the performance of vessels, and increase fuel consumption. The number of fouling organisms can be high, as highlighted by Farrapeira, *et al.*, (2007) who found 60 species of these marine organisms on hulls of ships moored in the harbor of Recife, Pernambuco State, Brazil.

Over time sips' hulls suffer mechanical and chemical wear, thus releasing substances from their surfaces (antifouling paints for example). Figs. (7) and (8) show an aggressive oxidation process in a ship stranded in the coast of Ceará State, Brazil.

Fig. (7). Front view of a ship's hull in advanced corrosion process (Personal archive).

Fig. (8). Enlarged view of a corrosive process of a ship's hull (Personal archive).

Tributyltin - TBT ($C_{12}H_{28}Sn$) is highly efficient in regard to the inhibition of fouling organisms, which can be very numerous. Nevertheless, this compound is also toxic to non-target organisms, such as molluscs, in which TBT provokes endocrine-disrupting effects (Tornero & Hanke, 2016). According to Antizar-Ladsilao (2008), these endocrine-disrupting effects are characterized by elevated testosterone production with the consequent phenomenon of imposex. This phenomenon is manifested when male sex characteristics are superimposed on normal females, thereby unleashing remarkable decreases in the population of these marine organisms. Additionally, TBT is able to disturb the function of mitochondria with severe consequences for the growth, development, reproduction, and survival of many marine species. In addition to TBT, antifouling paints have monobutyltin – MBT ($C_4H_{12}Sn$) and dibutyltin – DBT ($C_8H_{18}Sn$), but MBT and DBT are found in these paints as impurities or breakdown products. Other impurities containing organotin compounds, including triphenyltin, TPhT – (C_6H_5)$_3$Sn, were also detected in historic layers of antifouling paints on leisure boat hulls (Lagerström, *et al.*, 2017).

Pletsch, *et al.*, (2010) evaluated the accumulation of TBT and other organic forms of tin (monobutyltin and dibutyltin) in marine sediments and organisms from Todos os Santos Bay, Bahia State, Brazil. These authors found the following maximum concentrations (ng g^{-1}) of these organotin compounds in sediments: 438, 207, and 423 for tributyltin, dibutyltin, and monobutyltin, respectively. Analyses were also performed in a marine tunicate (*Phallusia nigra*), and the concentrations (ng g^{-1}) of dibutyltin and monobutyltin were, respectively, 294.1 and 148.5. According to these authors, it is possible to relate concentrations of tributyltin and dibutyltin to identify when organotin compounds were introduced into the marine environment. Of the 24 sampling sites, half of them had concentrations of tributyltin higher than those of dibutyltin, thus suggesting recent tributyltin introductions in the Todos os Santos Bay, which is the entrance to the port of the Salvador city. This port is one the most important Brazilian ports.

For many years, paints with TBT were successfully used to inhibit the growth of fouling organisms, but a global prohibition was ratified in 2008 (Tornero & Hanke, 2016). Since then, levels of TBT have declined, although significant amounts of this pollutant can still be found in sediments as discussed below.

Langston and Dope (1995) highlighted the strong affinity of TBT in relation to sediments collected in the Tamar River estuary (UK), thereby indicating potential risks to benthic species. This thermodynamically favorable association of TBT with sediments is due to its remarkable hydrophobicity, which causes a high affinity in relation to specific constituents of sediment particles (humified organic matter, for example). These authors used sediments with varied chemical and granulometric compositions, which were stirred with aqueous TBT solutions at different concentrations. The authors performed detailed experiments varying the pH and salinity of TBT aqueous solutions. It was observed that distribution coefficients (Kd), which are the ratio between TBT concentration in sediment (µg kg^{-1}) and water (µg L^{-1}), reached values as high as 30000 L kg^{-1}. These values of Kd highlighted the huge ability of TBT to associate with coast marine sediments. Moreover, kinetic studies (de Mora, *et al.*, 1995) showed that TBT has a relatively long half-life in marine sediments (2.5 years).

Compounds with copper in their structures, such as Cu_2O (copper oxide) and CuCHNS (copper thiocyanate), have been alternatively used as biocides to prevent encrustation of marine organisms on ships' hulls. These compounds are less toxic than tin-based paints. However, excessive concentrations of copper in marine environments can promote biological damages to marine life, as reported by Karlsson, *et al.*, (2010) and Ytreberg *et al.* (2010). Karlsson, *et al.*, (2010) evaluated the toxic effects of leachates of anti-fouling paints used on ships and leisure boats, and the authors observed that three marine organisms (bacterium – *Vibrio fischeri*, macroalga – *Ceramium tenuicorne*, and crustacean – *Nitroca spinipes*) were severely affected due to the elevated concentrations of copper. Although copper is essential for all forms of life, the excess of this element promotes the rupture of the cells (Nordeberg, *et al.*, 2007).

A study performed in an American port area (Florida) revealed alarming copper inputs in marine waters (Srinivasan & Swain, 2007). According to these authors, these copper inputs, which were derived from sailboats and powerboats, varied from 1.7 to 2.1 tons/year, respectively. Copper levels in the marine environments have a direct relationship with the intensity of vessel traffic (Lewis, 1998), and Schiff *et al.* (2004) classify vessel hull washing as one the most noxious activities regarding the release of copper into the marine water. Such activities have been responsible for increasing the levels of this metal in marinas.

CONCLUSIONS

This chapter provided an overview of the possibilities of interaction among chemical species and different marine environmental compartments. It was highlighted that humic organic matter plays important role in order to control the mobility and bioavailability of chemical species. This bioavailability is also controlled by several inorganic chemical species (different anions) that are found in marine water.

From the works that were cited and discussed in this chapter, it was possible to conclude that analyses of toxicity concerning the marine ecosystems require analytical methods able to perform chemical speciation. Besides the capacity of chemical speciation, these methods should have high sensitivity, because many metals and non-metals in marine environments (especially in water) are found at trace levels (μg L^{-1} or ng L^{-1}). In addition to chemical aspects involving speciation and sensitivity, environmental analyses in marine ecosystems should consider biological features, which are able to control kinetic aspects concerning the assimilation rates of chemical elements by living organisms.

Due to the high complexity of marine chemistry, the introduction of toxic substances, as well as the excessive supply of nutrients, must be closely monitored. Otherwise, the triggering of physical-chemical and biochemical processes can spread unpredictably for long periods of time, thus causing severe damages across the various marine food chains. In this chapter, we focused our attention on two major environmental problems, which are related to eutrophication and navigation. In the first case, it was highlighted that excesses of nitrogen and phosphorus have been responsible for the emergence of deoxygenated areas along coastal/estuarine areas. Moreover, eutrophication also promotes acidification and worrying consequences for serval forms of marine life, since this phenomenon avoids the normal development of organisms with shells of $CaCO_3$. Over long periods of time, these deleterious effects concerning the eutrophication can be responsible for imbalances in important biogeochemical cycles, including the oxygen cycle. In turn, navigation practices are/were sources of constant release of tin and copper. This source is derived from antifouling paints used on ships' hulls, which are composed by organotin compounds (tributyltin mainly) or copper-based biocides. In both situations, vital biochemical processes can be disrupted in many species of marine organisms.

It is an imperative need for the preservation of marine biodiversity, and all its biogeochemical implications, which are essential for keeping the life on our planet. Therefore, multidisciplinary efforts should be directed toward a deeper understanding of the interactions of metals and non-metals, as well as their different chemical forms, with living organisms and abiotic compartments. Additionally, sustainable practices concerning the treatment of domestic, agricultural and industrial wastes must be implanted to avoid large supplies of nutrients. Finally, the deleterious environmental effects concerning the growing movement of ships can be minimized through global regulations of toxic compounds in antifouling paints. This type of regulation was successfully applied to organotin compounds. However, more effort in needed to produce antifouling paints containing biocides more selective to the target organisms.

CONSENT FOR PUBLICATION

Not applicable.

CONFLICT OF INTERESTS

The authors confirm that this chapter contents have no conflict of interest.

ACKNOWLEDGEMENTS

The authors would like to thank the Conselho Nacional de Desenvolvimento Científico Tecnológico (CNPq), the Coordenação de Aperfeiçoamento de Pessoal de Nível Superior (CAPES), and the Fundação de Amparo à Pesquisa do Estado da Bahia (FAPESB) for fellowships and financial support (CNPq: 555522/2006-7, and 620041/2006-4 and FAPESB: APP0076/2009, RED APP0002/2014, and RED0017/2014).

REFERENCES

Allègre, C & Dars, R (2009) La géologie – passé, présent et avenir de la Terre, Belin: pour la science, Paris.

Antizar-Ladislao, B (2008) Environmental levels, toxicity and human exposure to tributyltin (TBT)-contaminated marine environment. a review. *Environ Int,* 34, 292-308.
[http://dx.doi.org/10.1016/j.envint.2007.09.005] [PMID: 17959247]

Aparicio, FL, Nieto-Cid, M, Borrull, E, Calvo, E, Pelejero, C, Sala, MM, Pinhassi, J, Gasol, JM & Marrasé, C (2016) Eutrophication and acidification: Do they induce changes in the dissolved organic matter dynamics in the coastal Mediterranean Sea? *Sci Total Environ,* 563-564, 179-89.
[http://dx.doi.org/10.1016/j.scitotenv.2016.04.108] [PMID: 27135581]

Bell, PRF (1992) Eutrophication and coral reefs – some examples in the great barrier reef lagoon. *Water Res,* 26, 553-68.
[http://dx.doi.org/10.1016/0043-1354(92)90228-V]

Boyce, DG, Petrie, B, Frank, KT, Worm, B & Leggett, WC (2017) Environmental structuring of marine plankton phenology. *Nat Ecol Evol,* 1, 1484-94.
[http://dx.doi.org/10.1038/s41559-017-0287-3] [PMID: 29185511]

Cockell, C (2008) *An introduction to the Earth-life system.* Cambridge University Press, Cambridge.

de Aguiar, MRMP (2002) Remoção de metais pesados de efluentes industriais por aluminossilicatos. *Quim Nova,* 25, 1145-54.
[http://dx.doi.org/10.1590/S0100-40422002000700015]

de Duve, C (1995) *Vital dust – the origin and evolution of life on Earth.* Basic Books, New York.

de Mora, SJ, Stewart, C & Phillips, D (1995) Sources and rate of degradation of tri(n-butyl)tin in marine sediments near Auckland, New Zeland. *Mar Pollut Bull,* 30, 50-7.
[http://dx.doi.org/10.1016/0025-326X(94)00178-C]

Emsley, J (2005) *The elements of murder – a history of poison.* Oxford University Press, Oxford.

Essington, ME (2004) *Soil and water chemistry – an integrative approach.* CRC Press, Boca Raton.
[http://dx.doi.org/10.1201/b12397]

Farrapeira, CMR, de Melo, AVOM, Barbosa, DF & da Silva, KME (2007) Ship hull fouling in the port of Recife, Pernambuco. *Braz J Oceanogr,* 55, 207-21.
[http://dx.doi.org/10.1590/S1679-87592007000300005]

Fergusson, JE (1990) *The heavy elements – chemistry, environmental impact and health effects.* Pergamon Press, Oxford.

Gao, G, Clare, AS, Chatzidimitriou, E, Rose, C & Caldwell, G (2018) Effects of ocean warming and

acidification, combined with nutrient enrichment, on chemical composition and functional properties of Ulva rigida. *Food Chem,* 258, 71-8.
[http://dx.doi.org/10.1016/j.foodchem.2018.03.040] [PMID: 29655756]

Gargaud, M, Martin, H, López-García, P, Montmerle, T & Pascal, R (2012) *Young sun, early Earth and the origins of life.* Springer-Verlag, Berlin.
[http://dx.doi.org/10.1007/978-3-642-22552-9]

Guimarães, GP & de Mello, WZ (2006) Estimativa do fluxo de amônia na interface ar-mar na Baía de Guanabara – estudo preliminar. *Quim Nova,* 29, 54-60.
[http://dx.doi.org/10.1590/S0100-40422006000100012]

Karlsson, J, Ytreberg, E & Eklund, B (2010) Toxicity of anti-fouling paints for use on ships and leisure boats to non-target organisms representing three trophic levels. *Environ Pollut,* 158, 681-7.
[http://dx.doi.org/10.1016/j.envpol.2009.10.024] [PMID: 19913342]

Lagerström, M, Strand, J, Eklund, B & Ytreberg, E (2017) Total tin and organotin speciation in historic layers of antifouling paint on leisure boat hulls. *Environ Pollut,* 220, 1333-41.
[http://dx.doi.org/10.1016/j.envpol.2016.11.001] [PMID: 27836476]

Langston, WJ & Pope, ND (1995) Determinants of TBT adsorption and desorption in estuarine sediments. *Mar Pollut Bull,* 31, 32-43.
[http://dx.doi.org/10.1016/0025-326X(95)91269-M]

Lewis, J A ((1998)) Marine biofouling and its prevention on underwater surfaces. *Mater forum,* 22, 41-61.

Miller, GT & Spoolman, SE (2013) *Environmental Science.* Brooks/Cole Cengage Learning, Independence.

Nascimento, JR, Sabadini-Santos, E, Carvalho, C, Keunecke, KA, César, R & Bidone, ED (2017) Bioaccumulation of heavy metals by shrimp (Litopenaeus schmitti): A dose-response approach for coastal resources management. *Mar Pollut Bull,* 114, 1007-13.
[http://dx.doi.org/10.1016/j.marpolbul.2016.11.013] [PMID: 27876373]

Nieboer, E & Richardson, DHS (1980) The replacement of the nondescript term 'heavy metals' by a biologically and chemically significant classification of metal ions. *Environ Pollut,* 1, 3-26. [Series B].
[http://dx.doi.org/10.1016/0143-148X(80)90017-8]

Nordberg, GF, Fowler, BA, Nordberg, M & Friberg, L (2007) *Handbook on the toxicology of metals.* Academic Press, Amsterdam.

Pavlaki, MD, Morgado, RG, van Gestel, CAM, Calado, R, Soares, AMVM & Loureiro, S (2017) Influence of environmental conditions on the toxicokinetics of cadmium in the marine copepod Acartia tonsa. *Ecotoxicol Environ Saf,* 145, 142-9.
[http://dx.doi.org/10.1016/j.ecoenv.2017.07.008] [PMID: 28732297]

Pereira, MdeG, Neta, LCS, Fontes, MPF, Souza, AN, Matos, TC, Sachdev, RdeL, dos Santos, AV, da Guarda Souza, MO, de Andrade, MV A S, Paulo, GMM, Ribeiro, JN & Ribeiro, AV F N (2014) An overview of the environmental applicability of vermicompost: from wastewater treatment to the development of sensitive analytical methods. *ScientificWorldJournal,* 2014917348
[http://dx.doi.org/10.1155/2014/917348] [PMID: 24578668]

Pletsch, AL, Beretta, M & Tavares, TM (2010) Distribuição especial de compostos orgânicos de estanho em sedimentos costeiros e em Phallusia nigra da Baía de Todos os Santos e litoral norte da Bahia – Brasil. *Quim Nova,* 33, 451-7.
[http://dx.doi.org/10.1590/S0100-40422010000200037]

Reinfelder, JR, Fisher, NS, Luoma, SN, Nichols, JW & Wang, WX (1998) Trace element trophic transfer in aquatic organisms: a critique of the kinetic model approach. *Sci Total Environ,* 219, 117-35.
[http://dx.doi.org/10.1016/S0048-9697(98)00225-3] [PMID: 9802246]

Reish, DJ, Kauwling, TJ, Mearns, AJ, Oshida, PS & Rossi, SS (1977) Marine and estuarine pollution. *J Water Pollut Control Fed,* 49, 1316-40.

[PMID: 328944]

Rothschild, LJ & Lister, AM (2003) *Evolution on Planet Earth – the impact of the physical environment.* Academic Press, London.

Russell, JB (1982) *Química Geral.* Editora Mc-Graw-Hill do Brasil Ltda, São Paulo.

Srinivasan, M & Swain, GW (2007) Managing the use of copper-based antifouling paints. *Environ Manage,* 39, 423-41.
[http://dx.doi.org/10.1007/s00267-005-0030-8] [PMID: 17253094]

Tchounwou, PB, Yedjou, CG, Patlolla, AK & Sutton, DJ D J (2014) Heavy metals toxicity and the environment. National Institutes of Health, 101, 133-64.

Tornero, V & Hanke, G (2016) Chemical contaminants entering the marine environment from sea-based sources: A review with a focus on European seas. *Mar Pollut Bull,* 112, 17-38.
[http://dx.doi.org/10.1016/j.marpolbul.2016.06.091] [PMID: 27444857]

vanLoon, GW & Duffy, SJ (2005) *Environmental Chemistry – a global perspective.* Oxford University Press, Oxford.

Vauk, GJM & Schrey, E (1987) Litter pollution from ships in the German bight. *Mar Pollut Bull,* 18, 316-9.
[http://dx.doi.org/10.1016/S0025-326X(87)80018-8]

Wallace, RB, Baumann, H, Grear, JS, Aller, RB & Gobler, CJ (2014) Coastal ocean acidification: The other eutrophication problem. *Estuar Coast Shelf Sci,* 148, 1-13.
[http://dx.doi.org/10.1016/j.ecss.2014.05.027]

Ytreberg, E, Karlsson, J & Eklund, B (2010) Comparison of toxicity and release rates of Cu and Zn from anti-fouling paints leached in natural and artificial brackish seawater. *Sci Total Environ,* 408, 2459-66.
[http://dx.doi.org/10.1016/j.scitotenv.2010.02.036] [PMID: 20347476]

CHAPTER 4

Sulfur, Aluminum, Arsenic, Cadmium, Lead, Mercury, and Nickel in Marine Ecosystems: Accumulation, Distribution, and Environmental Effects

Madson de Godoi Pereira[1,*], **Daniel Carneiro Freitas**[1], **Lourdes Cardoso de Souza Neta**[1], **Arnaud Victor dos Santos**[1], **Joselito Nardy Ribeiro**[2] and **Araceli Verónica Flores Nardy Ribeiro**[3]

[1] *Departamento de Ciências Exatas e da Terra – Universidade do Estado da Bahia, Rua Silveira Martins, 2555, CEP: 41.150-000, Cabula, Salvador-BA, Brazil*

[2] *Centro de Ciências da Saúde – Universidade Federal do Espírito Santo, Avenida Maruípe, S/N, CEP: 29.042-751, Vitória-ES, Brazil*

[3] *Coordenação de Licenciaturas – Instituto Federal de Educação, Ciência e Tecnologia do Espírito Santo, Avenida Vitória, 1729, CEP: 29.040-780, Vitória-ES, Brazil*

Abstract: This chapter deals with aspects concerning the presence of sulfur, aluminum, arsenic, cadmium, lead, mercury, and nickel in marine environments. For each of these elements, information about their natural and anthropic sources, as well as their toxicity and accumulation in different organisms, was provided. It was shown that the total accumulation of aluminum, cadmium, lead, mercury, and nickel in physical marine compartments (water, particulate matter, and sediments) is not necessarily related to the bioaccumulation of these elements, since many aspects concerning the bioavailability (including chemical speciation) should be considered.

Keywords: Chemical Speciation, Bioaccumulation, Human Health.

INTRODUCTION

In this chapter, we discuss chemical and environmental aspects concerning the presence of some key elements in marine ecosystems, choosing the following elements: sulfur, arsenic, aluminum, cadmium, lead, mercury, and nickel. These elements were chosen because they belong to specific groups, so that sulfur is an essential element (a secondary macroelement), while arsenic is classified as a

---
* **Corresponding author Madson de Godoi Pereira:** Departamento de Ciências Exatas e da Terra – Universidade do Estado da Bahia, Rua Silveira Martins, 2555, CEP: 41.150-000, Cabula, Salvador-BA, Brazil; Tel/Fax: +55-71-31-7-2372; E-mail: madson.pereira444@gmail.com

De-Sheng Pei & Muhammad Junaid (Eds.)
All rights reserved-© 2019 Bentham Science Publishers

non-metallic toxic element, and a microelement. In turn, aluminum, cadmium, lead, and mercury represent the group of toxic metals, which are commonly found in marine environments as microelements. Finally, nickel is a metallic microelement with some essential functions associated with different forms of life. In the next sections, we discussed the ways in which these seven elements are distributed, transformed, and accumulated in the different marine environmental compartments, as well as their natural and anthropic sources. This chapter also discusses toxicological aspects concerning the assimilation of some of these elements by humans. Fig. (**1**) shows marine environments that can integrate many compartments, including coastal vegetation and giant aquatic animals such as whales. In this photo, we have an adult humpback whale (Praia do Forte Beach, Bahia State, Brazil), whose length can reach 16 m.

Fig. (1). Humpback whale (*Megaptera novaeangliae*) in the Brazilian coast (Personal archive).

Sulfur

Sulfur is an essential element because it is found in the composition of amino acids as cysteine ($C_3H_7NO_2S$), thus participating in the chemical composition of proteins and/or enzymes. Much of the sulfur that reaches the oceans comes from leaching of soils, rocks, and minerals. This contribution of continental sulfur is delivered almost exclusively by rivers (Cardoso & Pitombo, 1992). The marine chemistry of several elements, including cadmium, lead, mercury, and nickel, is controlled by species containing sulfur since these microelements form insoluble sulfides, thus exhibiting direct impact on their bioavailability. As sulfur can assume many oxidation states (-2 to +6), this element presents several chemical forms, as discussed in Chapter 3, entitled *"Macroelements and microelements in marine ecosystems: an overview"*.

An important natural source of sulfur for marine water is the geological activities that occur in the oceanic floor (average depth of 4-5 km) where deep, cold marine water infiltrates through cracks in the rocks and is overheated by the proximity of hot rocks. This overheated water captures H_2S, which comes from magma, and becomes acidic, thus reacting with rocks and dissolving metals such as iron, copper, and zinc. Insoluble sulfides of these metals are formed, and their precipitation builds geological structures called fumaroles or smokers, the color of which tends to be black, due to the presence of iron sulfides, or white. Around these fumaroles, there is exuberant life for which chemosynthesis is the source of energy. Specialized bacteria promote the oxidation of H_2S, thus producing nutrients that sustain many forms of life (Martins and Nunes, 2009).

The transfer of sulfur from the atmosphere to the oceans occurs by means of the deposition of sulfate salts formed in atmospheric reactions in which H_2S is subsequently oxidized to SO_2, SO_3, and SO_4^{2-}. The average concentration of SO_4^{2-} in marine water is around 2.6 g L^{-1} (Garrison, 2006). The oceans also have elementary sulfur, but the occurrence of this species needs intermediate values of pE as well as small to intermediate values of pH, as discussed in Chapter 3 entitled *"Macroelements and microelements in marine ecosystems: an overview"*.

The flux of sulfur from the oceans to the atmosphere is by marine spray and production of $(CH_3)_2S$ by phytoplankton (Manham, 1994). Except for deep marine waters, in which there are large emissions of H_2S from geological activities, this weak acid is not found in large amounts in shallow water. However, it remains very important because of its dissociation, whose equilibria are indicated in the following Equations 1 and 2 with the respective constants (K_{a1} and K_{a2}) of 9.6 x 10^{-8} and 1.3 x 10^{-14}.

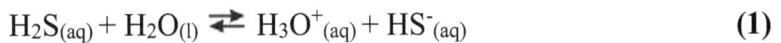

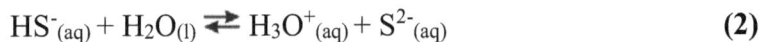

$$H_2S_{(aq)} + H_2O_{(l)} \rightleftharpoons H_3O^+_{(aq)} + HS^-_{(aq)} \qquad \textbf{(1)}$$

$$HS^-_{(aq)} + H_2O_{(l)} \rightleftharpoons H_3O^+_{(aq)} + S^{2-}_{(aq)} \qquad \textbf{(2)}$$

Equations 3 and 4 represent, respectively, the mass action law of the reactions indicated in Equations 1 and 2.

$$K_{a1} = [H_3O^+].[HS^-]/[H_2S] \qquad (3)$$

$$K_{a2} = [H_3O^+].[S^{2-}]/[HS^-] \qquad (4)$$

As can be noted, the concentrations of HS$^-$ and S^{2-} are strongly dependent on the pH of marine water, so it is possible to deduce an equation that relates [S^{2-}] and [H$_3$O$^+$]. For this purpose, we should multiply Equations 3 and 4 to obtain Equation 5.

$$K_{a1}.K_{a2} = [H_3O^+]^2.[S^{2-}]/[H_2S] \qquad (5)$$

Equation 5 shows that [S^{2-}] decreases quickly with an increase in [H$_3$O$^+$] for a fixed [H$_2$S]. Thus, the average pH (8.5) of marine water can assure sufficiently high [S^{2-}] to allow the precipitation of sulfides of many metals.

HS$^-$ can also be derived from the decomposition of organic matter, in which amino acids, such as cysteine, are microbiologically decomposed in the presence of the enzyme desulfhydrase, as indicated in Equation 6 (vanLoon & Duffy, 2005).

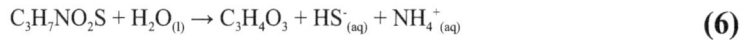

$$C_3H_7NO_2S + H_2O_{(l)} \rightarrow C_3H_4O_3 + HS^-_{(aq)} + NH_4^+{}_{(aq)} \qquad (6)$$

According to Ivanov, *et al.*, (1989), sulfate plays an important role in the oxidation of organic matter in marine sediments, thus highlighting how close the biogeochemical cycles of sulfur and carbon are. In marine sediments, sulfate-reducing bacteria are responsible for the oxidation of several organic compounds, including the fatty acids, as indicated in Equation 7.

$$4H(CH_2)_nCOO^- + (3n+1)SO_4^{2-} + H_2O \rightarrow (4n+4)HCO_3^- + (3n+1)HS^- + OH^- + nH^+ \qquad (7)$$

Arsenic

The toxicity of this element to humans has long been known since it was used in ancient China and India, respectively, to kill rodents and preserve paper from insect attack. Curiously, the vapor of this element was already employed to treat humans with coughs and asthma. According to the United States Environmental Protection Agency (USEPA), a human adult of 70 kg may support 1 mg of arsenic per day, and doses between 125 and 250 mg of arsenic would be enough to kill most adults. In terms of toxicity to humans, inorganic compounds of arsenic cause several illnesses such as diabetes, heart diseases, anemia, and disorders of the immune, nervous, and reproductive systems. Furthermore, these inorganic forms of arsenic are responsible for cancers of the skin, lungs, bladder, and prostate (Emsley, 2005).

Although the average arsenic concentration in seawater is only 24 µg L^{-1}, several marine organisms can concentrate this element because algae and cyanobacteria are able to methylate arsenic, thus potentiating its distribution through the food chain. As a result, some marine species may accumulate surprising amounts of arsenic (not informed if dry or wet weight), including oysters (4 mg kg^{-1}), mussels (120 mg kg^{-1}), and prawns (175 mg kg^{-1}). Due to its high capacity for accumulating arsenic, seafood is one of the most important sources of this element in countries such as Japan, where the consumption of fish and other marine animals is common (Emsley, 2005).

Approximately 200×10^6 kg y^{-1} of arsenic compounds can reach the oceans, and most of these compounds are derived from volcanic emissions and deposited from the atmosphere. Additionally, arsenic compounds that are applied to soils as insecticides can be brought to the oceans by rivers. Once in seawater, arsenic is predominantly distributed as $HAsO_4^{2-}$, $H_2AsO_4^{-}$, AsO_4^{3-}, $RAsO(OH)_2$, and $R_2AsO(OH)$, where R is an organic group. Large amounts of arsenic are immobilized in marine sediments, where concentrations of arsenic may reach 450 mg kg^{-1} (Fergusson, 1990).

Equations 8-10 represent three acid-base equilibria that occur among arsenic species.

$$H_3AsO_{4(aq)} + H_2O_{(l)} \rightleftharpoons H_3O^{+}_{(aq)} + H_2AsO_4^{-}{}_{(aq)}, K_{a1} = 5.6 \times 10^{-3} \qquad \textbf{(8)}$$

$$H_2AsO_4^{-}{}_{(aq)} + H_2O_{(l)} \rightleftharpoons H_3O^{+}_{(aq)} + HAsO_4^{2-}{}_{(aq)}, K_{a2} = 1.7 \times 10^{-7} \qquad \textbf{(9)}$$

$$HAsO_4^{2-}{}_{(aq)} + H_2O_{(l)} \rightleftharpoons H_3O^{+}_{(aq)} + AsO_4^{3-}{}_{(aq)}, K_{a3} = 4.0 \times 10^{-12} \qquad \textbf{(10)}$$

The mass action law concerning the reactions indicated in Equations 8-10 is indicated in Equations 11-13.

$$K_{a1} = [H_2AsO_4^{-}].[H_3O^{+}]/[H_3AsO_4] \qquad \textbf{(11)}$$

$$K_{a2} = [HAsO_4^{2-}].[H_3O^{+}]/[H_2AsO_4^{-}] \qquad \textbf{(12)}$$

$$K_{a3} = [AsO_4^{3-}].[H_3O^{+}]/[HAsO_4^{2-}] \qquad \textbf{(13)}$$

In Equation 11 if we consider $[H_3AsO_4] = [H_2AsO_4^{-}]$, we obtain Equation 14.

$$K_{a1} = [H_3O^{+}] \Rightarrow -\log K_{a1} = -\log[H_3O^{+}] \Rightarrow pK_{a1} = pH \qquad \textbf{(14)}$$

When pH = pK_{a1} or pH = $-\log(5.6 \times 10^{-3})$ = 1.25, it is possible to affirm that 50% of H_3AsO_4 is dissociated, thus forming 50% of $H_2AsO_4^-$. Similarly, we have Equations 15 and 16.

$$pK_{a2} = -\log(1.7 \times 10^{-7}) = 5.8 \qquad (15)$$

$$pK_{a3} = -\log(4.0 \times 10^{-12}) = 10.4 \qquad (16)$$

In this case, at pH = 5.8 $[H_2AsO_4^-]$ = $[HAsO_4^{2-}]$ and at pH = 10.4, $[HAsO_4^{2-}]$ = $[AsO_4^{3-}]$. Considering the average pH of marine water (8.5), $HAsO_4^{2-}$ is expected to predominate in this aquatic environment.

Many species of arsenic can be involved in redox reactions, as indicated in Equations 17 and 18 (Harris, 2008).

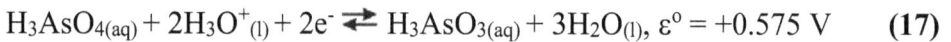

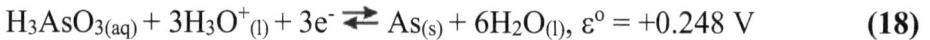

$$H_3AsO_{4(aq)} + 2H_3O^+_{(l)} + 2e^- \rightleftarrows H_3AsO_{3(aq)} + 3H_2O_{(l)}, \; \varepsilon^\circ = +0.575 \text{ V} \qquad (17)$$

$$H_3AsO_{3(aq)} + 3H_3O^+_{(l)} + 3e^- \rightleftarrows As_{(s)} + 6H_2O_{(l)}, \; \varepsilon^\circ = +0.248 \text{ V} \qquad (18)$$

Applying the same reasoning used to deduce the pE X pH diagram for sulfur (See the chapter titled "*Macroelements and microelements in marine ecosystems: an overview*"), it is possible to obtain Equations 19 and 20, which define the stability boundaries for all species involved in the equilibria indicated in equations 17 and 18. Here, the molar concentration of dissolved species is 10^{-2} mol L^{-1} and the partial pressure of gaseous species is 101,325 Pa, or 1 atm.

$$pE = 9.7 - pH \qquad (19)$$

$$pE = 3.5 - pH \qquad (20)$$

Accumulation and Distribution of Arsenic in the Marine Environment

In marine organisms, As(V) is absorbed into cells by means of phosphate transport systems with subsequent reduction to As(III), which is pumped out *via* membrane proteins to vacuoles. In turn, methylated forms of arsenic, including methylarsonous and dimethylarsinnous acids, are commonly recognized as products of detoxification for many marine organisms. More complex arsenic compounds, such as several arsenosugars and arsenolipids, were also identified in marine life including cyanobacteria that participate actively in the arsenic biogeochemical cycle. It was found that these bacteria can undertake, simultaneously, arsenic demethylation and arsenolipids biosynthesis (Xue, *et al.*, 2017).

Zhang, *et al*., (2016) investigated the biotransformation of As(III) and As(V) in a species of herbivorous fish (*Siganus fuscescens*). For this purpose, the cited fish was exposed to arsenic doses of up to 1,500 µg g^{-1} (dry weight) for 21 and 42 days. It was demonstrated that both As(III) and As(V) were biotransformed into the less toxic form of arsenobetaine, with a distribution percentage of 63.3 to 91.3% in the liver and 79.0 to 95.2% in muscle. The authors concluded that the biochemical transformation of inorganic arsenic forms includes oxidation of As(III) to As(V), reduction of As(V) to As(III), and methylation to monomethylarsonic and dimethylarsinic acids with subsequent conversion to arsenobetaine. Thus, it was observed that the biotransformation of inorganic arsenic species allows high rates of bioaccumulation of arsenic in the herbivorous fish evaluated.

A study performed in China (Li, *et al*. 2017) also evaluated the total concentrations of arsenic in 38 species of tropical fish, and the average concentration was 20.84 mg kg^{-1}, in dry weight. This level of arsenic was considered higher than the United State Environmental Protection Agency risk-based concentrations. The main form of organic arsenic was arsenobetaine ($C_5H_{11}O_2As$), while As(V) corresponded to less than 3.3% of the inorganic arsenic forms.

An important Chinese bay (Laizhou Bay) was evaluated for concentrations of arsenic in muscle samples of marine organisms (Liu, *et al*., 2017). It was found that concentrations of arsenic presented a large variation (0.48 ± 0.37 µg g^{-1}, wet weight) in the evaluated organisms, and the decreasing order of concentrations was: shell-fish > shrimp > crab > fish. According to China's assessment guidelines, the maximum recommended concentration of arsenic in marine organisms is 0.2 µg g^{-1}.

In 2012, an extensive study was performed in Cuba (Cienfuegos Bay) in order to quantify the total concentrations of arsenic in marine organisms (Alonso-Hernández, 2012). The average total concentrations of arsenic in fish (dry weight) varied from 2.09 to 28.4 µg g^{-1}, while concentrations varied from 21 to 23 µg g^{-1} in mollusks and from 15.6 to 36.2 µg g^{-1} in crustaceans. According to the authors, these results indicated that the Cienfuegos Bay has been impacted during the past 30 years by releases of discharge containing residual amounts of arsenic derived from a nitrogen fertilizer factory. However, it was not possible to conclude about the toxicological risks since a large part of the total arsenic is not available to cause biological damage to humans. Because of this, work dealing with speciation is very important to evaluate the real risks concerning human contamination.

Earlier in 2004, Cienfuegos Bay was evaluated in order to estimate the distribution of arsenic in fish (Fattorini *et al.*, 2004). When compared with the study of Alonso-Hernández, *et al.*, (2012), the total concentration of arsenic was much higher with values up to 500 μg g^{-1} (dry weight). In the muscle samples, inorganic species of arsenic were predominant, although many organic forms were also identified, including methylarsonate, dimethylarsinate, trimethylarsine oxide, and tetramethylarsonium.

The biogeochemical cycle of arsenic is closely interconnected with the sulfur cycle, so that many compounds containing arsenic also have sulfur in their structures. Seaweeds contain arsenic mainly as arsenoribosides, and arsenobetaine is the main arsenic compound found in marine animals. Nevertheless, more than 40 arsenic-containing compounds have already been identified in marine organisms, and the most complex of them belong to a class called thio-methylated arsenic compounds in which arsenic is bound to a sulfur atom and not to an oxygen atom. Recently, 14 arsenic-containing compounds were isolated from marine species that were collected in Australia (Maher *et al.* 2013).

The estuary of a South Korean river (Taehwa River) was monitored for different arsenic species (Hong, *et al.*, 2016). In this case, macroalgae, fish, crabs, bivalves, shrimps, gastropods, and water samples were collected and analyzed for inorganic and organic arsenic forms. Of the inorganic forms, As(V) was the most predominant dissolved form, while the main organic forms found were arsenobetaine (C$_5$H$_{11}$AsO$_2$), monomethylarsonic acid (C$_2$H$_5$AsO$_3$), dimethylarsinic acid (C$_2$H$_7$AsO$_2$), and arsenocholine (C$_5$H$_{14}$AsO$^+$). The predominance of As(V) was attributed to high oxygenation of the water samples evaluated. The Taehwa River estuary exhibited a predominance of fine particles (diameter range from 0.45 to 30 μm), which were responsible for effective adsorption of arsenic forms in the suspended particulate matter. Of all the organisms evaluated, a species of macroalgae (*Enteromorpha* spp.) exhibited the highest concentration of total arsenic, attaining a value of approximately 45 μg g^{-1}, dry weight, and thus highlighting the great bioaccumulation potential of algae for arsenic. This outcome is environmentally worrying because algae are one of the bases of marine food chains.

Speciation analyses for arsenic were also performed in marine organisms collected from the Aegean Sea, Turkey (Kucuksezgin *et al.* 2014). The average percentages of inorganic arsenic to total arsenic was 3.43 ± 3.38% (wet weight) for all biota samples. These analyses revealed that the more toxic arsenic form, which is the inorganic form, did not prevail in the analyzed samples, and based on the average consumption of fresh fish in the region of Izmir, there were no identified risks of human contamination with inorganic arsenic forms. Zhang, *et al.*, (2017) investigated how As(V) is bioaccumulated and biotransformed in marine benthic fish, including evaluations concerning biokinetic aspects. This kind of investigation is very interesting because sediments can accumulate arsenic and many other elements. It was verified that different diets and arsenic concentrations had no effect on the arsenic accumulations in the evaluated fish. Furthermore, the authors verified that bioaccumulation of inorganic arsenic in benthic fish was slow.

Price, *et al.*, (2016) evaluated the bioaccumulation and biotransformation of arsenic in deep-sea hydrothermal vent organisms in Manus Basin, Papua New Guinea. The authors evaluated the concentrations of cadmium in two species of gastropods (*Alviniconcha hessleri* and *Ifremeria nautilei*) and one species of mussel (*Bathymodiolus manunsensis*). In the tissues of these organisms, the highest arsenic concentration was found in the gills of *Alviniconcha hessleri* at 5,580 mg kg^{-1}, in dry weight. This high arsenic concentration in *Alviniconcha hessleri* was attributed to the presence of sulfur in this species as a consequence of a symbiotic association with microorganisms. The overall arsenic concentrations in the evaluated marine organisms decreased with increasing distance from diffuse vent areas, where arsenic concentrations in the water varied from 44.0 to 169.5 x 10^{-9} mol L^{-1}. From these low values of arsenic concentration in the water, we can see the remarkable ability of aquatic organisms to bioaccumulate arsenic. It was demonstrated that the high total concentration of arsenic in *Alviniconcha hessleri* was distributed in different arsenic species. Thus, in the digestive system and gills of this animal, the predominant forms of arsenic were As(V) and As(III). The latter represented almost 97% of all arsenic species, while arsenobetaine was found at a percentage of 42% in *Alviniconcha hessleri* muscles, thus highlighting that this organic arsenic form is a metabolite. Moreover, a rare type of sugar-containing arsenic was identified at trace levels only in digestive glands and gills.

Aluminum

Aluminum is a very abundant element in the Earth's crust, constituting approximately 8.1% by weight (Grotzinger & Jordan, 2010). However, due to its great thermodynamic tendency to precipitate as $Al(OH)_3$, cationic aluminum, which exists as a complex with water – $Al(H_2O)_6^{3+}$, it can be found only at trace concentrations (µg L^{-1} or less) in marine water, where the average pH is generally slightly higher than eight. At pH values between 9 and 10, aluminum initially precipitated as $Al(OH)_3$ can be solubilized as $Al(OH)_4^-$, but these values of pH are not commonly found in marine water, so the species $Al(OH)_3$ tends to have an important control on aluminum bioavailability. The existence of dissolved Al^{3+} is also limited by hydrolysis reactions, as indicated below in Equations 21-24.

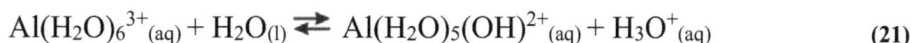

$$Al(H_2O)_6^{3+}{}_{(aq)} + H_2O_{(l)} \rightleftarrows Al(H_2O)_5(OH)^{2+}{}_{(aq)} + H_3O^+{}_{(aq)} \qquad \textbf{(21)}$$

$$Al(H_2O)_5(OH)^{2+}_{(aq)} + H_2O_{(l)} \rightleftarrows Al(H_2O)_4(OH)_2^{+}_{(aq)} + H_3O^{+}_{(aq)} \tag{22}$$

$$Al(H_2O)_4(OH)_2^{+}_{(aq)} + H_2O_{(l)} \rightleftarrows Al(H_2O)_3(OH)_{3(s)} + H_3O^{+}_{(aq)} \tag{23}$$

$$Al(H_2O)_3(OH)_{3(s)} + H_2O_{(l)} \rightleftarrows Al(H_2O)_2(OH)_4^{-}_{(aq)} + H_3O^{+}_{(aq)} \tag{24}$$

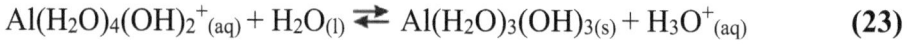

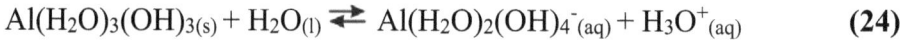

According to Gillmore, (2014), aluminum concentrations in ocean surface waters have large variations. Thus, the following concentrations (μg L^{-1}) were found: 0.5-5.5 (North Atlantic Ocean), 0.5-1.5 (Mediterranean Sea), 0.027-0.22 (Pacific Ocean), and 0.081-0.11 (Arctic Ocean).

It is very important to know the concentrations of aluminum in marine organisms because this element is neurotoxic to humans and it may be responsible for Alzheimer's disease (Mirza, *et al.*, 2017). A daily aluminum intake of up to 7 mg per kilogram body mass is tolerable for humans (WHO, 1996). Table **1** lists some results concerning the evaluation of aluminum concentrations in different marine samples.

Table 1. Research results concerning the concentration of aluminum in marine samples.

Marine Organism	Aluminum Concentration (μg g^{-1})	Local	Reference
Fish* (*Trichecus manatus*)	0.27±0.1**	Brazil	(Anzolin, 2011)
Fish (*Sebastes* spp.)	0.096±0.022**	Germany	(Ranau, *et al.*, 2001)
Mussel (*Mytilus edulis*)	71.9±1.1***	Germany	(Ranau, *et al.*, 1999)
Fishes (*Trachurus mediterraneus, Sadra sarda, Mullus barbatus, Engraulis encrasicolus, Gobius niger, Merlangius merlangus, Belone belone*, and *Pamatomus saltarix*)	7.84±1.14 to 19.2±3.0***	Turkey	(Küpeli, *et al.*, 2014)

*Blood samples **Wet weight ***Dry weight

Cadmium, Lead, and Mercury

Cadmium is an element extensively used in many industrial activities, including the production of batteries, polymers, and some insecticides. In this case, several sources of discharge can introduce cadmium into soils, rivers, and oceans. Marine sediments can accumulate cadmium at a very large range of concentrations, varying from 0.01 to 50,000 μg g^{-1}, while cadmium concentration in marine water is typically lower than 0.01 µg L^{-1}. The concentration of cadmium largely varies according to depth, such that the concentration increases until approximately 1,500 meters is reached. At greater depths, the concentration of cadmium stabilizes, and it is strongly correlated with concentrations of two nutrients (PO_4^{3-} and NO_3^-) with correlation coefficients found up to 0.9, a strong positive correlation. The increase in cadmium, phosphate, and nitrate concentrations with depth are due to the decomposition of organic matter and consequent release of both nutrients as well as cadmium (Fergusson, 1990).

In water, cadmium ions (Cd^{2+}) can be converted to different chemical species, since they can be involved in hydrolysis to generate $Cd(OH)^+$ and $Cd(OH)_2$. Species containing sulfur are formed, as indicated in Equation 25 in which it is possible to observe a reaction between a cadmium complex and an anion derived from H_2S. This is an interesting example because it considers the simultaneous occurrence of three chemical equilibria in marine water: complexation of Cd^{2+} with Cl$^-$, dissociation of H_2S, and formation of an insoluble salt.

$$CdCl^+_{(aq)} + HS^-_{(aq)} + H_2O_{(l)} \rightleftarrows CdS_{(s)} + H_3O^+_{(aq)} + Cl^-_{(aq)} \qquad (25)$$

pE X pH diagrams for cadmium show that the important species up to a pH of 7-8 is Cd^{2+}, while at higher values of pH (8 to10), the precipitation of $CdCO_3$ can control the concentration of Cd^{2+} in solution. In turn, in more strongly reducing environments, and at a wide range of pH, the predominant species is CdS. From pH 10 to 13, there is precipitation of $Cd(OH)_2$, and at pH higher than 13, the solubilization of cadmium as $Cd(OH)_3^-$ is observed. To form $CdCO_3$, it is necessary to have more oxidizing conditions, as well as pH values around 8 (Fergusson, 1990). Thus, at pH values commonly found in marine water (up to approximately 8.5), Cd^{2+} and its complexes with anions (Cl$^-$ for example) and/or humic organic matter predominate in marine waters, although precipitation of $CdCO_3$ may also occur.

It is very important to understand how cadmium is distributed and accumulated in marine organisms because the toxicity of this element for humans is high. Of the damage caused by cadmium, kidney failure, bone diseases, infertility, and different types of tumors should be highlighted (Nordberg, *et al.*, 2007). Table **2** lists some research results related to cadmium quantification in marine organisms.

Up to pH around 5, lead is mostly found in marine waters as Pb^{2+}, while from pH 5 to 9, insoluble $PbCO_3$ can be formed. Thus, if the amount of Pb^{2+} is sufficiently high, lead carbonate can precipitate. However, typical concentrations of total lead in the open ocean are around 0.015 µg L^{-1} or less, thus highlighting that this element can be kept as soluble form (Pb^{2+}) even when considering values of pH that are slightly alkaline (Fergusson, 1990). Nevertheless, like Cd^{2+}, Pb^{2+} can form soluble and stable complexes with several dissolved ligands in marine water.

Table 2. Research results concerning the concentration of cadmium in marine samples.

Marine Organism	Cadmium Concentration (µg g⁻¹)	Local	Reference
Algae (*Ulva* spp.)	0.39±0.20 to 0.48±0.16*	Romania (Black Sea)	(Jitar, *et al.*, 2015)
Mollusc (*Rapana venosa*)	1.10±1.23 to 1.64±0.76**	Romania (Black Sea)	(Jitar, *et al.*, 2015)
Fishes (*Thunnus albacores, Xiphias gladius*, and *Lutjanus*)	0.02±0.02 to 0.04±0.05*	Sri Lanka	(Jinadasa, *et al.*, 2010)
Fishes (*Oreochromis mossambicus, Aristichthys nobilis, Ctenopharyngodon idellus*, and *Siniperca chuatsi*)	0.10 to 0.17*	Estuary of Pearl River – China	(Cheung, *et al.*, 2008)
Fishes (*Macrodon ancilodon, Mugil liza, Micropogonias furnieri*, and *Sardinella brasiliensis*)	<0.01 to 0.287*	Brazil	(Morgano, *et al.*, 2011)
	0.14*	Malaysia	(Alam *et al.* 2012)

*Wet weight **Dry weight

Contrary to what is observed for cadmium, lead concentrations decrease with depth. This is because the largest input of lead into the oceans comes from atmospheric deposition. Thus, considering the strong relationship between anthropogenic activities and atmospheric concentrations of lead, it is possible to understand why the current surface lead concentrations in the oceans are much greater than those reported for prehistoric eras. For example, in the early 1980s, it was found that lead flow into the North Atlantic was approximately 170 ng cm^{-2} $year^{-1}$, while, for prehistoric eras, was estimated at 3 ng cm^{-2} $year^{-1}$ (Fergusson, 1990).

Because lead can be transferred to higher levels of the food chain and its toxicological effects in humans are very serious, there have been many investigations into the presence of this element in marine lifeforms (Table **3**). Several anthropogenic activities can emit lead to the atmosphere, thus resulting in subsequent deposition into the oceans. As an example of these activities, we can highlight metallurgy, which employs large amounts of lead due to its relatively low melting temperature, high resistance to corrosion, and notable flexibility (Bresciani Filho, 1997). Moreover, $Pb(CH_3COO)_2$ is used for producing some pesticides, while PbO and $Pb(NO_3)_2$ are largely used in the production of batteries and in the war industry, respectively (Johnson, 1998).

The ancient Greek poet and physician Nicander related some effects derived from lead poisoning, including hallucinations and paralysis (Emsley, 2005). This early description of symptoms already highlighted the devastating neurological effects of lead. Indeed, human intoxication by lead causes a severe psychiatric disturbance. In the case of chronic exposure to lead (commonly observed when contaminated marine organisms are ingested by humans), psychiatric disorders such as depression, anxiety, and irritability are observed (Mason, *et al.* 2014).

There is historical evidence that severe chronic contamination of Roman citizens by lead contributed to the fall of the Roman Empire. The ancient Romans had a substantial and systematic contact with lead, which was used to make several domestic utensils as well as ducts responsible for bringing drinking water from far areas to the residences. Moreover, ancient Roman cuisine had several recipes in which soaps that gave a sweetening taste to foods and drinks were used. These soaps were made by boiling sour wine in lead pans, thus resulting in the formation of lead acetate – $Pb(CH_3COO)_2$. As a result, high doses of lead were daily ingested by the Roman population, and it is estimated that a typical Roman aristocrat ingested 250 µg of lead per day while an ordinary Roman was exposed to approximately 35 µg of lead per day and in turn, a slave ingested around 15 µg of lead per day. This systematic contamination by lead can partially explain some of the serious psychiatric disturbances manifested by the emperors. In turn, these disturbances could be responsible for the ill-advised governmental acts that had disastrous geopolitical consequences for all aspects of Roman life. Besides its psychiatric and neurological consequences, lead provokes many other symptoms including recurrent attacks of abdominal cramps, as historically reported for Emperor Claudius who ruled from 41 to 54 of the Christian era (Emsley, 2005).

Table 3. Research results concerning the concentration of lead in marine samples.

Marine Organism	Lead Concentration ($\mu g\ g^{-1}$)	Local	Reference
Blue crab (*Callinects* sp.)	<0.3 to 12.6*	Brazil	(Virga, *et al.*, 2007)
Mussel (*M. galloprovincialis*)	0.25 to 0.75*	Morocco	(Moustaid, *et al.*, 2005)
Shrimp (*Pennaeus smithii*)	0.2 to 18.0**	Brazil	(Pfeifer, *et al.*, 1985)
Oyster (*Pinctata radiata*)	5.86 to 13.61*	Kuwait	(Bou-Olayan, *et al.*, 1995)
Blue crab (*Callinects danae*)	0.8 to 3.9**	Brazil	(Pfeifer, *et al.*, 1985)

*Dry weight **Wet weight

The chemistry of mercury in marine water is very complex because this element can assume many forms. Hg^{2+} hydrolyzes in the pH range from 2 to 6 according to Equations 26 and 27 with hydrolysis constants of 2.6×10^{-4} and 2.6×10^{-3}, respectively.

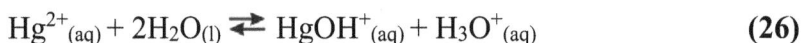

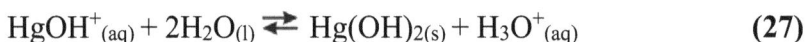

$$Hg^{2+}_{(aq)} + 2H_2O_{(l)} \rightleftharpoons HgOH^+_{(aq)} + H_3O^+_{(aq)} \qquad (26)$$

$$HgOH^+_{(aq)} + 2H_2O_{(l)} \rightleftharpoons Hg(OH)_{2(s)} + H_3O^+_{(aq)} \qquad (27)$$

The direct formation of insoluble $Hg(OH)_2$ is expected at pH higher than 4-5, and the pE X pH diagram for the system $Hg/Hg_2^{2+}/Hg^{2+}/HgO$ indicates that HgO predominates at pH lower than 3 and in oxidizing conditions. On the other hand, in a more reducing condition, metallic mercury predominates (Fergusson, 1990). The main reactions involving mercury in marine environments are indicated below in Equations 28-31.

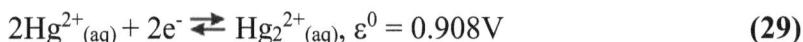

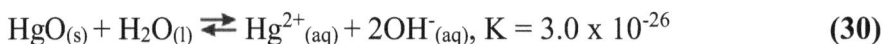

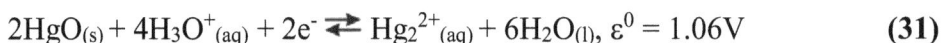

$$Hg_2^{2+}_{(aq)} + 2e^- \rightleftharpoons 2Hg_{(l)}, \; \varepsilon^0 = 0.793V \qquad (28)$$

$$2Hg^{2+}_{(aq)} + 2e^- \rightleftharpoons Hg_2^{2+}_{(aq)}, \; \varepsilon^0 = 0.908V \qquad (29)$$

$$HgO_{(s)} + H_2O_{(l)} \rightleftharpoons Hg^{2+}_{(aq)} + 2OH^-_{(aq)}, \; K = 3.0 \times 10^{-26} \qquad (30)$$

$$2HgO_{(s)} + 4H_3O^+_{(aq)} + 2e^- \rightleftharpoons Hg_2^{2+}_{(aq)} + 6H_2O_{(l)}, \; \varepsilon^0 = 1.06V \qquad (31)$$

Complexes can also be formed with Hg^{2+} and Cl^-, as indicated in Equations 32-35.

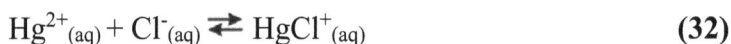

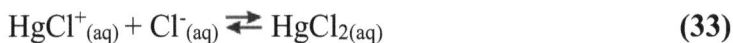

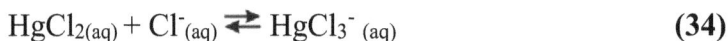

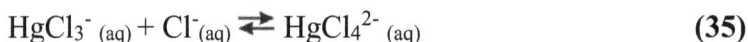

$$Hg^{2+}_{(aq)} + Cl^-_{(aq)} \rightleftharpoons HgCl^+_{(aq)} \qquad (32)$$

$$HgCl^+_{(aq)} + Cl^-_{(aq)} \rightleftharpoons HgCl_{2(aq)} \qquad (33)$$

$$HgCl_{2(aq)} + Cl^-_{(aq)} \rightleftharpoons HgCl_3^-{}_{(aq)} \qquad (34)$$

$$HgCl_3^-{}_{(aq)} + Cl^-_{(aq)} \rightleftharpoons HgCl_4^{2-}{}_{(aq)} \qquad (35)$$

Because of its several chemical forms in aquatic ecosystems, mercury is an element with a large potential to be transferred through environmental compartments. One possible transformation involves converting this element to methyl mercury, which is produced by bacteria. Furthermore, mercury can exist as Hg^+ (less soluble) and Hg^{2+} unless dissolved species of sulfur (HS^- and S^{2-}) are present to form HgS, which is insoluble. Sulfate-reducing bacteria naturally found in marine environments are responsible for generating methyl mercury, CH_3Hg^+, and dimethyl mercury, C_2H_6Hg. Due to its solubility in water, CH_3Hg^+ tends to be highly accumulated in animal tissues, especially in muscles and adipose layers. Despite its reasonable volatility, C_2H_6Hg can also be assimilated by several marine organisms, thus assuring its transfer through the food chain. If humans come in contact with contaminated water, CH_3Hg^+ and C_2H_6Hg can also be efficiently assimilated through the skin.

In marine water, total concentrations of mercury are around 0.4 µg L^{-1}, but due to the bioaccumulation of methylated forms of mercury, fishes can exhibit concentrations of mercury as elevated as 1 mg kg^{-1} (not informed if wet or dry weight). Animals that directly eat these fishes can exhibit very large concentrations of mercury. Examples of these animals include seals, bears, and some species of birds. Like lead, a remarkable part of the mercury that enters the oceans comes from the atmosphere, and it is estimated that 90% of all mercury in the air over the Atlantic Ocean comes from the burning of coal and waste incineration (Emsley, 2005).

Methyl mercury is many times more toxic than inorganic forms of mercury (Hg^0, Hg^+, and Hg^{2+}), so that while approximately 200 mg of this compound is enough to kill a human adult, the fatal dose for inorganic mercury can be up to 4 g (Emsley, 2005). Because methyl mercury has a positive charge, this compound is adsorbed on sediment particles that are negatively charged, although CH_3Hg^+ can also be formed directly on sediments after microbial conversion of inorganic forms of mercury to methyl mercury. Once absorbed by humans, methyl mercury is slowly eliminated from the body, and its biological half-life can be 80 days, considering urine, feces, and breast milk as the main sources of elimination. (Bisinoti & Jardim, 2004).

In fish, which are the main source of human contamination by methyl mercury, approximately 85% of the total mercury corresponds to CH_3Hg^+. The Brazilian limit for total mercury in fish is 0.5 mg kg^{-1}, but concentrations as elevated as 2.7 mg kg^{-1} were found in fish from some Brazilian areas where intense gold mining activities have been performed. As methyl mercury has a thermal stability up to 700°C and this temperature is not reached in any cooking practice, it is possible to affirm that methyl mercury can be efficiently assimilated from seafood (Bisinoti & Jardim, 2004).

Ikingura and Akagi (1999) performed a very interesting study in which fish exposed to inorganic mercury (salts of Hg^{2+}) exhibited a partition coefficient of fish/water varying from 5,000 to 7,000. On the other hand, when methylation occurred, these coefficients varied from 10,000 to 22,000, thus indicating that the mechanisms concerning the assimilation of inorganic and organic forms of mercury are very different. Besides its favorable partition coefficients, methyl mercury has a biological half-life in fish of approximately three years that favors constant augmentation of the methyl mercury concentrations (Bisinoti & Jardim, 2004).

One of the most serious known cases of marine pollution by methyl mercury happened in Japan, where the Minamata Bay was largely contaminated by effluents containing mercury. These effluents were discharged from 1932 to 1968 and were derived from an acetaldehyde-producing factory. Over these 36 years, approximately 80 tons of mercury were dumped into the bay, thus resulting in an accumulation of 2,000 mg kg^{-1} in the sediments. Bacteria were responsible for the biotransformation of inorganic mercury to CH_3Hg^+, thus ensuring very effective transfers of methyl mercury through the marine trophic chain (Tomiyasu, *et al.* 2006).

In early 1952, or 20 years from the beginning of the effluent discharges, several signals heralded the environmental disaster. Large quantities of fish were found dead and floating in the bay, and the birds that ate these dead fishes had erratic flights. Moreover, these birds fell down into the bay and subsequently drowned. Some cats that lived around the bay were observed to exhibit strange behavior, including very excessive salivation, uncoordinated movements, as well as falling into the water followed by death by drowning. Dogs and pigs were affected with deranged behavior, and after this tragic sequence of events, the next step was marked by severe human intoxication by methyl mercury. Daily, the people consumed large amounts of marine organisms containing frighteningly high amounts of methyl mercury. For example, crabs had up to 35 mg kg^{-1}, while some species of fish had 24 mg kg^{-1} of this powerful poison, so that in the summer of 1952, 21 humans died because of this methylated form of mercury (Emsley, 2005). It was not informed if these concentrations of mercury were expressed in dry or wet weight. When humans ingest toxic doses of methyl mercury, the set of symptoms is extremely serious with an emphasis on neurological disorders. Of these disorders, we can highlight the following: i) blurred vision, ii) visual field reduction, iii) low motor coordination, iv) skin insensitivity, v) nerve pain, vi) loss of hearing, vii) difficulty articulating words, viii) mental deterioration, and iv) muscle tremor. At extreme doses of intoxication, total paralysis and death are observed (Bisinoti & Jardim, 2004). These serious neurological problems are associated with the fact that mercury exhibits very strong affinity in relation to chemical groups containing sulfur. Thus, proteins/enzymes which have many sulfured amino acids and are involved in the working of the central nervous system react with chemical species of mercury, and their structures are sufficiently changed to cause their partial or total inactivation. One of the most important enzymes in this group is the Na/K-ATPase (Emsley, 2005). Table **4** lists some research results related to quantification of mercury in different marine organisms.

Table 4. Research results concerning the concentration of mercury in marine samples.

Marine Organism	Mercury Concentration ($\mu g\ g^{-1}$)	Local	Reference
Zooplankton (*Synchaeta*, and *Keratella*)	Up to 0.6*	Poland (Baltic Sea)	(Beldowska & Celgiolka, 2017)
Fishes (*Liza abu*, *Sparidentex hasta*, *Acanthopagrus latus*, *Thunnus tonggol*, and *Fenneropenaeus indicus*)	0.360 to 1.172**	Iran (Persian Gulf)	(Mortazavi & Sharifian, 2011)
Fishes (*Mugil seheli*)	0.00094 to 0.025**	Egypt	(El-Moselhy, 2006)
Fishes (*Thunnus thynnus*, *Pagrus pagrus*, *Centropomus* sp., and *Salmo solar*)	0.0077 to 0.9681***	Brazil	(CETESB, 2014)
Fish (*Cynoscion leiarchus*)	0.00526±0.00014 to 0.01399±0.00009**	Brazil	(Silva, 2014)
Mollusk (*Anadara granosa*)	6.85±0.95*	India	(Bhattacharya & Sarkar, 1996)
Oyster (*Saccostrea cucullata*)	0.11±0.004 to 2.090±0.17*	India	(Bhattacharya & Sarkar, 1996)
Sharks (*Brachaelurus waddi*)	1.18 to 1.22**	Australia	(Denton & Breck, 1981)
Fish (*Cyselurus comatus*)	<0.0004 to 0.01**	India	(Sanzgiry, *et al.*, 1988)
Crab (*Scylla serrata*)	0.004 to 0.009**	India	(Sanzgiry, *et al.*, 1988)

*Dry weight **Wet weight ***Uninformed

Nickel

Nickel is typically found in marine water at concentrations that vary from 0.1 to 0.5 $\mu g\ L^{-1}$ (Cempel & Nikel, 2006), and marine sediments can accumulate this metallic element at highly variable concentrations, as reported by Kljakovic-Gaspic, *et al.* (2009) who found total concentrations of nickel varying from 27.8 to 40.2 mg kg^{-1} in marine sediments collected in Soline Bay, Croatia.

Investigations into the distribution and accumulation of nickel in the abiotic and biotic marine environmental compartments are very important because almost all compounds of this element have significant carcinogenic effects on humans. Because of this, in 1990 the International Agency for Research on Cancer (IARC) classified the compounds of nickel as carcinogenic for humans. The molecular basis for the toxicology of nickel is related to the capacity of this element for producing oxidative damage to DNA (Denkhaus & Salnikow, 2002).

Sadiq (1992) performed an interesting study about the interaction of nickel and sediments from the northern Gulf of Arabia. It was observed that, after 75 days of contact between sediment particles and marine water (pH 8.1) enriched with Ni^{2+}, 60.1% of the nickel remained as Ni^{2+}, 16.9% was transformed into $NiCl^{+}$, thus highlighting the importance of chloride to form charged complexes with metallic cations, 5.0% of the initial amount of Ni^{2+} was converted to $NiCl_2$, and 0.4 and 17.5% were quantified as $NiOH^{+}$ and $NiSO_4$, respectively. The species $NiOH^{+}$ is derived from the first hydrolysis of Ni^{2+} (Equation 36). To a much lesser degree, the hydrolysis of nickel can continue to form $Ni(OH)_2$, according to Equation 37.

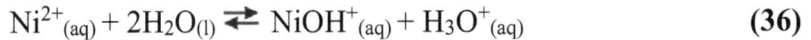

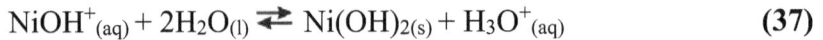

$$Ni^{2+}_{(aq)} + 2H_2O_{(l)} \rightleftarrows NiOH^{+}_{(aq)} + H_3O^{+}_{(aq)} \tag{36}$$

$$NiOH^{+}_{(aq)} + 2H_2O_{(l)} \rightleftarrows Ni(OH)_{2(s)} + H_3O^{+}_{(aq)} \tag{37}$$

pE X pH diagram for nickel indicates that Ni^{2+} domains all range in pE between the two lines of water stability from pH 0 to around 10. Metallic nickel is possible only at pH higher than 9 and in reducing conditions close to the reduction of water. Thus, in marine environments, the amount of metallic nickel is negligible (NIAIST 2005).

The assessment of nickel concentrations in marine organisms assumes essential importance in human toxicology, since this element is largely assimilated *via* daily diet. It is estimated that approximately 1 to 10% of all ingested nickel is absorbed by the gastrointestinal tract, but nickel assimilation rates are strongly dependent on the way it is ingested. Therefore, when large quantities of food are ingested, the toxicokinetics related to nickel assimilation tend to be decreased (Cempel and Nikel, 2006). Table **5** lists some relevant studies concerning the evaluation of nickel in marine organisms.

Hunt, *et al.*, (2002) evaluated the acute and chronic toxicity of nickel on three marine species: *Atherinops affinis* (fish), *Haliotis rufescens* (mollusk), and *Mysidopsis intii* (crustacean), which were collected on the U.S. west coast. Toxic effects of nickel were considered in terms of larval metamorphosis and juvenile growth. The authors found that the 96-h lethal concentration for 50% of the population (LC50 in µg L^{-1}) varied from 148.6 to 26,560, thus indicating large differences in nickel toxicity for the evaluated species. The acute-to-chronic ratios for nickel concentrations (µg L^{-1}) were 6.220, 5.505, and 6.727 for fish, mollusks, and crustaceans, respectively.

Table 5. Research results concerning the concentration of nickel in marine samples.

Marine Organism	Nickel Concentration (µg g^{-1})	Local	Reference
Fishes *(Bathyraja spinicauda, Polachius virens, Gadus morhua morhua, Raja radiata, Reinhardtius hippoglossoides, Sebastes marinus, Melanogrammus aeglefinus, Hippoglossoides platessoides, Cyclopterus lumpus, Pleuronectes platessa, Limanda limanda, Anarhichas minor, Anarhichas denticulatus, Anarhichas lupus, and Raja fyllae)*	< 1.0*	Russia	(Zauke, *et al.*, 1999)
Fishes *(Pampus argenteus, Epinephelus tauvina, Pomadasys argenteus, Acanthopagrus latus, Otolithus argentenus, Acanthopagrus cuvieri,* and *Hilsa ilisha)* and Shrimps *(Penaeus semisulcatus,* and *Metapenaeous affinis)*	0.96 to 1.33*	Kuwait	(Bou-Olayan, *et al.*, 1995)
Fish *(Mugil curema)*	4.59±1.38 to 12.56±1.3*	Brazil	(do Carmo, *et al.*, 2011)
Oyster *(S. cucullata)*	15.42 to 38.04*	Iran	(Bazzi, 2014)
Fish *(M. barbatus)*	Up to 1.73±1.86*	Greece	(Strogyloudi, *et al.*, 2014)

*Dry weight

CONCLUSIONS

This chapter provided an overview of the chemistry of seven elements (arsenic, cadmium, lead, mercury, nickel, and sulfur) in marine environments. For this purpose, aspects concerning the distribution and availability of these elements in different aquatic compartments were discussed.

The literature review showed that once chemical elements introduced into marine environments (from natural or anthropic sources), and they can transform several forms with different degrees of toxicity. Therefore, the pollution of these aquatic ecosystems can cause a severe imbalance in biogeochemical cycles with serious implications for aquatic life, as well as for human life.

CONSENT FOR PUBLICATION

Not applicable.

CONFLICT OF INTERESTS

Declare None.

ACKNOWLEDGEMENTS

The authors would like to thank the Conselho Nacional de Desenvolvimento Científico Tecnológico (CNPq), the Coordenação de Aperfeiçoamento de Pessoal de Nível Superior (CAPES), and the Fundação de Amparo à Pesquisa do Estado da Bahia (FAPESB) for fellowships and financial support (CNPq: 555522/2006-7 and 620041/2006-4 and FAPESB: APP0076/2009, and RED0017/2014).

REFERENCES

Alam, L, Mohamed, C A R & Mokhtar, MB (2012) Accumulation pattern of heavy metals in marine organisms collected from a coal burning power plant area of Malacca Strait. *Sci Asia,* 38, 331-9.
[http://dx.doi.org/10.2306/scienceasia1513-1874.2012.38.331]

Alonso-Hernández, CM, Gómez-Batista, M, Diáz-Asencio, M, Estévez-Alvares, J & Padilla-Alvares, R (2012) Total arsenic in marine organisms from Cienfuegos bay (Cuba). *Food Chem,* 130, 973-6.
[http://dx.doi.org/10.1016/j.foodchem.2011.07.087]

Anzolin, DG (2011) *Análise de contaminantes e biomarcadores em peixes-bois marinhos (Trichechus manatus).* Universidade Federal de Pernambuco, Recife.

Bazzi, AO (2014) Heavy metals in seawater, sediments and marine organisms in the Gulf of Chabahar, Oman Sea. *Journal of Oceanography and Marine Science,* 5, 20-9.
[http://dx.doi.org/10.5897/JOMS2014.0110]

Beldowska, M & Mudrak-Cegiolka, S (2017) Mercury concentration variability in the zooplankton of the Southern Baltic coastal zone. *Prog Oceanogr,* 159, 73-85.
[http://dx.doi.org/10.1016/j.pocean.2017.09.009]

Bhattacharya, B & Sarkar, SK (1996) Total mercury content in marine organisms of the Hooghly Estuary, West Bengal, India. *Chemosphere,* 33, 147-58.
[http://dx.doi.org/10.1016/0045-6535(96)00156-7]

Bisinoti, MC & Jardim, WF (2004) O comportamento do metilmercúrio (MetilHg) no ambiente. *Quim Nova,* 27, 593-600.
[http://dx.doi.org/10.1590/S0100-40422004000400014]

Bou-Olayan, AH, Al-Mattar, S, Al-Yakoob, S & Al-Hazeem, S (1995) Accumulation of lead, cadmium, copper and nickel by pearl oyster, *Pinctada radiata,* from Kuwait marine environment. *Mar Pollut Bull,* 30, 211-4.
[http://dx.doi.org/10.1016/0025-326X(94)00143-W]

Bresciani Filho, E (1997) *Seleção de metais não ferrosos.* Editora da UNICAMP, Campinas.

Cardoso, AA & Pitombo, LRM (1992) Contribuição dos compostos reduzidos de enxofre no balanço global do estoque de enxofre ambiental. *Quim Nova,* 15, 219-23.

Cempel, M & Nikel, G (2006) Nickel: a review of its sources and environmental toxicology. *Pol J Environ Stud,* 15, 375-82.

(2014) *Contaminação por mercúrio no Estado de São Paulo.* Secretaria do Meio Ambiente, São Paulo.

Cheung, KC, Leung, HM & Wong, MH (2008) Metal concentrations of common freshwater and marine fish from the Pearl River Delta, south China. *Arch Environ Contam Toxicol,* 54, 705-15.
[http://dx.doi.org/10.1007/s00244-007-9064-7] [PMID: 18080794]

Chiffoleau, JF, Chauvaud, L, Amouroux, D, Barats, A, Dufour, A, Pécheyran, C & Roux, N (2004) Nickel and vanadium contamination of benthic invertebrates following the *"Erika"* wreck. *Aquat Living Resour,* 17, 273-80.
[http://dx.doi.org/10.1051/alr:2004032]

Denkhaus, E & Salnikow, K (2002) Nickel essentiality, toxicity, and carcinogenicity. *Crit Rev Oncol Hematol,* 42, 35-56.
[http://dx.doi.org/10.1016/S1040-8428(01)00214-1] [PMID: 11923067]

Denton, GRW & Breck, WG (1981) Mercury in tropical marine organisms from North Queensland. *Mar Pollut Bull,* 12, 116-21.
[http://dx.doi.org/10.1016/0025-326X(81)90439-2]

Dhinamala, K, Shalini, R, Pushpalatha, M, Arivoli, S, Samuel, T & Raveen, R (2017) Bioaccumulation of nickel in gills and muscles of shellfish species from Pulicat Lake, Tamil Nadu, India. *International Journal of Fisheries and Aquatic Research,* 2, 1-5.

do Carmo, CA, Abessa, DMS & Machado Neto, JG (2011) Metais em águas, sedimentos e peixes coletados no estuário de São Vicente-SP, Brasil. *O. Mundo Saude,* 35, 64-70.
[http://dx.doi.org/10.15343/0104-7809.20113516470]

El-Moselhy, KM (2006) Bioaccumulation of mercury in some marine organisms from lake Timsah and Bitter Lakes (Suez Canal, Egypt). *Egyptian Journal of Aquatic Research,* 32, 124-34.

Emsley, J (2005) *The elements of murder – a history of poison.* Oxford University Press, Oxford.

Fattorini, D, Alonso-Hernandez, CM, Diaz-Asencio, M, Munoz-Caravaca, A, Pannacciulli, FG, Tangherlini, M & Regoli, F (2004) Chemical speciation of arsenic in different marine organisms: Importance in monitoring studies. *Mar Environ Res,* 58, 845-50.
[http://dx.doi.org/10.1016/j.marenvres.2004.03.103] [PMID: 15178123]

Fergusson, JE (1990) *The heavy elements – chemistry, environmental impact and health effects.* Pergamon Press, Oxford.

Garrison, T (2006) *Essentials of oceanography.* Thomson Brooks/Cole, Belmont.

Gillmore, M (2014) *Toxicity of aluminium in seawater: diatom sensitivity.* University of Wollongong, Wollongong.

Grotzinger, J & Jordan, TH (2010) *Understanding Earth.* W. H. Freeman & Company, New York.

Harris, DC (2008) Análise Química Quantitativa LTC Editora, Rio de Janeiro.

Hong, S, Kwon, HO, Choi, SD, Lee, JS & Khim, JS (2016) Arsenic speciation in water, suspended particles, and coastal organisms from the Taehwa River Estuary of South Korea. *Mar Pollut Bull,* 108, 155-62.
[http://dx.doi.org/10.1016/j.marpolbul.2016.04.035] [PMID: 27114086]

Hunt, JW, Anderson, BS, Phillips, BM, Tjeerdema, RS, Puckett, HM, Stephenson, M, Tucker, DW & Watson, D (2002) Acute and chronic toxicity of nickel to marine organisms: implications for water quality criteria. *Environ Toxicol Chem,* 21, 2423-30.
[http://dx.doi.org/10.1002/etc.5620211122] [PMID: 12389922]

Ikingura, JR & Akagi, H (1999) Methylmercury production and distribution in aquatic systems. *Sci Total Environ,* 234, 109-18.
[http://dx.doi.org/10.1016/S0048-9697(99)00116-3] [PMID: 10507152]

Ivanov, MV, Yu, A, Lein, MS, Reeburgh, MS & Skyring, GW (1989) In: Brimblecombe P & Lein, AY (Ed) *Evolution of the global biogeochemical sulphur cycle.* John Willey & Sons Ltd., New Jersey 125-79.

Kljakovic-Gaspic, Z, Bogner, D & Ujevic, I (2009) Trace metals (Cd, Pb, Cu, Zn and Ni) in sediment of the submarine pit Dragon ear (Soline Bay, Rogoznica, Croatia). *Environmental Geology,* 58, 751-60.
[http://dx.doi.org/10.1007/s00254-008-1549-9]

Kucuksezgin, F, Gonul, LT & Tasel, D (2014) Total and inorganic arsenic levels in some marine organisms from Izmir Bay (Eastern Aegean Sea): a risk assessment. *Chemosphere,* 112, 311-6.
[http://dx.doi.org/10.1016/j.chemosphere.2014.04.071] [PMID: 25048921]

Küpeli, T, Altundağ, H & Imamoğlu, M (2014) Assessment of trace element levels in muscle tissues of fish species collected from a river, stream, lake, and sea in Sakarya, Turkey. *ScientificWorldJournal,* 2014496107
[http://dx.doi.org/10.1155/2014/496107] [PMID: 24790570]

Jinadasa, BKKK, Rameesha, LRS, Edirisinghe, EMRKB & Rathnayake, RMUSK (2010) Mercury, cadmium and lead levels in three commercially important marine fish species in Sri Lanka, Sri Lanka. *J Aquat Sci,* 15, 39-43.

Jitar, O, Teodosiu, C, Oros, A, Plavan, G & Nicoara, M (2015) Bioaccumulation of heavy metals in marine organisms from the Romanian sector of the Black Sea. *N Biotechnol,* 32, 369-78.
[http://dx.doi.org/10.1016/j.nbt.2014.11.004] [PMID: 25500720]

Johnson, FM (1998) The genetic effects of environmental lead. *Mutat Res,* 410, 123-40.
[http://dx.doi.org/10.1016/S1383-5742(97)00032-X] [PMID: 9637233]

Li, J, Sun, C, Zheng, L, Jiang, F, Wang, S, Zhuang, Z & Wang, X (2017) Determination of trace metals and analysis of arsenic species in tropical marine fishes from Spratly islands. *Mar Pollut Bull,* 122, 464-9.
[http://dx.doi.org/10.1016/j.marpolbul.2017.06.017] [PMID: 28712770]

Liu, Y, Liu, G, Yuan, Z, Liu, H & Lam, PKS (2017) Presence of arsenic, mercury and vanadium in aquatic organisms of Laizhou Bay and their potential health risk. *Mar Pollut Bull,* 125, 334-40.
[http://dx.doi.org/10.1016/j.marpolbul.2017.09.045] [PMID: 28967412]

Maher, WA, Foster, S, Krikowa, F, Duncan, E, St John, A, Hug, K & Moreau, JW (2013) Thio arsenic species measurements in marine organisms and geothermal waters. *Microchem J,* 111, 82-90.
[http://dx.doi.org/10.1016/j.microc.2012.12.008]

Manham, SE (1994) *Environmental Chemistry.* Lewis Publishers, Boca Raton.

Martins, LR & Nunes, JC (2009) Escapes hidrotermais e bioprospecção. *GRAVEL,* 7, 57-65.

Mason, LH, Harp, JP & Han, DY (2014) Pb neurotoxicity: neuropsychological effects of lead toxicity. *BioMed Res Int,* 2014840547
[http://dx.doi.org/10.1155/2014/840547] [PMID: 24516855]

Mirza, A, King, A, Troakes, C & Exley, C (2017) Aluminium in brain tissue in familial Alzheimer's disease. *J Trace Elem Med Biol,* 40, 30-6.
[http://dx.doi.org/10.1016/j.jtemb.2016.12.001] [PMID: 28159219]

Morgano, MA, de Oliveira, APF, Rabonato, LC, Milani, RF, Vasconcellos, JP, Martins, CN, Citti, AL, Telles, EO & Balian, SC (2011) Avaliação de contaminantes inorgânicos (As, Cd, Cr, Hg e Pb) em espécies de peixes. *Rev Inst Adolfo Lutz,* 70, 497-506.

Mortazavi, MS & Sharifian, S (2011) Mercury bioaccumulation in some commercially valuable marine organisms from Mosa Bay, Persian Gulf. *Int J Environ Res,* 5, 757-62.

Moustaid, K, Nasser, B, Baudrimont, I, Anane, R, El Idrissi, M, Bouzidi, A & Creppy, EE (2005) Comparative evaluation of the toxicity induced by mussels (*Mytilus galloprovincialis*) from two sites of the Moroccan Atlantic coast in mice. *C R Biol,* 328, 281-9.
[http://dx.doi.org/10.1016/j.crvi.2005.01.004] [PMID: 15810552]

NIAIST- National Institute of Advanced Industrial Science and Technology (2005).

Nordberg, GF, Fowler, BA, Nordberg, M, Friberg, L (2007). *Handbook on the toxicology of metals.* Academic Press, Amsterdam.

Pfeiffer, WC, de Lacerda, LD, Fiszman, M & Lima, NRW (1985) Metais pesados no pescado da Baía de Sepetiba, Estado do Rio de Janeiro, RJ. *Cienc Cult,* 37, 297-302.

Price, RE, Breuner, C, Reeves, E, Bach, W & Pichler, T (2016) Arsenic bioaccumulation and biotransformation in deep-sea hydrothermal vent organisms from the PACMANUS hydrothermal field, Manus Basin, Papua New Guinea. *Deep Sea Res Part I Oceanogr Res Pap,* 117, 95-106.

Ranau, R, Oehlenschlager, J & Steinhert, H (1999) Determination of aluminium in the edible part of fish by GFAAS after sample pretreatment with microwave activated oxygen plasma. *Fresenius J Anal Chem,* 364, 599-604.

Ranau, R, Oehlenschlager, J & Steinhert, H (2001) Aluminium levels of fish fillets baked and grilled in aluminium foil. *Food Chem,* 73, 1-6.

Sadiq, M, Alam, IA & Al-Mohanna, H (1992) Bioaccumulation of nickel and vanadium by clams (Meretrix meretrix) living in different salinities along the Saudi coast of the Arabian Gulf. *Environ Pollut,* 76, 225-31. [PMID: 15091987]

Sanzgiry, S, Mesquita, A & Kureishy, TW (1988) Total mercury in water, sediments, and animals along the Indian coast. *Mar Pollut Bull,* 19, 339-43.

Silva, TS (2014) *Desenvolvimento de métodos espectrométricos para determinação de contaminantes inorgânicos (arsênio, cádmio, chumbo e mercúrio) em peixe.* Instituto Federal de Educação, Ciência e Tecnologia do Rio de Janeiro, Rio de Janeiro.

Strogyloud, E, Catsiki, VA & Bei, F (1998) Copper and nickel in marine fish from Greek waters. *Rapp. Comm. int. Mer Médit,* 35, 290-1.

Tomiyasu, T, Matsuyama, A, Eguchi, T, Fuchigami, Y, Oki, K, Horvat, M, Rajar, R & Akagi, H (2006) Spatial variations of mercury in sediment of Minamata Bay, Japan. *Sci Total Environ,* 368, 283-90. [PMID: 16293298]

vanLoon, GW & Duffy, SJ (2005) *Environmental Chemistry – a global perspective.* Oxford University Press, Oxford.

Virga, RHP, Geraldo, LP & dos Santos, FH (2007) Avaliação de contaminação por metais pesados em amostras de siris azuis. *Food Sci Technol (Campinas),* 27, 779-85.

Virga, RHP & Geraldo, LP (2008) Investigação dos teores de metais pesados em espécies de siris azuis do gênero Callinectes sp. *Food Sci Technol (Campinas),* 28, 943-8.

(1996) *Trace nlms in human nutrition and health.* WHO Library Cataloguing in Publication Data, Geneva.

Xue, XM, Yan, Y, Xiong, C, Raber, G, Francesconi, K, Pan, T, Ye, J & Zhu, YG (2017) Arsenic biotransformation by a cyanobacterium Nostoc sp. PCC 7120. *Environ Pollut,* 228, 111-7. [PMID: 28527322]

Zauke, GP, Savinov, VM, Ritterhoff, J & Savinova, T (1999) Heavy metals in fish from the Barents Sea (summer 1994). *Sci Total Environ,* 227, 161-73. [PMID: 10231981]

Zhang, W, Chen, L, Zhou, Y, Wu, Y & Zhang, L (2016) Biotransformation of inorganic arsenic in a marine herbivorous fish Siganus fuscescens after dietborne exposure. *Chemosphere,* 147, 297-304. [PMID: 26766368]

Zhang, W, Zhang, L & Wang, WX (2017) Prey-specific determination of arsenic bioaccumulation and transformation in a marine benthic fish. *Sci Total Environ,* 586, 296-303. [PMID: 28185737]

CHAPTER 5

Pollution Dynamics of Organic Contaminants in Marine Ecosystems

Donat-P. Häder[*]

Department of Biology, Friedrich-Alexander University Erlangen-Nürnberg, 91054 Erlangen, Germany

Abstract: While the biomass of marine ecosystems is only about 1% of their terrestrial counterparts, their productivity rivals that of all land-based ecosystems taken together. The structure and performance of these ecosystems are strongly affected by environmental factors, such as temperature, nutrients, transparency, solar visible and UV radiation. Increasing pollution, not only of coastal habitats but also of open ocean waters, results in changes in productivity and species composition. Persistent organic pollutants (POPs) are organic chemicals that are not degraded for long periods and include brominated flame retardants, perfluoroalkyl compounds, fluorotelomer alcohols, perfluoroalkylsulfonic acids (FPSAs), perfluorocarboxylic acids (PFCAs), fluorotelomer carboxylic acids, fluorotelomer sulfonic acids, and fluorinated polymers. Pesticides enter the aquatic ecosystems with terrestrial run-off, but are distributed not only in coastal areas and estuaries. Microplastics are of growing concern since they are concentrated in oceanic gyres. They are ingested by plankton and accumulated in the food chain. Accidental oil spills and catastrophic events are the reason for the pollution by crude oil and its products. Mineral oil pollution has been found to affect all the biota from plankton, *via* invertebrates to vertebrates.

Keywords: Organic Pollutants, Persistent Organic Pollutants, Pesticides, Microplastics, Mineral Oil, Coastal Ecosystems, Open Ocean Habitats.

INTRODUCTION

Marine ecosystems are major biomass producers. Prokaryotic and eukaryotic photosynthetic organisms generate about the same amount of biomass as all terrestrial ecosystems taken together and constitute important sinks for atmospheric CO_2 (Field, Behrenfeld, Randerson, & Falkowski, 1998). The biological pump in the oceans is a key element (Honjo *et al.*, 2014) mitigating fix dissolved CO_2 to produce organic biomass. This material in the form of dead

---

[*] **Corresponding author Donat-P. Häder:** Department of Biology, Friedrich-Alexander University Erlangen-Nürnberg, 91054 Erlangen, Germany; Tel/Fax: +49-9131 48730; E-mail: donat@dphaeder.de

De-Sheng Pei & Muhammad Junaid (Eds.)
All rights reserved-© 2019 Bentham Science Publishers

organisms and fecal pellets sinks to the deep sea sediment in the form of 'oceanic snow' and increases the largest carbon reservoir on Earth (IPCC, 2014).

Marine ecosystems are affected by a plethora of environmental stress factors threatening the sustainable development of resources for the rapidly increasing human population (2012). The primary producers are photosynthetic prokaryotes, such as cyanobacteria, eukaryotic phytoplankton, and macroalgae. In addition, many non-photosynthetic organisms are found in the water column ranging from viruses, heterotrophic bacteria to zooplankton and higher zoological taxa. The primary producers form the basis of the intricate food webs and therefore any disturbance at the basis is relayed to the primary and secondary consumers culminating in fish, birds and even humans for which the marine ecosystems contribute major resources for food production and technology.

Commencing in the 1970s, stratospheric ozone depletion by anthropogenic production and emission of chlorinated fluorocarbons (CFCs) and other trace gases resulted in an increase in the solar UV-B radiation (defined as 280-315 nm). Exposure to excessive solar UV-B radiation is detrimental for many organisms since it causes damage to the DNA (Sinha & Häder, 2002), reduces photosynthetic productivity (Jin, Duarte, & Agustí, 2017) and affects many other physiological and biochemical processes in the cell (Rastogi *et al.*, 2014). In addition to the damage of cellular targets, solar UV radiation produces reactive oxygen species (ROS) both inside the cell as well as outside *via* excitation of dissolved organic matter (DOM) in the water, which in turn impair cellular functions (Maraccini, Wenk, & Boehm, 2016).

Despite of the implementation of the Montreal Protocol and its amendments detrimental enhanced UV-B still persists because of the long lifetime of the CFCs in the stratosphere (Bais *et al.*, 2015; McKenzie *et al.*, 2011; Newman & McKenzie, 2011; Solomon *et al.*, 2016). Therefore, the ecophysiological effects of solar UV-B radiation are still a topic of increased interest (Häder & Gao, 2015).

Increasing temperatures due to global climate change (IPCC, 2014) are another stress factor for marine ecosystems. Fossil fuel burning, tropical deforestation, and altered land usage result in the increasing release of carbon dioxide into the atmosphere. The mean global temperature of the oceans has increased by ~1°C since 1900 (Fischetti, 2013), but the error bars are considerable. The temperature increase in Arctic and Antarctic waters are much higher than the mean values and amount to about 4°C depending on the region (Pithan & Mauritsen, 2014). One consequence of this ocean warming is an enhanced stratification and shoaling of the upper mixed layer (UML) (Boyce, Lewis, & Worm, 2010; G. Wang, Xie,

Huang, & Chen, 2015), which confines the organisms dwelling in this layer to a thinner water column and exposes them to higher solar visible and UV radiation (Gao *et al.*, 2012). In addition, the stronger thermocline, which defines the lower limit of the UML and separates it from the cooler deeper water, hinders the transport of dissolved macronutrients from below into the UML (Behrenfeld *et al.*, 2006).

Increasing pollution from a multitude of sources is a major problem for marine ecosystems, which affects primary producers, consumers and the intricate food webs (Gerlach, 2013; C. H. Walker & Livingstone, 2013). This has resulted in massive destruction of native populations and enhancing extinction of species in all taxa (Ceballos *et al.*, 2015). Organic pollutants include pesticides, detergents and surfactants, solvents and residues of oil spills. Many of the toxic materials are persistent organic pollutants (POPs). Pollution is more pronounced in coastal habitats than in the open ocean (Gómez & O'Farrell, 2014; Munir, Zaib-un-nisa Burhan, Morton, & Siddiqui, 2015). In contrast, the latter is affected by the accumulation of plastic materials, which amounts to about 250,000 tonnes collected in the oceanic gyres (Eriksen *et al.*, 2014). Coastal ecosystems accumulate pollutants from terrestrial run-off, which include heavy metals, industrial wastes, agricultural fertilizers, and pesticides, as well as surfactants and cleansing products from households (Nemerow, 1991; S.-L. Wang, Xu, Sun, Liu, & Li, 2013; Zaghden *et al.*, 2014).

This review provides an overview of the types and sources of organic pollutants in the oceans as well as on bioindicators and potential methods of remediation in order to preserve or restore the ecological integrity and providing the numerous services marine ecosystems offer.

PERSISTENT ORGANIC POLLUTANTS

Persistent organic pollutants (POPs) are organic chemicals that stay in the environment and are not degraded for long periods (Harrad, 2009). These pollutants include brominated flame retardants, perfluoroalkyl compounds, such as polyfluorinated sulfonamides (FSAs), fluorotelomer alcohols (FTOHs), perfluoroalkylsulfonic acids (FPSAs), perfluorocarboxylic acids (PFCAs), fluorotelomer carboxylic acids, fluorotelomer sulfonic acids, and fluorinated polymers. They are toxic wastes and are found in air, soil, water, and sediments. The United Nations Environment Programme (UNEP) listed the sources, behavior, fate and effects of POPs in the Stockholm Convention on 22 May 2001, which entered into force in 2004 when 50 countries had ratified it; in late 2008 180 parties had ratified it in order to protect human health and the environment from POPs. The Convention lists aldrin, chlordane, DDT, dieldrin, heptachlor, hexachlorobenzene, mirex, polychlorinated biphenyls, polychlorinated dibenzo-*p*-dioxins, polychlorinated dibenzofurans, and toxaphene, but allows adding further substances to the list (Xu, Wang, & Cai, 2013).

An international comparison of POPs in coastal waters collected polyethylene pellets at 30 beaches from 17 countries (Ogata *et al.*, 2009). The samples were analyzed for organochlorine compounds. Highest concentrations were found on US coasts, followed by Western Europe and Japan. Australia, tropical Asia, and Southern Africa had the lowest concentrations. The data on the collected samples showed a high correlation with data from another monitoring (Mussel Watch). In addition to the US west coast, the highest DDT concentrations were found on the coast in Vietnam. This may be due to the fact that DDT is currently used as a pesticide for malaria control. Likewise, high concentrations of HCHs were found in samples from South Africa where they are used as pesticides. Table 1 shows concentrations of various POPs measured over the open Atlantic Ocean (Jaward, Barber *et al.* 2004).

Table 1. Concentration of POPs measured over the open Atlantic Ocean (pg m^{-3}) (from Jaward, Barber *et al.* 2004).

Compound	Minimum	Mean	Maximum
PCB 28	<2.5	9.3	42.0
PCB 52	1.7	6.0	24.0
PCB 90/101	0.8	3.7	16.0
PCB 118	<0.6	1.1	6.0
PCB 138	<0.7	2.2	9.4
PCB 153/132	1.5	5.0	21.0
PCB 180	<0.3	1.0	3.6
Σ_{79} PCB	12.0	79.0	360.0
HCB	4.8	23.0	100.0
α-HCH	<0.1	3.0	11.0
γ-HCH	<3.6	22.0	100.0
p,p'-DDE	<1.5	20.0	47.0
p,p'-DDT	<2.2	2.2	5.4

In addition to sediments, persistent organochlorines were found in surface waters and porewater samples from Minjiang River Estuary, Fujian Province, China (Zhang, Hong, Zhou, Huang, & Yu, 2003). The most common ones were α-HCH, DDE, heptachlor, endosulfan and methoxychlor. The total concentrations of 18 organochlorine pesticides were found to be in the range of 214-1819 ng/L in surface water, 4541-13,699 ng/L in porewater and 29-52 ng/g in sediment. β-HCH was least degraded while DDT had been metabolized to DDE and DDE. DDE was least degradable. Concentrations of POPs have been compared in samples of surface sediments and near-bottom water collected on the Palos Verdes Shelf, California, USA and showed that the sediments had the highest concentrations from where the compounds diffused into the water (Fernandez, Lao, Maruya, & Burgess, 2014).

The effects of POPs on marine primary production have been evaluated in the Kattegat and the North Sea (Everaert, De Laender, Goethals, & Janssen, 2015). In comparison to other limiting factors, such as light, nutrient concentrations, temperature, and grazing, the effects of POPs were small, on the order of 10% inhibition, but there are large regional differences. Toxicity of organic pollutants was tested in Mediterranean, Atlantic and Southern Ocean phytoplankton communities (Echeveste, Galbán-Malagón, Dachs, Berrojalbiz, & Agustí, 2016). The results indicate that current concentrations affect marine phytoplankton. Small-sized picoplankton in temperate seawaters was found to be most affected. Larger Antarctic phytoplankton was more sensitive than Mediterranean picophytoplankton. In the ecologically and economically important oyster *Crassostrea gigas,* the concentrations of organic pollutants accumulated in the soft tissue of juvenile oysters was determined in the Marennes-Oléron Bay, which is the most important area for oyster production in the Gironde Estuary, France (Luna-Acosta, Bustamante, Budzinski, Huet, & Thomas-Guyon, 2015). Heavy polycyclic aromatic hydrocarbons (HPAHs), polychlorobiphenyls (PCBs), dichlorodipheyltrichlororethanes (DDTs), and lindane were detected, which compromised the antioxidant and immune-defense responses against infectious diseases. This may correlate with the observation of considerable oyster mortalities in this area. POPs were found in 10 marine species caught in the Gulf of Naples, Italy, including hexachlorobenzene (HCB), DDTs and 20 polychlorinated biphenyls (Naso, Perrone, Ferrante, Bilancione, & Lucisano, 2005). The PCB levels were the highest (from 57 ng to 48 mg/g on a lipid basis) followed by DDT and its metabolites. Highest concentrations were detected in resident species in shallow coastal waters indicating that the sources of these pesticides can be found in local agricultural, industrial and municipal activities. While the highest concentrations for HCB and DDTs were well below the allowed maximum residue limits (0.2 and 1 mg/kg fat weight), the concentrations for PCBs far exceeded the limit of 200 ng/g fresh weight recommended by the European Union for meat products. Even in Antarctic fish (Elephant Island), organochlorine compounds have been found (Weber & Goerke, 2003). Concentrations were found to be highest in bottom invertebrate feeders and least in krill feeders. Between 1987 and 1996, most POPs showed significant increases both in benthos and fish feeders, but afterwards, the concentrations remained almost constant. From this result, the authors concluded that the global distribution of HCB is close to equilibrium.

POPs accumulate in the food chain. These substances were analyzed in zooplankton, benthic invertebrates and fish collected in the Arctic Polynya. The results showed a strong bioaccumulation continuing into birds and mammals (Fisk, Hobson, & Norstrom, 2001). Fish oil was used as feed for growing Atlantic salmon; high concentrations of long-chain polyunsaturated fatty acids (LC-PUFA) and POPs were found in the flesh (Sprague *et al.*, 2015). Replacing the fish oil with a *Schizochytrium* algal meal reduced the pollutant concentrations, but the nutritional value was lower due to the low levels of EPA in the diet.

POPs are good markers for ocean fronts *e.g.,* in the Northeast Atlantic POP concentrations drop sharply south of the Canary Current, while elevated concentrations are found in the Gulf stream (Lohmann & Belkin, 2014). Large concentrations of perfluorinated compounds were found in the plumes of the Amazon River and Rio de la Plata. Concentrations of polycyclic aromatic hydrocarbons, such as polychlorinated biphenyls (PCBs, even though they are already phased out), have been found to be very high in Rhode Island Sound.

In order to monitor POPs and heavy metals in the Arctic Ocean, the Arctic Monitoring and Assessment Programme (AMAP) was established in 1991 to monitor identified pollution risks and their impacts on Arctic ecosystems. The first report was issued in 1997 (Arctic Pollution Issues: A State of the Arctic Environment Report) (Monitoring, 2004). It showed that the Arctic Ocean is not isolated but connected to the rest of the World's oceans. It greatly helped the negotiation of the protocols on persistent organic pollutants (POPs) and heavy metals to the United Nations Economic Commission for Europe's Convention on Long-Range Transboundary Air Pollution (LRTAP Convention) as well as the Stockholm Convention (see above) because POPs are long-lived, can be transported over long distances and are bioaccumulated. The Arctic Council decided to eliminate pollution of the Arctic also by identifying pollution from land-based sources (RPA). Since then the AMAP reports have included heavy metals, radioactivity, human health, and pathways.

The need for reliable methods for the detection of trace concentrations of POPs resulted in a review of 108 publications on the sources, distribution, accumulation, transformation, and toxicity of these substances (El-Shahawi, Hamza, Bashammakh, & Al-Saggaf, 2010). Especially for developing countries, there is an urgent need for low-cost methods, such as electrochemical techniques.

PESTICIDES

In addition to many other pollutants, various pesticides enter the marine environments by terrestrial run-off. The biocides cybutryn (Irgarol) and terbutryn, the herbicides aclonifen and bifenox, and the insecticides cypermethrin and heptachlor/heptachlor epoxide are new pollutants on the priority list of the Water Framework Directive of the European Union (Vorkamp, Bossi, Bester, Bollmann, & Boutrup, 2014). In seawater and fish samples from Denmark several pesticides were below the limits of detection, but cypermethrin and heptachlor were higher than the annual average environmental quality standards. However, no concentration exceeded the maximum allowable concentration. Sample collections in the Caribbean and Pacific indicated the presence of organochlorine and other cyclodiene pesticides in the surface of slicks at concentrations between <2 ng/L to 18.45 µg/L, which are many folds higher than in the water column (Menzies, Quinete, Gardinali, & Seba, 2013). Marine parasites, such as nematodes, cestodes, and acanthocephalans, can be used to monitor the environmental impact (Nachev & Sures, 2016). Parasites are suitable, because they bioconcentrate pollutants from their preys. In Europe eggs of herring and guillemot have been used to detect polychlorinated dibenzo-*p*-dioxins and dibenzofurans (PCDD/Fs) and dioxin-like polychlorinated biphenyls (dl-PCBs) (Miller *et al.*, 2014).

Since the concentrations of pesticides in the marine environment are low, new methods had to be developed for their detection, such as large-volume injection in gas chromatography-ion trap tandem mass spectrometry (Steen, Freriks, Cofino, & Brinkman, 1997). Organochlorine pesticides have been studied in the coastal marine environment of Mumbai, India using multi-compartment monitoring of residue levels (Pandit, Sahu, Sharma, & Puranik, 2006). Since these substances accumulate in the food chain it is not amazing that several organochlorine pesticides have been detected in the blubber in ringed seals in samples taken from various circumpolar regions in the Arctic with increasing concentrations from east to west with highest values in the Canadian Arctic. It is interesting to note that there are species-specific enantiomeric ratios in herring, grey seal, harbor seal and ringed seal as shown by enantioselective analysis of α-hexachlorocyclohexane (α-HCH), chlordanes, and chlordane metabolites taken from Swedish samples (Wiberg *et al.*, 1998). Chiral enrichment was also found for enantiomers of the organochlorine pesticide chlordane, which allows conclusions about the source, pathway, and fate of the pollutant (Singh, Hegeman, Laane, & Chan, 2016).

Also, ubiquitous herbicides, such as atrazine, diuron, Irgarol®1051, and isoproturon reach the marine environment by terrestrial run-off, however, in general, they are of low risk for the biota, but may reach occasional potential effect levels (Sjollema *et al.*, 2014). Using a pulse amplitude modulation fluorometry bioassay on marine, microalgae showed a clear species- and herbicide-specific toxicity. Therefore, it cannot be excluded that herbicide pollution may contribute to changes in phytoplankton species composition in coastal waters. However, in tropical environments, such as in Great Barrier Reef lagoons, a degradation of agricultural herbicides was detected, which is controlled by light and sediment (Mercurio *et al.*, 2016). Because of their ecological importance, wide geographic distribution and ease of handling bivalves, such as *Cerastoderma edule* and *Scrobicularia plana*, have been used in ecotoxicological bioassays to determine the toxic and biochemical effects of the widely used herbicide Primextra Gold TZ (Gonçalves *et al.*, 2016). A novel method for the determination of nine triazines in marine sediments has been developed based on Matrix Solid Phase Dispersion using ENI-Carb as dispersant and ethyl acetate as solvent. Recovery for analytical purposes was close to 100%, and the concentrations from 0.022 to 0.037 mg/kg could be detected (Rodríguez-González, González-Castro, Beceiro-González, & Muniategui-Lorenzo, 2017).

MICROPLASTICS

The uncontrolled use of plastic foils and bags since the 1940s has resulted in major pollution in many terrestrial habitats. In many developing countries in Africa and South America, one can see the countryside around villages and cities strewn with plastic debris. Luckily, some regions have prohibited the use of *e.g.* plastic bags. Still, millions of tons are drifted to the beaches where the material becomes brittle and cracks under the impact of solar UV radiation and mechanic forces, producing microparticles, commonly defined as plastic particles smaller than 5 mm in size (Thompson, 2015), which are carried into the oceans by wind and waves forming persistent organic pollutants. They can be collected and quantified by plankton samples, sandy, and muddy sediments, after ingestion by vertebrates and invertebrates and by pollutant interactions (do Sul & Costa, 2014).

The UV-induced fragmentation is augmented by increasing temperatures (Duis & Coors, 2016). Thus, climate change and UV radiation may enhance the number of microplastics in Easter Asia and the tropics where higher UV irradiances are expected (United Nations Environment Programme, 2017). These pollutants are found both in inland (Klein, Worch, & Knepper, 2015) and marine aquatic ecosystems (Lusher, 2016).

Other sources originate from tires, artificial turf, paint from boat hulls and laundry as listed in the updated report on "Swedish sources and pathways for microplastics to the marine environment - A review of existing data" from 2016 (Magnusson *et al.*, 2016). These microplastics reach the ocean *via* stormwater, wastewater or atmospheric deposition. While general quantification is difficult, the report estimates the yearly load of plastic particles from personal care products, synthetic fibers from laundry and household dust discharged to municipal wastewater in Sweden to amount to 67 - 927 tons. In addition, an estimated amount of 7670 tons is released from abrasion of tires; however, how much of this reaches the sea is not known. Most of these particles were >300 μm. These pollutants are abundant and ubiquitous not only in the coastal regions but also concentrated in mid-ocean gyres (Cole, Lindeque, Halsband, & Galloway, 2011).

Microplastics are of growing concern because of their persistence, ubiquity and their ability to transfer POPs to marine organisms (Ng & Obbard, 2006). Many microorganisms, such as phytoplankton consume material not according to taste but to size. Microplastics can inhibit biological processes after ingestion. Microplastics are in the size range where they are consumed by phyto- and zooplankton, which are then relayed through the food web reaching fish, birds, and mammals. In addition, they can leach toxic additives but also absorb waterborne contaminants, such as bisphenol A, phthalates, citrates, and Irgafos 168 phosphate (Bakir, Rowland, & Thompson, 2012; Suhrhoff & Scholz-Bottcher, 2016). Mussels and other passive filter-feeders ingest microplastics (Browne *et al.*, 2015; Van Cauwenberghe, Devriese, Galgani, Robbens, & Janssen, 2015; Wesch, Bredimus, Paulus, & Klein, 2016). However, the bioavailability and transfer efficiency across trophic levels still have to be quantified and modeled (Andrady, 2011) since most studies on the ecological effects of microplastics have been carried out in single species and in laboratory studies (Avio, Gorbi, & Regoli, 2016; Wesch *et al.*, 2016).

In order to identify and quantify microplastics from the marine ecosystems, three main collecting methods are used: selective, volume-reduced and bulk sampling (Hidalgo-Ruz, Gutow, Thompson, & Thiel, 2012). The samples are collected either on sandy beaches at the high tide mark while seawater samples are taken by neuston nets at the surface. The samples are then separated by density (floatation), filtration, sieving or visual sorting based on type, shape, degradation, and color, the latter being the most commonly used method. Chemical identification is done by (Fourier transform) infrared (FTIR) spectroscopy. The majority of the microplastics are composed of polyethylene, polypropylene, and polystyrene. Quantification is based on particles per area (surface and sediment) or particles per volume (water column). Most studies show two most abundant size ranges being 1- 500 μm and 500 μm - 5 mm.

CRUDE OIL

Spectacular disasters, such as the Deepwater Horizon blowout, have alerted the public about the ecologic impacts of crude oil spills (Camilli *et al.*, 2010; Smith, 1993). This accident resulted in the release of millions of barrels of crude oil (Forth, Mitchelmore, Morris, & Lipton, 2017). The Exxon Valdez spilled only about 20% (37,000 tonnes) of its cargo and the worst case scenario would be the complete loss of all the cargo being 150,000 tonnes from the very large crude carrier or about 306,000 tonnes from an ultra-large carrier (Fingas, 2016). However, the accidental spills due to storms, quakes, mechanical failure or human error accumulate too much larger total losses. The total cargo spills from US tanker accidents show that only in about 0.002% cases losses of more than 35% occurred. In 0.2% of the accidents about 11% was spilled, and in about 2.5% of the accidents, 20% were lost. Also spillings during operations on oil platforms and during loading and unloading add substantial amounts of crude oil to the environment. In addition, there is natural seepage of crude oil into the marine environment and it has been estimated that currently about 47% of crude oil released to the oceans is from natural sources, which amounts to about 600,000 metric tons per year and the rest from human-induced leaks (Kvenvolden & Cooper, 2003). However, the estimates have a very large range of uncertainty.

Crude oil components are toxic to the biota. In addition to the acute mortality after an oil spill, sublethal effects on the wildlife continue for a long time due to the persistence of toxic subsurface oil and chronic exposures resulting in population reductions and postponed recovery (Peterson *et al.*, 2003). Two years after an accident in Buzzards Bay, Massachusetts, which spilled about 600 metric tons of number 2 fuel oil the hydrocarbons still persisted both in the marsh and in offshore sediments (Blumer & Sass, 1972). Degradation is believed to proceed by microbial break-down of alkanes and partial dissolution of aromatic hydrocarbons. After the Deepwater Horizon accident, a cruise showed that water samples from 21% of the 14 stations were toxic to bacteria as indicated by the Microtox assay and 34% were toxic to phytoplankton (QwikLite assay) (Paul *et al.*, 2013). 43% of the samples induced DNA damage demonstrated by the L-Microscreen Prophage induction assay. The results suggest that the toxicity depends on the total petroleum hydrocarbon concentration in the water samples. Subsequent cruises showed that the mutagenicity measured with the Microscreen assay was present at least 1.5 years after capping the Macondo well. Also, sediment porewater samples in the vicinity of the well were highly genotoxic. Crude oil is toxic to fish as it impairs the development of the hearts in large pelagic predatory fish (Incardona *et al.*, 2014). It also affects the cardiac excitation-contraction coupling in fish (Brette *et al.*, 2014). Analysis of the molecular mechanisms of cardiotoxicity has shown that some polycyclic aromatic compounds are strong agonists of the aryl hydrocarbon receptor, which has also been found for dioxins (Incardona, 2017).

Birds are primarily affected by crude oil, which enters the plumage, restricts thermal insulation and prevents flying. When the birds try to clean their feathers they ingest the oil. This has been found to decrease the hemoglobin concentration, affect hepatic antioxidant enzymes activity and increase the liver weight, but the normally ingested oil quantity may not be sufficient to induce hemolytic anemia as demonstrated in sandpipers and cormorants (Bursian *et al.*, 2017; Harr *et al.*, 2017). High boiling petroleum components have been found to induce developmental deficits and some are mutagenic under *in vitro* conditions (Gray *et al.*, 2013).

In order to clean up accidental oil spills, dispersants are routinely applied in order to augment the dissolution of oil into the water. After this microbial biodegradation takes over (Kleindienst, Paul, & Joye, 2015). However, the application of dispersants is controversial. The effect of the dispersant is to break down the oil droplet size. In a test the initial droplet size was 5 to 10 µm with a few larger ones (Forth, Mitchelmore, Morris, Lay, & Lipton, 2017). After 4 days the size had decreased to 5 to 10 µm. Larger droplets rise to the surface. The dispersant Corexit 9500A used after the Deepwater Horizon blowup, as well as the dispersed crude oil itself, are highly toxic to microzooplankton including oligotrich ciliates (*Strombidium*), tintinnid ciliates (*e.g., Eutintinnus pectinis*) and heterotrophic dinoflagellates (Almeda, Hyatt, & Buskey, 2014). Almost 7 million liters of this dispersant were released in the Gulf of Mexico to clean up the oil spill after the Deepwater Horizon accident. Specifically, small ciliates were highly sensitive to the dispersant. In addition, the combination of the dispersant with the crude oil significantly enhances the toxicity. This effect on microzooplankton interferes with the energy transfer to higher trophic levels and affects the dynamics and structure of marine plankton communities. Natural oil and gas condensate are also toxic to coral reef larvae (Negri *et al.*, 2016). This was experimentally tested applying a fraction of light crude oil to larvae of the coral *Acropora tenuis* and the sponge *Rhopaloeides odorabile*. The metamorphosis of the coral larvae was completely blocked at a concentration of 103 µg/L. The sensitivity even increased when they were exposed to UV radiation. In contrast, the sponge larvae were less sensitive (>10 mg/L and additional UV exposure had no effect. The chemical dispersants used after the Gulf of Mexico accident were found to be cytotoxic and genotoxic to sperm whale skin cells (Wise, Wise, Wise, Thompson, & Wise, 2014).

Finally, bacteria degrade spilled oil. They often produce bioemulsifiers. Even though these microbes are ubiquitous in the oceans, it is advantageous to seed the spill site with hydrocarbonoclastic bacteria to speed up the process (Ron & Rosenberg, 2014). In addition, these prokaryotes require nitrogen and phosphorous, which are the rate-limiting factors for petroleum degradation in the water. Therefore, it is suggested to add these fertilizers. Chemical dispersants used after an accidental oil spill can suppress the activity of naturally occurring oil-degrading bacteria and alter the microbial community composition (Kleindienst, Seidel, *et al.*, 2015), *e.g.*, in an experiment the application of a dispersant selected the bacterium *Colwellia*, which has the potential to metabolize the dispersant. This bacterium was also found to bloom *in situ* in deep waters of the Gulf of Mexico during the dispersant application. The presence of oil but without adding a dispersant the natural hydrocarbon-metabolizing *Marinobacter* flourished. Oleophilic fertilizers, such as Inipol EAP 22 and urea-formaldehyde polymers, have been found to augment oil degradation on the shore, but seem to be less effective in the open water. Therefore, uric acid has been suggested as an effective fertilizer. Bioaccessibility of oil in the water and in the sediment depends on weathering, which is a natural process of degradation and dissipation as well as the hydrophobicity (Hong *et al.*, 2016).

DETERGENTS AND SURFACTANTS

Households and industry use vast amounts of detergents and surfactants for cleansing surfaces, dishes, laundry, and personal hygiene. They come as concentrated powders, suspensions or solutions. Detergents and soaps operate effectively because they are amphiphilic: part of the molecule is polar (hydrophilic) and another part is non-polar (hydrophobic). This characteristic facilitates the mixing of lipophilic substances, such as oil and grease with water. In addition, they tend to foam.

Chemically speaking we classify these substances according to their electrical charge (Rosen & Kunjappu, 2012). The first group comprises the anionic detergents, such as alkylbenzenesulfonates. The alkylbenzene part of the molecule is lipophilic while the sulfonate group is hydrophilic. Two forms are common, those with branched alkyl groups and others with linear alkyl groups. The former chemicals tend to be poorly biodegradable and are increasingly phased out. An estimated 6 million tonnes of anionic detergents are synthesized annually. The second group comprises cationic detergents, which are similar to the hydrophilic counterparts, but have a quaternary ammonium group at the polar head. The central ammonium surface center is positively charged. The third group of detergents is non-ionic and zwitterionic. They have hydrophilic, uncharged head groups. Typical examples are based on polyoxyethylene or glycosides (with a sugar as their uncharged hydrophilic head group) as marketed under the names Tween, Triton, and Brij. Zwitterionic detergents carry a net zero charge, *e.g.*,

CHAPS. They are also used in biochemistry, *e.g.* for analysis of membrane proteins by two-dimensional electrophoresis (Chevallet *et al.*, 1998).

Surfactants are used to lower the surface tension between liquids and solids. Therefore, they can be used as detergents, wetting agents, emulsifiers, foaming agents and dispersants. They are also amphiphilic with a hydrophilic head group and a hydrophobic tail (Kosswig, 2005). For this reason, they can diffuse in water and are adsorbed to interfaces between air or oil and water. The world production is estimated at about 15 million tons per year, half of which are soaps.

Because of the immense amounts of detergents dumped into the oceans the problem of toxicity to the marine biota is tackled in many publications (Gerlach, 2013). The environmental impact of the anionic detergent sodium dodecylbenzene sulfonate has been assayed using the marine diatom *Phaeodactylum tricornutum* isolated from an estuarine region in Canada (Aidar *et al.*, 1997). LC_{50} (concentration at which 50 percent inhibition or mortality is found) values were 1.94 and 1.80 mg/L after 48 and 96 h, respectively; NOEC (no observed effect concentration) values were 1.2 mg/L and 0.1 mg/L after the same times, respectively. In addition to mortality, effects on chlorophyll *a*, primary production and cell density have been found (Aidar *et al.*, 1997). Similar effects of detergents were found in green flagellates (Azizullah, Richter, & Häder, 2011, 2014; Azizullah, Richter, Jamil, & Häder, 2012). Because of its sensitivity, *Euglena gracilis* has been selected as a bioindicator to monitor toxicity in aquatic ecosystems (Azizullah *et al.*, 2014; Azizullah, Richter, Ullah, Ali, & Häder, 2013). The impact of anionic (ABS, LAS, LES 3 EO) and non-ionic (NP 10 EO, TAE 10 EO) detergents has been measured in marine fishes, crustaceans and bivalves in continuous-flow systems (Swedmark, Braaten, Emanuelsson, & Granmo, 1971). Fishes were found to be more sensitive (LC_{50} 0.8 - 6.5 ppm after 96 h) than bivalves (5 to > 100 ppm after 96 h). Crustaceans were even more resistant (25 to > 100 ppm after 96 h). Developmental stages were more sensitive than adults. Motile organisms reacted with avoidance, followed by inactivation and death. Sublethal effects are impaired locomotion and breathing rate. Another study confirmed the toxicity of NTA-containing detergents on adults and juveniles of 11 marine fish and invertebrate species (Eisler *et al.*, 1972). However, fairly high LD_{50} concentrations between 1.8 g/L and >10 g/L have been reported for grass shrimp, clams, crabs, starfish, lobster, mussels, snails, sandworm and two fishes. Corals have been found to be affected by household detergents. Growth and survival under the effect of the linear alkylbenzene sulfonates (LAS), the nonionic surfactant, and nonylphenol ethoxylate (NPE) were monitored in two branching coral species (*Stylophora pistillata* and *Pocillopora damicornis*) (Shafir, Halperin, & Rinkevich, 2014). Species-specific differences could result in a reduction of genetic diversity and shift species composition in the reef. A positive effect of the release of detergents into marine environments at low concentrations could be the enhancement of phosphorous availability (Jenkins & Ives, 2013). Phosphor is often a limiting nutrient for growth and reproduction in marine ecosystems (Ryther & Dunstan, 1971).

PHARMACEUTICALS

Advances in medicine are accompanied with an increase in the production and usage of pharmaceuticals. In addition to accidental spill into aqueous environments, many of these chemicals are excreted naturally either in their original chemical form or as metabolized substances. These drugs have been found to accumulate in the environment (Taylor *et al.*, 2015). Increasing concentrations of more than 1000 biologically active molecules used in human and veterinary medicine are detected in fresh water habitats, such as rivers and lakes, in drinking water as well as in the soil (Halm-Lemeille & Gomez, 2016; Li, 2014) where they pose a considerable risk for the environment, ecosystems, and wildlife (Arnold, Brown, Ankley, & Sumpter, 2014). Pharmaceuticals have been reported to occur in coastal ecosystems all over the globe from the Arctic to the Antarctic (Gaw, Thomas, & Hutchinson, 2015). There is growing evidence that these drugs at current concentrations can affect the biota including algae as primary producers with relayed effects for the aquatic food webs aggravated by bioaccumulation. The growing human population increasingly relies on aquatic food production *e.g.*, in coastal aquaculture. The main pathway for human pharmaceuticals as well as illicit drugs into the marine ecosystems is by sewage, *e.g.* an estimated 14.4 kg of antibiotics is deposited every day into Victoria Harbour, Hong Kong by seven wastewater treatment plants. 90% of the treated sewage is discharged into the ocean in New Zealand. Sewage from about 400 million people is collected in the Yangtze River, China, which includes an estimated 152 tonnes of pharmaceuticals every year.

Pharmaceuticals also reach marine ecosystems through terrestrial run-off and are detected in open water and sediments especially in coastal habitats at concentrations ranging from 0.01 ng/L to 2.4 µg/L. However, pharmaceutical contamination is only beginning to be investigated. But it goes undisputed that the potential risks for marine species need to be assessed because pharmaceuticals differ from conventional pollutants since they are designed to interact with physiological pathways at low doses (Fabbri & Franzellitti, 2016).

Pharmaceuticals found in marine sediments worldwide include analgesics, antibiotics, antiepileptics, antihistamines, anti-hypertensives, antimicrobial, bronchodilator, diuretics, estrogens, gastrointestinal drugs, hypolipidemic and nonsteroidal anti-inflammatory drugs (Fabbri & Franzellitti, 2016). In addition, drugs for use in animal husbandry and aquaculture are found in coastal and sea waters. A recent volume reviews the available data on pharmaceuticals in the Baltic Sea including the main sources and pathways collected by the Contracting Parties to the Convention on the Protection of the Marine Environment of the Baltic Sea Area (Helsinki Convention) (Vieno *et al.*, 2017), The report includes concentrations of pharmaceuticals, which are compared to threshold values and describes the environmental impact of 167 pharmaceutical substances. The main source of these drugs in the marine environment seems to be the excretion of pharmaceutical used by humans and animals, which have been calculated to amount to 1800 tons per year. Wastewater treatment was found to remove only nine out of 118 pharmaceuticals with an efficiency of over 95% and almost half of the substances were removed to less than 50%. The most common drugs belong to anti-inflammatory and analgesic compounds as well as cardiovascular and central nervous system agents and belong to the therapeutic groups of metabolic and gastrointestinal substances, such as clofibric acid, pyrimidone, and carbamazepine. In organisms, the highest concentrations were found in blue mussels. In addition, 20 pharmaceuticals have been found to bioaccumulate in snails and fish (golden grey mullet and black goby) in a coastal lagoon. (Moreno-González, Rodríguez-Mozaz, Huerta, Barceló, & León, 2016). A recent study also indicated that marine polychaetes (*Hediste diversicolor*) are affected when exposed to sediment spiked with five human pharmaceuticals (carbamazepine, ibuprofen, propranolol, fluoxetine, and 17α-ethynylestradiol). In addition to the energy status, neuroendocrine effects were found indicating that benthic communities may be at ecological risk from pharmaceuticals.

CONCLUSIONS

In conclusion, this chapter highlights that the marine ecosystems are strongly affected by environmental factors, such as temperature, nutrients, and transparency for solar visible and UV radiation. Increasing pollution, not only of coastal habitats but also of open ocean waters, results in changes in productivity and species composition. Various kinds of organic pollutants such as PAHs, PFCs and pesticides are entering the aquatic ecosystems with terrestrial run-off, but are distributed not only in coastal areas and estuaries. Moreover, microplastics are of growing concern since they are concentrated in oceanic gyres, where they are ingested by different marine animals and accumulate in the food chain. Accidental oil spills and catastrophic events are the reason for the pollution by crude oil and its products, which is causing hazardous impacts on marine biodiversity.

CONSENT FOR PUBLICATION

Not applicable.

CONFLICT OF INTEREST

Declare None.

ACKNOWLEDGEMENT

Declare None.

REFERENCES

Aidar, E, Sigaud-Kutner, TC, Nishihara, L, Schinke, KP, Braga, MCC, Farah, RE & Kutner, MB (1997) Marine phytoplankton assays: effects of detergents. *Mar Environ Res, 43*, 55-68.
[http://dx.doi.org/10.1016/0141-1136(96)00002-5]

Almeda, R, Hyatt, C & Buskey, EJ (2014) Toxicity of dispersant Corexit 9500A and crude oil to marine microzooplankton. *Ecotoxicol Environ Saf, 106*, 76-85.
[http://dx.doi.org/10.1016/j.ecoenv.2014.04.028] [PMID: 24836881]

Andrady, AL (2011) Microplastics in the marine environment. *Mar Pollut Bull, 62*, 1596-605.
[http://dx.doi.org/10.1016/j.marpolbul.2011.05.030] [PMID: 21742351]

Arnold, KE, Brown, AR, Ankley, GT & Sumpter, JP (2014) *Medicating the environment: assessing risks of pharmaceuticals to wildlife and ecosystems.* The Royal Society.

Avio, CG, Gorbi, S & Regoli, F (2016) Plastics and microplastics in the oceans: From emerging pollutants to emerged threat. *Mar Environ Res, 128*, 2-11.
[http://dx.doi.org/10.1016/j.marenvres.2016.05.012]

Azizullah, A, Richter, P & Häder, D-P (2011) Toxicity assessment of a common laundry detergent using the freshwater flagellate *Euglena gracilis*. *Chemosphere, 84*, 1392-400.
[http://dx.doi.org/10.1016/j.chemosphere.2011.04.068] [PMID: 21601907]

Azizullah, A, Richter, P & Häder, D-P (2014) Photosynthesis and photosynthetic pigments in the flagellate *Euglena gracilis* - as sensitive endpoints for toxicity evaluation of liquid detergents. *J Photochem Photobiol B, 133*, 18-26.
[http://dx.doi.org/10.1016/j.jphotobiol.2014.02.011] [PMID: 24658006]

Azizullah, A, Richter, P, Jamil, M & Häder, D-P (2012) Chronic toxicity of a laundry detergent to the freshwater flagellate *Euglena gracilis*. *Ecotoxicology, 21*, 1957-64.
[http://dx.doi.org/10.1007/s10646-012-0930-3] [PMID: 22644093]

Azizullah, A, Richter, P, Ullah, W, Ali, I & Häder, D-P (2013) Ecotoxicity evaluation of a liquid detergent using the automatic biotest ECOTOX. *Ecotoxicology, 22*, 1043-52.
[http://dx.doi.org/10.1007/s10646-013-1091-8] [PMID: 23783251]

Bais, AF, McKenzie, RL, Bernhard, G, Aucamp, PJ, Ilyas, M, Madronich, S & Tourpali, K (2015) Ozone depletion and climate change: impacts on UV radiation. *Photochem Photobiol Sci, 14*, 19-52.
[http://dx.doi.org/10.1039/C4PP90032D] [PMID: 25380284]

Bakir, A, Rowland, SJ & Thompson, RC (2012) Competitive sorption of persistent organic pollutants onto microplastics in the marine environment. *Mar Pollut Bull, 64*, 2782-9.
[http://dx.doi.org/10.1016/j.marpolbul.2012.09.010] [PMID: 23044032]

Behrenfeld, MJ, O'Malley, RT, Siegel, DA, McClain, CR, Sarmiento, JL, Feldman, GC, Milligan, AJ, Falkowski, PG, Letelier, RM & Boss, ES (2006) Climate-driven trends in contemporary ocean productivity. *Nature, 444*, 752-5.
[http://dx.doi.org/10.1038/nature05317] [PMID: 17151666]

Blumer, M & Sass, J (1972) Oil pollution: persistence and degradation of spilled fuel oil. *Science, 176*, 1120-2.

[http://dx.doi.org/10.1126/science.176.4039.1120] [PMID: 17775135]

Boyce, DG, Lewis, MR & Worm, B (2010) Global phytoplankton decline over the past century. *Nature,* 466, 591-6.
[http://dx.doi.org/10.1038/nature09268] [PMID: 20671703]

Brette, F, Machado, B, Cros, C, Incardona, JP, Scholz, NL & Block, BA (2014) Crude oil impairs cardiac excitation-contraction coupling in fish. *Science,* 343, 772-6.
[http://dx.doi.org/10.1126/science.1242747] [PMID: 24531969]

Browne, MA, Underwood, AJ, Chapman, MG, Williams, R, Thompson, RC & van Franeker, JA (2015) Linking effects of anthropogenic debris to ecological impacts. *Proceedings of the Royal Society B Biological Sciences,* 282, 20142929.
[http://dx.doi.org/10.1098/rspb.2014.2929]

Bursian, SJ, Dean, KM, Harr, KE, Kennedy, L, Link, JE, Maggini, I, Pritsos, C, Pritsos, KL, Schmidt, RE & Guglielmo, CG (2017) Effect of oral exposure to artificially weathered Deepwater Horizon crude oil on blood chemistries, hepatic antioxidant enzyme activities, organ weights and histopathology in western sandpipers (*Calidris mauri*). *Ecotoxicol Environ Saf,* 146, 91-7.
[http://dx.doi.org/10.1016/j.ecoenv.2017.03.045] [PMID: 28413080]

Camilli, R, Reddy, CM, Yoerger, DR, Van Mooy, BA, Jakuba, MV, Kinsey, JC, McIntyre, CP, Sylva, SP & Maloney, JV (2010) Tracking hydrocarbon plume transport and biodegradation at Deepwater Horizon. *Science,* 330, 201-4.
[http://dx.doi.org/10.1126/science.1195223] [PMID: 20724584]

Ceballos, G, Ehrlich, PR, Barnosky, AD, García, A, Pringle, RM & Palmer, TM (2015) Accelerated modern human-induced species losses: Entering the sixth mass extinction. *Sci Adv,* 1e1400253
[http://dx.doi.org/10.1126/sciadv.1400253] [PMID: 26601195]

Chevallet, M, Santoni, V, Poinas, A, Rouquié, D, Fuchs, A, Kieffer, S, Rossignol, M, Lunardi, J, Garin, J & Rabilloud, T (1998) New zwitterionic detergents improve the analysis of membrane proteins by two-dimensional electrophoresis. *Electrophoresis,* 19, 1901-9.
[http://dx.doi.org/10.1002/elps.1150191108] [PMID: 9740050]

Cole, M, Lindeque, P, Halsband, C & Galloway, TS (2011) Microplastics as contaminants in the marine environment: a review. *Mar Pollut Bull,* 62, 2588-97.
[http://dx.doi.org/10.1016/j.marpolbul.2011.09.025] [PMID: 22001295]

Ivar do Sul, JA & Costa, MF (2014) The present and future of microplastic pollution in the marine environment. *Environ Pollut,* 185, 352-64.
[http://dx.doi.org/10.1016/j.envpol.2013.10.036] [PMID: 24275078]

Duis, K & Coors, A (2016) Microplastics in the aquatic and terrestrial environment: sources (with a specific focus on personal care products), fate and effects. *Environ Sci Eur,* 28, 2.
[http://dx.doi.org/10.1186/s12302-015-0069-y] [PMID: 27752437]

Echeveste, P, Galbán-Malagón, C, Dachs, J, Berrojalbiz, N & Agustí, S (2016) Toxicity of natural mixtures of organic pollutants in temperate and polar marine phytoplankton. *Sci Total Environ,* 571, 34-41.
[http://dx.doi.org/10.1016/j.scitotenv.2016.07.111] [PMID: 27470667]

Eisler, R, Gardner, G, Hennekey, R, LaRoche, G, Wash, D & Yevich, P (1972) Acute toxicology of sodium nitrilotriacetic acid (NTA) and NTA-containing detergents to marine organisms. *Water Res,* 6, 1009-27.
[http://dx.doi.org/10.1016/0043-1354(72)90054-1]

El-Shahawi, MS, Hamza, A, Bashammakh, AS & Al-Saggaf, WT (2010) An overview on the accumulation, distribution, transformations, toxicity and analytical methods for the monitoring of persistent organic pollutants. *Talanta,* 80, 1587-97.
[http://dx.doi.org/10.1016/j.talanta.2009.09.055] [PMID: 20152382]

Eriksen, M, Lebreton, LC, Carson, HS, Thiel, M, Moore, CJ, Borerro, JC, Galgani, F, Ryan, PG & Reisser, J (2014) Plastic pollution in the world's oceans: more than 5 trillion plastic pieces weighing over 250,000 tons

afloat at sea. *PLoS One,* 9e111913
[http://dx.doi.org/10.1371/journal.pone.0111913] [PMID: 25494041]

Everaert, G, De Laender, F, Goethals, PL & Janssen, CR (2015) Relative contribution of persistent organic pollutants to marine phytoplankton biomass dynamics in the North Sea and the Kattegat. *Chemosphere,* 134, 76-83.
[http://dx.doi.org/10.1016/j.chemosphere.2015.03.084] [PMID: 25912805]

Fabbri, E & Franzellitti, S (2016) Human pharmaceuticals in the marine environment: Focus on exposure and biological effects in animal species. *Environ Toxicol Chem,* 35, 799-812.
[http://dx.doi.org/10.1002/etc.3131] [PMID: 26111460]

Fernandez, LA, Lao, W, Maruya, KA & Burgess, RM (2014) Calculating the diffusive flux of persistent organic pollutants between sediments and the water column on the Palos Verdes shelf superfund site using polymeric passive samplers. *Environ Sci Technol,* 48, 3925-34.
[http://dx.doi.org/10.1021/es404475c] [PMID: 24564763]

Field, CB, Behrenfeld, MJ, Randerson, JT & Falkowski, P (1998) Primary production of the biosphere: integrating terrestrial and oceanic components. *Science,* 281, 237-40.
[http://dx.doi.org/10.1126/science.281.5374.237] [PMID: 9657713]

Fingas, M (2016) *Oil spill science and technology.* Gulf professional publishing.

Fischetti, M (2013) Deep heat threatens marine life. *Sci Am,* 308, 92.
[http://dx.doi.org/10.1038/scientificamerican0413-92] [PMID: 23539796]

Fisk, AT, Hobson, KA & Norstrom, RJ (2001) Influence of chemical and biological factors on trophic transfer of persistent organic pollutants in the northwater polynya marine food web. *Environ Sci Technol,* 35, 732-8.
[http://dx.doi.org/10.1021/es001459w] [PMID: 11349285]

Forth, HP, Mitchelmore, CL, Morris, JM, Lay, CR & Lipton, J (2017) Characterization of dissolved and particulate phases of water accommodated fractions used to conduct aquatic toxicity testing in support of the Deepwater Horizon natural resource damage assessment. *Environ Toxicol Chem,* 36, 1460-72.
[http://dx.doi.org/10.1002/etc.3803] [PMID: 28328044]

Forth, HP, Mitchelmore, CL, Morris, JM & Lipton, J (2017) Characterization of oil and water accommodated fractions used to conduct aquatic toxicity testing in support of the Deepwater Horizon oil spill natural resource damage assessment. *Environ Toxicol Chem,* 36, 1450-9.
[http://dx.doi.org/10.1002/etc.3672] [PMID: 27805278]

Gao, KS, Xu, JT, Gao, G, Li, YH, Hutchins, DA & Huang, BQ (2012) Rising CO_2 and increased light exposure synergistically reduce marine primary productivity. *Nat Clim Chang,* 2, 519-23.
[http://dx.doi.org/10.1038/nclimate1507]

Gaw, S, Thomas, K & Hutchinson, TH (2015) *Pharmaceuticals in the marine environment Pharmaceuticals in the Environment.* Royal Society of Chemistry 70-91.
[http://dx.doi.org/10.1039/9781782622345-00070]

Gerlach, SA (2013) *Marine pollution: diagnosis and therapy.* Springer Science & Business Media.

Gómez, N & O'Farrell, I (2014) Phytoplankton from urban and suburban polluted rivers. *Adv Limnol,* 65, 127-42.
[http://dx.doi.org/10.1127/1612-166X/2014/0065-0038]

Gonçalves, A, Mesquita, A, Verdelhos, T, Coutinho, J, Marques, J & Gonçalves, F (2016) Fatty acids' profiles as indicators of stress induced by of a common herbicide on two marine bivalves species: *Cerastoderma edule* (Linnaeus, 1758) and *Scrobicularia plana* (da Costa, 1778). *Ecol Indic,* 63, 209-18.
[http://dx.doi.org/10.1016/j.ecolind.2015.12.006]

Gray, TM, Simpson, BJ, Nicolich, MJ, Murray, FJ, Verstuyft, AW, Roth, RN & McKee, RH (2013) Assessing the mammalian toxicity of high-boiling petroleum substances under the rubric of the HPV program. *Regul Toxicol Pharmacol,* 67 (Suppl.), S4-9.

[http://dx.doi.org/10.1016/j.yrtph.2012.11.014] [PMID: 23247262]

Häder, D-P & Gao, K (2015) Interactions of anthropogenic stress factors on marine phytoplankton. *Front Environ Sci,* 3, 14.
[http://dx.doi.org/10.3389/fenvs.2015.00014]

Halm-Lemeille, M-P & Gomez, E (2016) *Pharmaceuticals in the environment.* Springer.

Harr, KE, Cunningham, FL, Pritsos, CA, Pritsos, KL, Muthumalage, T, Dorr, BS, Horak, KE, Hanson-Dorr, KC, Dean, KM, Cacela, D, McFadden, AK, Link, JE, Healy, KA, Tuttle, P & Bursian, SJ (2017) Weathered MC252 crude oil-induced anemia and abnormal erythroid morphology in double-crested cormorants (*Phalacrocorax auritus*) with light microscopic and ultrastructural description of Heinz bodies. *Ecotoxicol Environ Saf,* 146, 29-39.
[http://dx.doi.org/10.1016/j.ecoenv.2017.07.030] [PMID: 28734789]

Harrad, S (2009) *Persistent organic pollutants.* John Wiley & Sons.
[http://dx.doi.org/10.1002/9780470684122]

Hidalgo-Ruz, V, Gutow, L, Thompson, RC & Thiel, M (2012) Microplastics in the marine environment: a review of the methods used for identification and quantification. *Environ Sci Technol,* 46, 3060-75.
[http://dx.doi.org/10.1021/es2031505] [PMID: 22321064]

Hong, S, Yim, UH, Ha, SY, Shim, WJ, Jeon, S, Lee, S, Kim, C, Choi, K, Jung, J, Giesy, JP & Khim, JS (2016) Bioaccessibility of AhR-active PAHs in sediments contaminated by the Hebei Spirit oil spill: Application of Tenax extraction in effect-directed analysis. *Chemosphere,* 144, 706-12.
[http://dx.doi.org/10.1016/j.chemosphere.2015.09.043] [PMID: 26408977]

Honjo, S, Eglinton, TI, Taylor, CD, Ulmer, KM, Sievert, SM & Bracher, A (2014) Understanding the role of the biological pump in the global carbon cycle: An imperative for ocean science. *Oceanography (Wash DC),* 27, 10-6.
[http://dx.doi.org/10.5670/oceanog.2014.78]

Incardona, JP (2017) Molecular mechanisms of crude oil developmental toxicity in fish. *Arch Environ Contam Toxicol,* 73, 19-32.
[http://dx.doi.org/10.1007/s00244-017-0381-1] [PMID: 28695261]

Incardona, JP, Gardner, LD, Linbo, TL, Brown, TL, Esbaugh, AJ, Mager, EM, Stieglitz, JD, French, BL, Labenia, JS, Laetz, CA, Tagal, M, Sloan, CA, Elizur, A, Benetti, DD, Grosell, M, Block, BA & Scholz, NL (2014) Deepwater Horizon crude oil impacts the developing hearts of large predatory pelagic fish. *Proc Natl Acad Sci USA,* 111, E1510-8.
[http://dx.doi.org/10.1073/pnas.1320950111] [PMID: 24706825]

IPCC (2014) Climate Change 2014: Impacts, Adaptation, and Vulnerability. *Part B: Regional Aspects Contribution of Working Group II to the Fifth Assessment Report of the Intergovernmental Panel on Climate Change,* Vol. 2014.

Jenkins, SH & Ives, KJ (2013) Phosphorus in Fresh Water and the Marine Environment. *Prog Water Technol,* 2 Elsevier, Pergamon Press.

Jin, P, Duarte, CM & Agustí, S (2017) Contrasting responses of marine and freshwater photosynthetic organisms to UVB radiation: A meta-analysis. *Front Mar Sci,* 4, 45.
[http://dx.doi.org/10.3389/fmars.2017.00045]

Klein, S, Worch, E & Knepper, TP (2015) Occurrence and spatial distribution of microplastics in river shore sediments of the Rhine-Main area in Germany. *Environ Sci Technol,* 49, 6070-6.
[http://dx.doi.org/10.1021/acs.est.5b00492] [PMID: 25901760]

Kleindienst, S, Paul, JH & Joye, SB (2015) Using dispersants after oil spills: impacts on the composition and activity of microbial communities. *Nat Rev Microbiol,* 13, 388-96.
[http://dx.doi.org/10.1038/nrmicro3452] [PMID: 25944491]

Kleindienst, S, Seidel, M, Ziervogel, K, Grim, S, Loftis, K, Harrison, S, Malkin, SY, Perkins, MJ, Field, J, Sogin, ML, Dittmar, T, Passow, U, Medeiros, PM & Joye, SB (2015) Chemical dispersants can suppress the

activity of natural oil-degrading microorganisms. *Proc Natl Acad Sci USA,* 112, 14900-5.
[http://dx.doi.org/10.1073/pnas.1507380112] [PMID: 26553985]

Kosswig, C (2005) *Surfactants Ullmann's Encyclopedia of Industrial Chemistry.* Wiley-VCH, Weinheim.

Kvenvolden, K & Cooper, C (2003) Natural seepage of crude oil into the marine environment. *Geo-Mar Lett,* 23, 140-6.
[http://dx.doi.org/10.1007/s00367-003-0135-0]

Landschützer, P, Gruber, N, Bakker, D & Schuster, U (2014) Recent variability of the global ocean carbon sink. *Global Biogeochem Cycles,* 28, 927-49.
[http://dx.doi.org/10.1002/2014GB004853]

Li, WC (2014) Occurrence, sources, and fate of pharmaceuticals in aquatic environment and soil. *Environ Pollut,* 187, 193-201.
[http://dx.doi.org/10.1016/j.envpol.2014.01.015] [PMID: 24521932]

Lohmann, R & Belkin, IM (2014) Organic pollutants and ocean fronts across the Atlantic Ocean: A review. *Prog Oceanogr,* 128, 172-84.
[http://dx.doi.org/10.1016/j.pocean.2014.08.013]

Luna-Acosta, A, Bustamante, P, Budzinski, H, Huet, V & Thomas-Guyon, H (2015) Persistent organic pollutants in a marine bivalve on the Marennes-Oléron Bay and the Gironde Estuary (French Atlantic Coast) - part 2: potential biological effects. *Sci Total Environ,* 514, 511-22.
[http://dx.doi.org/10.1016/j.scitotenv.2014.10.050] [PMID: 25666833]

Lusher, A (2016) Microplastics in the marine environment: distribution, interactions and effects. *Marine Anthropogenic Litter,* In: Bergmann, M., Gutow, L., Klages, M., (Eds.), Springer International Publishing 245-307.

Magnusson, K, Eliasson, K, Fråne, A, Haikonen, K, Hultén, J, Olshammar, M & Voisin, A (2016) *Microplastics to the marine environment.* Report Number C 183.

Maraccini, PA, Wenk, J & Boehm, AB (2016) Exogenous indirect photoinactivation of bacterial pathogens and indicators in water with natural and synthetic photosensitizers in simulated sunlight with reduced UVB. *J Appl Microbiol,* 121, 587-97.
[http://dx.doi.org/10.1111/jam.13183] [PMID: 27207818]

McKenzie, RL, Aucamp, PJ, Bais, AF, Björn, LO, Ilyas, M & Madronich, S (2011) Ozone depletion and climate change: impacts on UV radiation. *Photochem Photobiol Sci,* 10, 182-98.
[http://dx.doi.org/10.1039/c0pp90034f] [PMID: 21253660]

Menzies, R, Soares Quinete, N, Gardinali, P & Seba, D (2013) Baseline occurrence of organochlorine pesticides and other xenobiotics in the marine environment: Caribbean and Pacific collections. *Mar Pollut Bull,* 70, 289-95.
[http://dx.doi.org/10.1016/j.marpolbul.2013.03.003] [PMID: 23597795]

Mercurio, P, Mueller, JF, Eaglesham, G, O'Brien, J, Flores, F & Negri, AP (2016) Degradation of herbicides in the tropical marine environment: Influence of light and sediment. *PLoS One,* 11e0165890
[http://dx.doi.org/10.1371/journal.pone.0165890] [PMID: 27806103]

Miller, A, Nyberg, E, Danielsson, S, Faxneld, S, Haglund, P & Bignert, A (2014) Comparing temporal trends of organochlorines in guillemot eggs and Baltic herring: advantages and disadvantage for selecting sentinel species for environmental monitoring. *Mar Environ Res,* 100, 38-47.
[http://dx.doi.org/10.1016/j.marenvres.2014.02.007] [PMID: 24680644]

Monitoring, A (2004) AMAP Assessment 2002: Persistent Organic Pollutants in the Arctic.

Moreno-González, R, Rodríguez-Mozaz, S, Huerta, B, Barceló, D & León, VM (2016) Do pharmaceuticals bioaccumulate in marine molluscs and fish from a coastal lagoon? *Environ Res,* 146, 282-98.
[http://dx.doi.org/10.1016/j.envres.2016.01.001] [PMID: 26775009]

Munir, S (2015) Morphometric forms, biovolume and cellular carbon content of dinoflagellates from polluted waters on the Karachi coast, Pakistan. *Indian J Geo-Mar Sci,* 44, 1.

Nachev, M & Sures, B (2016) Environmental parasitology: Parasites as accumulation bioindicators in the marine environment. *J Sea Res,* 113, 45-50.
[http://dx.doi.org/10.1016/j.seares.2015.06.005]

Naso, B, Perrone, D, Ferrante, MC, Bilancione, M & Lucisano, A (2005) Persistent organic pollutants in edible marine species from the Gulf of Naples, Southern Italy. *Sci Total Environ,* 343, 83-95.
[http://dx.doi.org/10.1016/j.scitotenv.2004.10.007] [PMID: 15862838]

Negri, AP, Brinkman, DL, Flores, F, Botté, ES, Jones, RJ & Webster, NS (2016) Acute ecotoxicology of natural oil and gas condensate to coral reef larvae. *Sci Rep,* 6, 21153.
[http://dx.doi.org/10.1038/srep21153] [PMID: 26892387]

Nemerow, N L (1991) Stream, lake, estuary, and ocean pollution.

Newman, PA & McKenzie, R (2011) UV impacts avoided by the Montreal Protocol. *Photochem Photobiol Sci,* 10, 1152-60.
[http://dx.doi.org/10.1039/c0pp00387e] [PMID: 21455537]

Ng, KL & Obbard, JP (2006) Prevalence of microplastics in Singapore's coastal marine environment. *Mar Pollut Bull,* 52, 761-7.
[http://dx.doi.org/10.1016/j.marpolbul.2005.11.017] [PMID: 16388828]

Ogata, Y, Takada, H, Mizukawa, K, Hirai, H, Iwasa, S, Endo, S, Mato, Y, Saha, M, Okuda, K, Nakashima, A, Murakami, M, Zurcher, N, Booyatumanondo, R, Zakaria, MP, Dung, Q, Gordon, M, Miguez, C, Suzuki, S, Moore, C, Karapanagioti, HK, Weerts, S, McClurg, T, Burres, E, Smith, W, Van Velkenburg, M, Lang, JS, Lang, RC, Laursen, D, Danner, B, Stewardson, N & Thompson, RC (2009) International Pellet Watch: global monitoring of persistent organic pollutants (POPs) in coastal waters. 1. Initial phase data on PCBs, DDTs, and HCHs. *Mar Pollut Bull,* 58, 1437-46.
[http://dx.doi.org/10.1016/j.marpolbul.2009.06.014] [PMID: 19635625]

Pandit, GG, Sahu, SK, Sharma, S & Puranik, VD (2006) Distribution and fate of persistent organochlorine pesticides in coastal marine environment of Mumbai. *Environ Int,* 32, 240-3.
[http://dx.doi.org/10.1016/j.envint.2005.08.018] [PMID: 16213018]

Paul, JH, Hollander, D, Coble, P, Daly, KL, Murasko, S, English, D, Basso, J, Delaney, J, McDaniel, L & Kovach, CW (2013) Toxicity and mutagenicity of Gulf of Mexico waters during and after the deepwater horizon oil spill. *Environ Sci Technol,* 47, 9651-9.
[http://dx.doi.org/10.1021/es401761h] [PMID: 23919351]

Peterson, CH, Rice, SD, Short, JW, Esler, D, Bodkin, JL, Ballachey, BE & Irons, DB (2003) Long-term ecosystem response to the Exxon Valdez oil spill. *Science,* 302, 2082-6.
[http://dx.doi.org/10.1126/science.1084282] [PMID: 14684812]

Pithan, F & Mauritsen, T (2014) Arctic amplification dominated by temperature feedbacks in contemporary climate models. *Nat Geosci,* 7, 181-4.
[http://dx.doi.org/10.1038/ngeo2071]

Rastogi, RP, Sinha, RP, Moh, SH, Lee, TK, Kottuparambil, S, Kim, Y-J, Rhee, JS, Choi, EM, Brown, MT, Häder, DP & Han, T (2014) Ultraviolet radiation and cyanobacteria. *J Photochem Photobiol B,* 141, 154-69.
[http://dx.doi.org/10.1016/j.jphotobiol.2014.09.020] [PMID: 25463663]

Rodríguez-González, N, González-Castro, M-J, Beceiro-González, E & Muniategui-Lorenzo, S (2017) Development of a matrix solid phase dispersion methodology for the determination of triazine herbicides in marine sediments. *Microchem J,* 133, 137-43.
[http://dx.doi.org/10.1016/j.microc.2017.03.022]

Ron, EZ & Rosenberg, E (2014) Enhanced bioremediation of oil spills in the sea. *Curr Opin Biotechnol,* 27, 191-4.
[http://dx.doi.org/10.1016/j.copbio.2014.02.004] [PMID: 24657912]

Rosen, MJ & Kunjappu, JT (2012) *Surfactants and interfacial phenomena.* John Wiley & Sons.

[http://dx.doi.org/10.1002/9781118228920]

Ryther, JH & Dunstan, WM (1971) Nitrogen, phosphorus, and eutrophication in the coastal marine environment. *Science,* 171, 1008-13.
[http://dx.doi.org/10.1126/science.171.3975.1008] [PMID: 4993386]

Shafir, S, Halperin, I & Rinkevich, B (2014) Toxicology of household detergents to reef corals. *Water Air Soil Pollut,* 225, 1890.
[http://dx.doi.org/10.1007/s11270-014-1890-4]

Singh, K, Hegeman, WJ, Laane, RW & Chan, HM (2016) Review and evaluation of a chiral enrichment model for chlordane enantiomers in the environment. *Environ Rev,* 24, 363-76.
[http://dx.doi.org/10.1139/er-2016-0015]

Sinha, RP & Häder, D-P (2002) UV-induced DNA damage and repair: a review. *Photochem Photobiol Sci,* 1, 225-36.
[http://dx.doi.org/10.1039/b201230h] [PMID: 12661961]

Sjollema, SB, Martínezgarcía, G, van der Geest, HG, Kraak, MH, Booij, P, Vethaak, AD & Admiraal, W (2014) Hazard and risk of herbicides for marine microalgae. *Environ Pollut,* 187, 106-11.
[http://dx.doi.org/10.1016/j.envpol.2013.12.019] [PMID: 24463473]

Smith, C (1993) News sources and power elites in news coverage of the Exxon Valdez oil spill. *Journal Q,* 70, 393-403.
[http://dx.doi.org/10.1177/107769909307000214]

Solomon, S, Ivy, DJ, Kinnison, D, Mills, MJ, Neely, RR, III & Schmidt, A (2016) Emergence of healing in the Antarctic ozone layer. *Science,* 353, 269-74.
[http://dx.doi.org/10.1126/science.aae0061] [PMID: 27365314]

Sprague, M, Walton, J, Campbell, PJ, Strachan, F, Dick, JR & Bell, JG (2015) Replacement of fish oil with a DHA-rich algal meal derived from *Schizochytrium* sp. on the fatty acid and persistent organic pollutant levels in diets and flesh of Atlantic salmon (*Salmo salar*, L.) post-smolts. *Food Chem,* 185, 413-21.
[http://dx.doi.org/10.1016/j.foodchem.2015.03.150] [PMID: 25952887]

Steen, R, Freriks, I, Cofino, W & Brinkman, UT (1997) Large-volume injection in gas chromatography-ion trap tandem mass spectrometry for the determination of pesticides in the marine environment at the low ng/l level. *Anal Chim Acta,* 353, 153-63.
[http://dx.doi.org/10.1016/S0003-2670(97)87773-2]

Suhrhoff, TJ & Scholz-Böttcher, BM (2016) Qualitative impact of salinity, UV radiation and turbulence on leaching of organic plastic additives from four common plastics - A lab experiment. *Mar Pollut Bull,* 102, 84-94.
[http://dx.doi.org/10.1016/j.marpolbul.2015.11.054] [PMID: 26696590]

Swedmark, M, Braaten, B, Emanuelsson, E & Granmo, A (1971) Biological effects of surface active agents on marine animals. *Mar Biol,* 9, 183-201.
[http://dx.doi.org/10.1007/BF00351378]

Taylor, D, Roig, B, Gaw, S, Caldwell, D, Voulvoulis, N, Kostich, MS & Kolar, B (2015) *Pharmaceuticals in the Environment*. Royal Society of Chemistry.

Thompson, RC (2015) Microplastics in the marine environment: sources, consequences and solutions. *Marine Anthropogenic Litter*. In: Bergmann, M., Gutow, L., Klages, M., (Eds.), Springer International Publishing 185-200.
[http://dx.doi.org/10.1007/978-3-319-16510-3_7]

(2017) Environmental effects of ozone depletion and its interactions with climate change: Progress report, 2016. *Photochem Photobiol Sci,* 16, 107-45.
[http://dx.doi.org/10.1039/C7PP90001E] [PMID: 28124708]

Van Cauwenberghe, L, Devriese, L, Galgani, F, Robbens, J & Janssen, CR (2015) Microplastics in sediments: A review of techniques, occurrence and effects. *Mar Environ Res,* 111, 5-17.

[http://dx.doi.org/10.1016/j.marenvres.2015.06.007] [PMID: 26095706]

Vieno, N, Hallgren, P, Wallberg, P, Pyhälä, M, Zandaryaa, S & Commission, BMEP (2017) *Pharmaceuticals in the aquatic environment of the Baltic Sea region: a status report.* UNESCO Publishing.

Vorkamp, K, Bossi, R, Bester, K, Bollmann, UE & Boutrup, S (2014) New priority substances of the European Water Framework Directive: biocides, pesticides and brominated flame retardants in the aquatic environment of Denmark. *Sci Total Environ,* 470-471, 459-68.
[http://dx.doi.org/10.1016/j.scitotenv.2013.09.096] [PMID: 24148321]

Walker, B & Salt, D (2012) *Resilience thinking: sustaining ecosystems and people in a changing world.* Island Press, St. Louis.

Walker, CH & Livingstone, DR (2013) *Persistent pollutants in marine ecosystems.* Elsevier.

Wang, G, Xie, S-P, Huang, RX & Chen, C (2015) Robust warming pattern of global subtropical oceans and its mechanism. *J Clim,* 28, 8574-84.
[http://dx.doi.org/10.1175/JCLI-D-14-00809.1]

Wang, S-L, Xu, X-R, Sun, Y-X, Liu, J-L & Li, H-B (2013) Heavy metal pollution in coastal areas of South China: a review. *Mar Pollut Bull,* 76, 7-15.
[http://dx.doi.org/10.1016/j.marpolbul.2013.08.025] [PMID: 24084375]

Weber, K & Goerke, H (2003) Persistent organic pollutants (POPs) in antarctic fish: levels, patterns, changes. *Chemosphere,* 53, 667-78.
[http://dx.doi.org/10.1016/S0045-6535(03)00551-4] [PMID: 12962716]

Wesch, C, Bredimus, K, Paulus, M & Klein, R (2016) Towards the suitable monitoring of ingestion of microplastics by marine biota: A review. *Environ Pollut,* 218, 1200-8.
[http://dx.doi.org/10.1016/j.envpol.2016.08.076] [PMID: 27593351]

Wiberg, K, Oehme, M, Haglund, P, Karlsson, H, Olsson, M & Rappe, C (1998) Enantioselective analysis of organochlorine pesticides in herring and seal from the Swedish marine environment. *Mar Pollut Bull,* 36, 345-53.
[http://dx.doi.org/10.1016/S0025-326X(97)00187-2]

Wise, CF, Wise, JT, Wise, SS, Thompson, WD, Wise, JP, Jr & Wise, JP, Sr (2014) Chemical dispersants used in the Gulf of Mexico oil crisis are cytotoxic and genotoxic to sperm whale skin cells. *Aquat Toxicol,* 152, 335-40.
[http://dx.doi.org/10.1016/j.aquatox.2014.04.020] [PMID: 24813266]

Xu, W, Wang, X & Cai, Z (2013) Analytical chemistry of the persistent organic pollutants identified in the Stockholm Convention: A review. *Anal Chim Acta,* 790, 1-13.
[http://dx.doi.org/10.1016/j.aca.2013.04.026] [PMID: 23870403]

Zaghden, H, Kallel, M, Elleuch, B, Oudot, J, Saliot, A & Sayadi, S (2014) Evaluation of hydrocarbon pollution in marine sediments of Sfax coastal areas from the Gabes Gulf of Tunisia, Mediterranean Sea. *Environ Earth Sci,* 72, 1073-82.
[http://dx.doi.org/10.1007/s12665-013-3023-6]

Zhang, ZL, Hong, HS, Zhou, JL, Huang, J & Yu, G (2003) Fate and assessment of persistent organic pollutants in water and sediment from Minjiang River Estuary, Southeast China. *Chemosphere,* 52, 1423-30.
[http://dx.doi.org/10.1016/S0045-6535(03)00478-8] [PMID: 12867172]

CHAPTER 6

Monitoring of Organic Pollutants: PCBs in Marine Ecosystem

Rabeea Zafar[1,2], Muhammad Arshad[1,*], Kanza Naseer[1], Muhammad Arif Ali[3], Zaheer Ahmed[2] and Deedar Nabi[1]

[1] *Institute of Environmental Sciences and Engineering, School of Civil and Environmental Engineering, National University of Sciences and Technology, Sector H-12, Islamabad, 44000, Pakistan*

[2] *Department of Home and Health Sciences, Faculty of Sciences, Allama Iqbal Open University, Sector H-8, Islamabad, 44000, Pakistan*

[3] *Department of Soil Science, Faculty of Agricultural Sciences and Technology, Bahauddin Zakariya University, Multan, Pakistan*

Abstract: Marine ecosystem is rich and diverse, and plays a vital role in maintaining the natural balance of the planet. Though, the chemical revolution brought many benefits to human civilization but it also affected natural ecosystem due to chemical pollution. Unfortunately, oceans are one of environmental compartments that is at the most receiving end of the chemical pollution. There is a need to monitor chemical pollution in oceans for its normal functioning and providing a healthy habitat to marine biota. The chemical pollution of polychlorinated biphenyls (PCBs) is one of the most prominent types of organic contamination in the oceans. PCBs, comprising of 207 congeners, are considered legacy contaminants. PCBs are banned because of persistent, bioaccumulative and toxic attributes. Being hydrophobic in nature, they tend to bio-accumulate and bio-magnify, causing human health concerns that many of the sea organisms serve as food to human beings and other living organisms through food chain. Monitoring of PCBs in oceans can be done through various methods/techniques involving bio-indicators, biological monitoring, chemical monitoring, biomarkers and through isotopic analysis. The use of any single technique may not help in achieving the maximum control and monitoring of PCBs; so a use of combined approach is recommended to ensure proper monitoring of PCBs in the marine environment.

Keywords: Marine Ecosystem, PCBs Monitoring, Bioindicators, Control Measures, Isotopic analysis

---

* **Corresponding author Muhammad Arshad**: Institute of Environmental Sciences and Engineering, School of Civil and Environmental Engineering, National University of Sciences and Technology, Sector H-12, Islamabad, 44000, Pakistan; Tel: +92-51-90854309; Fax: +92-51-90854202; E-mail: marshad@iese.nust.edu.pk

De-Sheng Pei & Muhammad Junaid (Eds.)
All rights reserved-© 2019 Bentham Science Publishers

INTRODUCTION

Marine ecosystem is a complex system that is rich in biodiversity. The organisms in the marine environment are affected by variations in pH, temperature, dissolved oxygen, water circulations, and light. These changes can be natural but anthropogenic pollution load in the marine ecosystem is also contributing. The uptake of pollutants by the organisms is dependent upon the properties of pollutants such as partition coefficient, hydrophobicity, *etc*. The serious threat posed to the marine ecosystem is by pollutants, which are persistent/recalcitrant. PCBs are amongst these and are of major concern while dealing with the toxicities and disruption of living organisms functions. It is very important to continuously monitor the marine ecosystem for any toxic chemical or pollutant. The monitoring can be achieved through different mechanisms such as bioindicators, chemical monitoring, biological monitoring and through isotopes and biomarkers.

Marine pollution started when humans opened the doors of oceans for transport and from the industrialized era. Since the oceans are continually facing the disposal of wastes and pollution level increases day by day. Even if today we stop polluting oceans, persistent pollutants are already there in the ocean environment. Earth contains 70% of the water in oceans, therefore, everything ends up into oceans. Due to anthropogenic activities, hundreds of thousands of pollutants are introduced into the marine environment. Most of them are organic in nature and some are inorganics as well.

ORGANIC POLLUTANTS

Organic pollutants that have sufficiently long retention time in living organisms, pass through food chains and undergo the process of biomagnification and reach higher trophic levels. Less persistent compounds can be bioconcentrated or bioaccumulated at lower trophic levels, for example, polycyclic aromatic hydrocarbons (PAHs) (Potters, 2013). Organochlorines (dieldrin) and the PCBs, are those compounds that are present in high concentrations within the tissues of the highest trophic levels. Polychlorinated dibenzodioxin (PCDDs), polychlorinated dibenzofurans (PCDFs), and some organometallic compounds also fall in the same category (Gioia, *et al.*, 2011; Potters, 2013). The types and sources of various organic pollutants are listed in Table **1**.

Biomagnifcation along terrestrial food chains is principally due to bioaccumulation from food, the principal source of most of the pollutants. In a few instances, the major route of uptake may be from the air, from contact with contaminated surfaces, or from drinking water. Biomagnifcation along aquatic food chains may be the consequence of bioconcentration as well as bioaccumulation. Aquatic vertebrates and invertebrates can absorb pollutants from ambient water; bottom feeders can take up pollutants from sediments.

Urbanized runoff or wastewater discharge through point or non-point sources are the main cause of the presence of organochloride in water. In the ocean, organochlorine contaminants bound to marine sediments and are being continuously restored into the ecosystem by means of physical or biological disturbance (Sawyna, *et al.*, 2017). Benthic organisms are mostly exposed to organochlorine because of deposited organochlorine into sediments. Thus benthic organisms especially fish species can be a good indicator of the presence of organochlorine in marine ecosystems (Hinojosa-Garro, *et al.*, 2016). DDT, an organochlorine compound, is recognized as the pesticide, and known for its negative impacts on animals and humans. DDT is banned due to its several health and social issues such as accumulation and biomagnifcation properties in organisms (Rossi, *et al.*, 2017).

Table 1. Types and Sources of Organic Pollutants.

Compounds	Source	Available in the Marine Environment	Examples
Organometallic Compounds	Antifouling biocides for ships and fishing nets, agricultural fungicides, and rodent repellents	Bound to marine sediments and, bioaccumulated	Diethyl magnesium organolithium *e.g* butyllithium
Polycyclic Aromatic Hydrocarbons	Heating, burning, and pyrolysis of organic substances, *e.g* gas, coal, oil, garbage, wood, and *etc.*	Bioaccumulated, biomagnified	Pyrene, benzopyrene,
Polybrominated Compounds	Electronics, airplanes, motor vehicles, textiles, foams, and plastics	Bioaccumulated, biomagnified	decabromodiphenyl ethers and its isomers *etc.*
Plastics	Cosmetics, personal care products, packaging products *e.g* Plastic bags, storage containers, Rope, bottle caps, and *etc.*	Bioaccumulated, present in water and sediments	Polyethylene, polypropylene
Organochloride compounds	Industrial and wastewater	Bound to marine sediments	DDT

Organometallic compounds are being used in industry and agriculture. They are toxic for the marine environment. Their toxicity not only depends on the nature of the metallic atom, but also on the organic compounds bound to the metal. Phenylsilatrane is a very toxic example of organometallic compounds (Egorochkin, *et al.*, 2013). The toxicity of metal (mercury, vanadium, chromium, iron, rhodium, cobalt, iridium, phosphorus, boron, selenium, molybdenum, silicon, germanium, tin, tungsten, manganese and platinum) bearing compounds is determined not only by the composition and inductive resonance along with steric characteristics but also by polarizability (Egorochkin, *et al.*, 2013; Kuznetsova, *et al.*, 2015). Mercury is also widely present in marine ecosystem both from natural earth processes and anthropogenic activities (Glibert, 2017). Methyl mercury (MeHg) is an extremely toxic organometallic compound and is produced by marine microorganisms by using inorganic mercury. It has the potential to highly bioaccumulate and magnified along the food web at higher levels (Cavalheiro, *et al.*, 2016).

Polybrominated compounds are ubiquitous in environment and biota. Artificially synthesized polybrominated compounds had been used widely as flame retardants since the 1970s. Widely present polybrominated compounds are persistent in nature and they accumulate into marine biota. They have the quality of long-distance transportation. Due to this property, they can be present into the diverse and large environments (Jang, *et al.*, 2017).

Polycyclic aromatic hydrocarbons (PAHs) are included in persistent organic pollutants. PAH compounds are primarily derived from incomplete combustion *e.g* transport and industries and also through pyrolysis of organic materials and added to the environment through several processes including deposition through the atmosphere, industrial sewage, transport (marine ships), petroleum spills and terrestrial runoff. Compounds that are present in PAHs groups are mainly carcinogenic, mutagenic and teratogenic (Sun, *et al.*, 2016). PAHs are hydrophobic in nature. Due to this property, they easily and rapidly absorbed within the water either in suspended form or deposited on sediments. Through a connection with water and sediments, PAHs can easily transfer to marine organisms through bioaccumulation. These compounds affect the growth and reproduction of marine organisms and also human health by consuming seafood (Liu, *et al.*, 2017). Many studies show that coastal ecosystems are under threat due to large concentrations of PAHs (Zhang, *et al.*, 2016; Wang, *et al.*, 2012). Therefore, from last few decades, concerns about PAHs pollution have increased.

Plastics are extraordinarily useful materials due to their low cost and durability. Polyethylene, polyethylene terephthalate, polyvinyl chloride, polystyrene, and polypropylene are the famous and frequently used plastics types. Sources of marine plastic contamination include but not limited to cosmetics and personal care products, packaging products such as plastic bags, storage containers, rope, bottle caps, gear, strapping, cups, pipe, fishing nets, rope, bottles, strapping, cigarette filters and glitters. From the last 60 years, plastic production over the world is increased tremendously, nowadays known as a serious risk to the marine ecosystems. Microplastics have been now found around the world in almost every marine ecosystem (Lusher, 2015). Plastic debris size varies in marine ecosystem form macro-plastics to micro and nano-plastics. Larger particles of plastics break down into smaller particles, and thus degrade extremely slow. Microplastics and nanoplastics have the potential to accumulate and persist into organisms for very larger time periods in the marine environment (Hidalgo-ruz, *et al.*, 2012). Microplastics also act as a pathway to transfer other organic pollutants into marine organisms by adsorption onto their surface (Andrady, 2011).

RELEASE OF PCBS INTO MARINE ECOSYSTEM

Anthropogenic activities, the industrial revolution, and the increased agricultural production have introduced synthetic organic pollutants into the marine ecosystems (Dachs and Méjanelle, 2010). A small amount of these chemicals is toxic, persistent, and bioaccumulative. Recently much attention has been given to PCBs, a class of POPs. Despite being present infractions, they are of concern due to their presence widely in abiotic and biotic matrices. The release of PCBs by anthropogenic sources is divided into two stages or eras (Jones and de Voogt, 1999).

a. Industrial: 1800-1945
b. Acceleration: 1945 to present.

The organic chemical pollution started from industrial time but increased at an alarming rate after the second stage, causing an increased pressure to ecosystems and humans. Some chemicals, such as PAHs are in the environment since the beginning due to wood fire and forest fires. The importance of marine pollutants is not because of their occurrence but due to the fact that the amount of pollutants in any marine ecosystem is huge. A fraction of these pollutants is persistent; thus, they could stay in the marine environment for longer periods of time. Due to climate change, biodiversity loss, disturbance in the hydrological cycle, and socio-economic changes the marine ecosystems are being modified, which is another reason for the increased marine pollutions and enhanced pollutant loads. The main sources of these chemicals are herbicides used in agriculture, detergents, flame retardants, pharmaceutical waste, combustion by-products and other applications (Wania and Mackay, 1999). Mostly the degradation products are more polar than the original parent compound.

DISTRIBUTION AND BEHAVIOR OF PCBS

Biotic Movement

In migratory birds, the amount of PCBs is very small. The only reason for significant environmental transfer can be the movement of these migratory seabirds from one hemisphere to another and this totals the amount of PCBs transferred equally to a ton per year (Zhang, *et al.*, 2003)

Isolation and Transformation of PCBs

The composition of PCBs changes after they enter the environment. There are different mechanisms involved in changing the PCBs composition. Some of the important factors are illustrated in Fig. (**1**).

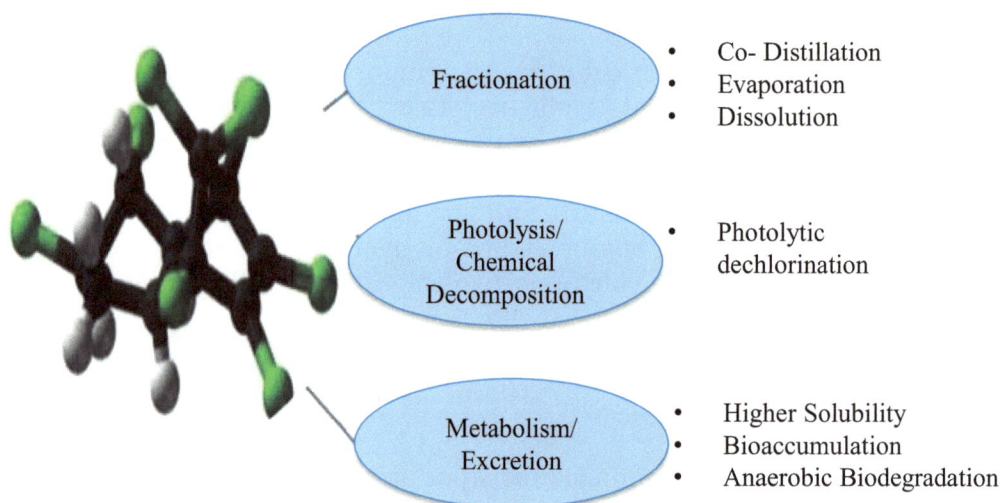

Fig. (1). Factors responsible for the change of PCBs Isomers.

Water solubility and vapor pressure of PCBs decreases with an increase in chlorine though the pattern is not constant. The movement through different environmental matrices is governed by the above-mentioned factors. They only affect the lower isomers' proportion and not the higher isomers (Paasivirta, *et al.*, 1999). Higher isomers are partially affected by photolytic breakdown. Due to photolysis, lower isomers are formed (which may not be present in parent mixture), and the process is very slow (Potters, 2013).

Research has shown that higher isomers are retained easily in rats and birds. This could be due to low solubility. The lower isomers are not retained due to differential excretion/metabolism and higher water solubility. So, we can suggest that lower isomers are decomposed easily in marine sediments. In bottom sediments, it is likely that anaerobic decomposition is the main reason for the decreased proportion of lower isomers of PCBs. PCBs with 5, 6 or fewer chlorine atoms can be decomposed easily and the biodegradability of PCB isomers depends on the position of chlorine atoms on the isomers (Gioia, *et al.*, 2011).

Bio-Magnification of PCBs

A large proportion of PCBs is reported in fish and other organisms that are at a high level in the food chain. The exact mechanism is not clear, that whether the concentrations of PCBs increasing with increasing trophic levels in a uniform pattern. However, researches show that PCB levels are magnified by the factor of 10-100 at every trophic level. These levels can be used to estimate the PCB concentrations in spatial gradients and in the same species but cannot be used as a direct measurement in the environment (Wania and Mackay, 1996; Dachs and Méjanelle, 2010).

Uptake of Organic Contaminants/PCBs by Organisms

The uptake of Organic Contaminants/PCBs involves two steps:

PHASE 1: These reactions hydrolyze/oxidize the molecule to increase water solubility.

PHASE 2: These reactions involve the combination of products of phase 1 and a substance that reduces its bioactivity and is easily and readily excreted out. But sometimes the product of Phase 1 is more toxic than the parent compound/original contaminant (Fig. **2**).

Highly chlorinated compounds have high accumulation due to the reason that they are metabolized slowly, whereas less chlorinated compounds can be readily eliminated as they are metabolized faster.

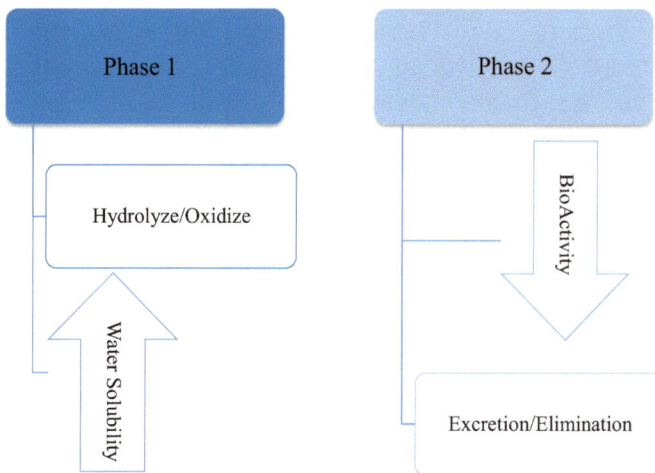

Fig. (2). Potential uptake schemes of PCBs by the organisms.

FATE OF PCBS

PCBs are well-known carcinogens, causing toxicity and being persistent. Due to their carcinogenic properties, the congress in America voted for stopping the production in 1976 and made amendments to the Toxic Substances Control Act (TSCA). Being hydrophobic in nature means they are lipophilic so, uptake by the organisms is of serious concern. The uptake by the organisms is through skin, digestive system and gills. To explain the uptake, we use the term Bioaccumulation and it refers to all the uptakes from all the sources in all the environmental compartments (Gioia, *et al.*, 2011).

Characteristics Determining the Fate of PCBs

Different characteristics of PCBs important for their environmental fate were identified as:

1. Multiphase Chemicals
2. Long-lived/Persistent Chemicals
3. Difficult to Measure
4. Higher Trophic Level Organisms are more vulnerable to PCBs

Multiphase Chemicals

PCBs have a strong tendency to partition between the organic, air, and water phases. This partitioning property helps in the rapid movement of PCBs from one environmental compartment to another. This movement is facilitated by evaporation and deposition. The important parameter for modeling the uptake by organisms from water is Kow (Octanol+Water). For PCBs, the range for Kow varies from 10^4 to 10^7 (Paasivirta, *et al.*, 1999). In the environment, the air phase is the largest by volume, then comes water and the last is the organic phase. There is a need of modeling which aims to quantify the PCB pathways from atmosphere to terrestrial surface to water and then to fish and benthic sediments.

Long-lived/Persistent Chemicals

PCBs are known for their persistent nature. The processes involved in their breakdown, (discussed in section 3.2) depend on the half-lives. The half-life of PCBs in water should not be more than 6±2 months (Ogata, *et al.*, 2009; Gioia, *et al.*, 2011). If the half-life is more, it tends to accumulate and results in long-range transport.

Difficult to Measure

Measurement of PCBs is expensive and difficult. Determination of concentrations and fluxes at high spatial and temporal resolutions is very difficult. The data received from various laboratories and from various time-periods are never comparable. A very common example is; earlier PCBs were always reported as mixtures but recently they are reported as isomers. To date, no hydrodynamic model of very high resolution for determining the fate of PCBs in the marine ecosystem has been reported (Bremner, *et al.*, 2003; Zhang, *et al.*, 2003)

Vulnerability of Higher Trophic Level Organisms towards PCBs

As already discussed in section 4.1.1, PCBs have Kow of 10^7 or more. This Kow value is an indicator of bioaccumulation and vulnerability of higher trophic level organisms. Because of their partitioning in lipid-phase and bioaccumulative property, severe detrimental effects are reported in marine organisms/mammals. Consumption of seafood contaminated with PCBs especially fish can result is severe health damages in humans (Jones and de Voogt, 1999; Potters, 2013).

PCBS POLLUTION MONITORING

Degradation of oceans and marine environment (coastal) should not have harmful effects on human life, socio-economic development and decline of biodiversity. Environmental monitoring explains processes of occurrence, distribution and fate of chemicals in different environmental compartments. It also describes the spatial and temporal trends and helps in determining and maintaining the quality of the environment.

It is achieved by using different strategies, some of which are depicted in Fig. (**3**).

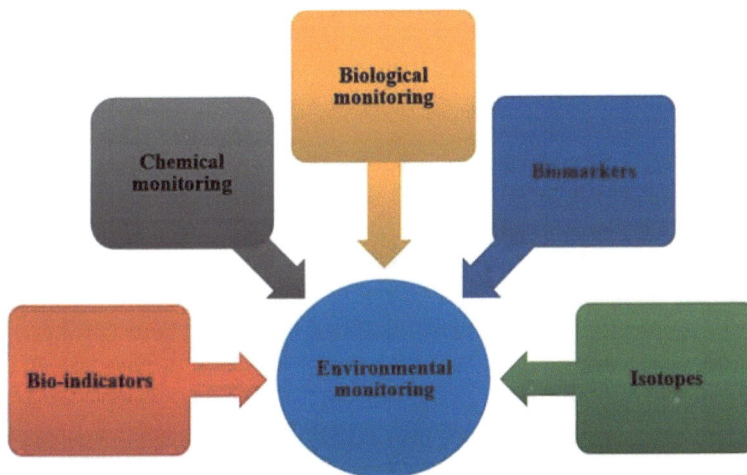

Fig. (3). PCBs Pollution Monitoring Strategies.

Bio-indicators

Bioindicators are the species that determine the overall health and quality of an ecosystem. The examples include molluscs, amphibians, helminths, algae, zooplanktons, and fish. In scientific research, photosynthetic organisms such as algae are being used as bio-detectors for detection of xenobiotics in addition to marine fauna. Micro-algae are named as "GREEN LIVER" of marine ecosystem as they are key determinants for xenobiotics. The reason for their increasing use in research is their substantial biomass and a large surface-volume ratio (Okamoto and Colepicolo, 1998).

Perfect Bio-indicator has:

  i. High tolerance of pollutants without death.
 ii. Wide spread abundance and distribution.
iii. Long lifecycle.
iv. Key importance in food chain.
 v. Easy to sample.

There are two principal rationales for using bio-indicator approach:

  i. The most important is health and safety.
 ii. The use of organisms as indicators of existing pollutants levels in their environment.

Both reasons, demand a detailed information on the physiology and the ecology of the indicator species. In any biological monitoring the most important task is to measure the changes occurred due to contaminants in a population of organisms (Mohan, *et al.*, 2016).

The two approaches used for environmental monitoring through bio-indicators are:

Critical Pathway Approach

If the marine organisms are monitored for ensuring health and safety of the people in the area under study, as they will be consuming seafood, then the main route of PCB transfer to the consumer can be identified. This approach is termed "critical pathway", and it uses only those species as bio-indicators which are consumed by people as part of their diet rather than those which readily accumulate the PCBs or organic pollutants. The key species can be identified through survey of population and their eating habits, after identifying key species sampling and analysis can be done through standard procedures (Lusher, 2015).

Bio-Indicator Approach

Second approach is using marine organisms as environmental matrices to provide information about the level of PCBs in the surrounding environment. This approach is termed as biological monitoring (discussed in detail in next section) and the organism use to quantify the bio-available fraction of the pollutant is called Bio-indicator or Sentinel Species (Hidalgo-Ruz, *et al.*, 2012). The key points to be considered in using bio-indicator technique for pollution monitoring is summarized in the Table **2**.

Table 2. Parameters for Bio-indicator Technique.

Key Points for Bio-Indicator Technique	
Key Parameters	**Advantages**

Use of Bio-indicators	• Provide time-integrated information about bio-available fraction and concentration.
	  • Tendency to bioconcentrate at low concentrations.
	  • No need of measuring pollutants directly in sea water.
	  • Provide time-integrated picture of bio-available.

Selection of species	• Sufficiently depict contaminant concentrations in the surrounding water/environment.
Contaminant/Pollutant Interactions	• Provide enough information about the uptake mechanism. • Helps in identifying the suitable bio-indicator.

Transplanted Organisms	• Helpful in areas where suitable bio-indicators are not available.
	  • Deployed through buoy or cage system.
	  • Clearly depicts the age variability of pollutants.
	  • Provide information on time-integrated uptake/loss kinetics.

Despite having limitations, many marine organisms can be used as bio-indicators to get information about pollutants (spatial and temporal) in the marine environment. The most significant information they provide is about bio-concentration factor and bio-available fractions in the marine environment which reduces the tedious and lengthy chemical analysis procedures.

Chemical Monitoring

Quantitative analysis of toxins in benthic zones and surface water is done through "chemical monitoring." The qualitative and quantitative chemical analysis of contaminants is achieved by coupling instrumental techniques such as Liquid Chromatography-Quadruple Time-Of-Flight (LC-QTOF), Gas Chromatography-Mass Spectrometry (GC-MS). The concentration monitoring of organic compounds in the abiotic environment cannot correlate to real effect on biota, thus bio-monitoring is required (Mohan, *et al.*, 2016).

PCBs are hydrophobic in nature, semi permeable membrane device (SPMD), passive samplers, biological samples and other environmental compartments have higher concentration as compared to water. These matrices are more suitable for chemical monitoring of PCBs. (Tanabe and Subramanian, 2005). Various chemical methods are available for monitoring an analysis of PCBs but the important consideration is of acceptance of method as international standards (see Table 3). Use of high-resolution mass spectrometry and isotope-labeled standards are the new methods for monitoring and analysis of PCBs makes it relatively an expensive monitoring method (Ogata, *et al.*, 2009).

Table 3. Chemical monitoring methods for PCBs.

Technique	Purpose	Advantage	Reference
Passive Sampler	Chemical potentials of OCBs in two media	Contaminants from episodic events are collected not possible through spot sampling. Gives pollutant concentrations over extended time-period with single sample	(Greenberg, *et al.*, 2014)
GC-ECD/MS	Separation and Quantification	Can determine 100 individual components at low concentrations of ng/g. Can monitor PCBs in food and tissues residues	(Greenberg, *et al.*, 2014; Binelli, *et al.*, 2009; Van Emon, *et al.*, 1998)
ELISA	Rapid assessment of PCBs	Well-suited in developing countries relying on spectrophotometric techniques but not having GC	(Muir & Sverko, 2006; Ogata, *et al.*, 2009)
HRMS	Determination of all PCBs	Congener-specific determination of non-ortho and mono-ortho substituted PCBs	EPA method (8290A, 1613 and 1668); (Ogata, *et al.*, 2009)
Time-of-Flight MS (ToF-MS	Determination of all PCBs especially non-ortho PCBs	High resolution than HRMS	(Robertson, *et al.*, 2018; Greenberg, *et al.*, 2014)

Data on concentrations of PCBs in marine environment is required to assess the quality of water and food, contamination of sea-food and to determine the toxicological effects of PCBs on marine biota and its transfer from marine to other environmental matrices. Information on loss of analytes form extraction till final analysis of PCBs is done through surrogate spiking and their recovery standards. Ideal surrogates for PCBs are isotopically labeled surrogates, which are then analyzed through LRMS and HRMS (Cavalheiro, *et al.*, 2016).

Techniques like 2D-GC, fast GC (GC-ECD) are widely used in developing countries due to their high resolution and cost-effectiveness. Refinement in monitoring and analytical for PCBs has helped in achieving better understanding of PCB sources, fate, behavior, effects and exposure. Quadrupole mass spectrometry and triple quadrupole mass spectrometry has helped in detecting low-level concentrations of PCBs at picograms with low costs. Quantification of chiral compounds through 2D-GC with high resolution ToF-MS helped in analyzing PCBs in complex mixtures (Robertson, *et al.*, 2018; Kania-Korwel and Lehmler, 2016).

Biological Monitoring

Human exposure to natural and man-made chemicals can be evaluated through biological monitoring or bio-monitoring. Sampling and analysis is based upon test organism's tissues and fluids. The most commonly used methods for marine and aquatic pollution are Acute Toxicity Assay, Chronic Toxicity Assay, Bacterial Culture Tests, Residue Analysis, *etc.* (Zhou, *et al.*, 2008).

Techniques for Biological Monitoring

Fig. (**4**). highlighted the summarized perspectives of *In situ* Bio-Monitoring in terms of community effects, bioconcentration of pollutants, and effects of pollutants. However, the specific biomonitoring techniques related PCBs and other hazardous pollutants are listed below:

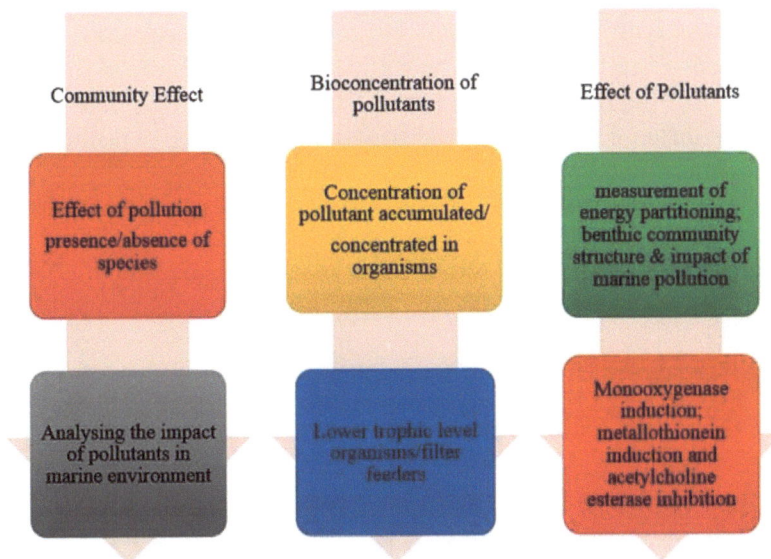

Community Effect

Effect of pollution presence/absence of species

Analysing the impact of pollutants in marine environment

Bioconcentration of pollutants

Concentration of pollutant accumulated/concentrated in organisms

Lower trophic level organisms/filter feeders

Effect of Pollutants

measurement of energy partitioning; benthic community structure & impact of marine pollution

Monooxygenase induction; metallothionein induction and acetylcholine esterase inhibition

Fig. (4). *In situ* bio-monitoring adapted from Hopkin (1993).

a. Bioaccumulation

It occurs when the absorbance of toxic substance/chemicals by the organism is at a higher rate than at which it is lost. This results in increase in chemical concentration in organisms over time. Bioaccumulation is a dynamic process resulting due to equilibrium between external exposure and uptake, storage, degradation and excretion in an organism. It is usually calculated through bio-concentration factor (BCF), higher BCF indicates that the uptake and storage is higher than the degradation/metabolism and excretion. BCF is affected by gender, age, size of organism, reproductive & nutritional status of an organism. In addition to these, temperature, salinity, pH, light, seasonal variations, presence of another chemical also play an important role in determining the bioaccumulation of PCBs (Zhou, *et al.*, 2008).

b. Biochemical Changes/Alterations

Action mechanism of organic pollutants such as PCBs can be determined by studying the interaction of pollutants with protein, enzymes, nucleic acids and other macromolecules. Selection of biomarkers should be carefully done while observing biochemical changes, keeping in view the target pollutant, test organism and area of investigation (Hinojosa-Garro *et al.* 2016; Zhou, *et al.*, 2008).

c. Morphological and Behavior Changes

PCBs direct effect on organism can be assessed through morphological and behavior changes. The techniques are standardized for the evaluation of toxicity (individual or combined) and also for the risk assessment. The information is provided on the target cells, tissues and organelles exposed to pollutants.

Acute toxicity assays can be used and the result of lethality/toxicity can refer to the selection of sub-lethal levels of pollutants. The parameter used in this is LC_{50}. For more sensitive analysis, sub-lethal toxic effects are evaluated. The parameter of EC_{50} and LOEC are used. (Zhou, *et al.*, 2008; Dachs and Méjanelle, 2010).

d. Population and Community Level Responses

To evaluate the ecological balance, with reference to water quality, population and community response is of significant importance. It may involve size density, distribution of population and/or species richness and community composition. Change in population and community level may indicate the disturbance at ecological level which requires bio-monitoring (Hidalgo-Ruz, *et al.*, 2012).

e. Modeling in Bio-Monitoring

To understand the different biological changes due to environmental pollution stress, mechanistic modes can be developed based on published data, prediction of chemical toxicity& mechanism for toxicity and metabolism of pollutants can be achieved. Modeling methods is helpful in risk assessment with low cost technique (Moore, *et al.*, 2007).

Application of Biological Monitoring

1. Assessment & evaluation of organic pollutants in marine environment

2. Evaluation of metal pollution

3. Bioremediation

4. Toxicity prediction

5. Risk and hazard assessment

6. Toxicological Mechanism

Biomarkers

Biomarkers are extended to sub-organism monitoring. It includes biochemical responses, histopathological and physiological modifications/alteration due to chemicals (Shugart, 1996). Biomarkers are used for early warning signals in human toxicology. They are specific and thus reducing the parameters to be monitored significantly. Biomarker EROD (ethoxyresorufin O-deethylase) gives information about planar PCBs, few PAHs, dioxins and other chemicals of similar structure. Thus, many chemical measurements are replaced by one single analysis/measurement. Table **4** shows different types of biomarkers and the specific purpose of its use.

Table 4. Types and Specific Use of Different Biomarkers.

Biomarkers of Marine Toxic Pollutants		
Types of Biomarkers	**Purpose of Studies**	**Reference**
Lysosomal Membrane Stability (LMS)	General Stress Bio-monitoring	(Domouhtsidou and Dimitriadis, 2001)
Neutral Red Retention Assay (NRR) of the hemocyte lysosomes	Lysosomal content efflux in the cytosol (stressed mussels); membrane damage & capacity of the cellular process to stress	(Lowe and Pipe, 1994)

(Table 4) cont.....

Biomarkers of Marine Toxic Pollutants		
Types of Biomarkers	**Purpose of Studies**	**Reference**
Acetylcholinesterase (AChE) activity (Biochemical Biomarker)	Effects of pesticides; Thermal Stress; High Critical Lethal Ambient Temperatures (irreversible cellular damage in tissues of marine bivalve)	(Domouhtsidou and Dimitriadis, 2001; Bocchetti and Regoli, 2006; Moore, *et al.*, 2007)
N-acetyl-b-hexosaminidase (Hex), acid phosphatase (AcP) and b-glucuronidase (b-Gus)	Histochemical Localization	(Raftopoulou and Dimitriadis, 2012)
Cytochrome P 4501A, DNA Integrity, Acetyl Cholinesterase (AChE) and Metallothionein	Molecular Biomarkers for Marine Pollution Monitoring; effect of metals; Genotoxic effects of PCBs	(Sarkar, *et al.*, 2006)

Biomarkers with high ecological relevance are often found at higher hierarchical levels. Biomarkers with high specificity are often found at lower levels of biological organization (Fig. **5**).

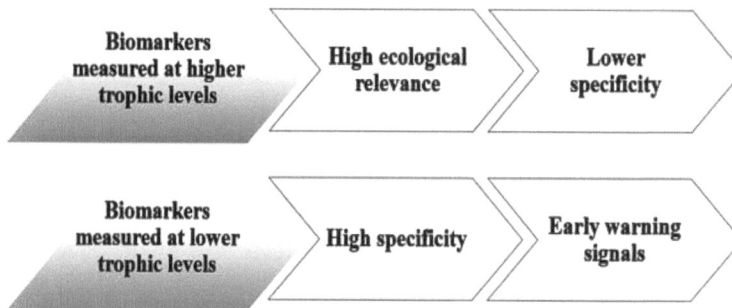

Fig. (5). Biomarkers measured at different trophic levels.

Properties of Biomarkers

Extraction of information that can be beneficial for environmental management is one of the main goals of environmental monitoring. It is very important that effects are analyzed and seen as they are caused by the chemicals. Following are the properties of biomarkers, which help in the selection of biomarkers (Cajaraville, *et al.*, 2000; Viarengo, *et al.*, 2007).

i. Specificity for chemicals

ii. Ecological relevance

iii. Early warning signal

iv. Mixture effects

v. Status/Adoption in Environmental Monitoring Programs

vi. Analysis/International Standard Protocols

vii. Temporal relevance

viii. Confounding biotic and abiotic factors

Integrated response of an organism can be monitored through multi-marker approach. In environmental pollution monitoring, cause and effect relationship can be effectively determined with the help of biomarker. Organic pollutants such as PCBs are genotoxic and damage or alter the structure of DNA. Biomarker for genotoxicity include DNA damage, chromosomal damage, DNA-DNA, DNA-Protein cross linking and presence of micronuclei. For the health status of marine environment, the sensitive biomarker is histopathological alterations in tissues (Liu, *et al.*, 2017; Viarengo, *et al.*, 2007).

Isotopes

The technique which provides diagnostic and dynamic determination of the source of contaminants, accumulation history, environmental pathways and effects on marine ecosystem using isotopes known as nuclear or isotopic technique. Identification of sources of PCBs, PAHs, hydrocarbons in waste disposal is achieved by stable C, N and H isotope ratios. Carbon 14 isotope is used for studying the fate of toxic chemicals and to understand the fate of contaminants. For tracking of biomagnification of PBTs, the stable carbon and nitrogen isotopes are used (Aubail, *et al.*, 2011; Cardona-Marek, *et al.*, 2009). This is achieved by the difference in fractionation within different trophic levels. The increase of N isotope concentration by 3-4%/trophic level is determined by increase instable nitrogen isotope ratios (15N/14N) at every trophic level (every step of the food chain). Whereas the increase in stable carbon isotope is about 1‰/trophic level providing information about carbon source in the food webs (Michener and Kaufman, 2007).

LIMITATIONS OF MONITORING TECHNIQUES

Limitations of Chemical Monitoring

Although, it is the most widely used method for monitoring of organic compounds like PCBs, but it has certain limitations:

i. It provides information about the chemicals only which are being analyzed.

ii. The cost is increased drastically if a large number of chemicals are to be analyzed.

iii. Information about unexpected or unknown chemicals is very limited due to the transformation of chemicals from one form to another.

iv. The synergistic and antagonist effects of different chemicals in a diffused mixture cannot be determined by chemical monitoring. This is due to the range of factors (salinity, temperature, diversity, age, interaction with environmental factors, *etc.*), which affect the toxicity and uptake of substances under analysis (Whitfield, 2001).

Limitation of Biological Monitoring

Following are the limitation of bio-monitoring:

i. Difficulty in the determination of the cause of alteration in nature of chemicals.

ii. Identification of cost-effective route to risk reduction.

iii. Bio-monitoring methods/techniques are often hard to be standardized, thus interpretation of results is subjective/biased.

iv. Early warning system cannot be achieved through biological monitoring, unless and until the population or community is badly affected. The effects cannot be seen and analyzed in advance (Wania and Mackay, 1996; Cajaraville, *et al.*, 2000).

CONCLUSIONS

Uncontrolled use of pesticides to increase agricultural production has not only affected water but has also caused severe harm in disturbing trophic levels and ecosystems' balance. Other industrial processes releasing organic PCBs are the major reason for depleted water quality and the decline of marine biodiversity. The monitoring of these and other toxic chemicals is very essential in order to reduce the hazardous effects on the marine ecosystem and saving the biodiversity. The monitoring of toxic substances provides information about the fate and distribution in the environment. There are a number of ways to monitor these chemicals including chemical monitoring, biological monitoring, using Isotopes, through bio-indicators and the latest through biomarkers. In future, the important characteristics of PCBs determining the fate and transport in different environmental matrices need to be researched. The emphasis should be on coordinated approaches to sampling and modeling rather than a statistical accumulation of data.

CONSENT FOR PUBLICATION

Not applicable.

CONFLICT OF INTEREST

Authors have no conflict of interest to declare for the present work.

ACKNOWLEDGEMENTS

Financial support to Ms. Rabeea Zafar for PhD studies from the Higher Education Commission of Pakistan under HEC 5000 Scheme is duly acknowledged.

REFERENCES

Andrady, AL (2011) Microplastics in the marine environment. *Mar Pollut Bull,* 62, 1596-605.
[http://dx.doi.org/10.1016/j.marpolbul.2011.05.030] [PMID: 21742351]

Aubail, A, Teilmann, J, Dietz, R, Riget, FF, Harkonen, T, Karlsson, O, Rosing-Asvid, A & Caurant, F (2011) Investigation of mercury concentrations in fur of phocid seals using stable isotopes as tracers of trophic levels and geographical regions. *Polar Biol,* 34, 1411-20.
[http://dx.doi.org/10.1007/s00300-011-0996-z]

Binelli, A, Sarkar, SK, Chatterjee, M, Riva, C, Parolini, M, Bhattacharya, Bd, Bhattacharya, AK & Satpathy, KK (2009) Congener profiles of polychlorinated biphenyls in core sediments of Sunderban mangrove wetland (N.E. India) and their ecotoxicological significance. *Environ Monit Assess,* 153, 221-34.
[http://dx.doi.org/10.1007/s10661-008-0351-1] [PMID: 18563606]

Bocchetti, R & Regoli, F (2006) Seasonal variability of oxidative biomarkers, lysosomal parameters, metallothioneins and peroxisomal enzymes in the Mediterranean mussel *Mytilus galloprovincialis* from Adriatic Sea. *Chemosphere,* 65, 913-21.
[http://dx.doi.org/10.1016/j.chemosphere.2006.03.049] [PMID: 16678235]

Bremner, J, Rogers, SI & Frid, CLJ (2003) Assessing functional diversity in marine benthic ecosystems: a comparison of approaches. *Inter-Res Mar Ecol Prog Ser,* 254, 11-25.
[http://dx.doi.org/10.3354/meps254011]

Cajaraville, MP, Bebianno, MJ, Blasco, J, Porte, C, Sarasquete, C & Viarengo, A (2000) The use of biomarkers to assess the impact of pollution in coastal environments of the Iberian Peninsula: a practical approach. *Sci Total Environ,* 247, 295-311.
[http://dx.doi.org/10.1016/S0048-9697(99)00499-4] [PMID: 10803557]

Cardona-Marek, T, Knott, KK, Meyer, BE & O'Hara, TM (2009) Mercury concentrations in Southern Beaufort Sea polar bears: variation based on stable isotopes of carbon and nitrogen. *Environ Toxicol Chem,* 28, 1416-24.
[http://dx.doi.org/10.1897/08-557.1] [PMID: 19226182]

Cavalheiro, J, Sola, C, Baldanza, J, Tessier, E, Lestremau, F, Botta, F, Preud'homme, H, Monperrus, M & Amouroux, D (2016) Assessment of background concentrations of organometallic compounds (methylmercury, ethyllead and butyl- and phenyltin) in French aquatic environments. *Water Res,* 94, 32-41.
[http://dx.doi.org/10.1016/j.watres.2016.02.010] [PMID: 26921711]

Dachs, J & Méjanelle, L (2010) Organic pollutants in coastal waters, sediments, and biota: a relevant driver

for ecosystems during the Anthropocene? *Estuaries Coasts,* 33, 1-14.
[http://dx.doi.org/10.1007/s12237-009-9255-8]

Domouhtsidou, GP & Dimitriadis, VK (2001) Lysosomal and lipid alterations in the digestive gland of mussels, *Mytilus galloprovincialis* (L.) as biomarkers of environmental stress. *Environ Pollut,* 115, 123-37.
[http://dx.doi.org/10.1016/S0269-7491(00)00233-5] [PMID: 11586767]

Egorochkin, AN, Kuznetsova, OV, Khamaletdinova, NM & Domratcheva-Lvova, LG (2013) Toxicity of organometallic compounds: Correlation analysis via substituent constants. *J Organomet Chem,* 735, 88-92.
[http://dx.doi.org/10.1016/j.jorganchem.2013.03.029]

Gioia, R, Dachs, J, Nizzetto, L, Berrojalbiz, N, Galban, C, Vento, SD, Mejanelle, L & Jones, KC (2011) Sources, Transport and Fate of Organic Pollutants in the Oceanic Environment. *Persistent Pollution – Past, Present and Future.* Springer-Verlag Berlin Heidelberg.
[http://dx.doi.org/10.1007/978-3-642-17419-3_8]

Glibert, PM (2017) Eutrophication, harmful algae and biodiversity - Challenging paradigms in a world of complex nutrient changes. *Mar Pollut Bull,* 124, 591-606.
[http://dx.doi.org/10.1016/j.marpolbul.2017.04.027] [PMID: 28434665]

Greenberg, MS, Chapman, PM, Allan, IJ, Anderson, KA, Apitz, SE, Beegan, C, Bridges, TS, Brown, SS, Cargill, JG, IV, McCulloch, MC, Menzie, CA, Shine, JP & Parkerton, TF (2014) Passive sampling methods for contaminated sediments: risk assessment and management. *Integr Environ Assess Manag,* 10, 224-36.
[http://dx.doi.org/10.1002/ieam.1511] [PMID: 24343931]

Hidalgo-Ruz, V, Gutow, L, Thompson, RC & Thiel, M (2012) Microplastics in the marine environment: a review of the methods used for identification and quantification. *Environ Sci Technol,* 46, 3060-75.
[http://dx.doi.org/10.1021/es2031505] [PMID: 22321064]

Hinojosa-Garro, D, Burgos Chan, AM & Rendón-von Osten, J (2016) Organochlorine Pesticides (OCPs) in Sediment and Fish of Two Tropical Water Bodies Under Different Land Use. *Bull Environ Contam Toxicol,* 97, 105-11.
[http://dx.doi.org/10.1007/s00128-016-1828-1] [PMID: 27209546]

Jang, M, Shim, WJ, Han, GM, Rani, M, Song, YK & Hong, SH (2017) Widespread detection of a brominated flame retardant, hexabromocyclododecane, in expanded polystyrene marine debris and microplastics from South Korea and the Asia-Pacific coastal region. *Environ Pollut,* 231, 785-94.
[http://dx.doi.org/10.1016/j.envpol.2017.08.066] [PMID: 28865384]

Jones, KC & de Voogt, P (1999) Persistent organic pollutants (POPs): state of the science. *Environ Pollut,* 100, 209-21.
[http://dx.doi.org/10.1016/S0269-7491(99)00098-6] [PMID: 15093119]

Kuznetsova, OV, Egorochkin, AN, Khamaletdinova, NM & Domratcheva-Lvova, LG (2015) Reactivity of organometallic compounds and polarizability effect. *J Organomet Chem,* 779, 73-80.
[http://dx.doi.org/10.1016/j.jorganchem.2014.12.004]

Liu, N, Li, X, Zhang, D, Liu, Q, Xiang, L, Liu, K, Yan, D & Li, Y (2017) Distribution, sources, and ecological risk assessment of polycyclic aromatic hydrocarbons in surface sediments from the Nantong Coast, China. *Mar Pollut Bull,* 114, 571-6.
[http://dx.doi.org/10.1016/j.marpolbul.2016.09.020] [PMID: 27663644]

Lowe, DM & Pipe, RK (1994) Contaminant induced lysosomal membrane damage in marine mussel digestive cells: an *in vitro* study. *Aquat Toxicol,* 30, 357-65.
[http://dx.doi.org/10.1016/0166-445X(94)00045-X]

Lusher, A (2015) Microplastics in the Marine Environment: Distribution, Interactions and Effects. *Marine Anthropogenic Litter.* Springer, Cham.
[http://dx.doi.org/10.1007/978-3-319-16510-3_10]

Michener, RH & Kaufman, L (2007) Stable isotope ratios as tracers in marine food webs: an update. *Stable Isotopes in Ecology and Environmental Science.* In: Michener, R., Lajtha, K., (Eds.), Blackwell Publishing

Ltd, Oxford, UK.
[http://dx.doi.org/10.1002/9780470691854.ch9]

Mohan, M, Jyothy, S, Cherian, N, Augustine, T, Sreedharan, K & Gopikrishna, VG (2016) Monitoring of toxic pollutants in the marine environment- a review. *Int J Mater Sci, 6,* 1-20.

Moore, MN, Viarengo, A, Donkin, P & Hawkins, AJ (2007) Autophagic and lysosomal reactions to stress in the hepatopancreas of blue mussels. *Aquat Toxicol, 84,* 80-91.
[http://dx.doi.org/10.1016/j.aquatox.2007.06.007] [PMID: 17659356]

Muir, D & Sverko, E (2006) Analytical methods for PCBs and organochlorine pesticides in environmental monitoring and surveillance: a critical appraisal. *Anal Bioanal Chem, 386,* 769-89.
[http://dx.doi.org/10.1007/s00216-006-0765-y] [PMID: 17047943]

Ogata, Y, Takada, H, Mizukawa, K, Hirai, H, Iwasa, S, Endo, S, Mato, Y, Saha, M, Okuda, K, Nakashima, A, Murakami, M, Zurcher, N, Booyatumanondo, R, Zakaria, MP, Dung, Q, Gordon, M, Miguez, C, Suzuki, S, Moore, C, Karapanagioti, HK, Weerts, S, McClurg, T, Burres, E, Smith, W, Van Velkenburg, M, Lang, JS, Lang, RC, Laursen, D, Danner, B, Stewardson, N & Thompson, RC (2009) International Pellet Watch: global monitoring of persistent organic pollutants (POPs) in coastal waters. 1. Initial phase data on PCBs, DDTs, and HCHs. *Mar Pollut Bull, 58,* 1437-46.
[http://dx.doi.org/10.1016/j.marpolbul.2009.06.014] [PMID: 19635625]

Okamoto, OK & Colepicolo, P (1998) Response of superoxide dismutase to pollutant metal stress in the marine dinoflagellate *Gonyaulax polyedra. Comp Biochem Physiol C Pharmacol Toxicol Endocrinol, 119,* 67-73.
[http://dx.doi.org/10.1016/S0742-8413(97)00192-8] [PMID: 9568375]

Paasivirta, J, Sinkkonen, S, Mikkelson, P, Rantio, T & Wania, F (1999) Estimation of vapor pressures, solubilities and Henry's law constants of selected persistent organic pollutants as functions of temperature. *Chemosphere, 39,* 811-32.
[http://dx.doi.org/10.1016/S0045-6535(99)00016-8]

Potters, G (2013) Marine Pollution, Oxford University Press: 12.

Raftopoulou, EK & Dimitriadis, VK (2012) Aspects of the digestive gland cells of the mussel *Mytilus galloprovincialis,* in relation to lysosomal enzymes, lipofuscin presence and shell size: contribution in the assessment of marine pollution biomarkers. *Mar Pollut Bull, 64,* 182-8.
[http://dx.doi.org/10.1016/j.marpolbul.2011.12.017] [PMID: 22225914]

Robertson, LW, Weber, R, Nakano, T & Johansson, N (2018) PCBs risk evaluation, environmental protection, and management: 50-year research and counting for elimination by 2028. *Environ Sci Pollut Res Int, 25,* 16269-76.
[http://dx.doi.org/10.1007/s11356-018-2467-3] [PMID: 29934860]

Rossi, M, Scarselli, M, Fasciani, I, Maggio, R & Giorgi, F (2017) Dichlorodiphenyltrichloroethane (DDT) induced extracellular vesicle formation: a potential role in organochlorine increased risk of Parkinson's disease. *Acta Neurobiol Exp (Warsz), 77,* 113-7.
[http://dx.doi.org/10.21307/ane-2017-043] [PMID: 28691715]

Sarkar, A, Ray, D, Shrivastava, AN & Sarker, S (2006) Molecular Biomarkers: their significance and application in marine pollution monitoring. *Ecotoxicology, 15,* 333-40.
[http://dx.doi.org/10.1007/s10646-006-0069-1] [PMID: 16676218]

Sawyna, JM, Spivia, WR, Radecki, K, Fraser, DA & Lowe, CG (2017) Association between chronic organochlorine exposure and immunotoxicity in the round stingray (*Urobatis halleri*). *Environ Pollut, 223,* 42-50.
[http://dx.doi.org/10.1016/j.envpol.2016.12.019] [PMID: 28153417]

Shugart, LR (1996) Molecular markers to toxic Agents. *Ecotoxicology, a hierarchical treatment. Boca Ranton.* In: Newman, M.C., Jagoe, C.H., (Eds.), CRC Press 133-61.

Sun, RX, Lin, Q, Ke, CL, Du, FY, Gu, YG, Cao, K, Luo, XJ & Mai, BX (2016) Polycyclic aromatic

hydrocarbons in surface sediments and marine organisms from the Daya Bay, South China. *Mar Pollut Bull,* 103, 325-32.
[http://dx.doi.org/10.1016/j.marpolbul.2016.01.009] [PMID: 26778499]

Tanabe, S & Subramanian, A (2005) *Bioindicators suitable for POPs monitoring in developing countries.* Kyoto University Press, Kyoto, Japan.

Van Emon, JM, Gerlach, CL & Bowman, K (1998) Bioseparation and bioanalytical techniques in environmental monitoring. *J Chromatogr B Biomed Sci Appl,* 715, 211-28.
[http://dx.doi.org/10.1016/S0378-4347(98)00261-8] [PMID: 9792512]

Viarengo, A, Lowe, D, Bolognesi, C, Fabbri, E & Koehler, A (2007) The use of biomarkers in biomonitoring: a 2-tier approach assessing the level of pollutant-induced stress syndrome in sentinel organisms. *Comp Biochem Physiol C Toxicol Pharmacol,* 146, 281-300.
[http://dx.doi.org/10.1016/j.cbpc.2007.04.011] [PMID: 17560835]

Wang, Z, Liu, Z, Yang, Y, Li, T & Liu, M (2012) Distribution of PAHs in tissues of wetland plants and the surrounding sediments in the Chongming wetland, Shanghai, China. *Chemosphere,* 89, 221-7.
[http://dx.doi.org/10.1016/j.chemosphere.2012.04.019] [PMID: 22578517]

Wania, F & Mackay, D (1996) Peer reviewed: tracking the distribution of persistent organic pollutants. *Environ Sci Technol,* 30, 390A-6A.
[http://dx.doi.org/10.1021/es962399q] [PMID: 21649427]

Wania, F & Mackay, D (1999) The evolution of mass balance models of persistent organic pollutant fate in the environment. *Environ Pollut,* 100, 223-40.
[http://dx.doi.org/10.1016/S0269-7491(99)00093-7] [PMID: 15093120]

Whitfield, J (2001) Vital signs. *Nature,* 411, 989-90.
[http://dx.doi.org/10.1038/35082694] [PMID: 11429567]

Zhang, D, Liu, J, Jiang, X, Cao, K, Yin, P & Zhang, X (2016) Distribution, sources and ecological risk assessment of PAHs in surface sediments from the Luan River Estuary, China. *Mar Pollut Bull,* 102, 223-9.
[http://dx.doi.org/10.1016/j.marpolbul.2015.10.043] [PMID: 26616744]

Zhang, ZL, Hong, HS, Zhou, JL, Huang, J & Yu, G (2003) Fate and assessment of persistent organic pollutants in water and sediment from Minjiang River Estuary, Southeast China. *Chemosphere,* 52, 1423-30.
[http://dx.doi.org/10.1016/S0045-6535(03)00478-8] [PMID: 12867172]

Zhou, Q, Zhang, J, Fu, J, Shi, J & Jiang, G (2008) Biomonitoring: an appealing tool for assessment of metal pollution in the aquatic ecosystem. *Anal Chim Acta,* 606, 135-50.
[http://dx.doi.org/10.1016/j.aca.2007.11.018] [PMID: 18082645]

CHAPTER 7

An Overview of Pollution Dynamics along the Pakistan Coast with Special Reference of Nutrient Pollution

Noor U. Saher[*, 1], Asmat S. Siddiqui[1], Nayab Kanwal[1], Altaf Hussain Narejo[1], Ayesha Gul[2], Muhammed A. Gondal[2] and Fakhar I. Abbass[3]

[1] *Centre of Excellence in Marine Biology, University of Karachi, Karachi, Pakistan*

[2] *Department of Biosciences, COMSATS Institute of Information Technology, Park Road, Islamabad, Pakistan*

[3] *Bioresource Research Centre, Bazar road, Islamabad, Pakistan*

Abstract: Pollution in coastal waters is quickly becoming a conspicuous problem throughout the world and the coastal areas of Pakistan are also included in severely affected and therefore no exception. Anthropogenic activities are generally accountable for the deprivation of the marine environment along with their resources across the ocean bodies. The oceans economy not only offers significant development opportunities but also raise some challenges. Not only marine sources, the land-based sources are the prominent contributor of pollution as add in the pollution through direct and indirect wastes discharge as well as effluents in the adjacent coastal waters from untreated domestic and industrial sources. In this chapter, the magnitude of pollution (organic and inorganic) in coastal environments of Pakistan was discussed including plastic pollution as in recent days, it's a hot issue and a detailed topic itself. The weathering material, river runoff, industrial and domestic waste water enter through different channels and take part in coastal pollution. Most of the pollutants like pesticides, herbicides, heavy metals and macro-nutrients, presented intensification in a marine environment. Nutrient dynamics and their cycling influence the process of eutrophication in the adjacent coastal waters and an enrichment of macro-nutrients in coastal waters reveals an increment in the explosion frequency of harmful algal blooms were reported. The animal manure, sewage treatment, runoff of fertilizers, storm water runoff, plant discharges, and power plant emissions, and failing septic tanks are the primary sources of nutrient pollution. The algal blooms are responsible to produce algal toxins or red-tide toxins and these naturally-derived toxins harm the organisms, including humans. These bloom toxins initially contaminated the fish or seafood species, then responsible for significant loss of fish and shellfish species and ultimately economy damage.

[*] **Corresponding author Noor U. Saher:** Centre of Excellence in Marine Biology, University of Karachi, Karachi, Pakistan; Tel: +92-21-9261397, +92-21-9261551; Fax: +92-21-9261398; E-mail: noorusaher@gmail.com

De-Sheng Pei & Muhammad Junaid (Eds.)
All rights reserved-© 2019 Bentham Science Publishers

Keywords: Marine Environment, Coastal Pollution, Nutrient Dynamics, Heavy metal Contamination, Pollution Impacts.

INTRODUCTION

Coastal and estuarine ecosystems always remain sturdily subjective by the activities of mankind through pollution and habitat loss throughout the world. The environmental degradation, climate change, over-exploitation, pollution, poverty and lack of basic (health, water as well as education) facilities are the conspicuous issues for the coastal areas including the associated population. Pollution, now become one of the most significant challenges to the health of coastal ecology and systems. The pollution sources mainly include the land affected by agricultural or industrial activities, livestock or domestic waste discharge and also from coastal waters by aquaculture as well as other anthropogenic activities. The direct and untreated discharge of industrial and agricultural effluents and domestic sewage are the main contributor of pollution for the 990 km long coastline of Pakistan. The impacts of coastal pollution appeared as a consequence of various environmental issues mainly includes; the enrichment of organic matter leading to eutrophication, pollution through chemicals (metals and oil), sea level rise due to the global climate change and sedimentation as a result of land-based activities. According to preliminary estimation, the fisheries and allied resources are the primary livelihood for 80% of the coastal population of Pakistan as fishery-related exports acquiesce per year on average sum of PKR 8.8 billion (US$ 838 million) for the country. but this trade benefits are significantly dependent on the sustainable utilization of these marine resources. Over 75% of all marine pollution originates from land-based sources, which are primarily industrial, agricultural and urban. Point and non-point source pollutions continue globally, resulting in the steady degradation of coastal and marine ecosystems. There are various means of pollution incorporated through various human activities, including offshore oil and gas production and marine oil transportation. Other contaminants produced either naturally or anthropogenically ultimately flow into marine waters. Pharmaceuticals are also an important pollution source, mostly due to overproduction and incorrect disposal. Ship breaking and recycling industries (SBRIs) also releases various pollutants and substantially deteriorates habitats and marine biodiversity of adjacent coastal areas of Sindh and Balochistan coast.

AN OVERVIEW OF POLLUTION AND POLLUTANTS IN MARINE ECOSYSTEM OF PAKISTAN

In Pakistan, marine pollution is primarily restricted to the areas, which receive waste from the industrial, municipal, agriculture and oil spill sources. The coast of Pakistan faces semi-diurnal tides, therefore, washed twice a day and taking away the pollutants, however inside the harbours or creeks; the pollutants are oscillating for several days until they dispersed, washed or settle down at the bottom (Rizvi *et al.*, 1988; Sayied, 2007; Saher and Siddiqui, 2016). The 800 Km coastline of Balochistan is almost free from marine pollution from land-based activities as it is sparingly populated. Sonmiani Bay is one of the most populated city along the Balochistan coast, located about 90 km away from Karachi (Saifullah and Rasool, 1995; Gondal *et al.*, 2012; Saher and Siddiqui, 2016). The sources of fresh water are the seasonal runoff of the Porali and Windor Rivers (Rasool *et al.*, 2002). These rivers receive effluents of around 122 industries, functioning at the Hub and Windor Industrial Trading Estate, which include textile weaving, plastic, chemical, food preservation, engineering, paper and paper product industries, *etc.*, and mainly contribute to coastal contamination (LGB, 2008; Saleem *et al.*, 2013; Saher and Siddiqui, 2016). The close adjoining area is from industrial sites of Karachi city and the few locations at the Hub industrial areas of Balochistan are the major waste receiving areas. Karachi is the most urbanized and industrialized city along the coast of Pakistan that has about 167 km long shoreline along the Sindh coast. By the virtue of the biggest city of Pakistan, it has the highest risk towards the environmental pollution, which insert from the diverse point and non-point sources (Rizvi *et al.*, 1988; Saher and Siddiqui, 2016). Solid waste discharged into the marine environment is also a conspicuous serious threat to marine life. A substantially large quantity of solid waste from the coastal towns enters the sea on a regular basis, which is accumulated on beaches as well as in the shallow coastal waters, making the coastal area polluted (MFF, 2016). It is estimated that Karachi produces approximately 8000 tons of solid waste per day and a substantial portion, *i.e.* around 60% of uncollected solid waste mixes up with wastewater and enter in the sea at Karachi Harbour. It along with the ships, jetties, and inlets, thus compound the existing problem. The entered amount of solid wastes spreads in the harbour and accumulates in different points of the harbour and is expected to substantially increase with the rapid growth of population and economic activity. According to an estimation by the year 2020, solid waste generation in Karachi may come up to 16,000 to 18,000 tons each day and therefore, there is a need to improve present solid waste management practices and make them more effective and modernized according to acquired demand (MFF, 2016).

There are four main (Karachi harbour, located on the Lyari River, Port Qasim on the Indus deltaic region, Gizri creek near the Malir River and a Cape Monze area a side of Hub River) coastal areas of Karachi which continuously receive the land-based pollution. More than six thousand functional industries present in six different industrial estates and their untreated effluent along with 300 MGD municipal wastewater continuously discharged into coastal waters of Karachi (WWF, 2002; Mashiatullah *et al.*, 2016). Two rivers (Malir and Lyari) are the seasonal rivers and flow during southwest monsoon and main point sources of coastal contamination. Both rivers act as an exposed sewage channel that loaded with highly polluted industrial and domestic wastewater. The Lyari river annually releases the 90 tons of phosphate compounds, 160,000 tons organic matter, 800 tons of nitrogenous compounds, 130,000 tons of solid nitrogen, and 12,000 tons of suspended solids into the Manora Channel (JICA, 2007; Mashiatullah *et al.*, 2016). It is estimated that more than 430 (MGD) of untreated sewage dumped into the adjacent coastal environment of Karachi through the tributaries of Malir and Lyari rivers. These rivers are polluted through organic and inorganic pollutants which come from diverse sources such as agricultural runoff, industrial effluents, urban and domestic waste that ultimately drain into the Arabian Sea through the beaches of Karachi (Rizvi *et al.*, 1988; Sayied, 2007; Siddique *et al.*, 2009; Saher and Siddiqui, 2016).

According to an estimation >1,500 tons/day of Biochemical Oxygen Demand (BOD) is added into the Karachi coastal waters through associated industries. In addition, along with inorganic pollutant the sewage generated quantity has increased up to 350 MGD during the previous years. The combined treatment capacity of the treatment plants is approximately less than 50% of the total sewage generated (Sayied, 2007). The adjacent coastal waters and various creeks in Karachi show evidence of eutrophication due to elevated levels of organic pollution. Due to the discarding of a huge quantity of untreated effluents and sewage, the few habitats mainly Manora Channel and Gizri creek has been near to completely destroy as mostly areas of these water bodies are now devoid of any marine fauna (Fig. **1**).

In addition, a major portion of the creek system near to Port Qasim is also observed as heavily polluted due to receive of heavy metal discharge from the steel mill and tanneries, thermal discharges from power plants, and organic discharge of the cattle colony (MFF, 2016).

The coastal (Fig. **2**) development or the unscientific infrastructural development along the shoreline also considered a threat towards the coastal environments. Karachi is the largest industrial and commercial estate situated near the proximity of the Indus delta along the coastal belt of Pakistan and continues to expand rapidly. However, rapid establishment of new housing sectors, industrial estates and construction activities contribute to waste generation.

A navigational channel of about 45 km in length is the connection between the Phitti creek connects Port Bin Qasim to the open Arabian Sea. The widening and deepening of this channel also a cause of severe coastal erosion problem and therefore resulted in the substantial destruction of barrier islands which used to act as the nature's first line of defense against coastal erosion (Muzaffar *et al.*, 2017). Indications are that erosional forces continue to alter the hydraulic regime in the area even though the remedial measures to stabilize coastal sediment erosion through the plantation of mangrove have been taken. Prior to developmental activities in this area, all coastal processes were naturally controlled (Muzaffar *et al.*, 2017). For instant economic profit, the developers also use sand as excavated from the adjacent beach as infill material as lack of awareness that beach sand provides protection from high energy waves (Muzaffar *et al.*, 2017).

Fig. (1). The solid waste material distribution along the Korangi Gizri creek area and in adjacent Korangi Harbour waters.

Fig. (2). Some ongoing construction activities along the coastal areas of Sindh.

The coastal erosion over the centuries is a result of natural processes and sea level change. In recent times, the rate of erosion appears to have increased at certain specific points along the coast. The coastal erosion hotspot areas in Balochistan coast include Jiwani, ShadiKor, and Damb, where intensity is very severe. In the Sindh coast, severe erosion intensity has estimated at various points of Karachi (Phitti and Gizri creeks, Hawksbay, Clifton, DHA phase-8 and RasMurai), Thatta (KharoChann, KetiBundar, Mirpursakro, Ghorabari, Gharo creek), Sujjawal (Jati and Shah Bunder) and Badin. These areas are display severe to very severe erosion issues resulting either by natural processes or human activities, which is mainly attributed to reduced freshwater flows and seawater intrusion (MFF, 2016).

Pollution Hotspots along with Pollution Types

The biggest port of Pakistan, Port Bin Qasim is located at the South of Karachi in the Indus Delta Region. The port related activities and associated industries are the main contributors of pollution in the deltaic region. The shipping activities at the Port Qasim and contaminated water from Power Generation Plants and Pakistan Steel Mill are the major sources of pollution. However, the industrial effluent, domestic sewage, wastes from the Landhi cattle colony and the Korangi fish Harbour. Hub Industrial Trading Estate (HITE) and the HUBCO power plant are also the major sources of marine pollution in the vicinity of Malir River and Gizri creek area. The Industries operating in HITE and the HUBCO power plant continuously discharged effluent into the Hub River that ultimately dumped into the Arabian Sea (Sayied, 2007).

Radioactive Waste

Nuclear Power Plant (KANUPP) is the causative source of artificial radioactivity along the Karachi coast. The periodic monitoring of radioactivity also done through survey of the adjacent vegetation and soil material and shown that radioactivity has not increased in the water environment. Another larger, nuclear power plant is being built upcountry near Chashma in the district Mianwali. Through this source some radioactivity may reach the coastal waters of Pakistan *via* the river Indus, but the level is expected to be insignificant.

Oil Pollution, Shipping, and Accidents

The oil pollution can be evidenced through the tar-balls on different beaches along the coast. Oil spills can be caused by release of crude oil from drilling rigs and wells, offshore platforms, oil tankers, and spills of refined petroleum products and their by-products. The intertidal areas are also blackened with oil in the vicinity of the oil refineries along Korangi and Gizri creeks. The Indus delta is located at a safe distance from another source of oil pollution: mechanized fishing boats as relatively small in number (Akhter, 1995). Annually more than 200 oil tankers and 2,500 ships visit the Karachi Harbour through the Manora channel. There is large scale shipping traffic at Port Qasim. The bilges, washings from engine rooms of vessels, discharges, leaks and small spills occurring during loading and unloading at oil piers are the sources of oil pollution in Manora channel. Although a great deal of oil pollution is present in the Manora channel (flushed daily by the tide) virtually no trace can be found along the beaches in the vicinity of the channel. This may be due to the large-scale dilution because of the direction of the current as occurs along the watercourse when the effluent reaches to the sea. Oily waste from city-based sources, including service stations, also ends up in the Harbour area (MFF, 2016).

No information is available for the southeast coast. The sandy and rocky intertidal zones at Gadani are smeared with oil, which flow from the oil tanks of vessels scrapped at the ship-breaking yard. The country's largest ship breaking industry is located at Gadani only 50 km northwest of the Karachi. The ships are brought for scrap since last 50 years and causing all sorts of organic and inorganic pollution along the coast. A gradual building of heavy metals has been reported in the sediments from the Gadani coastal area. Light refined oil is known to be more toxic to adult fish than crude and heavy fuel oil. It is generally believed that oil spilled at sea, in small or large quantities, does not immediately affect stocks of fish and shellfish, but oil pollution may be gradual and chronic, resulting in the long-term reduction of fish production and can be hazardous for inland and coastal water fauna. The new naval Harbour at Ormara and the upcoming Harbour in Gwadar would contribute towards pollution if strict measures are not taken to prevent pollution (Sayied, 2007). The recent oil spillage from the tanker Tasman Spirit, which went aground on 30 July 2003 as spilled approximately 35,000 tons of oil into the Arabian Sea, attracted a great deal of criticism from medical circles concerning the effects, livelihood of 90,000 registered fishermen of Sindh were affected as the oil slick in the fishing zone led to a sharp decline in the sale of seafood in the city markets. Prices of different fish species came down by 60-70% and the both short-term and long-term on the health of people, living or working in the vicinity of the affected area (Sayied, 2007, Khurshid *et al.*, 2008).

Thermal Pollution

Hot water discharges from power plants such as the Karachi Nuclear Power Plant (KANUPP), and plants owned by the Steel Mills and the Karachi Electric Supply Corporation affect the fauna and flora immediately around the outfall; but since the emission is relatively small compared to the receiving environment, the damage is localized (Akhter, 1995). In addition, a major part of the creek system adjacent to Port Qasim is also heavily polluted because of thermal discharges from power plants, heavy metal discharge from the Steel Mill and tanneries, and organic discharge from the cattle colony (MFF, 2016).

Marine Debris

Marine debris along the beach fronts is another foremost constituent of marine pollution and major portion of debris enters from the adjacent land, inland water channel and through beach visitors. According to the United Nations Joint Group of Experts on the Scientific aspects of Marine Pollution (GESAMP, 1991); land-based sources accounted for about 80% of the worldwide marine pollution. However, significant quantities of plastic trash are buoyant, resilient and slow to degrade (Williams and Simmons, 1996). Apart from municipal sewage and industrial wastes, the Lyari River also brings in a heavy load of floating garbage. It is estimated that the 50% pollution of the Karachi Harbour is brought in by the Lyari River. Ali and Shams (2015) quantified and characterized the debris along the most popular recreational sandy beach front beach Clifton (Karachi). The study reveals the high abundance and composition of various man-induced debris (Fig. **3**).

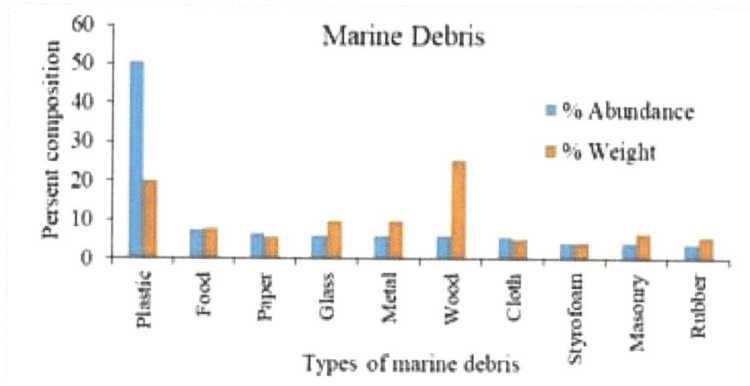

Fig. (3). Mean composition of debris km^{-1} of beach front (Clifton) Karachi (modified from Ali and Shams, 2015).

MAJOR POLLUTANTS

Pakistan has been struggling to develop and improve its industrial as well as agricultural sectors and the rapid urbanization and indiscriminate industrialization pose an impact and numerous environmental issues have been originating related to the ecological integrities. About 80% of the industrial growth is restricted to major cities of the country like Hyderabad, Multan, Faisalabad, Karachi, Lahore, Sialkot, Gujranwala, Rawalpindi, Peshawar, and Kasur (Aftab *et al.*, 2000; Mehmood *et al.*, 2017).

Persistent Organic Pollutants/Polycyclic Aromatic Hydrocarbons/ Pesticides

Thousands of persistent organic pollutants (POP) chemicals remain migrating to the environment for certain families of chemicals (for example, about 209 different polychlorinated biphenyls with a wide range of difference in chlorination and substitution position) and having long half-lives in the air, soil, sediments, and biota. POPs are lipophilic and hydrophobic chemicals and have the capability to be stored in fatty tissues. These POPs persistence in the biota and can easily accumulate them in the food chain (Jones and de-Voogt, 1999; Mehmood *et al.*, 2017). Karachi is one of the highest emerging sites, for POPs along the coastal zone followed by Thatta in the Deltic zone (Fig. **4**), which pose a serious threat to the marine environment (Mehmood *et al.*, 2017).

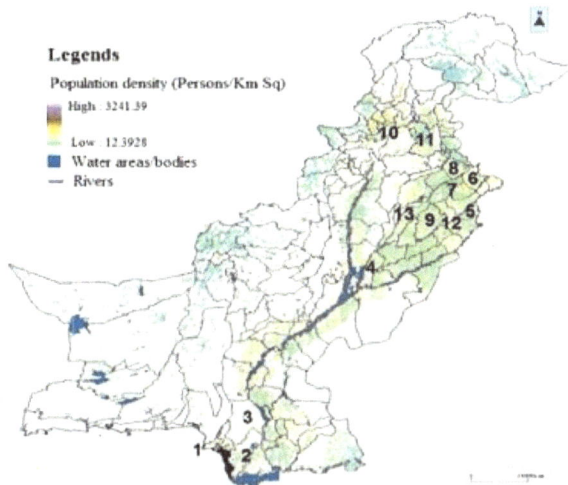

Fig. (4). Map of Pakistan showing density of population and locations of studies reported for newly emerging POPs (1: Karachi, 2: Thatta, 3: Haiderabad, 4: Multan, 5: Lahore, 6: Sialkot, 7: Gujranwala, 8: Gujrat, 9: Faisalabad, 10: Nowshehra, 11: Islamabad, 12: River Ravi, and 13: River Chenab) (modify from Mehmood *et al.*, 2017).

Polycyclic aromatic hydrocarbons (PAHs) represent another group of conspicuous persistent organic pollutants (more than 100 congeners) renowned due to mutagenic, carcinogenic and immunotoxic properties (Din *et al.*, 2013; Callen *et al.*, 2011; Jang *et al.*, 2013; Hamid *et al.*, 2015). The occurrence of PAHs in various environments is recognized as through natural (volcanoes and forest fires) and anthropogenic sources (coal and coke incineration, purifying and emission of petroleum and its related products, exhaust of motor vehicles, *etc.*) (Ravindera *et al.*, 2007; Zhang and Tao, 2009; Hamid *et al.*, 2015). The thousands of glaciers of the Himalayas are also known as the third Pole in the world and also act as a condenser for PAHs similarly in the Arctic region (Kang *et al.*, 2009; Hamid *et al.*, 2015).

In Pakistan, the chemical pesticides are also used specifically the use of agro-chemical pesticides started in 1954 with 254 metric tons (MT) however, the formulations and its consumption increased over 7000 tons/year. Pesticide consumption has been increasing from 16,226 MT in 1976–77 and increase each year, reached to a maximum of 20,648 MT in 1986–87 (Baloch, 1985) and further increased up to 78,132 tons/annum (Syed and Malik, 2011). The use of pesticide has been increased 100 times in Pakistan during 1980–2002 (Khan *et al.*, 2002; Mehmood *et al.*, 2017). Insecticides (74%) are the commonly used pesticides in the country followed by 14% of herbicides, 9% of fungicides, 2% and 1% of acaricides and fumigants respectively. In Pakistan about 69% of total pesticides used for cotton crop and the rest are used for wheat, maize, rice, *etc.* (Economic Survey of Pakistan, 2005–2006). Presently, about 108 types of insecticides, 30 types of fungicides, 39 types of weedicides, 5 types of acaricides and 6 different types of rodenticides, are commonly used pesticides in Pakistan (PPSGDP, 2002). The use of pesticide has almost increased by 1169% in the country and the number of sprays has reached 10/crop that can pose the drastic effect to human health (Mehmood *et al.*, 2017).

In 2003, Pakistan imported 78,133 metric tons (MT) of pesticides belong to organochlorine, organophosphate, and pyrethroid groups, while in the same year sales of Endosulfan were 117.98 MT in Pakistan (Khooharo *et al.*, 2008). Despite the importance, there only few studies are reported regarding the chlorinated pesticides in the coastal environment of Pakistan (Saleem *et al.*, 2013). In Fig. (**5**), the analysis of 16 organochlorine pesticides (OCP's) was presented in coastal waters of Pakistan (Saleem *et al.*, 2016). Traces of these pesticides were also reported in human foods and marine biota of Karachi coast (Munshi *et. al.* 2004; Sanpra *et al.*, 2003). Khan *et al.*, (2010) reported the environmental safe limits of organochlorine pesticides (OCPs) in Karachi Harbour and adjoining areas as compared to regional countries. The impact and toxicity of pesticides on marine organisms in Pakistan has been studied and well documented (Naqvi *et al.*, 2017; Shoaib *et al.*, 2013; Shoaib *et al.*, 2012; Shoaib and Siddiqui, 2015).

Heavy Metals

Heavy metals are toxic, persistence and non-biodegradable in nature, therefore, considered as potentially hazardous materials in marine environments (Ali *et al.*, 2014; Saher and Siddiqui, 2016; Çoğun *et al.*, 2017). They originate from natural (weathering of parent rock) as well as anthropogenic (mining, agricultural activities, industrial applications, and atmospheric deposition) sources into coastal environments (Bryan and Langston, 1992; Callender, 2005; Saher and Siddiqui, 2016). Few metals (for instance Se, Fe, Cu, Cr, Zn, Co, Mo, V, Mn, and Ni) play a role as essential micronutrients for organisms in several metabolic processes but become toxic, if in excess level. On the other hand, some metals (Hg, Cd, Ag, and Pb) are considered as nonessential and can cause toxicity in small dosage (Bryan, 1979; Eisler, 2010; Saher and Siddiqui, 2016; Duarte *et al.*, 2017). Due to their nature, they easily accumulate in organisms and biomagnified in the food web and possess the severe threats to aquatic fauna (Ahearn *et al.*, 2004; Rainbow, 2007; Vilhena *et al.*, 2012; Duarte *et al.*, 2017; Kanwal and Saher 2018).

Organochlorine Pesticides in coastal
water

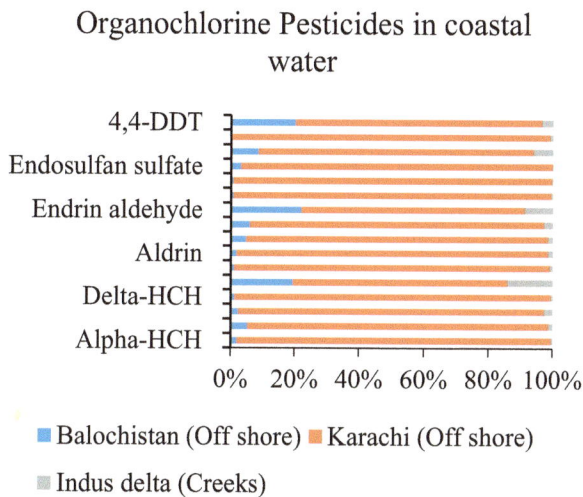

Fig. (5). Percent composition of organochlorine pesticides from coastal waters of Pakistan (modified from Saleem *et al.*, 2016).

Heavy metals could be circulated through blood within the body and reside or accumulate in the organs of consumer and then appear in toxic forms, both for the organisms and humans that consume contaminated seafood (Wang *et al.*, 2005; Çoğun *et al.*, 2017). The concentration levels of heavy metals in water, sediment and biota are studied from a coastal environment in Pakistan (Table **1**). Heavy metals concentrations in sediments of Pakistan extensively evaluated in recent decades as the sediment considered as the primary indicator for pollution monitoring (Saifullah *et al.*, 2002; Qari *et al.*, 2005; Siddique *et al.*, 2009; Mashiatullah *et al.*, 2013; Chaudhary *et al.*, 2013a, b; Ali *et al.*, 2014; Saher and Siddiqui, 2016).

Table 1. The literature review on the study of heavy metals and Heavy metal concentrations in water, sediments and fauna along the coastal areas of Pakistan.

	Co	Ag	Cd	Cr	Cu	Fe	Mn	Ni	Pb	Zn	Hg	Mg	Ref.
COASTAL WATER	-	15	4	29	38	93	161	19	109	1215	87	-	[1]
	-	-	0.0006	-	0.0014	-	-	-	0.0007	0.130	-	-	[2]
MARINE SEDIMENTS													[1]
	-	0.53	0.44	32.3	24.4	23, 259	238	76.4	7.9	65.3	0.95	-	
	0.07	-	0.01	0.03	0.01	6.75	5.19	0.03	0.07	0.06	-	10.95	[3]
	22.18	-	2.27	83.27	30.93	1.85%	39.88	27.91	24.80	65.91	-	-	[4]
	-	-	-	96.75	-	3.07	500	31.39	23.24	204.79	-	-	[5]
	1.1	-	0.4	171	64.2	-	-	34	45	68	-	-	[6]
	-	-	1.43	-	52.23	-	-	-	59.99	111.87	-	-	[2]
Monitoring Year I (MY-I)	9.225	-	0.958	31.66	57.43	865.6	-	34.50	40.38	100.67	-	-	[7]
Monitoring Year II (MY-II)	8.51	-	1.20	107.2	82.6	978.13	-	46.23	53.40	100.7	-	-	[7]
SEAWEED													
Cladophora	-	0.76	2.94	4.58	4.73	592	3.30	6.13	5.70	12.36	0.24	-	[1]
Dictyota sp.	-	0.40	0.32	7.93	0.99	124	7.15	19.24	1.87	2.87	0.21	-	[1]
Lyngeriastellata	-	0.45	0.32	7.93	0.99	124	7.15	19.01	0.15	2.61	0.21	-	[1]
Padinapavonia	-	0.42	0.76	9.12	1.70	373	10.03	19.16	0.44	4.10	0.23	-	[1]
MANGROVES													
Pneumatophores	-	-	1.04	-	11.78	-	-	-	14.75	90.40	-	-	[2]
Bark	-	-	0.70	-	15.37	-	-	-	9.956	80.08	-	-	[2]
Leaves	-	-	0.46	-	5.36	-	-	-	0.681	61.28	-	-	[2]
Flowers	-	-	0.40	-	1.87	-	-	-	0.178	18.08	-	-	[2]
Fruits	-	-	0.38	-	1.80	-	-	-	0.433	37.76	-	-	[2]
Twigs	-	-	0.24	-	4.89	-	-	-	0.59	39.54	-	-	[2]
FISH													
Chactadonjayakeri	-	0.29	0.35	5.10	0.89	937	4.87	12.12	0.59	4.99	0.09	-	[1]
Rastrelligerkanagurta	-	0.53	0.36	8.51	1.56	1791	7.68	18.28	0.14	19.83	0.16	-	[1]
Pornadysismaculaturn	-	0.30	0.26	5.20	0.83	889	6.65	12.09	11.63	7.22	0.15	-	[1]
Oreochromismossambicus	-	-		25.4–33.7		6.8–6.94			2.87–3.6	27.42–66.2	2.24–2.69		[8]
Cyprinuscarpio	-	-		1.12–4.82		1.65–2.09			1.08–1.52	36.6–39.7	4.98–8.72		[9]
Megalaspiscordyla													[10]
Liver	-	-	1.86	1.90	-	439.17	8.92	-	1.62	-	-	-	
Kidney	-	-	1.81	1.90	-	31.51	3.31	-	1.72	-	-	-	
Gills	-	-	1.84	1.33	-	31.12	4.40	-	1.71	-	-	-	
Muscle	-	-	0.45	0.32	-	41.23	2.49	-	0.41	-	-	-	
Sardinellaalbella													[11]
Liver	-	-	1.835	-	14.74	433.1	9.77	-	2.08	51.78	-	-	
Gills	-	-	1.897	-	2.97	41.48	3.42	-	1.62	14.86	-	-	
Muscles	-	-	0.81	-	1.97	6.16	1.77	-	0.48	2.5	-	-	
CRAB													
Macrophthalmusdepressus	11.02	-	1.76	14.36	85.26	654.4	-	27.21	68.8	88.2	-	-	[12]

(Table 1) cont.....

	Co	Ag	Cd	Cr	Cu	Fe	Mn	Ni	Pb	Zn	Hg	Mg	Ref.
Ilyoplax frater	14.54	-	2.22	10.8	122.06	373.8	-	26.93	72.7	94.91	-	-	[13]
Opusiaindica	22.1	-	1.94	9.2	69.1	409.8	-	19.91	25.56	78.07	-	-	[14]
Austrucasindensis	6.77	-	1.54	7.3	58.2	33.1	-	25.92	94.13	39.3	-	-	[15]
SHRIMP													
Metapeneaus monocerous		0.29	0.47	5.90	1.20	806	6.42	12.64	1.71	4.21	0.15	-	[1]
Penaeus japonicus		0.25	0.47	0.18	4.55	837	3.14	13.10	1.60	7.11	0.13	-	[1]

Reference: [1] Tariq *et al.*, (1993); [2] Ismail *et al.*, (2014); [3] Qari *et al.*, (2005); [4] Siddique *et al.*, (2009); [5] Mashiatullah *et al.*, (2013); [6] Ali *et al.*, (2014); [7] Saher and Siddiqui, 2016; [8] Jabeen and Chaudhry (2009); [9] Jabeen and Chaudhry (2010); [10] Ahmed *et al.*, (2014a); [11] Ahmed *et al.*, (2014b); [12] Saher and Siddiqui (2019); [13] Siddiqui and Saher (2019); [14] Siddiqui and Saher (2016); [15] Saher and Siddiqui (2017)

The severity level of metal toxicity in organisms, in both cases (essential and nonessential), primarily depends on the concentrations, persistence, mobility and bioavailability of metals in water or/and sediments (Ahearn *et al.*, 2004; Luoma and Rainbow, 2008; Vilhena *et al.*, 2012; Duarte *et al.*, 2017). During the past decades, the severity of heavy metal contamination in the marine environment of Pakistan has been intensified mainly due to anthropogenic activities and lack of management towards the dumping of contaminants in river bank and coastal areas (Fig. **6**).

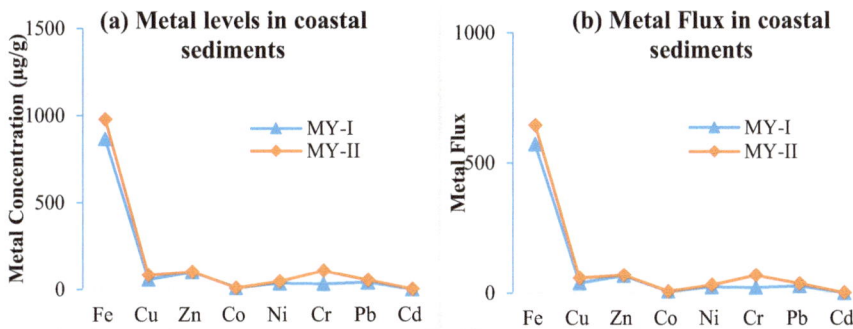

Fig. (6). Comparison of heavy metals pollution in coastal sediments between the two monitoring years by single metal pollution indices.

Heavy metal pollution in coastal sediment evaluated by means of various multiple pollution indices Geo-accumulation index (I*geo*), contamination factor (CF), the enrichment factor (EF), and potential ecological risk factor (ER) each based on single metal to illuminate the pollution elevation or relegation during two monitoring years. The Geo - accumulation index showed negative or less than zero Igeo values of Zn, Co, Fe, Cu, Cr, and Ni specify overall an unpolluted condition. However, few sites presented specific contamination, such as moderate pollution of Zn and Cu evaluated in the Sandspit mangrove area during both years (Fig. 7). The enrichment factor indicates the extremely high enrichment (>50) for Cu, Pb, Zn, and Cd during both monitoring years, suggesting potential threats by these metals to the marine environment and biodiversity. The contamination factor showed the low contamination of Fe, Co and Ni perceived during the last decade. The ecological risk factor estimated to scrutinize the heavy metal eco-toxicology and their influence on benthic fauna. Low risk of all metals was perceived along the coastal sediment of Pakistan, except Cd, that showed the considerable risk in both monitoring years.

Fig. (7). The industrial pollution traces evidenced in Sandspit backwater areas during some ongoing ecological studies.

For MY-I, contamination degree was ranged from 1.53 to 20.8, indicating low to a considerable degree of contamination along the coastal sediment of Pakistan. Whereas, it varied from 7.21 to 33.21 for MY-II, indicated low to very high degree of contamination along the coastal sediment of Pakistan (Fig. **8e**). However, PERI varied from 55 to 272 during the MY-II, indicated low to moderate risk of heavy metals in the coastal sediments of Pakistan (Fig. **8f**).

Nutrients Dynamics in Marine Environment of Pakistan

The nutrients (Phosphate, Nitrate, Nitrite, and Ammonia) play an imperative role in the marine food chain while to consider and act fertilizers of the sea chiefly in primary production of marine waters. Nutrient cycling is essential for the maintenance of any ecosystem as it's generally known as the chemical alteration of nutrients and/or the flux of nutrients between different compartments or stages, such as organisms, habitats or even ecosystems (Vanni, 2002). These nutrient alterations sustain and stimulate the growth of phytoplankton (primary productivity in the marine habitat and basic constituent of the marine food chain) primarily serve as the food for primary consumers (zooplankton community, larval stages of various invertebrates and vertebrates organism, juveniles of fish and crustaceans) and for filter feeders (EMC, 2013). They need nutrients for their growth and nourishment thus the seasonal changes in nutrients may effect on phytoplankton production (Shoaib *et al.*, 2017). Nutrients can also be pulsed during storm-mediated increases in mixing (Wetz and Paerl, 2008), transport (Lovelock *et al.*, 2011) and upwelling events (Field *et al.*, 1980). The changes in the phytoplankton community size and structure influence the providence of Net primary production NPP (Malone, 1980; Legendre and Rassoulzadegan, 1996; Pomeroy *et al.*, 2007; Marañón, 2009).

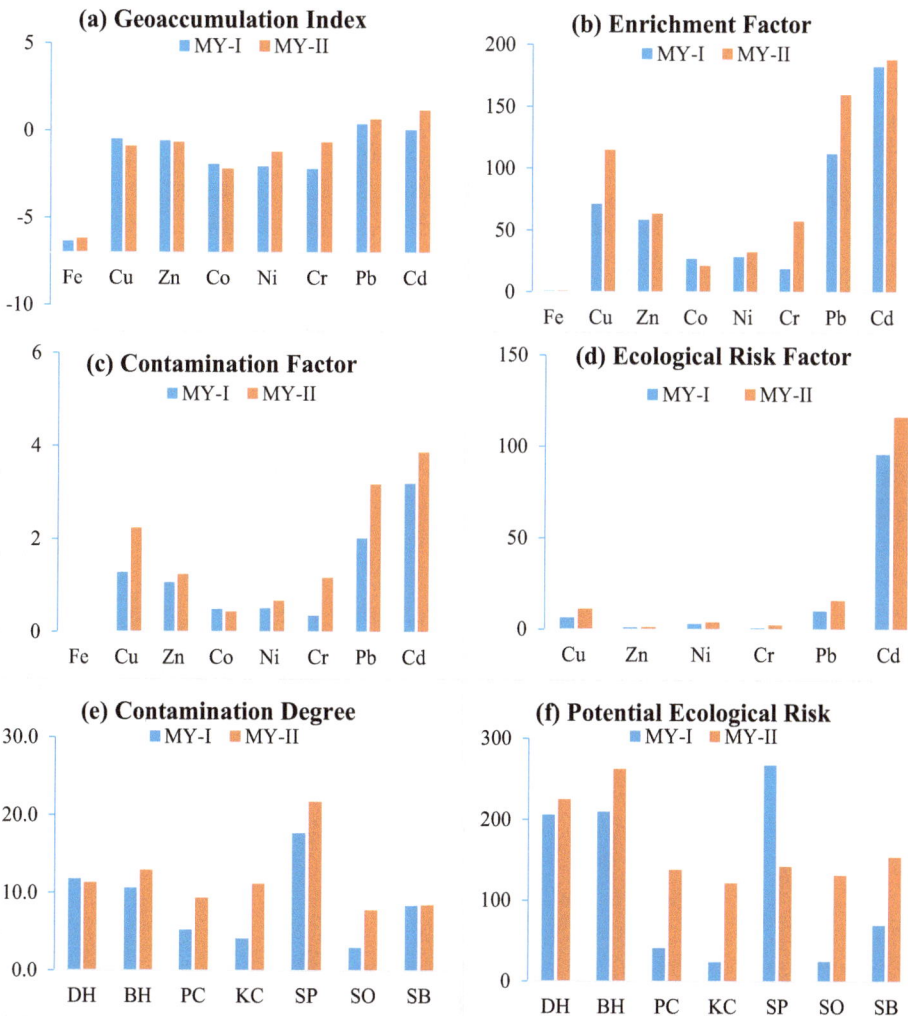

Fig. (8). Comparison of heavy metals pollution in coastal sediments between the two monitoring years by single and combined metal pollution indices.

The NPP is the entire photosynthetic fixation of inorganic carbon into autotrophic biomass. NPP sustains the life and energizes the global cycles of carbon, nitrogen, phosphorus, and other nutrients, and is also a main factor of atmospheric CO_2 and O_2 levels. An estimated Global NPP is about ~105 Pg C yr^{-1} and half of this includes marine plants (Field *et al.*, 1998; Falkowski and Raven, 1997; West berry *et al.*, 2008). In the upper ocean, euphotic zone the macrophytes and phytoplankton, respectively account for ~6 percent (~3.0 Pg C yr-1) and ~94 percent (~50 ± 28 Pg C yr-1) of NPP (Duarte *et al.*, 2005; Carr *et al.*, 2006; Schneider *et al.*, 2008; Chavez *et al.*, 2011; Ma *et al.*, 2014; Rousseaux and Gregg, 2014). In the marine environment, the bioavailability of nitrogen (N) is necessary for biological production and is mostly depleted at the surface of the world ocean (Montoya *et al.*, 2004; Ahmed *et al.*, 2017). Therefore, the supply of nitrate from deep water is the major source of new nitrogen sustaining primary productivity, marine biological N_2 fixation accounts for ~50 percent of N_2 fixation globally (Ward, 2012). The Arabian Sea always remains identified as site for active N_2 fixation dynamics (Devassy *et al.*, 1978; Capone *et al.*, 1998; Ahmed *et al.*, 2017). The excess level of phosphate in surface waters arising through denitrification is the main factor that favors the growth of nitrogen fixers in the Arabian Sea (Bange *et al.*, 2005; Ahmed *et al.*, 2017) as the bloom and a recurring phenomenon of filamentous diazotrophic cyanobacteria (*Trichodesmium* sp.) are well documented in this region (Desa *et al.*, 2005; Parab *et al.*, 2006; Basu *et al.*, 2011; Ahmed *et al.*, 2017). The incorporation of nutrient through river runoff increase the NPP in coastal waters throughout the growing season (Seitzinger *et al.*, 2005; Seitzinger *et al.*, 2010) therefore, account for the largest part of NPP during spring blooms at high latitudes (Malone, 1980). The nutrient supply pattern and variation in sunlight duration predominantly related to annual cycles of NPP and are likely to raise the amplitude. In addition, the seasonal amplification in NPP in general follow the winter mixing and when nutrient concentrations amplify, then the cycles are also more evident in coastal waters in accordance to seasonal upwelling. The Arabian sea considered as a high productive zone due to the intense upwelling phenomenon caused by the southwest monsoon (Qasim, 1982). The high production of organic matter leads to an intense oxygen minimum zone (OMZ) at 150–1200 m water depth in the Arabian Sea (Wyrtki, 1973; DeSousa *et al.*, 1996; Linsy *et al.*, 2018). Higher concentrations of nutrients result in overproduction and utilization of dissolved oxygen in the seawater (Amjad and Rizvi, 1999). In addition, the high phytoplankton and zooplankton productivities lead to high fish production in the Arabian sea (Schenau and De-Lange, 2000; Linsy *et al.*, 2018). The huge amount of nutrients supplied to the coastal areas by the Indus River, Hub River and seasonal rivers such as Malir river, Lyari river, Windor river, Porali river, Hingol river, ShadiKhor and Dasht river (Amjad and Rizvi, 1999). The coastal the mixing and nutrient-rich water is the major factor for phytoplankton abundance in the northern Arabian sea during northeast (winter) monsoon period (Banse and Mc Clain, 1986; Levy *et al.*, 2007; Shoaib *et al.*, 2017). However, these phenomena were opposed in the backwater area due to high load of domestic and industrial effluents (Harrison *et al.*, 1997; Mashiatullah *et al.*, 2004; Saifullah *et al.*, 2004; Iftikhar *et al.*, 2015).

Phosphorus (P) is another vital nutrient for life, found in forms of phosphate and organic phosphate and plays a key role in regulating primary productivity over geological time scales in both terrestrial and aquatic ecosystems (Correll, 1998; Elser *et al.*, 2007; Civan *et al.*, 2018; Defforey and Paytan, 2018). The conspicuous role of Phosphorus is also well documented in the regulation of bio-community structure and act as a limiting factor (affects the primary production rate, species structure and distribution) in primary productivity (Smith, 1984; Bastami *et al.*, 2018) therefore, determine the productivity of aquatic ecosystems (Karl *et al.*, 2001; Paytan and McLaughlin, 2010; Bastami *et al.*, 2018). Biologically active Phosphorous (BAP) in natural waters occurs in the form of phosphate (PO_4^{-3}); an organic form (organically bound phosphates) and in dissolved inorganic form (*i.e.* orthophosphates and polyphosphates). The longer time scales BAP entrance in the ecosystem occur through weathering of rocks followed by multifaceted biogeochemical interactions with as compared to anthropogenic P inputs (Benitez-Nelson, 2000). An industrial fertilizer, animal wastes and sewage are the main Primary anthropogenic sources of BAP (Jarvie *et al.*, 2006). However, at the global scale the particulate form of Phosphorus mainly contributes in the eutrophication and impairment of surface water quality which comes from point and diffuse sources (Hecky and Kilham, 1988; Mainstone and Parr, 2002; Civan *et al.*, 2018). An estimated BAP, which reaches to the open ocean from rivers, ranges from a few tenths to perhaps 1 Tg P yr-1 (Seitzinger *et al.*, 2005; Meybeck, 1982; Sharpies *et al.*, 2013). Mahowald *et al.*, (2008) estimated that atmospheric inputs of BAP are ~0.1 Tg P yr-1. Together all inputs would support ~0.1 percent of NPP annually. Therefore, like N, all NPP is propping up by BAP as recycled within the ocean on a global scale.

The nitrogen (N) and phosphorus (P) is used as fertilizer and the major nutrients responsible for aquatic eutrophication, their providence and consequence are site-specific (Finnveden and Potting 1999; Henryson *et al.*, 2018). The physio-chemical characteristics of the Arabian Sea make it an appropriate site to study the pathways of P accretion in sediments thus, extensive studies has been done on the variation and dynamics of total P in the sediments of Arabian Sea (Murty *et al.*, 1968; Setty and Rao, 1972 and Rao *et al.*, 1978, 1987; Linsy *et al.*, 2018) as well as P fractionation and cycling in the Eastern Arabian Sea (Babu and Nath, 2005; Acharya *et al.*, 2016), the Oman Margin and northern part of the Eastern Arabian sea (Schenau and De Lange, 2001), and the western Arabian sea

(Kraal *et al.*, 2012; 2015). The high rate of benthic flux of phosphate is accounted with the degradation of organic matter and dissolution of fish debris and may escort the P to authigenic phases (Schenau and De Lange, 2001; Babu and Nath, 2005; Linsy *et al.*, 2018). In addition to organic matter, iron also was found to play a major role in the P cycling and authigenesis in the Arabian sea sediments (Kraal *et al.*, 2012; Linsy *et al.*, 2018).

Some areas of Karachi creek receive large quantities of nutrients (such as Karachi harbour, Gharo creek, Gizri creek, and Korangi creek) in the form of liquid as well as solid waste. There has been a severe reduction observed in the nutrients load as brought by Indus River for the last fifty years.

This reduction in Indus discharge has a negative impact on the Indus estuary and the productivity of the mangrove forest and fisheries (Amjad and Rizvi, 1999; Inam *et al.*, 2017). The nutrient distribution in the Karachi harbour area is also presented in Table **2**, show that nutrient enrichment, particularly of inorganic phosphate over the entire Harbour area, which is the influence of industrial and sewage discharges on nutrient salt concentrations in surface and near-bottom layers. Excessive nutrient in coastal waters may cause eutrophication in coastal areas (EMC, 2013).

Table 2. The concentrations of various nutrients in coastal areas of Pakistan.

Locations	Ammonia (NH$_3$)	Nitrite (NO$_2$)	Nitrate (NO$_3$)	Phosphate (PO$_4$)	References
Sandspit (µgL^{-1})	0.05-18.07	0.03-0.05	0.16-5.64	0.11-2.95	(Shoaib *et al.*, 2017)
Korangi creek (µgL^{-1})	1.81–12.30	>0.20	0.45–1.69	1.00–23.60	(Saleem *et al.*, 2014)
Kemari (µgL^{-1})	17-33	0.56-1.1	7.8-27.0	11-48	NIO Data Archive
Clifton (µgL^{-1})	31-41	1.4-4.9	25-26	12-68	NIO Data Archive
Clifton beach (µgL^{-1})	59-60	3.9-5.0	8.4-28.0	11-70	NIO Data Archive
Sea view (µgL^{-1})	57-110	3.6-11.0	16-59.0	16-140	NIO Data Archive
Gizri creek (µgL^{-1})	36-160	9.8-14	30-81.0	36-77	NIO Data Archive
West wharf (gmL^{-1})	-	2.44	3.81	10.35	(Khan and Saleem, 1988)
Fish harbour (gmL^{-1})	-	0.1	0.18	23.25	(Khan and Saleem, 1988)
Lyari river mouth (gmL^{-1})	-	0.83	1.35	4.6	(Khan and Saleem, 1988)
Chari Kund (gmL^{-1})	-	5.16	8.74	6.7	(Khan and Saleem, 1988)
Manora channel (gmL^{-1})	-	1.02	2.05	14.05	(Khan and Saleem, 1988)
Mausa channel (gmL^{-1})	-	1.29	1.19	4.4	(Khan and Saleem, 1988)

Some areas of Karachi creek receive higher nutrients (phosphate, silicate, nitrate, and ammonia) values occurred during low tides and low values occurred during high tide. Moreover, the concentrations of nitrite were relatively higher in the bottom waters, but in general, nitrate concentration was not high *i.e.* 0.45-1.69 µM and its concentration inversely correlated with tidal fluctuations. Results of dissolved nutrients (phosphate, silicate, nitrate, and ammonia) showed a positive correlation with salinity, while the negative correlation with the tide (Saleem *et al.*, 2014). A substantial variation in nutrient concentrations was recorded at different sites during different time periods as high surface nitrate concentration was recorded at French Beach (2.657 µM/L) and low at MV (0.743 µM/L). High surface phosphate concentration was noted at Hawkes Bay (0.203 µM/L) and low (0.135 µM/L) at French Beach (Ali *et al.*, 2017).

Nutrients from the urban wastes and land run-off also reach the coastal waters, however, the sewage from the urban wastes bring a sizable amount of nutrients from Karachi city (Amjad and Rizvi, 1999). Recently, Inam *et al.*, 2017 reported major nutrients (nitrite, ammonia, phosphate and silicate) along the Indus river water over a distance of about 1000 km from Terbella Dam to Khobar creek through which the river water discharges into the Arabian sea and also reported the excessive loads of silicates along the Indus river as compared to nitrite and phosphate levels. According to Malone (1980), larger phytoplankton (diatoms), mainly prefer nitrate, whereas smaller cells favor utilizing ammonium ion. Shoaib *et al.,* (2017) shown a negative correlation of phytoplankton abundance with ammonium and confirms the above statement *Cyclotella cf. meneghiniana*, planktonic species were recorded in both seasons (summer and winter) and showed a preference for higher nutrient levels. *C. cf. meneghiniana* has also been reported as bio-indictor of many metal pollutants in aquatic ecosystems. The high abundance of this species was also recorded from both marine and freshwater area and their association with low dissolved oxygen concentration suggests poor water quality (Bestawy, 2000; El-Kassas and Gharib, 2016; Shoaib *et al.*, 2017).

Nutrient Pollution and Belonging Hazards

Whilst nitrogen and phosphorus are essential for maintenance and sustainable Net Primary production as well as important components of a healthy ecosystem; an excess amount of these nutrients can result to an overgrowth of algae and known as Nutrient pollution. The emergence of blooms and eutrophication are the two main conspicuous outcomes of Nutrient pollution as can alter the whole ecosystem structure and function in almost all types of aquatic habitats. The Pakistan Agriculture sector is the backbone of Pakistan's economy and accounts for 24% of the GDP and employs 48.4% of the entire labor demand. The percent of the total population depends upon the fishery sector. The coastal zones already confronted the several concerned issues that can be resulted in intense variation in the marine environment. The global warming, changes in biochemical cycles, acidification, hypoxia and alterations in the physicochemical stipulation and circulation patterns are conspicuous under the climate change regime; however, the habitat loss and degradation also associated with climate change and result in thrashing of intertidal areas, terrestrial shoreline areas and decline in freshwater body likely respond in biodiversity changes. The nutrients found in excessive amounts in underground waters, lakes, rivers, and coastal waters, and in the Indus Basin of Pakistan, water pollution is most important problem associated with agriculture and the contamination of various chemical substances is another immense challenge faced by the marine environment of country; Metropolitan municipal sewage and industrial effluent are two major sources of coastal water pollution. Nutrient input can be increased by humans or can occur naturally and termed eutrophication. Because of the disposal of a huge quantity of sewage, the habitat in Manora channel and Gizri creek has been completely destroyed and most of the area in these two water bodies is devoid of any marine life. The eutrophication likely due to an increase of nutritional resources to meticulous waters includes the supply of mineral nutrients (nitrogen, phosphorus, silicon, trace elements) as well as organic carbon. In the current era, the eutrophication has been most pronounced in the developed world, but it has to be expected that it will become more significant in the developing countries of Africa, Asia, and Latin America in the near future (Dolbeth *et al.*, 2007). The accession of nutrients leads to anoxic conditions, which triumphs approximately 40% of the regions of Karachi harbour, whereas 60% areas of the Gizri creek (Rizvi *et al.,* 1999; Amjad and Rizvi, 1999). Municipal and industrial sewage is one of the major sources of eutrophication in these adjacent coastal areas, which includes domestic wastewater that contains a wide variety of dissolved and suspended impurities, this nutrient-rich sewage water becomes a source of excessive algal growth. As algae die, decomposition of organic matter is enhanced through the bacterial activity, which consumes dissolved oxygen in the water, leading to depletion of oxygen in the water, making it difficult for other aquatic organisms to survive (MFF, 2016). This nutrient enrichment, eutrophication, initially stimulate growth of phytoplankton, microalgae, and macroalgae, and ultimately cause algal bloom, hypoxia which in turn can lead to other impacts such as: decrease and loss of other vegetation, variability and reduction in biodiversity, dominance of gelatinous organism, coral reef damage or reef growth inhibition, change in phytoplankton species composition, Low dissolved oxygen and formation of hypoxic or "dead" zones (oxygen-depleted waters), which in turn can lead to ecosystem collapse due to mass fish kills.

In the Arabian sea, inorganic nutrients (P, S, and N) and sunlight are the auspicious aspects for the increase of flagellates and dinoflagellates during high upwelling as during the Northeast monsoon season, the nutrients lead to the upper-column building suitable condition for the growth of dinoflagellates (Banse, 1987; Smith and Bottero, 1997; Landry, 2000; Brown *et al.*, 2002; Tang *et al.*, 2002). The water recirculation in spring monsoon increases the growth of phytoplankton to some extent (Gomes *et al.*, 2000). Some recent work on seasonal variations in abundance, diversity, and growth of phytoplankton community including diatoms and dinoflagellates has been reported from coastal waters of Karachi, Pakistan (Munir *et al.*, 2012, 2013a, b, 2015a, b; Naz *et al.*, 2010, 2012, 2013a, b, 2014; Khokhar *et al.*, 2016). The highest amount of nitrogen can enhance the production of dinoflagellates and the cells can multiply and strengthen the toxin synthesis or change in cells chemical composition; result as the immediate explosion of harmful algal bloom specifically dinoflagellates known as Red tides.

The unexpected and colossal growth of Algae or phytoplankton is known as Red tide or Harmful Algal Bloom (HAB). Production of algal toxins or red-tide toxins during algal blooms has been increasing worldwide mainly because of the effects of organic matter (OM) pollution and (GW) global warming (Prince *et al.*, 2008; Castle and Rodgers Jr., 2009; Yates and Rogers, 2011; Mostofa *et al.*, 2013b) Marine surface waters are undergoing acidification (Doney *et al.*, 2009; Beaufort *et al.*, 2011; Cai *et al.*, 2011; Xiao *et al.*, 2011), which is well known to instigate the changes in marine chemistry and ultimately production of algal toxins (Gao *et al.*, 2012). The occurrence, abundance and geographical distribution of toxin-producing algae or cyanobacterial blooms have substantially increased during the last few decades, because of increased anthropogenic contribution of organic matter pollution, nutrients and because of global warming (Yan and Zhou, 2004; Luckas *et al.*, 2005; McCarthy *et al.*, 2007; Mostofa *et al.*, 2013). According to D'Silva *et al.*, (2012) total 101 harmful algal blooms (39 spp.) were recorded during 1908 to 2009, and Padmakumar (2012) described 15% increment in HAB frequency during last twelve years in the Indian Ocean. Algal toxins or red tide toxins produced during algal blooms in surface waters are responsible for various ecological, physiological and environmental adverse effects *e.g.*: Decline of water quality, Depletion of dissolved oxygen below the pycnocline, loss of benthos, loss of phytoplankton, mortality of fish, coral reefs, livestock and wildlife Inhibition of enzymes and photosynthesis, cell and membrane damage (Howarth, 2008; Castle and Rodgers Jr., 2009; Bricelj and Lonsdale, 1997; Imai and Kimura, 2008; Southard *et al.*, 2010; Yates and Rogers, 2011). In addition, the shellfish or finfish poisoning by neurotoxic compounds (brevetoxins) produced by red-tide dinoflagellates (Backer *et al.*, 2005, 2008; Moore *et al.*, 2008). Illness or even death of higher organisms or humans, associated with consumption of contaminated fish, seafood and water as well as Adverse health effects (*e.g.* eczema or acute respiratory illness) from direct contact with, ingestion, or inhalation of cyanobacteria or various toxins (Fleming *et al.*, 2005; Moore *et al.*, 2008; Backer *et al.*, 2005, 2008; Mostofa *et al.*, 2013b and references therein).

Currently, more than two thousand dinoflagellate species have been recorded worldwide and approximately 200 species are toxic and accountable for harmful algal blooms in various coastal areas (Banse, 1987; Eppley *et al*, 1970; Smayda, 1997, 2002; Landry, 2000; Brown *et al.*, 2002; Gómez, 2005). In Pakistan since the 1970s, many researchers have been studying about the effects causing by the red tide and various species have been recognized, such as Saifullah, 1973; 1978 (*Gonialax*), Saifullah, 1990 (*Noctiluca scintillans*), Chaghtai, 1997 (*Phaeocystis*) and Hassan, 1973 (*Peridinium*). However, extensive work has been done on the distribution and taxonomy of dinoflagellates in Pakistani coastal waters such as Korangi creek and Manora channel (Taylor, 1976; Hassan and Saifullah, 1971, 1972; Hassan, 1976; and the Balochistan coast (Ghazala *et al.*, 2006). A few putative bloom-forming species have been previously reported from the Pakistan coastal water, such as *Prorocentrum micans, Ceratium shurunk, Gonyaulax diesing*, and *Noctiluca scintillans* (Chaghtai *et al.*, 2006). According to Munir *et al.*, (2013), most of the planktonic and toxic epiphytic dinoflagellates species belonging to a *Prorocentrales* group (genus *Prorocentrum* and genus *Mesoporos*) cause potentially harmful blooms in the coastal water of Pakistan. The species of the genus *Gambierdiscus* recently reported in coastal waters of Karachi, Pakistan, which produces the toxin ciguatera (Munir *et al.*, 2011; Munir *et al.*, 2011, 2013). A recent study on the distribution and abundance of dinoflagellates from Manora channel reported the sixty-six bloom forming (non-toxic) and 28 toxic species (Khokar *et al.*, 2018).

In the Arabian Sea, the Algal blooms explosion is a normal practice and has been observed during the monsoon seasons and usually emerges in two (Red or Orange and Green) different colors (Fig. **9**) but it was also observed that all blooms are not severely harmful to biodiversity. However, in the last few years, numerous incidences of mass-mortality of fishes by red tides or bloom have been reported in the coastal areas of Pakistan. Type and source of bloom monitored at distinct sites along the coast differs: as it may be set up locally or may have developed at some remote site and later transferred by coastal currents.

The fish mortalities and economic damages are reported due to the bloom of *Prorocentrum minimum* Gwadar Bay, Balochistan during Nov. 1987 (Rabbani *et al.*, 1990). It is further recognized that remarkable mortality of fish in the Arabian Sea experiences risen in the last few years, which yet adversely affects fish marketing and fisheries production and is of growing social involvement.

These blooms are common in Korangi as well as Gizri creeks (Harrison, 1997), and in coastal waters adjacent to Clifton beach that reported *Asterionella japonica* bloom (Khan, 1986), and dinoflagellates bloom reported in the waters of Sandspit and Hawks bay beaches (Saifullah and Chaghtai, 1990).

Fig. (9). The harmful algal bloom (Green) observed along the coastal areas of Karachi.

An incidence of red tide from 1999 – 2000 year in Pakistan's coastal waters where different species of the wild fish population were affected by red tide including Bottlenose and Rough-toothed Dolphins, Filefish, Speckled Siderial Moray, Parrot fishes, and baleen whale were found dead. The water sample during these years showed that the concentration of *Noctiluca* and *Gymnodinium* species are responsible for the emergence of bloom. The fish mortalities and economic damages are reported due to the bloom of *Prorocentrum minimum* Gwadar Bay, Balochistan and in Kuwait Bay (Gilbert, 2007). Where as in the Arabian Sea, a bloom of *Cochlodinium polykrikoides* has been generated in UAE (August 2008), and extended to coasts of Qatar and Iran and the killing of tons of fish (Anon., 2008) as well as destruction of the coral reefs ecosystem and fish farm were recorded (Richlen *et al.*, 2010).

In Gawader bay, Pakistan, mass mortality of *Terapon puta, Congresox sp* and *Pomadasys maculatus* were reported due to toxic dinoflagellate *Prorocentrum minimum* (Rabbani *et al.*, 1990). The *Noctiluca scintillans*, which forms both green and red tides, have been reported in Pakistan by various researchers (Subrahmanyan, 1954; Saifullah and Chaghtai, 1990; Chaghtai and Saifullah, 2006) and some toxic species has also been reported from the coast of Pakistan (Chaghtai and Saifullah, 2001; Gul and Saifullah, 2010; 2011; Munir *et al.*, 2011). The algal bloom of *Synedra acus* was observed in Karachi coastal waters (Luqman *et al.*, 2015). Massive blooms of stinging Jellyfish has recorded in the Karachi offshore waters in the sample area at the mouth of the River Indus (about 140 km south of Karachi) and Ormara along Sindh and Balochistan coast which influenced the fishing activities (DAWN, 2016). Jellyfish blooms are yet not well known in Pakistan. It is expected that climate change and ocean condition are two constituents, which are supporting components of most jellyfish blooms increasing in wide areas. The release of untreated sewage having nutrient enter coastal waters can lead to eutrophication make possible better feed for jellyfish polyps and supporting to form bloom. In 2013, along with open coastal waters of Clifton beach, sea breams, mullets, John croaker, Tiger tooth croaker, Tiger tooth croaker, and other small fish species were noticed to have been influenced by red tide. In 2013, along with open coastal waters of Clifton beach, sea bream, mullets, John croaker, Tiger tooth croaker, Tiger tooth croaker, and other small fish species were noticed to have been influenced by red tide. In August and September 2015, the red algal bloom appeared along the Sindh-Balochistan coast that killed a number of marine fishes. Except for fish mortality by the red tide in Pasni and Ormara in Balochistan, coast and coral reefs were also severely affected in Churna Island, Karachi. (DAWN, 2015). The most recent incidence of algal bloom of *Noctiluca scintillans* was observed during May-June 2017 and the bloom was noticed by the foul smell as felt in various adjacent areas however no mass mortality was observed during the stench decaying of bloom. Over the previous twenty years or so, harmful algal blooms have raised around the world in their number, intensity, and geographic scope. Due to the rise of the bulk of nutrients (P and N) that are brought into coastal waters or estuaries through sewage and industrial discharge, agricultural runoff and climate change the frequency of algal bloom, which has been increasing in coastal areas of Pakistan.

CONCLUSION

Oceans and coastal areas have unique interactions of numerous anthropogenic (Urbanization, Industrialization and Pollution discharge) and natural progressions. Unfortunately, in Pakistan, not much detailed surveys and research work have been done to understand the pollution impact and interactions in chemical, physical, hydrological and biological processes in estuaries, marine habitats, coastal waters and the inter-dependency among various marine resources.

There is a paucity of important information such as run-off from inorganic fertilizer and pesticides to water bodies, economic valuation of important features such as wetlands, seasonal and annual variations in pollution loads from land-based activities, impact and expected consequence of contamination in coastal sediments, waters and associated living web, bioaccumulation, bio-concentration, biomagnification and trophic transfer of hazardous pollutants and the recycling and biosorption of nutrients and other pollutant at their early discharge stage *etc*.

An integrated research-based approach is required for the assessment, evaluation, regular monitoring and sustainability of ocean space and resources. The development of a more consistent, integrated and structured framework that takes account of the economic potential of all exploited marine natural resources, which include energy sources from the oceans. Various pressures resulting from urban and industrial development include pollution from the disposal of solid waste and sewage are the widespread concerns along the coast. There is a factual lack of information about pollution hotspots and on how and where pollution is attenuate on land, in rivers, and in the sea as essential for managing and reducing pollution along the coast. There is no structural baseline for the estimation of coastal pollution types, pollutant and level of intensity in various hotspots and vulnerability to available living resources of adjacent coastal areas and the earlier initiatives and ongoing efforts are still not sufficient for the development and betterment of the coastal waters including resources.

The clean technologies need to be introduced for industry and treatment of wastes at earliest discharge source with best practices for agriculture and lack of uniform standards for the establishment of permissible limits of toxic discharge by the relationships between specific activities, the types and amount of discharge pollutants and discharge sources and attenuation rates or carrying capacity of receiving environments

CONSENT FOR PUBLICATION

Not applicable.

CONFLICT OF INTERESTS

The authors confirm that this chapter contents have no conflict of interest.

ACKNOWLEDGEMENT

Declare None.

REFERENCES

Acharya, SS, Panigrahi, MK, Kurian, J, Gupta, AK & Tripathy, S (2016) Speciation of phosphorus in the continental shelf sediments in the Eastern Arabian Sea. *Cont Shelf Res,* 115, 65-75.
[http://dx.doi.org/10.1016/j.csr.2016.01.005]

Adachi, R & Fukuyo, Y (1979) The thecal structure of a marine toxic dinoflagellate *Gambierdiscus toxicus* gen. et sp. nov. collected in a ciguatera-endemic area. *Nippon Suisan Gakkaishi,* 45, 67-71.
[http://dx.doi.org/10.2331/suisan.45.67]

Aftab, Z, Ali, L & Khan, AM (2000) *Industrial Policy and the Environment in Pakistan United Nations Industrial Development Organization United Nations Industrial Development Organization-UNIDO, (2000).* Industrial Policy and the Environment in Pakistan, Pakistan.

Agawin, NSR, Duarte, CM & Agustí, S (2000) Nutrient and temperature control of the contribution of picoplankton to phytoplankton biomass and production. *Limnol Oceanogr,* 45, 591-600.

[http://dx.doi.org/10.4319/lo.2000.45.3.0591]

Ahearn, GA, Mandal, PK & Mandal, A (2004) Mechanisms of heavy-metal sequestration and detoxification in crustaceans: a review. *J Comp Physiol B,* 174, 439-52.
[http://dx.doi.org/10.1007/s00360-004-0438-0] [PMID: 15243714]

Ahmed, Q, Khan, D & Ali, QM (2014) Heavy metals (Fe, Mn, Pb, Cd, and Cr) concentrations in muscles, liver, kidneys and gills of Torpedo Scad [*Megalaspis cordyla* (Linnaeus, 1758)] from Karachi waters of Pakistan. *Int J Biol Biotechnol,* 11, 517-24.

Ahmed, Q, Khan, D & Elahi, N (2014) Concentrations of heavy metals (Fe, Mn, Zn, Cd, Pb, and Cu) in muscles, liver and gills of adult *Sardinella albella* (Valenciennes, 1847) from Gwadar water of Balochistan, Pakistan. *FUUAST Journal of Biology,* 4, 195-204.

Ahmed, A, Gauns, M, Kurian, S, Bardhan, P, Pratihary, A, Naik, H, Shenoy, DM & Naqvi, SWA (2017) Nitrogen fixation rates in the eastern Arabian Sea. *Estuar Coast Shelf Sci,* 191, 74-83.
[http://dx.doi.org/10.1016/j.ecss.2017.04.005]

Khooharo, AA, Memon, RA & Mallah, MU (2008) An empirical analysis of pesticide marketing in Pakistan. *Pak Econ Soc Rev,* 46, 57-74.

Akhtar, M, Ahmad, N & Hussain, SP (2005) Climate change in upper Indus basin of Pakistan, A case study In: Muhammad, A, Hussain, SS, (Eds.), *Proceedings of National workshop on global change perspective in Pakistan-Challenges, Impacts, Opportunities and Prospects.*

Ali, S, Begum, F, Hussain, SA, Khan, AS, Ali, H, Khan, T, Raza, G, Ali, K & Karim, R (2014) Biomonitoring of heavy metals availability in the marine environment of Karachi, Pakistan, using oysters (*Crassostrea* sp). *Int J Biosci,* 4, 249-57.

Ali, U, Malik, RN, Syed, JH, Mehmood, ChT, Sánchez-García, L, Khalid, A & Chaudhry, MJI (2014) Mass burden and estimated flux of heavy metals in Pakistan coast: sedimentary pollution and eco-toxicological concerns. *Environ Sci Pollut Res Int,* 22, 4316-23.
[http://dx.doi.org/10.1007/s11356-014-3612-2] [PMID: 25296937]

Ali, A, Siddiqui, PJA & Aisha, K (2017) Characterization of macro algal communities in the coastal waters of Sindh (Pakistan), a region under the influence of reversal monsoons. *Reg Stud Mar Sci,* 14, 84-92.
[http://dx.doi.org/10.1016/j.rsma.2017.05.008]

Ali, R & Shams, ZI (2015) Quantities and composition of shore debris along Clifton Beach, Karachi, Pakistan. *J Coast Conserv,* 19, 527-35.
[http://dx.doi.org/10.1007/s11852-015-0404-x]

Amjad, S & Rizvi, SHN (1999) *Pakistan's national program of action under the global program of action for the protection of the marine environment from land based activities.* Ministry of Environment and Local Government and Rural Development, Government of Pakistan and National Institute of Oceanography, Clifton, Karachi, Pakistan.

Anderson, DM & White, AW (1989) Toxic dinoflagellates and marine mammal mortalities. *Proceedings of an Expert Consultation held at the Woods Hole Oceanographic Institution.*

Babu, CP & Nath, BN (2005) Processes controlling forms of phosphorus in surficial sediments from the eastern Arabian Sea impinged by varying bottom water oxygenation conditions. *Deep Sea Res Part II Top Stud Oceanogr,* 52, 1965-80.
[http://dx.doi.org/10.1016/j.dsr2.2005.06.004]

Backer, LC, Carmichael, W, Kirkpatrick, B, Williams, C, Irvin, M, Zhou, Y, Johnson, TB, Nierenberg, K, Hill, VR, Kieszak, SM & Cheng, YS (2008) Recreational exposure to microcystins during a Microcystis aeruginosa bloom in a small lake. *Marine Drugs,* 6, 389e406.

Banse, K (1987) Seasonality of phytoplankton chlorophyll in the central and northern Arabian Sea. *Deep-Sea Res,* 34, 713-23.
[http://dx.doi.org/10.1016/0198-0149(87)90032-X]

Banse, K & McClain, CR (1986) Winter blooms of phytoplankton in the Arabian Sea as observed by the Coastal Zone Color Scanner. *Ecological Progress Series,* 34, 201-11.
[http://dx.doi.org/10.3354/meps034201]

Bastami, KD, Neyestani, MR, Raeisi, H, Shafeian, E, Baniamam, M, Shirzadi, A, Esmaeilzadeh, M, Mozaffari, S & Shahrokhi, B (2018) Bioavailability and geochemical speciation of phosphorus in surface sediments of the Southern Caspian Sea. *Mar Pollut Bull,* 126, 51-7.
[http://dx.doi.org/10.1016/j.marpolbul.2017.10.095] [PMID: 29421132]

Brown, SL, Landry, MR, Christensen, S, Garrison, D, Gowing, MM, Bidigare, RR & Campbell, L (2002) Microbial community dynamics and taxon specific phytoplankton production in the Arabian Sea during 1995 monsoon seasons. *Deep-Sea Res,* 49, 2345-76.

Bryan, GW & Langston, WJ (1992) Bioavailability, accumulation and effects of heavy metals in sediments with special reference to United Kingdom estuaries: a review. *Environ Pollut,* 76, 89-131.
[http://dx.doi.org/10.1016/0269-7491(92)90099-V] [PMID: 15091993]

Bryan, GW (1979) Bioaccumulation of marine pollutants. *Philos Trans R Soc Lond B Biol Sci,* 286, 483-505.
[http://dx.doi.org/10.1098/rstb.1979.0042] [PMID: 40274]

Buitenhuis, ET, Li, WKW, Vaulot, D, Lomas, MW, Landry, MR, Partensky, F, Karl, DM, Ulloa, O, Campbell, L, Jacquet, S, Lantoine, F, Chavez, F, Macias, D, Gosselin, M & McManus, GB (2012) Picophytoplankton biomass distribution in the global ocean. *Earth Syst Sci Data,* 4, 37-46.
[http://dx.doi.org/10.5194/essd-4-37-2012]

Cai, WJ, Hu, X, Huang, WJ, Murrell, MC, Lehrter, JC, Lohrenz, SE, Chou, WC, Zhai, W, Hollibaugh, JT & Wang, Y (2011) Acidification of subsurface coastal waters enhanced by eutrophication. *NatureGeosciences,* 4, 766-70.
[http://dx.doi.org/10.1038/ngeo1297]

Callen, MS, De la Cruz, MT, Lopez, JM & Mastral, AM (2011) PAH in airborne particulate matter. Carcinogenic character of PM10 samples and assessment of the energy generation impact. *Fuel Process Technol,* 92, 176-82.

Callender, E (2005) Heavy metals in the environment-historical trends. *Journal of Treatise on Geochemistry* Elsevier, Pergamon Oxford 67-106.

Chaghtai, F & Saifullah, SM (2001) Harmful Algal bloom (HAB) organisms of the northern Arabian Sea bordering Pakistan-1 *Gonyaulax* Diesing. *Pak J Bot,* 33, 69-75.

Chaghtai, F & Saifullah, SM (2006) On the occurrence of green *Noctiluca scintillans* blooms in the coastal waters of Pakistan, Northern Arabian Sea. *Pak J Bot,* 38, 893-8.

Chaghtai, F & Saifullah, SM (1997) Occurrence of Phaeocystis in mangrove creeks near Karachi. *Pakistan Journal of Marine Sciences,* 6, 105-8.

Chapman, PM (2017) Assessing and managing stressors in a changing marine environment. *Mar Pollut Bull,* 124, 587-90.
[http://dx.doi.org/10.1016/j.marpolbul.2016.10.039] [PMID: 27760713]

Chaudhary, MZ, Ahmad, N, Mashiatullah, A, Ahmad, N & Ghaffar, A (2013) Geochemical assessment of metal concentrations in sediment core of Korangi creek along Karachi Coast, Pakistan. *Environ Monit Assess,* 185, 6677-91.
[http://dx.doi.org/10.1007/s10661-012-3056-4] [PMID: 23279880]

Chaudhary, MZ, Ahmad, N, Mashiatullah, A, Munir, S & Javed, T (2013) Assessment of metals concentration and ecotoxicology of the sediment core of Rehri creek, Karachi Coast, Pakistan. *Acta Geol Sin,* 87, 1434-43.
[http://dx.doi.org/10.1111/1755-6724.12140]

Chaudhry, Q & Sheikh, MM (2003) Climatic change and its impact on the water resources of mountain regions of Pakistan. *Journal of Pakistan Meteorology,* 1, 28-34.

Clark, R (2001) *Marine pollution.* Oxford University Press.

Codispoti, LA, Brandes, JA, Christensen, JP, Devol, AH, Naqvi, SWA, Paerl, HW & Yoshinari, T (2001) The oceanic fixed nitrogen and nitrous oxide budgets: Moving targets as we enter the anthropocene? *Sci Mar,* 65, 85-105.
[http://dx.doi.org/10.3989/scimar.2001.65s285]

Çoğun, HY, Firat, Ö, Aytekin, T, Firidin, G, Firat, Ö, Varkal, H, Temiz, Ö & Kargin, F (2017) Heavy Metals in the Blue Crab (*Callinectes sapidus*) in Mersin Bay, Turkey. *Bull Environ Contam Toxicol,* 98, 824-9.
[http://dx.doi.org/10.1007/s00128-017-2086-6] [PMID: 28409194]

Comber, S, Gardner, M, Georges, K, Blackwood, D & Gilmour, D (2013) Domestic source of phosphorus to sewage treatment works. *Environ Technol,* 34, 1349-58.
[http://dx.doi.org/10.1080/09593330.2012.747003] [PMID: 24191467]

Correll, DL (1998) The role of phosphorus in the eutrophication of receiving waters: a review. *J Environ Qual,* 27, 261-6.
[http://dx.doi.org/10.2134/jeq1998.00472425002700020004x]

Cullen, JJ (2008) Observation and prediction of harmful algal blooms. *Real-Time Coastal Observing Systems for Marine Ecosystem Dynamics and Harmful Algal Blooms: Theory, Instrumentation and Modelling* Monographs on Oceanographic Methodology Series UNESCO 1-41.

https://www.dawn.com/news/1304629

D'Silva, MS, Anil, AC, Naik, RK & D'Costa, PM (2012) Algal blooms: a perspective from the coasts of India. *Nat Hazards,* 63, 1225-53.
[http://dx.doi.org/10.1007/s11069-012-0190-9]

Ud Din, I, Rashid, A, Mahmood, T & Khalid, A (2013) Effect of land use activities on PAH contamination in urban soils of Rawalpindi and Islamabad, Pakistan. *Environ Monit Assess,* 185, 8685-94.
[http://dx.doi.org/10.1007/s10661-013-3204-5] [PMID: 23595691]

Dodds, WK, Bouska, WW, Eitzmann, JL, Pilger, TJ, Pitts, KL, Riley, AJ, Schloesser, JT & Thornbrugh, DJ (2009) Eutrophication of U.S. freshwaters: analysis of potential economic damages. *Environ Sci Technol,* 43, 12-9.
[http://dx.doi.org/10.1021/es801217q] [PMID: 19209578]

Dolbeth, M, Cardoso, PG, Ferreira, SM, Verdelhos, T, Raffaelli, D & Pardal, MA (2007) Anthropogenic and natural disturbance effects on a macrobenthic estuarine community over a 10-year period. *Mar Pollut Bull,* 54, 576-85.
[http://dx.doi.org/10.1016/j.marpolbul.2006.12.005] [PMID: 17240405]

GESAMP (IMO/FAO/UNESCO/WMO/WHO/IAEA/UN/UNEP Joint Group of Experts on the Scientific Aspects of Marine Pollution) (1991) *Global strategies for marine environmental protection. Rep Stud GESAMP,* 36. *Global strategies for marine environmental protection.*

Ghazala, B, Ormond, R & Hannah, F (2006) Phytoplankton communities of Pakistan: I. Dinophyta and Bacillariophyta from the coast of Sindh (Pakistan). *Int J Phycol Phycochem,* 2, 183-96.

Glibert, PM (2007) Eutrophication and harmful algal blooms: a complex global issue, examples from the Arabian Seas including Kuwait Bay, and an introduction to the global ecology and oceanography of harmful algal blooms (GEOHAB) programme. *International Journal of Oceans and Oceanography,* 2, 157-69.

Glibert, PM & Bouwman, L (2012) *Land-based Nutrient Pollution and the Relationship to Harmful Algal Blooms in Coastal Marine Systems,* 2, 5-7.

Glibert, PM, Anderson, DM, Gentien, P, Granéli, E & Sellner, KG (2005) The global, complex phenomena of harmful algal blooms. *Oceanography (Wash DC),* 18, 136-47.
[http://dx.doi.org/10.5670/oceanog.2005.49]

Glibert, PM (2016) *Why nutrient ratios matter: global and regional changes and consequences for phytoplankton community structure. 8th International Conference on Marine Pollution and Ecotoxicology*

University of Hong Kong, Hong Kong.

Gomes, HR, Goes, JI & Saino, T (2000) Influence of physical processes and freshwater discharge on the seasonality of phytoplankton regime in the Bay of Bengal. *Cont Shelf Res,* 20, 313-30.
[http://dx.doi.org/10.1016/S0278-4343(99)00072-2]

Gómez, F (2005) A list of free-living dinoflagellates species in the world's oceans. *Acta Bot Croat,* 64, 129-212.

Gondal, MA, Saher, NU & Qureshi, NA (2012) Diversity and biomass distribution of intertidal fauna in Sonmiani Bay (Miani Hor), Balochistan (Pakistan). *Egyptian Academic Journal of Biological Sciences,* 4, 219-34.
[http://dx.doi.org/10.21608/eajbsz.2012.14304]

González-De Zayas, R, Merino-Ibarra, M, Soto-Jiménez, MF & Castillo-Sandoval, FS (2013) Biogeochemical responses to nutrient inputs in a Cuban coastal lagoon: runoff, anthropogenic, and groundwater sources. *Environ Monit Assess,* 185, 10101-14.
[http://dx.doi.org/10.1007/s10661-013-3316-y] [PMID: 23856810]

Greene, RM, Lehrter, JC & Hagy, JD, III (2009) Multiple regression models for hindcasting and forecasting midsummer hypoxia in the Gulf of Mexico. *Ecol Appl,* 19, 1161-75.
[http://dx.doi.org/10.1890/08-0035.1] [PMID: 19688924]

Gul, S & Saifullah, SM (2010) Taxonomic and ecological studies of three marine genera of *Dinophysiales* from Arabian Sea shelf of Pakistan. *Pak J Bot,* 42, 2647-60.

Gul, S & Saifullah, SM (2011) The Dinoflagellate Genus *Prorocentrum* (Prorocentrales, Prorocentracea) from the northern Arabian Sea. *Pak J Bot,* 43, 3061-5.

Hallegraeff, GM (2010) Ocean climate change, phytoplankton community responses, and harmful algal blooms: a formidable predictive challenge. *J Phycol,* 46, 220-35.
[http://dx.doi.org/10.1111/j.1529-8817.2010.00815.x]

Hamid, N, Syed, JH, Kamal, A, Aziz, F, Tanveer, S, Ali, U, Cincinelli, A, Katsoyiannis, A, Yadav, IC, Li, J, Malik, RN & Zhang, G (2015) A Review on the Abundance, Distribution and Eco-Biological Risks of PAHs in the Key Environmental Matrices of South Asia. *Rev Environ Contam Toxicol,* 240, 1-30.
[http://dx.doi.org/10.1007/398_2015_5007] [PMID: 26809717]

Harrison, PJ, Khan, N, Yin, K, Saleem, M, Bano, N, Nisa, M, Ahmed, SI, Rizvi, N & Azam, F (1997) Nutrient and phytoplankton dynamics in two mangrove tidal creeks of the Indus River Delta, Pakistan. *Mar Ecol Prog Ser,* 157, 13-9.
[http://dx.doi.org/10.3354/meps157013]

Hassan, D & Saifullah, SM (1971) Some thecate Dinophyceae from inshore waters of Karachi. *Pak J Bot,* 3, 61-70.

Hassan, D & Saifullah, SM (1974) The genus Ceratium Schrank from coastal waters of Karachi: I. The subgenera *Amphiceratium* and *Biceratium. Bot Mar,* 17, 82-7.
[http://dx.doi.org/10.1515/botm.1974.17.2.82]

Hassan, D (1976) The genus *Ceraitum* schrank from coastal water of Karachi. Part II. The subgenus *Euceratium. Bot Mar,* 19, 287-94.
[http://dx.doi.org/10.1515/botm.1976.19.5.287]

Hassan, D & Saifullah, SM (1972) Genus *Peridinium* from inshore waters of Karachi. *Pak J Bot,* 4, 157-70.

Hecky, RE & Kilham, P (1988) Nutrient limitation of phytoplankton in freshwater and marine environments: a review of recent evidence on the effects of enrichment. *Limnol Oceanogr,* 33, 796-822.
[http://dx.doi.org/10.4319/lo.1988.33.4part2.0796]

Iftikhar, M, Ayub, Z & Siddiqui, G (2015) Impact of marine pollution in green mussel *Perna viridis* from four coastal sites in Karachi, Pakistan, Northern Arabian Sea: Hisopathological observation. *Indian J Exp Biol,* 3, 222-7.

Imai, I & Kimura, S (2008) Resistance of the fish-killing dinoflagellate *Cochlodinium polykrikoides* against algicidal bacteria isolated from the coastal Sea of Japan. *Harmful Algae,* 7, 360-7.
[http://dx.doi.org/10.1016/j.hal.2007.12.010]

Imai, I, Yamaguchi, M & Hori, Y (2006) Eutrophication and occurrences of harmful algal blooms in the Seto Inland Sea, Japan. *Plankton Benthos Res,* 1, 71-84.
[http://dx.doi.org/10.3800/pbr.1.71]

Inam, A, Muzaffar, M, Awan, KM, Hasany, SI, Zia, I & Hashmi, M (2017) An assessment of nutrient, sediment and carbon fluxes to the Indus Delta. *J Biodivers Environ Sci,* 10, 27-37.

Ismail, S, Saifullah, SM & Khan, SH (2014) Bio-geochemical studies of Indus Delta mangrove ecosystem through heavy metal assessment. *Pak J Bot,* 46, 1277-85.

Jang, E, Alam, MS & Harrison, RM (2013) Source apportionment of polycyclic aromatic hydrocarbons in urban air using positive matrix factorization and spatial distribution analysis. *Atmos Environ,* 79, 271-85.
[http://dx.doi.org/10.1016/j.atmosenv.2013.06.056]

Jarvie, HP, Neal, C & Withers, PJ (2006) Sewage-effluent phosphorus: a greater risk to river eutrophication than agricultural phosphorus? *Sci Total Environ,* 360, 246-53.
[http://dx.doi.org/10.1016/j.scitotenv.2005.08.038] [PMID: 16226299]

Jarvie, HP, Sharpley, AN, Flaten, D, Kleinman, PJA, Jenkins, A & Simmons, T (2015) The pivotal role of phosphorus in a resilient water-energy-food security nexus. *J Environ Qual,* 44, 1049-62.
[http://dx.doi.org/10.2134/jeq2015.01.0030] [PMID: 26437086]

JICA (2007) *Study on Water Supply and Sewerage System in Karachi, JICA, February 2007.*

Jones, KC & de Voogt, P (1999) Persistent organic pollutants (POPs): state of the science. *Environ Pollut,* 100, 209-21.
[http://dx.doi.org/10.1016/S0269-7491(99)00098-6] [PMID: 15093119]

Kamal, T, Tanoli, MAK, Mumtaz, M, Ali, N & Ayub, S (2015) Bioconcentration potential studies of heavy metals in *Fenneropenaeus penicillatus* (Jaira or Red Tail Shrimp) along the littoral states of Karachi City. *J Basic Appl Sci,* 11, 611-8.
[http://dx.doi.org/10.6000/1927-5129.2015.11.82]

Khan, N, Muller, J, Khan, SH, Amiad, S, Nizamani, S & Bhanger, MI (2010) Organochlorine pesticides (OCPs) contaminants from Karachi Harbour, Pakistan. *J Chem Soc Pak,* 32, 542-9.

Khan, SH (1986) Nontoxic bloom of *Asterionella japonica* on Clifton beach. *Pak J Bot,* 18, 361-3.

Khan, TMA, Razzaq, DA, Chaudhry, QZ, Quadir, DA, Kabir, A & Sarker, MA (2002) Sea Level Variations and Geomorphological Changes in the Coastal Belt of Pakistan. *Mar Geod,* 25, 159-74.
[http://dx.doi.org/10.1080/014904102753516804]

Khattak, MI & Khattak, MI (2013) Evaluation of heavy metals (As and Cd) contamination in the edible fish along Karachi-Makran Coast. *Pak J Zool,* 45, 219-26.

Khurshid, R, Sheikh, MA & Iqbal, S (2008) Health of people working/living in the vicinity of an oil-polluted beach near Karachi, Pakistan. *East Mediterr Health J,* 14, 179-82.
[PMID: 18557466]

Legendre, L & Rassoulzadegan, F (1996) Food-web mediated export of biogenic carbon in oceans: hydrodynamic control. *Mar Ecol Prog Ser,* 145, 179-93.
[http://dx.doi.org/10.3354/meps145179]

Lehninger, AL, Nelson, DL & Cox, MM (1993) *Principles of Biochemistry.* Worth Publishers, New York.

Levy, L, Shankar, D, Andre, J-M, Shenoi, SSC, Durand, F & Montegut, CB (2007) Basin wide seasonal evolution of the Indian Ocean's phytoplankton blooms. *J Geophys Res,* 112, 1-14.
[http://dx.doi.org/10.1029/2007JC004090]

Lewis, MR, Hebert, D, Harrison, WG, Platt, T & Oakey, NS (1986) Vertical nitrate fluxes in the oligotrophic

ocean. *Science,* 234, 870-3.
[http://dx.doi.org/10.1126/science.234.4778.870] [PMID: 17758109]

Linsy, P, Nath, NB, Mascarenhas-Pereira, MBL, Vinitha, PV, Ray, D, Babu, CP, Rao, BR, Kazip, A, Sebastian, T, Kocherla, M & Miriyala, P (2018) Benthic cycling of phosphorus in the Eastern Arabian Sea: Evidence of present day phosphogenesis. *Mar Chem,* 199, 53-66.
[http://dx.doi.org/10.1016/j.marchem.2018.01.007]

Luoma, SM & Rainbow, PS (2008) *Metal Contamination in Aquatic Environments: Science and Lateral Management* 573.

Luqman, M, Javed, MM, Yousafzai, A, Saeed, M, Ahmad, J & Chaghtai, F (2015) Blooms of pollution indicator micro-alga (*Synedra acus*) in northern Arabian Sea along Karachi, Pakistan. *Indian J Geo-Mar Sci,* 44, 1377-81.

Marañón, E (2009) Phytoplankton Size Structure. *Encyclopedia of Ocean Science.* In: Steele, J.H., Turekian, K.K., Thorpe, S.A., (Eds.), Academic Press.

Mashiatullah, A, Ahmad, N & Mahmood, R (2016) Stable Isotope Techniques to Address Coastal Marine Pollution.*Applied Studies of Coastal and Marine Environments.* InTech.
[http://dx.doi.org/10.5772/62897]

Mashiatullah, A, Chaudhary, MZ, Ahmad, N, Javed, T & Ghaffar, A (2013) Metal pollution and ecological risk assessment in marine sediments of Karachi Coast, Pakistan. *Environ Monit Assess,* 185, 1555-65.
[http://dx.doi.org/10.1007/s10661-012-2650-9] [PMID: 22580789]

Mashiatullah, A, Qureshi, LM, Ahmad, N, Javed, T & Shah, Z (2004) Distribution of trace metals in intertidal sediment along Karachi coast, Pakistan. *Geological Bulletin. University of Peshawar,* 37, 215-23.

McCarthy, MJ, Lavrentyev, PJ, Yang, LY, Zhang, L, Chen, YW, Qin, BQ & Gardner, WS (2007) Nitrogen dynamics and microbial food web structure during a summer cyanobacterial bloom in a subtropical, shallow, well-mixed, eutrophic lake Lake Taihu, China. *Hydrobiologia,* 581, 195-207.
[http://dx.doi.org/10.1007/s10750-006-0496-2]

Mehmood, A, Mahmood, A, Eqani, SAMAS, Ishtiaq, M, Ashraf, A, Bibi, N, Qadir, A, Li, J & Zhang, G (2017) A review on emerging persistent organic pollutants: Current scenario in Pakistan. *Hum Ecol Risk Assess,* 23, 1-13.
[http://dx.doi.org/10.1080/10807039.2015.1133241]

Mangroves for the Future, MFF (2016) *A Handbook on Pakistan's Coastal and Marine Resources* 78.

Mohiuddin, S & Naqvi, II (2014) Marine Sediment's Profile for Mercury as Pollution Indicator at Karachi Coast. *International Journal of Economic and Environmental Geology,* 5, 15-24.

Montoya, JP, Holl, CM, Zehr, JP, Hansen, A, Villareal, TA & Capone, DG (2004) High rates of N_2 fixation by unicellular diazotrophs in the oligotrophic Pacific Ocean. *Nature,* 430, 1027-32.
[http://dx.doi.org/10.1038/nature02824] [PMID: 15329721]

Moore, SK, Trainer, VL, Mantua, NJ, Parker, MS, Laws, EA, Backer, LC & Fleming, LE (2008) Impacts of climate variability and future climate change on harmful algal blooms and human health. *Environ Health,* 7 (Suppl. 2), S4.
[http://dx.doi.org/10.1186/1476-069X-7-S2-S4] [PMID: 19025675]

Morrice, JA, Valett, HM, Dahm, CN & Campana, ME (1997) Alluvial characteristics, groundwater-surface water exchange and hydrological retention in headwater streams. *Hydrol Processes,* 11, 253-67.
[http://dx.doi.org/10.1002/(SICI)1099-1085(19970315)11:3<253::AID-HYP439>3.0.CO;2-J]

Mostofa, KMG, Liu, CQ, Vione, D, Gao, K & Ogawa, H (2013) Sources, factors, mechanisms and possible solutions to pollutants in marine ecosystems. *Environ Pollut,* 182, 461-78.
[http://dx.doi.org/10.1016/j.envpol.2013.08.005] [PMID: 23992682]

Mostofa, KMG, Liu, CQ, Li, S & Mottaleb, A (2013) Impacts of global warming on biogeochemical cycles in natural waters. *Photobiogeochemistry of Organic Matter: Principles and Practices in Water Environment*

Springer, New York 851-914.
[http://dx.doi.org/10.1007/978-3-642-32223-5_10]

Mostofa, KMG, Liu, CQ, Pan, XL, Yoshioka, T, Vione, D, Minakata, D, Gao, K, Sakugawa, H & Komissarov, GG (2013) Photosynthesis in nature: a new look. *Photobiogeochemistry of Organic Matter: Principles and Practices in Water Environment.* In: Mostofa, K.M.G., Yoshioka, T., Mottaleb, A., Vione, D., (Eds.), Springer, New York 561-686.
[http://dx.doi.org/10.1007/978-3-642-32223-5_7]

Chaghtai, F (2001) Harmful algal bloom (HAB) organisms of the north Arabian Sea bordering Pakistan-I. Gonyaulax diesing. *Pakistan Journal of Botany,* 33, 69-75. 34.

Chaghtai, F & Saifullah, S M (2006) On the occurrence of green Noctiluca scintillans blooms in coastal waters of Pakistan, North Arabian Sea. *Pakistan Journal of Botany,* 38, 893-8. 35.

Munir, S, Naz, T, Burhan, Z, Siddiqui, PJA & Morton, SL (2012) Potentially harmful dinoflagellates (Dinophyceae) from the coast of Pakistan. Proceedings of the 14th International Conference on Harmful Algae. *International Society for the Study of Harmful Algae and Intergovernmental Oceanographic Commission of UNESCO,* 36. ISBN 978- 87-990827-3-5.

Munir, S, Naz, T, Burhan, Z, Siddiqui, PJA & Morton, SL (2013) Seasonal abundance, biovolume and growth rate of the heterotrophic Dinoflagellate (Noctiluca scintillans) from coastal waters of Pakistan. *Pak J Bot,* 45, 1109-13.

Munir, S, Naz, T, Burhan, Z, Siddiqui, PJA & Morton, SL (2016) Species composition and abundance of dinoflagellates from the coastal waters of Pakistan. *J Coast Life Med,* 4, 448-57.
[http://dx.doi.org/10.12980/jclm.4.2016J6-58]

Munir, S, Siddiqui, PJA & Morton, LS (2011) The occurrence of ciguatera fish poisoning producing dinoflagellate genus *Gambierdiscus* in Pakistan waters. *Algae,* 26, 317-25.
[http://dx.doi.org/10.4490/algae.2011.26.4.317]

Munir, S, Burhan, Z, Naz, T, Siddiqui, PJA & Morton, SL (2013) Morphotaxonomy and seasonal distribution of planktonic and benthic Prorocentrales in Karachi waters, Pakistan Northern Arabian Sea. *Chin J Oceanology Limnol,* 31, 267-81.
[http://dx.doi.org/10.1007/s00343-013-2150-y]

Munshi, AB, Detlef, SB, Schneider, R & Zuberi, R (2004) Organochlorine concentrations in various fish from different locations at Karachi Coast. *Mar Pollut Bull,* 49, 597-601.
[http://dx.doi.org/10.1016/j.marpolbul.2004.03.019] [PMID: 15476838]

Murty, PSN, Reddy, CVG & Varadachari, VVR (1968) Distribution of total phosphorus in the shelf sediments off the west coast of India. *Proceedings of the National Academy of Sciences,* India 134-141.

Muzaffar, M, Inam, A, Hashmi, MA, Mehmood, K, Zia, I & Hasaney, IM (2017) Climate change and role of anthropogenic impact on the stability of Indus deltaic Eco-region. *J Biodivers Environ Sci,* 10, 164-76.

Naqvi, G, Shoaib, N & Ali, AM (2017) Pesticides impact on protein in fish (*Oreochromis mossambicus*) tissues. *Indian J Geo-Mar Sci,* 46, 1864-8.

Nasira, K, Shahina, F & Kamran, M (2010) Response of free-living marine nematode community to heavy metal contamination along the coastal areas of Sindh and Balochistan, Pakistan. *Pak J Nematol,* 28, 263-78.

Naz, T, Burhan, Z, Munir, S & Siddiqui, PJA (2014) Growth rate of diatoms in natural environment from the coastal waters of Pakistan. *Pak J Bot,* 46, 1129-36.

Naz, T, Munir, S, Burhan, Z & Siddiqui, PJA (2013) Seasonal abundance and morphological observations of a raphid pennate diatom *Asterionella glacialis castracane* from the coastal waters of Karachi, Pakistan. *Pak J Bot,* 45, 677-80.

Parab, SG, Matondkar, SP, Gomes, HDR & Goes, JI (2006) Monsoon driven changes in phytoplankton populations in the eastern Arabian Sea as revealed by microscopy and HPLC pigment analysis. *Cont Shelf Res,* 26, 2538-58.

[http://dx.doi.org/10.1016/j.csr.2006.08.004]

Paytan, A & McLaughlin, K (2007) The oceanic phosphorus cycle. *Chem Rev,* 107, 563-76.
[http://dx.doi.org/10.1021/cr0503613] [PMID: 17256993]

Pomeroy, LR (2007) The microbial loop. *Oceanography (Wash DC),* 20, 28-33.
[http://dx.doi.org/10.5670/oceanog.2007.45]

Prince, EK, Myers, TL & Kubanek, J (2008) Effects of harmful algal blooms on competitors: allelopathic mechanisms of the red tide. *Limnol Oceanogr,* 53, 531-41.
[http://dx.doi.org/10.4319/lo.2008.53.2.0531]

Punjab Private Sector Groundwater Development Project, PPSGDP (2002) Environmental Assessment and Water Quality Monitoring Program. *Irrigation and Power Department, Government of the Punjab, Pakistan Technical Report 54*

Qari, R, Olufemi, A, Rana, M & Rahim, AA (2015) Seasonal variation in occurrence of heavy metals in *Perna Viridis* from Manora Channel of Karachi, Arabian Sea. *Int J Mater Sci,* 5, 1-13.

Qari, R & Ahmed, S (2014) Heavy metal distribution in *Avicennia marina* from Sonmiani, Pakistan Coast. *Journal of Shipping and Ocean Engineering,* 4, 38-42.

Qari, R, Siddiqui, SA & Qureshi, NA (2005) A comparative study in surficial sediment from coastal areas of Karachi, Pakistan. *Mar Pollut Bull,* 50, 583-608.
[http://dx.doi.org/10.1016/j.marpolbul.2005.01.024]

Qasim, SZ (1982) Oceanography of the northern Arabian Sea. *Deep-Sea Res A, Oceanogr Res Pap,* 29, 1041-68.
[http://dx.doi.org/10.1016/0198-0149(82)90027-9]

Rabbani, MM, Rehman, AU & Harms, CE (1990) Mass mortality of fishes caused by dinoflagellate blooms in Gawader Bay, Southwest Pakistan. *Toxic Marine Phytoplankton* Elsevier, New York 209-14.

Rabbani, MM, Inam, A, Tabrez, AR, Sayed, NA & Tabrez, SM (2008) The impact of sea level rise on Pakistan's coastal zones–in a climate change scenario. National Institute of Oceanography, Clifton, Karachi. *Conference Paper* March 2008.

Rainbow, PS (2007) Trace metal bioaccumulation: models, metabolic availability and toxicity. *Environ Int,* 33, 576-82.
[http://dx.doi.org/10.1016/j.envint.2006.05.007] [PMID: 16814385]

Rao, CM, Paropkari, AL, Mascarenhas, A & Murty, PSN (1987) Distribution of phosphorus and phosphatisation along the western continental margin of India. *J Geol Soc India,* 30, 423-38.

Rao, CM, Rajamanickam, GV, Paropkari, AL & Murty, PSN (1978) Distribution of phosphate in sediments of the northern half of the western continental shelf of India. *Indian J Geo-Mar Sci,* 7, 146-50.

Rasool, F, Tunio, S, Hasnian, SA & Ahmad, E (2002) Mangrove conservation along the coast of Sonmiani, Balochistan, Pakistan. *Pak J Bot,* 16, 213-7.

Ravindra, K, Sokhi, R & Vangrieken, R (2008) Atmospheric polycyclic aromatic hydrocarbons: source attribution, emission factors and regulation. *Atmos Environ,* 2895-921.
[http://dx.doi.org/10.1016/j.atmosenv.2007.12.010]

Richlen, MA, Morton, SL, Jamali, EA, Rajan, A & Anderson, DA (2010) The catastrophic 2008-2009 red tide in the Arabian Gulf region, with observations on the identification and phylogeny of fish-killing dinoflagellate *Cochlodinium polykrikoides. Harmful Algae,* 9, 163-72.
[http://dx.doi.org/10.1016/j.hal.2009.08.013]

Rizvi, SHN, Nisa, M & Haq, SM (1999) Environmental degradation and Marine Pollution along the Karachi coast and adjacent creeks in the Indus Delta. *The Indus River (Biodiversity, Resources, human kind)* Oxford University Press, Karachi 77-90.

Rizvi, SHN, Saleem, M & Baquer, J (1988) Steel mill effluents: influence on the Bakran creek environment.

Proceedings of Marine Science of the Arabian Sea American Institutes of Biological Sciences 549-69.

Rousseaux, CS & Gregg, WW (2014) Inter annual variations in phytoplankton primary production at a global scale. *Remote Sens,* 6, 1-19.
[http://dx.doi.org/10.3390/rs6010001]

Saher, NU & Siddiqui, AS (2016) Comparison of heavy metal contamination during the last decade along the coastal sediment of Pakistan: Multiple pollution indices approach. *Mar Pollut Bull,* 105, 403-10.
[http://dx.doi.org/10.1016/j.marpolbul.2016.02.012] [PMID: 26876559]

Saher, NU & Kanwal, N (2018) Heavy metal contamination and human health risk indices assessment in Shellfish species from Karachi coast Pakistan. *Academia Journal of Food Research,* 6, 012-20.

Saher, NU & Kanwal, N (2018) Some biomonitoring studies of heavy metals in commercial species of crustacean along Karachi coast, Pakistan. *Int J Biol Biotechnol,* 15, 269-75.

Saher, NU & Siddiqui, AS (2017) Evaluation of heavy metals contamination in mangrove sediments and their allied fiddler crab species (*Austruca sindensis* Alcock 1900) from Hawks Bay, Karachi, Pakistan. *Int J Biol Biotechnol,* 14, 411-7.

Saifullah, SM & Chaghtai, F (1990) Incidence of *Noctiluca scintillans* (Mac Cartney) Ehernb blooms along Pakistan's Shelf. *Pak J Bot,* 22, 94-9.

Saifullah, SM & Hassan, D (1973) Planktonic dinoflagellates from inshore waters of Karachi: 1. *Gonyaulax* Diessing. *Pak J Zool,* 5, 143-7.

Saifullah, SM & Hassan, D (1973) Planktonic dinoflagellates from inshore waters of Karachi: II. *Amphisolenia* Stein. *Pak J Zool,* 5, 149-55.

Saifullah, SM (1973) A preliminary survey of the standing crop of seaweed form Karachi coast. *Bot Mar,* 16, 139-44.
[http://dx.doi.org/10.1515/botm.1973.16.3.139]

Saifullah, SM (1978) Inhibitory effects of copper on marine dinoflagellates. *Mar Biol,* 44, 299-308.
[http://dx.doi.org/10.1007/BF00390893]

Saifullah, SM, Ismail, S, Khan, SH & Saleem, M (2004) Land use iron pollution in mangrove habitat of Karachi, Indus Delta. *Earth Interact,* 8, 1-9.
[http://dx.doi.org/10.1175/1087-3562(2004)8<1:LUPIMH>2.0.CO;2]

Saifullah, SM, Khan, SH & Ismail, S (2002) Distribution of nickel in a polluted mangrove habitat of the Indus Delta. *Mar Pollut Bull,* 44, 570-6.
[http://dx.doi.org/10.1016/S0025-326X(02)00088-7] [PMID: 12146841]

Saifullah, SM & Rasool, F (1995) A Preliminary Survey of Mangroves of Balochistan. *WWF-Pakistan Project Report.*

Saleem, M, Rizvi, SHN, Aftab, J, Kahkashan, S, Khan, AA & Qammaruddin, M (2013) Short Assessment Survey of Organochlorine Pesticides in Marine Environment of Damb (Sonmiani) Balochistan. *Pakistan Journal of Chemical Society,* 3, 107-15.
[http://dx.doi.org/10.15228/2013.v03.i03.p03]

Saleem, M, Aftab, J, Kahkashan, S, Kalhoro, NA & Ahmad, W (2014) Diurnal variation of nutrients, water quality and plankton composition in the Hajamro creek (Indus delta) during north east monsoon period. *Nucleus,* 51, 51-61.

Sanpera, C, Ruiz, X, Llorente, GA, Jover, L & Jabeen, R (2002) Persistent organochlorine compounds in sediment and biota from the Haleji lake: a wildlife sanctuary in South Pakistan. *Bull Environ Contam Toxicol,* 68, 237-44.
[http://dx.doi.org/10.1007/s001280244] [PMID: 11815794]

Sattar, A, Kroeze, C & Strokal, M (2014) The increasing impact of food production on nutrient export by rivers to the Bay of Bengal 1970-2050. *Mar Pollut Bull,* 80, 168-78.

[http://dx.doi.org/10.1016/j.marpolbul.2014.01.017] [PMID: 24467860]

Savage, C, Thrush, SF, Lohrer, AM & Hewitt, JE (2012) Ecosystem services transcend boundaries: estuaries provide resource subsidies and influence functional diversity in coastal benthic communities. *PLoS One,* 7e42708
[http://dx.doi.org/10.1371/journal.pone.0042708] [PMID: 22880089]

Sayied, N (2007) *Environmental issues in coastal waters, Pakistan as a case study.* http://commons.wmu.se/all_dissertations/201

Sharpies, J, Middelburg, JJ, Fennel, K & Jickells, TD (2013) Riverine delivery of nutrients and carbon to the oceans. *Nature,* 504, 61-70.
[http://dx.doi.org/10.1038/nature12857]

Shoaib, N & Siddiqui, PJA (2015) Acute toxicity of organophosphate and synthetic pyrethroid pesticides to juveniles of the penaeid shrimps, *Metapenaeus Monoceros. Pak J Zool,* 47, 1655-61.

Shoaib, M, Burhan, Z, Shafique, S, Jabeen, H & Siddiqui, PJA (2017) Phytoplankton composition in a mangrove ecosystem at Sandspit, Karachi, Pakistan. *Pak J Bot,* 49, 379-87.

Shoaib, N, Siddiqui, PJA & Khalid, H (2012) Acute toxic effect of pesticides on brine shrimp and opossum shrimp. *Pak J Zool,* 44, 1753-7.

Shoaib, N, Siddiqui, PJA & Khalid, H (2013) Toxicity of synthetic pyrethroid pesticides, fenpropathrin and fenvalerate, on killifish *Aphanius dispar* juveniles. *Pak J Zool,* 45, 1160-4.

Siddique, A, Mumtaz, M, Zaigham, NA, Mallick, KA, Saied, S, Zahir, E & Khwaja, HA (2009) Heavy metal toxicity levels in the coastal sediments of the Arabian Sea along the urban Karachi (Pakistan) region. *Mar Pollut Bull,* 58, 1406-14.
[http://dx.doi.org/10.1016/j.marpolbul.2009.06.010] [PMID: 19616812]

Siddiqui, AS & Saher, NU (2015) Heavy metals distribution in sediments and their transfer rate to benthic fauna in mangrove area near Hawks Bay Karachi, Pakistan. *Pakistan Journal of Marine Science,* 24, 09-17.

Siddiqui, AS & Saher, NU (2016) Assessment of heavy metal accumulation in *Opusiaindica* (Alcock, 1900) (Ocypodoidea: Camptandriidae) with Reference to sediment contamination from Coastal areas of Pakistan. *International Journal of Biology Research,* 4, 56-61.

Smayda, TJ (2002) Adaptive ecology, growth strategies and global expansion of dinoflagellates. *J Oceanogr,* 58, 281-94.
[http://dx.doi.org/10.1023/A:1015861725470]

Smayda, TJ (1997) Harmful algal blooms: their ecophysiology and general relevance to phytoplankton blooms in the sea. *Limnol Oceanogr,* 42, 1137-53.
[http://dx.doi.org/10.4319/lo.1997.42.5_part_2.1137]

Syed, JH & Malik, RN (2011) Occurrence and source identification of organochlorine pesticides in the surrounding surface soils of the Ittehad Chemical Industries Kalashah Kaku, Pakistan. *Environ Earth Sci,* 62, 1311-21.
[http://dx.doi.org/10.1007/s12665-010-0618-z]

Tang, DL, Kawamura, H & Luis, AJ (2002) Short-term variability of phytoplankton blooms associated with a cold eddy in the northwestern Arabian Sea. *Remote Sens Environ,* 81, 82-9.
[http://dx.doi.org/10.1016/S0034-4257(01)00334-0]

Tariq, J, Jaffar, M, Ashraf, M & Moazzam, M (1993) Heavy metal concentrations in fish, shrimp, seaweed, sediment, and water from the Arabian Sea, Pakistan. *Mar Pollut Bull,* 26, 644-7.
[http://dx.doi.org/10.1016/0025-326X(93)90504-D]

Taylor, FJR (1976) Dinoflagellates from International Indian Ocean Expedition. *Bibl Bot,* 132, 1-234.

Ulloa, O, Canfield, DE, DeLong, EF, Letelier, RM & Stewart, FJ (2012) Microbial oceanography of anoxic oxygen minimum zones. *Proc Natl Acad Sci USA,* 109, 15996-6003.

[http://dx.doi.org/10.1073/pnas.1205009109] [PMID: 22967509]

Van-Drecht, G, Bouwman, A, Harrison, J & Knoop, JM (2009) Global nitrogen and phosphate in urban waste water for the period 1970–2050. *Global Biogeochem Cycles,* 23GB0A03
[http://dx.doi.org/10.1029/2009GB003458]

Vilhena, MSP, Costa, ML & Berredo, JF (2013) Accumulation and transfer of Hg, As, Se, and other metals in the sediment-vegetation-crab-human food chain in the coastal zone of the northern Brazilian state of Pará (Amazonia). *Environ Geochem Health,* 35, 477-94.
[http://dx.doi.org/10.1007/s10653-013-9509-z] [PMID: 23334486]

Voss, M, Bange, HW, Dippner, JW, Middelburg, JJ, Montoya, JP & Ward, B (2013) The marine nitrogen cycle: recent discoveries, uncertainties and the potential relevance of climate change. *Philos Trans R Soc Lond B Biol Sci,* 36820130121
[http://dx.doi.org/10.1098/rstb.2013.0121] [PMID: 23713119]

Walsh, PJ, Smith, SL, Fleming, LE, Solo-Gabriele, HM, Gerwick, WH (2008). *Oceans and Human Health: Risk and Remedies from the Sea* Elsevier 644.

Wang, X, Sato, T, Xing, B & Tao, S (2005) Health risks of heavy metals to the general public in Tianjin, China *via* consumption of vegetables and fish. *Sci Total Environ,* 350, 28-37.
[http://dx.doi.org/10.1016/j.scitotenv.2004.09.044] [PMID: 16227070]

Ward, BB (2012) The Global Nitrogen Cycle. *Fundamentals of Geomicrobiology.* In: Knoll, A.H., Canfield, D.E., Konhauser, K.O., (Eds.), Wiley-Blackwell, Chichester, UK 36-48.
[http://dx.doi.org/10.1002/9781118280874.ch4]

Ward, BB (2013) Oceans. How nitrogen is lost. *Science,* 341, 352-3.
[http://dx.doi.org/10.1126/science.1240314] [PMID: 23888027]

Wetz, MS & Paerl, HW (2008) Estuarine phytoplankton responses to hurricanes and tropical storms with different characteristics (trajectory, rainfall, winds). *Estuaries Coasts,* 31, 419-29.
[http://dx.doi.org/10.1007/s12237-008-9034-y]

Williams, AT & Simmons, SL (1996) The degradation of plastic litter in rivers: implications for beaches. *J Coast Conserv,* 2, 63-72.
[http://dx.doi.org/10.1007/BF02743038]

Wu, J, Sunda, W, Boyle, EA & Karl, DM (2000) Phosphate depletion in the western North Atlantic Ocean. *Science,* 289, 759-62.
[http://dx.doi.org/10.1126/science.289.5480.759] [PMID: 10926534]

WWF (2002) Study of heavy metal pollution level and impact on the fauna and flora of the Karachi and Gwadar coast. *Final Project Report, No. 50022801.*

WWF (2016) *Unusual Jellyfish blooms affecting fishing activities in the Arabian Sea.* http://www.wwfpak.org/ newsroom/261216_Jellyfish.php

Wyrtki, K (1973) Physical oceanography of the Indian Ocean. *Ecological Studies: Analysis and Synthesis* Springer, Berlin 18-36.
[http://dx.doi.org/10.1007/978-3-642-65468-8_3]

Wyrtki, K (1973) *Physical Oceanography of the Indian Ocean* Springer, Berlin, Heidelberg 18-36.

Xiao, M, Wu, F, Zhang, R, Wang, L, Li, X & Huang, R (2011) Temporal and spatial variations of low-molecular-weight organic acids in Dianchi Lake, China. *J Environ Sci (China),* 23, 1249-56.
[http://dx.doi.org/10.1016/S1001-0742(10)60567-0] [PMID: 22128530]

Yan, T & Zhou, M-J (2004) Environmental and health effects associated with Harmful Algal Bloom and marine algal toxins in China. *Biomed Environ Sci,* 17, 165-76.
[PMID: 15386942]

Yates, BS & Rogers, WJ (2011) Atrazine selects for ichthyotoxic *Prymnesium parvum,* a possible explanation for golden algae blooms in lakes of Texas, USA. *Ecotoxicology,* 20, 2003-10.

[http://dx.doi.org/10.1007/s10646-011-0742-x] [PMID: 21809122]

Zehra, I, Kauser, T, Zahir, E & Naqvi, II (2003) Determination of Cu, Cd, Pb and Zn concentration in edible marine fish *Acanthopagurus berda* (Dandya) along Baluchistan Coast, Pakistan. *Int J Agric Biol,* 5, 80-2.

Zhang, Y & Tao, S (2009) Global atmospheric emission inventory of polycyclic aromatic hydrocarbons (PAHs) for 2004. *Atmos Environ,* 812-9.
[http://dx.doi.org/10.1016/j.atmosenv.2008.10.050]

Zinia, NJ & Kroeze, C (2015) Future trends in urbanization and coastal water pollution in the Bay of Bengal: the lived experience. *Environ Dev Sustain,* 17, 531-46.
[http://dx.doi.org/10.1007/s10668-014-9558-1]

CHAPTER 8

Ecotoxicology of Heavy Metals in Marine Fish

Lizhao Chen, Sen Du, Dongdong Song, Peng Zhang and **Li Zhang**[*]

Guandong Provincial Key Laboratory of Applied Marine Biology, South China Sea Institute of Oceanology, Chinese Academy of Sciences, Guangzhou, China

Abstract: Heavy metal pollution in the marine environment has been realized and developed to an important environmental problem since the 1950's. In the polluted areas, marine organisms are exposed to high level of heavy metals *via* different routes, accumulate them in the body, and may have harmful effects from molecular level to population level. Heavy metals in marine fish have been taken much attention due to human consumption and health. Marine fish accumulate heavy metals depending on the concentration and species of metals in water and food, and trophic level, ionic physiology, feeding habits (carnivorous, herbivorous or omnivorous), habitats (demersal, pelagic, or bento-pelagic), growing of fish, and other factors. Consequently, the concentrations of heavy metals in marine fish vary considerably among species and different sites, which can be well explained by the biokinetic model. High levels of heavy metals in marine fish can induce various acute and chronic toxic effects, including behavioral changes, organ pathological changes, biochemical and physiological changes, hematological changes, and so on. Heavy metal-contaminated fish consumption will pose threats to organisms at higher trophic level and humans. Here, we review the occurrence and chemistry of heavy metals in the marine environment, bioaccumulation, and toxicity of heavy metals in marine fish, and the general risk assessment of heavy metal in fish to human health.

Keywords: Heavy Metals, Bioaccumulation, Toxicology, Risk Assessment.

OCCURRENCE AND CHEMISTRY OF HEAVY METALS

Heavy metals are elements having atomic weights between 63.5 and 201, and a specific gravity greater than 5.0 (Fu and Wang, 2011). They include copper (Cu), zinc (Zn), iron (Fe), nickel (Ni), cobalt (Co), selenium (Se), silver (Ag), aluminum (Al), chromium (Cr), cadmium (Cd), lead (Pb), mercury (Hg), arsenic (As) and so on. Among multitudinous classifications proposed, heavy metals are popularly divided into essential and non-essential metals for life by aquatic toxicologists (Wood *et al.*, 2011, Wood *et al.*, 2012). Indeed, the small quantities

[*] **Corresponding author Li Zhang:** Guandong Provincial Key Laboratory of Applied Marine Biology, South China Sea Institute of Oceanology, Chinese Academy of Sciences, Guangzhou, China; Tel/Fax: +86-20-89221322; E-mail: zhangli@scsio.ac.cn

De-Sheng Pei & Muhammad Junaid (Eds.)
All rights reserved-© 2019 Bentham Science Publishers

of these heavy metals, like Cu, Fe, Co, Mo, Zn, Ni, Mn, and Cr, are essential for organisms of proteins owing to their participation in metabolic reactions as cofactors or integral parts especially enzymes, but above the permissible limit, they can be hazardous to organisms. Some other heavy metals including Al, Cd, Pb, As, and Hg are not essential but toxic to organisms due to their interaction with biomolecules and interfere with corresponding functions. Heavy metals in the environment come from both natural and anthropogenic sources (Morel and Price, 2003). For many metals, anthropogenic inputs have exceeded the natural inputs currently. Heavy metals enter the seawater *via* river runoff, wind-blown dust, diffusion from sediments, hydrothermal vent inputs, and many anthropogenic activities (Fu and Wang, 2011). With large-scale industrial activities and fast urbanization processes, anthropogenic activities have released very substantial amounts of heavy metal into seawater and exerted tremendous pressure on marine ecosystems. Metal pollution in estuaries, bays, and coastal areas is often considered as a "traditional" environmental problem, but with such rapid industrialization and often "uncontrolled" releases of industrial wastes, it has led to further deterioration in marine environments and become a new challenge (Li *et al.*, 2012, Pan and Wang, 2012).

The concentrations of heavy metals vary both horizontally and vertically through the world's oceans, determined by the relative rates of supply and removal (Donat and Dryden, 2001, Morel and Price, 2003). Beside concentrations, the chemical speciation of heavy metals is vital for physiology and toxicology (Donat and Dryden, 2001, Fu and Wang, 2011). The metals in seawater are mainly in the dissolved or particulate forms. It is generally recognized that the particulate metals exhibit negligible toxicity and bioavailability to aquatic organism relative to the dissolved metals. Dissolved metals can exist in different oxidation states, such as Fe(II)/Fe(III), Mn(II)/Mn(IV), Cr(III)/Cr(VI), Cu(I)/Cu(II), and As(III)/As(V), and chemical forms, such as free ions, organometallic compounds, organic complexes (*e.g.* metals bound to proteins or humic substances), and inorganic complexes (*e.g.*, metals bound to Cl^-, OH^-, HCO_3^-, SO_4^{2-}, *etc.*), depending on redox potential, pH and biological processes (Wood *et al.*, 2011, Wood *et al.*, 2012). The toxicity and bioavailability of heavy metals have been found to be proportional to the concentrations of their free metal ions but not their total concentrations. Complexation of metals by organic ligands will decrease the concentration of the free ion, thereby decreasing its toxicity or bioavailability. Seawater contains high levels of major ions, such as Na^+, Ca^{2+}, Cl^-, and HCO_3^-. For many metals, complexation with Cl^- and dissolved organic matters (DOMs) and the protective effects of competition by high concentrations of Na^+, Mg^{2+}, and Ca^{2+} lower the toxicity and availability of heavy metals in seawater (Grosell *et al.*, 2007).

Marine animals can accumulate, retain, and transform heavy metals inside their bodies when exposed to them through different routes/sources, such as diet, water, sediments, particles, and *etc*. Therefore, marine organisms are mostly good bioindicators for long-term monitoring of metal accumulation (Zhou *et al.*, 2008). Fish are usually considered as an organism of choice for assessing the effects of heavy metal pollution on aquatic ecosystems (van der Oost *et al.*, 2003). They are continuously exposed to heavy metals through their gills, skin, and intestine, resulting in high bioaccumulation and potential toxicity due to acute and chronic effects. The high bioaccumulation of heavy metals can affect biochemical and physiological systems, including behavior, organ histopathology, material/energy metabolism, enzyme activities, immune function, and gene expression. Fish also appear to have evolved different mechanisms for detoxification of heavy metals to counter the ambient heavy metal contamination. Heavy metal contaminated fish consumption could result in heavy metal exposure to humans and lead to an adverse health effect. Therefore, fish are good bioindicators of heavy metal toxicity and can be used as sentinels for biomonitoring of food safety.

BIOACCUMULATION MECHANISMS

Bioaccumulation is typically defined as the increase of concentrations of contaminants in aquatic organisms following uptake from the ambient environmental medium. Bioaccumulation is usually considered as a good integrative indicator of the chemical exposures of organisms in polluted ecosystems (Wang, 2016). Fish are at the high trophic levels of the aquatic food chains. Their metabolic activities allow them to accumulate the major, essential, and non-essential elements from water, food, or sediment (Castro-Gonzalez and Mendez-Armenta, 2008). Given the importance of fish both as food and bio-monitors, numerous studies have therefore determined the bioaccumulation of heavy metals in various fish over the past few decades. Several reviews have summarized metal accumulation in fish and concluded that the accumulation of heavy metals in fish is mostly depending on different feeding habits (carnivorous, herbivorous or omnivorous), differences in the aquatic environmental lives (demersal, pelagic, or bento-pelagic), growing rates of the species, types of tissues analyzed, and other factors (Neff, 2002, Varjani *et al.*, 2018, Yilmaz *et al.*, 2017, Yilmaz *et al.*, 2018).

Table **1** summarizes the concentration range of some essential metals (Cu, Zn, Fe, Cr, Mn, Mo, Ni) and non-essential metals (Al, As, Cd, Pb) in different fish species collected from different regions of the world. Among these data, metal concentrations were either quantified based on tissue dry weights or wet weights which were specified in the table and text. Typically, the wet weight to dry weight ratio of the fish would be in the range of 4-5 (Neff, 2002, Onsanit *et al.*, 2010);

thus, the readers can do a simple conversion of these concentrations based on either dry or wet tissue weights. It is rather difficult to summarize some commonalities among numerous bioaccumulation data because of the differences of regions, interspecies, and intraspecies, when compared the bioaccumulation potential of metals in fish. Here, we firstly simply summarize the metal concentrations and some differences in different fish species from different regions; we then try to use the biodynamic model to explain these observed differences.

Copper (Cu) is an essential cofactor of several enzymes and necessary for the synthesis of hemoglobin. Most marine organisms have evolved mechanisms to regulate the Cu levels in their tissues in the presence of variable concentrations in the ambient water, sediments, and food. A range of potential cellular targets for Cu was manifested in altering physiology and toxicity at the organ and organism level during Cu exposure (Wood *et al.*, 2011). The Cu concentrations in the muscles of different fish collected from different regions put forward by many researchers, *e.g.*Zhang and Wang (2012): 0.34-7.35 µg/g d.w. in the coastal area of China, Gu *et al.* (2015a, 2017): 0.12-1.13 µg/g w.w. in the South China Sea, Huang *et al.* (2011): 1.74-4.08 µg/g in Yangtze River Estuary, Cui *et al.* (2011): 2.32-11.2 µg/g d.w. in Yellow River Estuary, Du *et al.* (2017): 0.36-6.99 µg/g d.w. in Daya Bay (China) and 0.6-11.6 µg/g d.w. in Jiaozhou Bay (China), Gu *et al.* (2015b): 0.01-0.62 µg/g w.w. in Qinzhou Bay (China), Yilmaz *et al.* (2017): nd-6.24 µg/g w.w. in Iskenderun Bay (Turkey), Uluozlu *et al.* (2007): 0.73-1.83 µg/g d.w. in Black and Aegean Seas, Tuzen (2009): 0.65-2.78 µg/g w.w. in Black Sea, Türkmen *et al.* (2009b): 0.07-1.48 µg/g w.w. in Aegean and Mediterranean, Bilandžić *et al.* (2011): 0.001-57.3 µg/g w.w. in Adriatic Sea. Generally, Cu accumulation levels depend on many factors, such as geological regions, fish species, organs, feeding habits, and habitats. In all previous studies, the levels of Cu in the muscle of carnivore and demersal fish, such as *Chelidonichthys lucernus*, *Trigla lucerna*, *Epinephelus aeneus*, *Merluccius merluccius*, *Mullus surmuletus*, *Mullus barbatus*, *Saurida undosquamis*, *Solea solea*, *Sparus aurata*, and *Upeneus molluccensis*, were 1.11-1.50 µg/g w.w. (Ersoy and Çelik, 2010), 0.95-4.19 µg/g w.w. (Ateş *et al.*, 2015, Yılmaz *et al.*, 2010), 0.1-4.58 µg/g w.w., 0.98-2.04 µg/g w.w. (Ateş *et al.*, 2015, Ersoy and Çelik, 2010), 0.2-1.21 µg/g w.w. (Dural *et al.*, 2010, Tepe, 2008), nd-4.19 µg/g w.w. (Çiçek *et al.*, 2008, Çoğun *et al.*, 2006, Dural *et al.*, 2010, Kalay *et al.*, 1999, Kargin, 1996, Manaşirli *et al.*, 2016, Tepe *et al.*, 2008, Türkmen *et al.*, 2005), nd-2.61 µg/g w.w. (Ateş *et al.*, 2015, Çiçek *et al.*, 2008, Ersoy and Çelik, 2010, Manaşirli *et al.*, 2016, Türkmen *et al.*, 2005), 0.94-1.66 µg/g w.w. (Çoğun *et al.*, 2005, Ersoy and Çelik, 2010), nd-6.24 µg/g w.w. (Çoğun *et al.*, 2005, Ersoy and Çelik, 2010, Kargin, 1996, Türkmen *et al.*, 2005, Yilmaz, 2005), and nd-1.7 µg/g w.w. (Ateş *et al.*, 2015, Dural and BïCkïCï, 2010, Ersoy and Çelik, 2010), respectively. The Cu concentrations in the muscles of *Etrumeus teres*, *Scomber japonicus*, and *Trachurus mediterraneus*, carnivore and pelagic, were 0.36-0.48 µg/g w.w. (Ersoy and Çelik, 2009), 1.18-1.94 µg/g w.w. (Ersoy and Çelik, 2009, Türkmen *et al.*, 2009a), and 0.66-1.91 µg/g w.w. (Ersoy and Çelik, 2009, Yilmaz, 2003), respectively. The Cu levels in the muscles of *Mugil cephalus* and *Pagellus erythrinus*, omnivore and bento-pelagic, were 0.15-2.58 µg/g w.w. (Çoğun *et al.*, 2006, Kalay *et al.*, 1999, Tepe, 2008, Türkmen *et al.*, 2006, Yilmaz, 2003, Yilmaz, 2005) and nd-4.54 µg/g w.w. (Çiçek *et al.*, 2008, Dural and BïCkïCï, 2010, Manaşirli *et al.*, 2016), respectively. The Cu concentrations in the muscle of *Etrumeus teres* (carnivore and pelagic) were 0.36-0.48 µg/g w.w. (Ersoy and Çelik, 2009). Thus, the Cu levels in the carnivore and demersal fish species were always higher than those found in omnivore and pelagic species due to their variations in feeding habits and habitats. Moreover, in most previous studies, the mean value of Cu levels in organs is in the following order: liver (nd-66 µg/g w.w.) (Çoğun *et al.*, 2006, Dural and BïCkïCï, 2010, Ersoy and Çelik, 2009, Kargin, 1996) > gonad (nd-59.4 µg/g w.w.) (Yilmaz, 2003, Yilmaz, 2005) > skin (nd-25.6 µg/g w.w.) (Dural and BïCkïCï, 2010, Yilmaz, 2003, Yilmaz, 2005) > spleen (9.38-20.1 µg/g w.w.) (Kargin, 1996) > intestine (3.25-16.9 µg/g w.w.) (Dural and BïCkïCï, 2010, Türkmen *et al.*, 2013) > gill (0.96-8.26 µg/g w.w.) (Çoğun *et al.*, 2005, Çoğun *et al.*, 2006, Kargin, 1996, Türkmen *et al.*, 2013) > kidneys (3.62-7.26 µg/g w.w.) (Kargin, 1996) > muscle (nd-6.24 µg/g w.w.) (Çoğun *et al.*, 2006, Dural and BïCkïCï, 2010, Ersoy and Çelik, 2009, Kargin, 1996). In general, biomagnification, defined as increasing tissue burdens over three trophic levels, is not considered a major factor for Cu, presumably owing to the relatively strong homeostatic control of this essential element (DeForest *et al.*, 2007).

Zinc (Zn) is an essential cofactor in nearly 300 metalloenzymes (Vallee and Auld, 1990) and other metabolic compounds, such as DNA. Aquatic animals can regulate Zn levels in their tissues that come from seawater and diet. It follows that when Zn availability is low, fish will take up as much of it as needed from their environment and in cases where environmental concentrations are in excess of the requirement, they will attempt to avoid further accumulation (Wood *et al.*, 2011). The Zn concentrations in the muscles of fish ranged from 10.4 to 79.1 µg/g d.w. in the coastal area of China (Zhang and Wang, 2012), 2.34 to 6.88 µg/g w.w. in the South China Sea (Gu *et al.*, 2017, Gu *et al.*, 2015a), 2.2 to 52.3 µg/g d.w. in Pearl River Estuary (Ip *et al.*, 2005, Zhang and Wang, 2012), 1.85 to 15.4 µg/g in Yangtze River Estuary (Huang *et al.*, 2011), 21.6 to 79.4 µg/g d.w. in Yellow River Estuary (Cui *et al.*, 2011), 11.7 to 40.9 µg/g d.w. in Daya Bay (China) (Du *et al.*, 2017), 14.5 to 49.2 µg/g d.w. in Jiaozhou Bay (China) (Unpublished data), 5.73 to 7.18 µg/g w.w. in Qinzhou Bay (China) (Gu *et al.*, 2015b), 0.016 to 51.1 µg/g w.w. in Iskenderun Bay (Turkey) (Yilmaz *et al.*, 2017), 35.4 to 106 µg/g d.w. in Black and Aegean Seas (Uluozlu *et al.*, 2007), 38.8 to 93.4 µg/g w.w. in Black Sea (Tuzen, 2009), 3.51 to 53.5 µg/g w.w. in Aegean and Mediterranean Seas (Türkmen *et al.*, 2009b). The Zn levels in the muscle of carnivore and demersal fish, carnivore and pelagic fish, omnivore and bento-pelagic fish, and omnivore and pelagic fish were 0.11-39.4 µg/g w.w. (Çiçek *et al.*, 2008, Çoğun *et al.*, 2006, Dural *et al.*, 2006, Ersoy and Çelik, 2010, Kargin, 1996, Manaşirli *et al.*, 2016), 2.02-30.4 µg/g w.w. (Ersoy and Çelik, 2009, Türkmen *et al.*, 2009a, Yilmaz, 2003), 0.64-51.3 µg/g w.w. (Çiçek *et al.*, 2008, Çoğun *et al.*, 2006, Dural *et al.*, 2009, Dural *et al.*, 2006, Manaşirli *et al.*, 2016, Tepe, 2008, Türkmen *et al.*, 2006, Yilmaz, 2003, Yilmaz, 2005), and 2.37-8.04 µg/g w.w. (Ersoy and Çelik, 2009, Tepe, 2008, Türkmen *et al.*, 2009a), respectively. These results indicate that variations in feeding habits and habitats lead to the difference of Zn bioaccumulation. In wild marine fish, the mean values of Zn concentrations in organs are in the following order: gonad > gill > skin > liver > muscle in nearly all studies. Typical Zn concentrations in gonad, gill, skin, liver, and muscle among different fish species were 10.7-347 µg/g w.w. (Dural *et al.*, 2006, Yilmaz, 2003, Yilmaz, 2005), 7.63-142 µg/g w.w. (Çoğun *et al.*, 2005, Çoğun *et al.*, 2006, Dural *et al.*, 2011, Dural *et al.*, 2006, Kalay *et al.*, 1999, Kargin, 1996, Türkmen *et al.*, 2013), 0.21-133 µg/g w.w. (Dural and BïCkïCï, 2010, Dural *et al.*, 2010, Dural *et al.*, 2011, Dural *et al.*, 2009, Yilmaz, 2003, Yilmaz, 2005, Yılmaz *et al.*, 2010), 0.23-112 µg/g w.w. (Çoğun *et al.*, 2006, Dural *et al.*, 2006, Ersoy and Çelik, 2010, Ersoy and Çelik, 2009, Kargin, 1996), and 0.022-51.1 µg/g w.w. (Çiçek *et al.*, 2008, Dural *et al.*, 2011, Kalay *et al.*, 1999, Manaşirli *et al.*, 2016, Yilmaz, 2003), respectively. This distribution of Zn in fish tissues undoubtedly reflects the distribution of requirements for Zn as a cofactor in several important enzymes. Much of Zn in tissues of fish appear to be bound to metallothionein (MT) (Hamilton and Mehrle, 1986). MTs are non-enzymatic proteins with low molecular weight that play a role in the homeostatic control of essential metals, such as Zn and Cu (Krezel and Maret, 2007). The variations of MTs in organs may explain the tissue-specific accumulation. Similar to Cu, there is also no evidence for biomagnification in the food chain because Zn concentrations in higher trophic levels are not higher than those in lower levels (Besser *et al.*, 2007a).

Iron (Fe) is probably the most well-known metal in biologic systems, and is essential for life, being involved in oxygen transfer, respiratory chain reactions, DNA synthesis, and immune function (Papanikolaou and Pantopoulos, 2005). Owing to the essentiality and requirement of Fe for hemoglobin function, tissue Fe values are typically high. Tissue Fe concentrations vary considerably with species and geographic regions. The Fe levels in the muscles of different fish were reported by Zhang and Wang (2012) in the coastal area of China (473-11788 µg/g d.w.), Gu *et al.* (2015a, 2017) in the South of China Sea (2510-22990 µg/g w.w.), Zhang and Wang (2012) in the Pearl River Estuary (3441-11788 µg/g d.w.), Yilmaz *et al.* (2017) in Iskenderun Bay (Turkey) (0.03-217 µg/g w.w.), Uluozlu *et al.* (2007) in Black and Aegean Seas (68.6-163 µg/g d.w.), Tuzen (2009) in Black Sea (36.2-145 µg/g w.w.), Türkmen *et al.* (2009b) in Aegean and Mediterranean Seas (0.51-7.05 µg/g w.w.). The Fe levels in the muscle of carnivore and demersal fish, carnivore and pelagic fish, omnivore and bento-pelagic fish, and omnivore and pelagic fish were 0.21-217 µg/g w.w. (Ateş *et al.*, 2015, Çiçek *et al.*, 2008, Çoğun *et al.*, 2005, Çoğun *et al.*, 2006, Ersoy and Çelik, 2010, Manaşirli *et al.*, 2016, Tepe, 2008, Türkmen *et al.*, 2005, Türkmen *et al.*, 2013), 0.03-57.5 µg/g w.w. (Ersoy and Çelik, 2009, Türkmen *et al.*, 2009a, Yilmaz, 2003), 1.08-98.1 µg/g w.w. (Çiçek *et al.*, 2008, Çoğun *et al.*, 2006, Dural *et al.*, 2009, Dural *et al.*, 2006, Manaşirli *et al.*, 2016, Tepe, 2008, Türkmen *et al.*, 2006, Yilmaz, 2003, Yilmaz, 2005), and 1.18-61.6 µg/g w.w. (Ersoy and Çelik, 2009, Tepe, 2008), respectively. In addition, nearly in all previous studies, the mean rate of Fe accumulation is in the following order: gonad > gill > skin > liver > muscle, like Zn concentrations. In most species and most conditions, the gonad, gill, skin, liver, and muscle contained 1.8-478 µg/g w.w. (Dural *et al.*, 2006, Yilmaz, 2003, Yilmaz, 2005), 15-137 µg/g w.w. (Çoğun *et al.*, 2006, Dural *et al.*, 2011, Dural *et al.*, 2006, Kalay *et al.*, 1999, Kargin, 1996, Türkmen *et al.*, 2013), 3.17-194 µg/g w.w. (Dural and BïCkïCï, 2010, Dural *et al.*, 2010, Dural *et al.*, 2009, Yilmaz, 2003, Yilmaz, 2005, Yılmaz *et al.*, 2010), 10.3-520 µg/g w.w. (Çoğun *et al.*, 2006, Dural *et al.*, 2009, Dural *et al.*, 2006, Ersoy and Çelik, 2010, Ersoy and Çelik, 2009, Kargin, 1996, Türkmen *et al.*, 2013), and 0.03-217 µg/g w.w. (Çoğun *et al.*, 2006, Dural *et al.*, 2009, Dural *et al.*, 2006, Ersoy and Çelik, 2010, Ersoy and Çelik, 2009, Kargin, 1996, Türkmen *et al.*, 2013), respectively. The considerably higher Fe concentrations in the liver of fish relative to muscle tissue are expected due to the physiological role of this organ in blood synthesis. The higher Fe values observed in spleen and intestine range from 442 to 638 µg/g w.w. (Kargin, 1996) and from 51.9 to 262 µg/g w.w. (Dural *et al.*, 2011, Dural *et al.*, 2009, Türkmen *et al.*, 2013). Biomagnification also appears unlikely under most environmental conditions because the combination of homeostatic regulation and abiotic factors limits aqueous and dietary Fe bioaccumulation in fish (Nfon *et al.*, 2009, Winterbourn *et al.*, 2000).

Chromium (Cr) is known to enhance the action of insulin, a hormone critical to the metabolism and storage of carbohydrate, fat, and protein in the body. It primarily has two valence states in the ocean: particle-active trivalent form, Cr(III), and more soluble hexavalent form, Cr(VI). Cr(III) is possibly required in trace amounts in human metabolism for sugar and lipid metabolism, but Cr(VI) is very toxic and mutagenic when inhaled (Vincent, 2000). Marine organisms have evolved mechanisms for accumulating and regulating Cr from seawater and food (Wood *et al.*, 2011). The Cr concentrations data of previous studies are given in

Table **1**, and ranged from 17.6 to 121 ng/g d.w. in the coastal area of China (Zhang and Wang, 2012), 20 to 1260 ng/g w.w. in the South China Sea (Gu *et al.*, 2017, Gu *et al.*, 2015a), 110 to 4270 ng/g w.w. in Pearl River Estuary (Ip *et al.*, 2005, Zhang and Wang, 2012), 3600 to 5130 ng/g d.w. in Yellow River Estuary (Cui *et al.*, 2011), 60 to 650 ng/g d.w. in Daya Bay (China) (Du *et al.*, 2017), 35 to 156 ng/g d.w. in Jiaozhou Bay (China) (Unpublished data), 2310 to 2750 ng/g w.w. in Qinzhou Bay (China) (Gu *et al.*, 2015b), nd to 6566 ng/g w.w. in Iskenderun Bay (Turkey) (Yilmaz *et al.*, 2017), 950 to 1980 ng/g d.w. in Black and Aegean Seas (Uluozlu *et al.*, 2007), 630 to 1740 ng/g w.w. in Black Sea (Tuzen, 2009), <10 to 450 ng/g w.w. in Aegean and Mediterranean Seas (Türkmen *et al.*, 2009b). Previous studies have shown that the Cr accumulation in fish might be related to many factors, such as fish species, feeding habits, habitats, physio-chemical parameters of the aquatic environment, and bioavailability of chemicals in food and water. The Cr concentrations in fish generally decreased in the order: carnivore and demersal fish species (nd-6566 ng/g w.w.) (Ateş *et al.*, 2015, Ersoy and Çelik, 2010, Kalay *et al.*, 1999, Türkmen *et al.*, 2009a, Türkmen *et al.*, 2013) > carnivore and pelagic fish species (40-1870 ng/g w.w.) (Ersoy and Çelik, 2009, Türkmen *et al.*, 2009a, Yilmaz, 2003) > omnivore and bento-pelagic fish species (120-1790 ng/g w.w.) (Dural *et al.*, 2009, Ersoy and Çelik, 2010, Tepe, 2008, Türkmen *et al.*, 2006, Yilmaz, 2003, Yilmaz, 2005) > omnivore and pelagic fish species (40-740 ng/g w.w.) (Ersoy and Çelik, 2009, Tepe, 2008, Türkmen *et al.*, 2009a). The Cr concentrations in tissues generally decreased in the order: liver (nd-30139 ng/g w.w.) (Dural *et al.*, 2009, Ersoy and Çelik, 2010, Ersoy and Çelik, 2009, Kalay *et al.*, 1999, Türkmen *et al.*, 2009a) > skin (44-27260 ng/g w.w.) (Dural *et al.*, 2009, Yilmaz, 2003, Yilmaz, 2005) > gill (nd-19400 ng/g w.w.) (Dural *et al.*, 2011, Kalay *et al.*, 1999, Türkmen *et al.*, 2013) > gonad (nd-16820 ng/g w.w.) (Yilmaz, 2003, Yilmaz, 2005) > intestine (456-6123 ng/g w.w.) (Dural *et al.*, 2011, Dural *et al.*, 2009, Türkmen *et al.*, 2013) > muscle (nd-6566 ng/g w.w.) (Dural *et al.*, 2009, Ersoy and Çelik, 2010, Ersoy and Çelik, 2009, Kalay *et al.*, 1999, Türkmen *et al.*, 2009a). According to the existed literature, no significant biomagnification of Cr in the aquatic food web has been reported (Seenayya and Prahalad, 1987). The lack of any biomagnification of Cr is likely due to the conversion of Cr(III) and the virtual indigestibility of Cr(III).

Table 1. Comparison of heavy metal concentration in fish from different areas. dw: dry weight; ww: wet weight.

Locations		Al (µg/g)	As (µg/g)	Cd (ng/g)	Cr (ng/g)	Cu (µg/g)	Fe (µg/g)	Mo (ng/g)	Ni (ng/g)	Pb (ng/g)	Zn (µg/g)	Mn (µg/g)	Co (ng/g)	Reference
		Different Trace Element Concentrations												
Qin Huang Dao	dw	5.68-37.0	4.0-48.4	9.68-59.9	48.3-121	1.43-5.88	1329-4185	8.75-33.8	39.8-149	10.9-290	21.6-79.0			Zhang et al. (2012)
Da Lian	dw	7.79-10.6	3.41-7.04	19.4-42.0	55.9-81.7	1.0-2.87	1472-2571	10.8-16.6	43.7-75.8	9.65-46.4	18.4-24.6			
Qing Dao	dw	9.02-17.2	6.72-14.3	17.0-30.5	59.3-74.4	1.15-5.78	1033-6329	14.1-34.1	46.1-153	5.43-90.2	27.2-41.8			
Shang Hai	dw	6.19-22.3	2.09-17.6	5.49-71.3	55.1-89.9	0.76-1.27	797-2609	13.3-35.4	36-141	12.5-53.3	18.5-79.1			
Hui Lai	dw	6.87-11.1	17.9-36.1	4.09-6.75	50.4-56.0	0.82-1.29	849-2963	9.69-12.6	62.9-92.3	29.0-70.0	16.4-18.7			
Zhan Jiang	dw	1.19-21.8	4.81-134	2.37-54.1	17.6-87.0	0.34-1.94	474-9937	9.27-38.7	22.3-125	6.91-119	10.4-42.4			
Hai Nan	dw	2.54-8.97	3.43-24.8	2.43-69.1	55.2-98.2	0.43-1.74	550-1983	11.1-18	26.7-95.9	2.69-51.2	14.8-25.4			
The South China Sea	ww			0.51-116	20-1260	0.12-1.13	2510-22990		8.32-250	0.54-680	2.34-6.88	0.04-0.81		Gu et al. (2015a, 2017)
Pearl River Estuary	dw	18.0-47.6	2.69-14.8	2.54-29.6	71.3-109	1.09-7.35	3441-11788	10.8-81.4	126-461	54.2-128	19.6-52.3			Zhang et al. (2012)
Pearl River Estuary	ww			10-130	110-4270	0.15-7.55			170-2080	90-30700	8.78-30.3		20-480	Ip et al. (2005)
Yangtze River Estuary			0.29-0.99	80-310		1.74-4.08				0-2200	1.85-15.4			Huang et al. (2011)
Yellow River Estuary	dw		0.23-8.97	nd	3600-5130	2.32-11.2			1350-3520	nd-910	21.6-79.4	2.41-4.29		Cui et al. (2011)

182 Marine Ecology: Current and Future Developments, Vol. 1

(Table 1) cont......

Different Trace Element Concentrations

Locations		Al (µg/g)	As (µg/g)	Cd (ng/g)	Cr (ng/g)	Cu (µg/g)	Fe (µg/g)	Mo (ng/g)	Ni (ng/g)	Pb (ng/g)	Zn (µg/g)	Mn (µg/g)	Co (ng/g)	Reference
Daya Bay	dw		2.65-36.8	10-80	60-650	0.36-6.99			10-1350	30-580	11.7-40.9		2-620	Du et al. (2017), Gu et al. (2016), Wang et al. (2009)
Jiaozhou Bay	dw		6.1-28.7	11-90	35-156	0.6-11.6		7-60	19-732	34-138	14.5-49.2		10-46	Unpublished data
Qinzhou Bay	ww			1-154	2310-2750	0.01-0.62			410-810	64-299	5.73-7.18			Gu et al. (2015b)
Iskenderun Bay (Turkey)	ww	0.1-17.0	0.98-1.74	nd-7600	nd-6566	nd-6.24	0.03-217		nd-4720	nd-10870	0.016-51.1	nd-2.01	nd-571	Yilmaz et al. (2017, 2018)
Black and Aegean Seas	dw			450-900	950-1980	0.73-1.83	68.6-163		1920-5680	330-930	35.4-106			Uluozlu et al. (2007)
Black Sea	ww		0.11-0.32	100-350	630-1740	0.65-2.78	36.2-145		1140-3600	280-870	38.8-93.4			Tuzen (2009)
Aegean and Mediterranean	ww			<10-390	<10-450	0.07-1.48	0.51-7.05		180-2780	30-1720	3.51-53.5			Türkmen et al. (2009b)
Adriatic Sea	ww		0.01-70.9	1-850		0.001-57.3				1-340				Bilandžić et al. (2011)

Manganese (Mn) is an essential element for both animals and plants with some identified functions in enzymes and proteins. The Mn deficiency results in severe skeletal and reproductive abnormalities in mammals (Schramm and Wedler, 1986). The Mn concentrations data of previous studies are listed in Table 1. Gu *et al.* (2015a, 2017), Cui *et al.* (2011) and Yilmaz *et al.* (2017) reported that the Mn concentrations were 0.04-0.81 µg/g w.w. in the South China Sea, 2.41-4.29 µg/g d.w. in Yellow River Estuary, and nd-2.01 µg/g w.w in Iskenderun Bay (Turkey), respectively. Nearly in all previous studies, the Mn concentrations are in the following order: gill (nd-20.4 µg/g w.w.) (Dural *et al.*, 2011, Türkmen *et al.*, 2013) > liver (nd-15.1 µg/g w.w.) (Dural *et al.*, 2009, Ersoy and Çelik, 2010, Ersoy and Çelik, 2009, Türkmen *et al.*, 2009a, Türkmen *et al.*, 2013) > intestine (2.63-14.8 µg/g w.w.) (Dural *et al.*, 2011, Dural *et al.*, 2009, Türkmen *et al.*, 2013) > skin (nd-6.37 µg/g w.w.) (Dural *et al.*, 2011, Dural *et al.*, 2009, Yılmaz *et al.*, 2010) > muscle (nd-2.01 µg/g w.w.) (Dural *et al.*, 2009, Ersoy and Çelik, 2010, Ersoy and Çelik, 2009, Türkmen *et al.*, 2009a, Türkmen *et al.*, 2013) > gonads (nd-0.77 µg/g w.w.) (Yilmaz *et al.*, 2017). The Mn values in the carnivore and demersal fish species, carnivore and pelagic fish species, omnivore and bento-pelagic fish species, and omnivore and pelagic fish species were nd-2.01 µg/g w.w. (Ersoy and Çelik, 2010, Türkmen *et al.*, 2005, Türkmen *et al.*, 2013), 0.08-0.85 µg/g w.w. (Ersoy and Çelik, 2009, Türkmen *et al.*, 2009a), 0.26-1.21 µg/g w.w. (Dural *et al.*, 2009, Tepe, 2008, Türkmen *et al.*, 2006), and 0.02-0.58 µg/g w.w. (Ersoy and Çelik, 2009, Tepe, 2008), respectively.

Nickel (Ni) is well established as an essential nutrient for plants and terrestrial animals, but not in aquatic animals (Muyssen *et al.*, 2004). However, the evidence is mounting to suggest that Ni is probably essential in fish. The Ni bioaccumulation in fish has received relatively little research attention compared to other more toxic metals, such as Cu or Cd. Table 1 shows the Ni concentration in different fish from different regions, such as the coastal area of China (22.3-153 ng/g d.w.) (Zhang and Wang, 2012), the South China Sea (8.32-250 ng/g w.w.) (Gu *et al.*, 2017, Gu *et al.*, 2015a), the Pearl River Estuary (126-461 ng/g d.w. and 170-2080 ng/g w.w.) (Ip *et al.*, 2005, Zhang and Wang, 2012), the Yellow River Estuary (1350-3520 ng/g d.w.) (Cui *et al.*, 2011), the Daya Bay (China) (10-1350 ng/g d.w.) (Du *et al.*, 2017), and the Jiaozhou Bay (China) (19-732 ng/g d.w.) (Unpublished data), the Qinzhou Bay (China) (410-810 ng/g w.w.) (Gu *et al.*, 2015b), the Iskenderun Bay (Turkey) (nd-4720 ng/g w.w.) (Yilmaz *et al.*, 2018), the Black and Aegean Seas (1920-5680 ng/g d.w.) (Uluozlu *et al.*, 2007), the Black Sea (1140-3600 ng/g w.w.) (Tuzen, 2009), and the Aegean and Mediterranean Seas (180-2780 ng/g w.w.) (Türkmen *et al.*, 2009b). Many researchers have dealt with carnivore and demersal fish, carnivore and pelagic fish, omnivore and bento-pelagic fish, and omnivore and pelagic fish species, and found the Ni values were nd-4720 ng/g w.w. (Ersoy and Çelik, 2010, Türkmen *et al.*, 2005, Türkmen *et al.*, 2013, Yilmaz, 2005), 100-1630 ng/g w.w. (Ersoy and Çelik, 2009, Turan *et al.*, 2009, Türkmen *et al.*, 2009a, Türkmen *et al.*, 2009b, Yilmaz, 2003), 130-1720 ng/g w.w. (Kalay *et al.*, 1999, Tepe, 2008, Türkmen *et al.*, 2006, Yilmaz, 2003, Yilmaz, 2005), and 120-550 ng/g w.w. (Ersoy and Çelik, 2009, Tepe, 2008, Türkmen *et al.*, 2009b), respectively. Typical Ni levels in gonad (0.43-24.4 ng/g w.w.) (Yilmaz, 2003, Yilmaz, 2005), gill (nd-25.3 ng/g w.w.) (Dural *et al.*, 2010, Yilmaz, 2003, Yılmaz *et al.*, 2010), liver (nd-9.99 ng/g w.w.) (Ersoy and Çelik, 2010, Ersoy and Çelik, 2009, Türkmen *et al.*, 2005, Türkmen *et al.*, 2013), skin (0.026-7.91 ng/g w.w.) (Kalay *et al.*, 1999, Türkmen *et al.*, 2013, Yilmaz, 2005), muscle (nd-4.72 ng/g w.w.) (Ersoy and Çelik, 2010, Ersoy and Çelik, 2009, Türkmen *et al.*, 2005, Türkmen *et al.*, 2013), and intestine (2.15-3.78 ng/g w.w.) (Türkmen *et al.*, 2013) were reported in previous studies. With regards to the results in wild marine fish, the pattern of Ni occurrence in the selected tissue can be written in descending order as follows: gonad > gill > liver > skin > intestine > muscle. There is no evidence for either biomagnification or bioconcentration of Ni in aquatic ecosystem. Muyssen *et al.* (2004) reported a negative relationship between exposure concentration and bioconcentration factors in fish after reviewing the available literature. They suggested that this negative relationship can be explained by fish actively regulating Ni uptake and elimination processes.

Cobalt (Co) is an essential element to fish and other organisms. It is the active center of coenzymes called cobalamins, the most common example of which is vitamin B12 (Wood *et al.*, 2011). It has also been implicated in blood pressure regulation and been found to be necessary for proper thyroid function. The Co accumulations in fish have received less attention when compared to other metals, and the reported values show that Co is not strongly accumulated in most fish species and as is the case with several other metals, Co does not appear to biomagnify in the food web extensively (Wood *et al.*, 2011). Table **1** shows the data of Co concentrations in some areas from previous studies. The Co levels in wild fish ranged from 20 to 480 ng/g w.w. in Pearl River Estuary (Ip *et al.*, 2005), 2 to 620 ng/g d.w. in Daya Bay (China) (Du *et al.*, 2017), 10 to 46 ng/g d.w. in Jiaozhou Bay (China) (Unpublished data), and nd to 571 ng/g w.w. in Iskenderun Bay (Turkey) (Yilmaz *et al.*, 2018). The Co levels in the muscles of carnivore and demersal fish, carnivore and pelagic fish, and omnivore and bento-pelagic fish species, were nd-571 ng/g w.w. (Tepe, 2008, Türkmen *et al.*, 2005, Türkmen *et al.*, 2013, Türkmen *et al.*, 2009b, Yılmaz *et al.*, 2010), 30-420 ng/g d.w. (Türkmen *et al.*, 2009a, Türkmen *et al.*, 2009b), and 66-3824 ng/g d.w. (Tepe, 2008, Türkmen *et al.*, 2006), respectively. A comparison of the ranges of tissue-specific Co accumulation shows that the gills (nd-590 ng/g w.w.) (Türkmen *et al.*, 2013) and skins (nd-210 ng/g w.w.) (Yılmaz *et al.*, 2010) are the strongest accumulators of Co, followed by the liver (70-1240 ng/g w.w.) (Ateş *et al.*, 2015,

Tepe, 2008, Türkmen *et al.*, 2009a, Türkmen *et al.*, 2013, Türkmen *et al.*, 2009b, Yılmaz *et al.*, 2010) and the intestine (190-270 ng/g w.w.) (Türkmen *et al.*, 2013), and that muscle tissue (nd-571 ng/g w.w.) (Ateş *et al.*, 2015, Tepe, 2008, Türkmen *et al.*, 2009a, Türkmen *et al.*, 2013, Türkmen *et al.*, 2009b, Yılmaz *et al.*, 2010) is not a strong accumulator of Co. The relative higher accumulation in certain tissues has been related to the exposure pathway (*i.e.* gills and gut), the physiological role of the organ (*i.e.* hematopoietic function), and the excretion pathway (*i.e.* gut and kidney) (Baudin and Fritsch, 1989).

Aluminum (Al) has no established biological function but can be extremely toxic to fish when solubilized under acidic (pH<6) or alkaline conditions (pH>8) (Gensemer and Playle, 1999). It can be taken with drinking water as well as diet, and water-soluble forms are more active for biological accumulation. Dietary uptake has only been reported at very high oral doses and biomagnification *via* the food chain does not occur (Wood *et al.*, 2012). Table **1** shows the data of Al concentration from previous studies in the coastal area of China (1.19-37.0 μg/g d.w.) (Zhang and Wang, 2012), the Pearl River Estuary (18.0-47.6 μg/g d.w.) (Zhang and Wang, 2012), and Iskenderun Bay (Turkey) (0.51-84.8 μg/g d.w.) (Yilmaz *et al.*, 2018). The Al levels in the muscle of *M. barbatus*, *M. surmuletus*, *S. undosquamis*, and *S. aurata*, carnivore and demersal, were 1.61-8.38 μg/g d.w. (Dural *et al.*, 2010, Turan *et al.*, 2009, Türkmen *et al.*, 2005), 7.52- 37.6 μg/g d.w. (Dural *et al.*, 2010), 0.51-1.23 μg/g d.w. (Türkmen *et al.*, 2005), and 0.69-1.23 μg/g d.w. (Türkmen *et al.*, 2005), respectively. The Al concentrations in the muscle of *M. cephalus* (omnivore and bento-pelagic) ranged from 0.61 to 1.71 μg/g d.w. (Türkmen *et al.*, 2006). However, the higher Al values were observed in *Engraulis encrasicholus* (carnivore and pelagic) (24.8±4.80 μg/g d.w.) and *Merlangius merlangus* (carnivore and bento-pelagic) (84.8±7.84 μg/g d.w.) (Turan *et al.*, 2009). Although the levels of Al accumulation vary with such factors as fish species, habitats, feeding habits, *etc.*, the gill is the target organ of rapid adsorption from waterborne exposure to Al, whereas cellular uptake from the water is slow, but gradual accumulation in internal organs does occur over time (Wood *et al.*, 2012). The mean values of Al in the liver, skin, and muscle ranged from 2.65 to 14.0 μg/g w.w. (Dural and BiCkiCi, 2010, Dural *et al.*, 2010, Yılmaz *et al.*, 2010), 1.88 to 16.6 μg/g w.w. (Dural and BiCkiCi, 2010, Dural *et al.*, 2010, Yılmaz *et al.*, 2010), and 0.1 to 17.0 μg/g w.w. (Dural and BiCkiCi, 2010, Dural *et al.*, 2010, Turan *et al.*, 2009, Türkmen *et al.*, 2005, Yılmaz *et al.*, 2010).

Arsenic (As) is a moderately toxic, naturally abundant metalloid with no known nutritional or metabolic roles (Neff, 1997). It can occur in several oxidation states, but in natural waters and sediments mostly in the inorganic forms as oxyanions of trivalent arsenite [As(III)] or pentavalent arsenate [As(V)], with organic forms as monomethylarsonous acid (MMA), dimethylarsinic acid (DMA), arsenobetaine (AsB) and arsenocholine (AsC) being major components in marine fish (Akter *et al.*, 2005). It has been acknowledged that the accumulation of As in marine fish, influenced by the concentration and chemical form of As in water, trophic level, species, and diet, vary considerably between species and among different site (Rahman *et al.*, 2012). In wild marine fish, the As concentrations in the muscle of different fish varied from 2.09 to 134 µg/g d.w. in the coastal area of China (Zhang and Wang, 2012), 2.69 to 14.8 µg/g d.w. in the Pearl River Estuary (Zhang and Wang, 2012), 0.29 to 0.99 µg/g in Yangtze River Estuary (Huang *et al.*, 2011), 0.23 to 8.97 µg/g d.w. in Yellow River Estuary (Cui *et al.*, 2011), 2.65 to 36.8 µg/g d.w. in Daya Bay (China) (Du *et al.*, 2017) and 6.1 to 28.7 µg/g d.w. in Jiaozhou Bay (China) (Unpublished data), 0.98 to 1.74 µg/g w.w. in Iskenderun Bay (Turkey) (Yilmaz *et al.*, 2018), 0.11 to 0.32 µg/g w.w. in Black Sea (Tuzen, 2009), and 0.01 to 70.9 µg/g w.w. in Adriatic Sea (Bilandžić *et al.*, 2011). In addition, the As levels in planktivores (alewife and killifish; approximately 0.07-0.075 µg/g w.w.) were higher than omnivores (black crappie, bluegill sunfish and yellow perch; approximately 0.03-0.04 µg/g w.w.) and a piscivore (largemouth bass; approximately 0.04 µg/g w.w.), presumably due to the differences in their feeding strategies (Chen and Folt, 2000). Yılmaz *et al.* (2010) found the highest As level in the liver of *Solea lascaris* (1.98 µg/g w.w), while mean value of As levels in *Lophius budegassa* and *S. lascaris* were liver > muscle > skin, and in *T. lucerna*, muscle > liver > skin. The highest level of As belonged to the muscle of a demersal fish *M. merluccius*, living in the Adriatic Sea, while lower levels were found in tissues of pelagic species like *Trachurus trachurus* and *Sardina pilchardus* (Jureša and Blanuša, 2003). Biomagnification does not occur with As. Conversely, As diminishes through increasing trophic levels (Chen and Folt, 2000, Ikemoto *et al.*, 2008).

Cadmium (Cd), a nonessential and toxicologically-significant metal, is considered to have no any specific uptake mechanism and appears to behave adventitiously, following existing pathways for essential metals (Hilmy *et al.*, 1985). Cd can displace Cu and Zn from the cells' enzyme pools, thus disrupting Cu and Zn homeostasis (Wood *et al.*, 2012). Not surprisingly, high concentrations of Cd may result in enzyme inhibition, and enzymes from different tissues may also be affected differently by Cd. Cd has received considerable attention due to its high toxicity and the fact that there is little knowledge concerning its function in the cells. Table 1 shows the concentration data for Cd from previous studies in the coastal area of China (2.37-71.3 ng/g d.w.) (Zhang and Wang, 2012), in the South China Sea (0.51-116 ng/g w.w.) (Gu *et al.*, 2017, Gu *et al.*, 2015a), in the Pearl River Estuary (2.54-29.6 ng/g d.w.) (Ip *et al.*, 2005, Zhang and Wang, 2012), in Yangtze River Estuary (80-310 ng/g) (Huang *et al.*, 2011), in Yellow River Estuary (not detected) (Cui *et al.*, 2011), in Daya Bay (China) (10-80 ng/g d.w.)

(Du *et al.*, 2017), in Jiaozhou Bay (China) (11-90 ng/g d.w.) (Unpublished data), in Qinzhou Bay (China) (1-154 ng/g w.w.) (Gu *et al.*, 2015b), in Iskenderun Bay (Turkey) (nd-7600 ng/g w.w.) (Yilmaz *et al.*, 2018), in Black and Aegean Seas (450-900 ng/g d.w.) (Uluozlu *et al.*, 2007), in Black Sea (100-350 ng/g w.w.) (Tuzen, 2009), in Aegean and Mediterranean Seas (<10-390 ng/g w.w.) (Türkmen *et al.*, 2009b), and in Adriatic Sea (1-850 ng/g w.w.) (Bilandžić *et al.*, 2011). The Cd concentrations in carnivore and demersal fish, carnivore and pelagic fish, omnivore and bento-pelagic fish, and omnivore and pelagic fish were nd-7600 ng/g w.w. (Çoğun *et al.*, 2006, Ersoy and Çelik, 2010, Ersoy and Çelik, 2009, Kargin, 1996, Türkmen *et al.*, 2005), <10-380 ng/g w.w. (Ersoy and Çelik, 2009, Türkmen *et al.*, 2009a, Türkmen *et al.*, 2009b), 12-440 ng/g w.w. (Çoğun *et al.*, 2006, Dural *et al.*, 2006, Tepe, 2008, Türkmen *et al.*, 2006), and 40-170 ng/g w.w. (Ersoy and Çelik, 2009, Türkmen *et al.*, 2009b), respectively. In marine fish, Cd accumulated maximally in kidney (3880-25300 ng/g w.w.) (Kargin, 1996), liver (nd-19900 ng/g w.w.) (Dural *et al.*, 2006, Ersoy and Çelik, 2010, Ersoy and Çelik, 2009, Türkmen *et al.*, 2009a), gill (76-94000 ng/g w.w.) (Çoğun *et al.*, 2006, Dural *et al.*, 2006, Kargin, 1996, Türkmen *et al.*, 2006, Türkmen *et al.*, 2013), intestine (100-3170 ng/g w.w.) (Dural *et al.*, 2009, Türkmen *et al.*, 2013), to a lesser extent in muscle (nd-7600 ng/g w.w.) (Dural *et al.*, 2006, Ersoy and Çelik, 2010, Ersoy and Çelik, 2009, Türkmen *et al.*, 2009a), skin (nd-2044 ng/g w.w.) (Dural *et al.*, 2011, Dural *et al.*, 2009, Yılmaz *et al.*, 2010), and gonad (76-90 ng/g w.w.) (Dural *et al.*, 2006), although the pattern of accumulation varied with exposure route. As reviewed by Suedel *et al.* (1994), little evidence existed to suggest biomagnification in aquatic systems, however, a few examples of Cd biomagnification can be found in the peer-reviewed literature (Croteau *et al.*, 2005, Wang, 2002).

Lead (Pb) has no known biological function and there exists no evidence that it is required, or otherwise beneficial for life. Furthermore, Pb is toxic even at low doses, which apparently exerts its toxic effects by binding to cellular binding sites and biomolecules, such as enzymes and hormones (Demayo *et al.*, 1982). It is well known that fish are able to bioaccumulate Pb from the water and diet within various tissues. However, the degree to which Pb will bioaccumulate depends on a number of factors including developmental stage, water quality, diet type, Pb concentration, and variability within and among population and/or species. The Pb concentrations in different fish are shown in Table **1** from previous studies, such as 2.69-290 ng/g d.w. in the coastal area of China (Zhang and Wang, 2012), 0.54-680 ng/g w.w. in the South China Sea (Gu *et al.*, 2017, Gu *et al.*, 2015a), 54.2-128 ng/g d.w. (Zhang and Wang, 2012) and 90-30700 ng/g w.w. (Ip *et al.*, 2005) in Pearl River Estuary, 0-2200 ng/g in Yangtze River Estuary (Huang *et al.*, 2011), nd-910 ng/g d.w. in Yellow River Estuary (Cui *et al.*, 2011), 30-580 ng/g d.w. in Daya Bay (China) (Du *et al.*, 2017), 34-138 ng/g d.w. in Jiaozhou Bay

(China) (Unpublished data), 64-299 ng/g w.w. in Qinzhou Bay (China) (Gu *et al.*, 2015b), nd-10870 ng/g w.w. in Iskenderun Bay (Turkey) (Yilmaz *et al.*, 2018), 330-930 ng/g d.w. in Black and Aegean Seas (Uluozlu *et al.*, 2007), 280-870 ng/g w.w. in Black Sea (Tuzen, 2009), 30-1720 ng/g w.w. in Aegean and Mediterranean Seas (Türkmen *et al.*, 2009b), and 1-340 ng/g w.w. in Adriatic Sea (Bilandžić *et al.*, 2011). Generally, the Pb concentrations show significant differences among the seasons, stations, fish species, and tissues in all samples. Many researchers have put forward the Pb values in the muscles of carnivore and demersal fish (nd-5700 ng/g w.w.) (Çoğun *et al.*, 2006, Ersoy and Çelik, 2010, Ersoy and Çelik, 2009, Türkmen *et al.*, 2005), carnivore and pelagic fish (140-1380 ng/g w.w.) (Ersoy and Çelik, 2009, Türkmen *et al.*, 2009a, Türkmen *et al.*, 2009b, Yilmaz, 2003), omnivore and bento-pelagic fish (250-10870 ng/g w.w.) (Çoğun *et al.*, 2006, Dural *et al.*, 2009, Kalay *et al.*, 1999, Tepe, 2008, Türkmen *et al.*, 2006, Yilmaz, 2003), and omnivore and pelagic fish (140-640 ng/g w.w.) (Ersoy and Çelik, 2009, Tepe, 2008, Türkmen *et al.*, 2009b). In wild marine fish, the mean values of Pb concentration were gonads (470-90970 ng/g w.w.) (Yilmaz, 2003, Yilmaz, 2005) > spleen (32860-63700 ng/g w.w.) (Kargin, 1996) > skin (1010-57910 ng/g w.w.) (Dural *et al.*, 2010, Dural *et al.*, 2009, Yilmaz, 2003, Yilmaz, 2005, Yılmaz *et al.*, 2010) > kidney (23520-41680 ng/g w.w.) (Kargin, 1996) > gill (550-9540 ng/g w.w.) (Çoğun *et al.*, 2005, Çoğun *et al.*, 2006, Kalay *et al.*, 1999, Türkmen *et al.*, 2013) > intestine (670-9125 ng/g w.w.) (Dural *et al.*, 2011, Dural *et al.*, 2009, Türkmen *et al.*, 2013) > liver (140-14800 ng/g w.w.) (Çoğun *et al.*, 2006, Dural *et al.*, 2009, Ersoy and Çelik, 2010, Ersoy and Çelik, 2009, Türkmen *et al.*, 2006, Türkmen *et al.*, 2009b) > muscle (nd-10870 ng/g w.w.) (Çoğun *et al.*, 2006, Dural *et al.*, 2009, Ersoy and Çelik, 2010, Ersoy and Çelik, 2009, Türkmen *et al.*, 2006, Türkmen *et al.*, 2009b), true for nearly all previous studies. It is reported that Pb does not biomagnify along the food web, which is likely due to the fact that most accumulated Pb is sequestered in calcified tissues (*e.g.* bone) and is scarcely found in the large and far more readily assimilated tissue masses of the muscles (Demayo *et al.*, 1982, Farag *et al.*, 1998, Settle and Patterson, 1980). Furthermore, at the subcellular level, Pb is portioned to metal-rich granules, which putatively represent a largely non-trophically available fraction in potential prey (Goto and Wallace, 2010). Although Pb does not biomagnify within the food web, some trophic transfer assuredly takes place for some species (Farag *et al.*, 1994).

In summary, the bioaccumulation data recorded in marine fish have shown large variations in geographic regions, interspecies and intraspecies. To explain why such complicated and variable concentrations of heavy metals are observed in marine fish requests to evaluate both the flows into the body (influx) and out of the body (efflux). Wang *et al.* (1996) developed a biokinetic model to study the influx and efflux of heavy metals in marine mussels, and then this biokinetic model was widely used to study the bioaccumulation in other organisms. In this model, the bioaccumulation of heavy metals in the organism can be quantified by the following equation (Wang *et al.*, 1996):

$$\frac{dC_t}{dt} = k_u \times C_w + AE \times IR \times C_f - (k_e + g) \times C_t \tag{1}$$

where C_t is the metal concentration in the marine fish at time t, k_u is the uptake rate constant from the dissolved phase (L/g/d), C_w is the metal concentration in the dissolved phase (µg/L), AE is the metal assimilation efficiency in the dietary phase, IR is the weight-specific ingestion rate of the fish (g/g/d), C_f is the metal concentration in the dietary phase (µg/g), k_e is the efflux rate constant (/d), and g is the growth rate constant (/d).

The first term describes the first-order dissolved uptake and the second term describes the first order dietary uptake. The third term describes the loss of heavy metals from organisms. To explain the observed geographic variations and inter/intra-species differences, it is noted that heavy metals are accumulated from both dissolved and dietary sources in marine fish. In general, almost all heavy metals in fish are from ingestion of food particles (Wang, 2002). Thus, both the differences of heavy metals in the ambient environment (C_w, C_f) and the biological processes of the fish (k_u, AE, k_e, g) could result in these differences in heavy metal concentrations.

Under steady-state conditions, the concentrations of heavy metals in the aquatic animals (C_{ss}) can be calculated as:

$$C_{ss} = [(k_u \times C_w + AE \times IR \times C_f)]/[(k_e + g)] \tag{2}$$

Wang (2016) reported that total metal concentration in the water could be calculated from the C_w, C_f, and total suspended solids (TSS), whereas C_f could be calculated from C_w and partitioning coefficient of metals in the particles (k_d):

$$C_t = C_w + (C_f \times TSS) \tag{3}$$

$$C_t = C_w + (C_w \times k_d \times TSS) \tag{4}$$

$$C_w = C_t/(1 + k_d \times TSS) \tag{5}$$

$$C_{ss} = [(k_u + AE \times IR \times k_d) \times C_t]/[(1 + TSS \times k_d) \times (k_e + g)] \qquad (6)$$

Thus, the bioaccumulation factor of metals in the aquatic animals, which reflects the bioaccumulation potential of metals, can be derived as:

$$BAF = C_{ss}/C_t \qquad (7)$$

$$BAF = [(k_u + AE \times IR \times k_d)]/[(1 + TSS \times k_d) \times (k_e + g)] \qquad (8)$$

Clearly, the bioaccumulation potential, which can be quantified by BAF, is determined by many biokinetic parameters (k_u, AE, k_e), aquatic animals' physiology (IR, g), metal geochemistry (k_d), and environmental conditions (TSS). Comparison of bioaccumulation potentials among species and metals needs to address the differences among these parameters.

Numerous studies have demonstrated that differences in diets of fish are the key factor affecting heavy metal levels between and within fish species. For example, variations in fish Hg levels can occur with varied food web structure (Ferriss and Essington, 2014), habitat specific foraging (Eagles-Smith *et al.*, 2008, Karimi *et al.*, 2016), as well as food availability (quality/quantity) (Wang and Wang, 2012). Different diets may affect the heavy metal biodynamic processes in fish, especially the AE. Considerable variations in the AEs of metals (*e.g.*, Zn, Cd, As and inorganic Hg) in both marine and freshwater fish have been reported when feeding with different diets, probably regulated by both physiological (gut passage time and ingestion rate) and biochemical mechanisms (subcellular distribution and metal partitioning) (Wang and Wong, 2003, Zhang and Wang, 2006), and may be concentration-dependent (Guan and Wang, 2004). Thus, the different feeding habits and living habitats significantly affect the intake, bio-assimilation, and subsequent bioaccumulation of heavy metals in different fish.

There are also major differences in the k_us of heavy metals among different fish species. For example, Veltman *et al.* (2008) demonstrated the significant relationships between the covalent index representing the binding affinity of metals/biotic ligand and the k_u of 10 metals in 17 aquatic species. Zhang and Wang (2007b) reported that relatively small animals displayed a higher k_u than larger animals. Environmental factors can also considerably affect the k_u of contaminants. Among these environmental factors, salinity, temperature, DOM, other competing ions, such as H^+, Ca^{2+}, Mg^{2+}, and dissolved oxygen, have received the most attention (Wang and Rainbow, 2008). Most of such influences are due to changes in the speciation of contaminants as well as the physiological and biochemical processes of the animals. Zhang and Wang (2007a) illustrated the complexity of metal chemistry and fish physiology in affecting metal uptake in fish at different salinities. Thus, the difference of k_us may explain the inter/intra-species differences.

Differences in k_e of heavy metal are also important in determining interspecies differences in accumulated metal concentrations among organisms. A common consensus for efflux is that the difference among metals is smaller than the difference among organisms (Wang, 2016). For example, the k_e of different metals in fish ranges between 0.001/d and 0.09/d, whereas for marine bivalves the k_e of different metals ranges from 0.01/d to 0.03/d. Thus, the lower metal efflux rates of fish lead to the higher metal accumulation in fish, and vice versa. The difference of k_e may be caused by the subcellular distribution of heavy metals in fish (Wang and Rainbow, 2008). Wallace *et al.* (2003) fractionated metals into 5 operationally defined subcellular pools, consisting of metal-rich granules (MRG), cellular debris (mainly cellular membrane fragments), organelles (metals bound with mitochondria, lysosomes, endoplasmic reticulum), heat-sensitive proteins (HSP, including enzymes), and heat-resistant proteins (generally considered to be metallothionein or more correctly metallothionein-like proteins MTLP). The higher efflux of heavy metals might be related to a higher partitioning in the soluble fraction (HSP+MTLP) and a lower partitioning in the insoluble fraction (MRG+ cellular debris+ organelles) (Ng *et al.*, 2007, Pan and Wang, 2008a). Efflux is also influenced by environmental conditions, such as temperature, tissue concentrations of the heavy metals, routes of exposure, as well as food conditions or internal sequestration (Wang, 2016).

Differences in growth rate (g) in accounting for the differences in metal accumulation among different organism have been well documented (Wang and Schrenk, 2012, Wang and Rainbow, 2008). Generally, an increase in the growth rate may result in a reduced body concentration of metals due to the growth dilution effect. Several previous studies have attempted to attribute such allometry to changes in the metal biokinetics (Pan and Wang, 2008b). Thus, the importance of the growth rate of fish in controlling metal concentrations needs to be considered, especially when examining the difference in metal concentrations in fish from different locations.

TOXICITY OF HEAVY METALS

Acute and Chronic Toxicity

According to the usage of terms for Ambient Water Quality Criteria (AWQC), acute toxicity for fish refers to mechanisms that are operative in causing lethality at concentrations effective in 96 h tests, whereas chronic toxicity refers to mechanisms causing pathology or performance decrements in trials lasting 21-30 days (or longer, *i.e.* up to lifetime) (Wood *et al.*, 2011, Wood *et al.*, 2012). In general, the chronic threshold values are at least 10-fold lower than the acute LC_{50} values for the same species (Hunt *et al.*, 2002, Marcantonio *et al.*, 2011). Both the acute and chronic toxicity of heavy metals are influenced by water hardness, pH, total suspended solids, salinity, metal speciation, fish species, and developmental stage (Al-Reasi *et al.*, 2011, Chakoumakos *et al.*, 1979, Green *et al.*, 1986, Richards *et al.*, 2001, Spry and Wiener, 1991, Zitko *et al.*, 1973, Zitko and Carson, 1976). Most of the accepted endpoints of acute toxicity can be death, reduced growth, reduced reproductive output, or behavioral changes, while the chronic toxicity is sub-lethal, including biochemical disturbances, hematological changes, immunotoxicity, and other organic, cellular, developmental, and gene expressive effects (Annabi *et al.*, 2011, Atli and Canli, 2013, Besser *et al.*, 2007b). In fact, the boundaries of acute and chronic toxicity are not strict, the essential difference may be the mechanisms of toxicity. Unlike acute toxicity, where lethality can often be attributed to only one or two mechanisms (ionoregulatory and respiratory disturbances), there is often a plethora of chronic toxicity mechanisms. It is probably more realistic to assume that the fish's health gradually "runs down" owing to the combined load of many disturbances, with the eventual result of one or more of decreased survival, growth, or reproductive output. These disturbances include costs of acclimation ("damage repair") (Mcdonald and Wood, 1993), detoxification (metallothionein, glutathione synthesis) (Chiaverini and De, 2010, Krezel and Maret, 2007, Polec-Pawlak *et al.*, 2007), immune suppression (Mushiake *et al.*, 1984), and the "burning out" of an ability to mount a corticosteroid stress response (Hontela, 1998), impacts that are common to many metals.

Behavioral Effects

Heavy metals have long been known to affect normal behavior in fish, often at levels that are close to or even below AWQC (Atchison *et al.*, 1987, Scott and Sloman, 2004). However, the effects of heavy metal on behavioral endpoints have been ignored or discounted by most regulatory authorities, such that behavioral disturbance cannot be used as an endpoint in deriving AWQCs, and such information is usually overlooked in ecological risk assessments (Melvin and Wilson, 2013). Behavioral responses relevant to chronic, sublethal toxicant exposure can be categorized into three broad categories: (1) attraction-avoidance behaviors to contaminants, (2) locomotory responses, and (3) appropriate responses to social information (Atchison *et al.*, 1987, Sabullah *et al.*, 2015). Attraction-avoidance responses are characterized as a fish's response to an environmental stimulus. If the stimulus is a toxic contaminant, an appropriate behavior would be to avoid the contaminated area (Moreira-Santos *et al.*, 2008, Svecevičius, 1999). If the stimulus is food or the presence of an essential nutrient, then the appropriate response would be an attraction to the stimulus (Kasumyan, 2001). This is troubling when attraction to unfavorable areas, or displacement from otherwise favorable areas occur in wild fish due to the considerable ecological cost (Melvin and Wilson, 2013). Furthermore, reduction in the ability to detect and avoid predators, locate prey, maintain social hierarchies, find suitable mates and spawning grounds, or undertake directional migrations could result in "ecological death" (Mirza *et al.*, 2009, Sloman, 2007). However, such behavior impairment would not have been detected or verified in typical laboratory tests, because the results derived under controlled laboratory conditions may not reflect metal effects under natural field conditions owing to several biotic and abiotic factors that can influence metal toxicity.

Such effects of heavy metals on complex behaviors in fish are likely caused by interference with a combination of many physiological systems, such as sensory, hormonal, neurological, and metabolic system, so the impacts of heavy metals on all of them need to be considered (Atchison *et al.*, 1987). Many metals could enter the olfactory system of fish, where they can potentially cause cell death or sublethal damage of the olfactory system, and subsequently disrupt the electrical transmission of sensory information from the olfactory epithelium to higher levels of the brain, finally influencing the behavior response in fish (Tierney *et al.*, 2010). Some metals could also agonize or antagonize endogenous hormones, disrupting the synthesis or metabolism of endogenous hormones and their receptors (Green and Planchart, 2017). Many metals have been shown to disrupt the hypothalamic-pituitary-interrenal (HPI) axis that controls the cortisol response to stress, which may consequently alter normal fish behavior. Changes in brain function of metal exposed fish, including alterations of cholinesterase (ChE) activity, neurotransmitter (*e.g.* acetylcholine, serotonin, dopamine) levels, enzyme function, or electrophysiological properties, has the potential to alter numerous different behavior systems of fish (Sabullah *et al.*, 2015). Heavy metals could disrupt various aspects of metabolism in fish, from whole-organism responses (*e.g.* metabolic rate and swim performance) to tissue responses (*e.g.* metabolic substrate availability and enzyme activity). By altering metabolism and thus food requirements and assimilation in fish, aquatic toxicants could alter optimal foraging strategies, which could have potential implications for numerous aspects of fish behavior. However, behavioral responses are highly variable, and can be site specific and context specific, and resulting data from behavioral testing can be difficult to interpret.

Histopathology Effects on Organs

Heavy metals show toxicity in many organs of fish, such as skin, kidney, liver, lung, muscle, and gastrointestinal tract. Among these, gill, gut, or skin are the major uptake organs for waterborne and diet borne exposure; liver and kidney serve as scavenging and clearance organs; muscle might be the storage organ. Acute and chronic toxic effects have been largely reported in such organ system, details are summarized as follows.

Skin

Skin is the first physical barrier to protect the body from damage by chemical hazards in the environment. Numerous skin changes occur due to acute and chronic heavy metal exposure. The toxic effects to the skin of *Clarias batrachus* Linn, including wear and tear, sloughing of the epithelial cells (ECs), hyperplasia of mucous cells (MCs) and club cells (CCs), along with severe degenerative changes, have been observed after 1 mg/L disodium arsenate heptahydrate exposure by Singh and Banerjee (2008). Poleksic *et al.* (2010) have analyzed the skin histopathology in Danube sterlet (*Acipenser ruthenus* Linnaeus, 1758) collected in heavy metal contaminated drainage basin, and observed pyknotic nuclei in the matrix layer of the epidermis, erosion with desquamation of epithelium, rupture (excoriation) of parts of epidermis, hyperplasia of the epidermal cells (MCs and CCs), and even a leucocytes infiltration in the epidermis. A thick layer of slime on the skin surface secreted by ECs, containing sulphated, acidic or a mixture of neutral and acidic/sulphated glycoproteins, is a common effort to protect the skin from the toxic stress of heavy metals (Kumari *et al.*, 2013). Metals can bind to the glycoproteins that have electronegative charges at neutral by their histidine and cysteine residues; histidine binding through its imidazole nitrogen, and cysteine through its thiol groups (Luckey and Venugopal, 1977). Secretion of the ECs also contributes actively to playing a protective but thin covering of slime on the outer surface of the epidermis. However, the protective role played by the slimy coating does not last long, perhaps due to extensive loss and the altered nature of the slime following prolonged exposure, thus leading to wear and tear, and sloughing of the superficial cells (Singh and Banerjee, 2008). The precise role played by the CCs is still a matter of great debate. However, the protective role played by these protein-rich cells against the stress of various physical as well as chemical hazards has been well accepted (Kumari *et al.*, 2017). The CCs of exposed fish thus also help to prevent quick penetration of the toxicants *via* the skin and withstand the stress of xenobiotics more effectively, even though the CCs are also badly damaged by the ambient heavy metal. Continuation of exposure often leads to further destruction, followed by uncontrolled regeneration of the epidermis, causing significant alteration in its histomorphology and cellular architecture. An increased mucous production by MCs is a first defense mechanism to protect the skin from the toxic stress of heavy metal. However, it is often followed by lack of production manifested by emptied, exhausted mucous cells, and finally complete lack of this cell type, showing continual degradation of the environmental conditions that lead to chronic changes (Devi and Banerjee, 2007, Sunita and Banerjee, 2003).

Gill

The gill is a multifunctional organ performing vital functions, such as respiration, osmoregulation, acid-base balance, and nitrogenous waste excretion, and playing a vital role in the overall protection against harmful substances by acting as a first barrier to lower the total uptake of toxic molecules by other organs (Topal *et al.*, 2017). Due to their large surface area and poor detoxification system, gill is highly vulnerable to toxic chemicals. Metals, in particular, are one of the most deleterious environmental toxicants affecting the general morphology and ultrastructure of gills of fish. The histopathological changes in the gill, such as desquamation, edema, epithelial necrosis, lifting of the lamellar epithelium, fusion of secondary lamellae, hemorrhage at filaments, hypertrophy of epithelial cells, and sloughing off of epithelial surface, are the major effects reported in gills of fish exposed to various types of metals in laboratory and field studies (Pandey *et al.*, 2008, Sonne *et al.*, 2014). Mallatt (1985) listed some of the frequently recorded histopathologic lesions as lifting, necrosis, hyperplasia, hypertrophy, rupture of gill epithelium, bulbing or fusing of gill lamellae, hypersecretion and proliferation of mucocytes, and changes in chloride cells and gill vasculature. Fonseca *et al.* (2017) uncovered several alterations in the gills of fish, such as lamellar fusion, filament epithelium proliferation, laminar epithelium proliferation, vasodilation, aneurisms, edema, lifting, and necrosis. Most of these histopathologic lesions are non-specific and can be considered as an adaptive response against metal exposure. For example, cell proliferation probably occurred as the defense mechanism (Ventura and Paperna, 1985) towards heavy metal insult, leading not only to increased epithelial thickness to prevent further chemical absorption but also, in extreme cases, to induce lamellar fusion. Such damages on the gills may imply the hindering of key physiological functions, such as gas exchange and osmotic balance, and likely result in hypoxia, respiratory failure problems with ionic and acid-base balance. Lifting and swelling could be related to a decrease in the gill Na^+ and K^+ activated ATPase and/or a decline in blood Na^+ and Cl^- concentration (Nieboer and Richardson, 1980). Fusion of secondary lamellae could cause a decrease in free gas exchange thus affecting the general health of fish (Skidmore and Tovell, 1972). Cell proliferation of secondary lamellar filaments and lamellar cell hypertrophy decreased the space between lamellae and causes fusion. Such lesions would increase the thickness of water-blood barrier and decrease the oxygen uptake, causing capillary hemorrhage (Jiraungkoorskul *et al.*, 2002, Nowak, 1992).

Muscle

The muscles provide the power for swimming and constitute up to 80% of the fish itself. Myomeres are the elementary units of the muscles arranging in multiple directions that allow fish to move in any direction (Gaworecki *et al.*, 2012). Muscle is the most commonly consumed portion of fish and contributes most to the mass of fish. However, muscle tissue in fish does not play an active role in the metal accumulation in laboratory and field study when compared to other soft tissues due to its indirect contact of heavy metals and inactive role of detoxification, thus lowering the transport of heavy metals from other tissue to muscles. In spite of relative low bioaccumulation of heavy metals in muscle tissue, muscle histopathology, such as focal necrosis, cellular dissolution, intermyofibrillar oedema, and a decline or loss of striation in muscle fibers, has been found as a result of heavy metal exposure (Adali and Koca, 2016, Koca *et al.*, 2005, Mughal. *et al.*, 2004). Degenerative features, including muscular atrophy, broken myofibril, swollen sarcolemmal nucleus, and sarcoplasmic reticulum, have been reported in muscular injuries. Barillet *et al.* (2010) observed degeneration and disorganization of myofibrillar sarcomeric pattern (myofibrils being twisted, tangled, or split), abnormal localization of mitochondria within muscle and altered endomysial sheaths after waterborne uranium exposure. Barillet *et al.* (2010) speculated that the degenerative effects of uranium on muscle structure might be linked to its ability to modulate acetylcholinesterase (AChE) activity since several authors noticed a close relationship between AChE activity and muscular structure and function. Moreover, Lerebours *et al.* (2009) reported that genes encoding detoxication, apoptosis or inflammatory processes were transcriptionally strongly down-regulated after 3 d waterborne uranium exposure in muscles of fish, while genes involved in mitochondrial mechanisms were up-regulated. Thus, disorganizations of the muscle tissue might, therefore, be related to severe modulations of gene expression. Arsenic exposure during embryogenesis altered expression of myosin light chain 2 (MLC2), myosin heavy chain 2 (MHC2), and actin filament capping protein Z (CapZ), and resulted in a significant reduction in muscle fiber size in the hatchlings of killifish, thus leading to aberrant muscle formation. Heavy metal exposure also alters or prevents muscle regeneration after injury, partly due to a reduction in myogenin, the transcription factor that controls differentiation from myoblasts (Gaworecki *et al.*, 2012).

Liver

Liver is a detoxification organ and is essential for both the metabolism and biotransformation of toxic substances in fish (Khoshnood *et al.*, 2010, Poleksic *et al.*, 2010, Sonne *et al.*, 2014). The liver of fish is sensitive to environmental contaminants because many contaminants tend to accumulate in the liver, making this organ exposed to much higher levels (several orders of magnitude) than in the environment or in other organs (Feist *et al.*, 2015, Vasanthi *et al.*, 2013). There were numerous reports of histopathology changes in the livers of fish exposed to a wide range of heavy metals. Salamat *et al.* (2017) divided the liver histopathology into three stages, stage I, slight damage, stage II, moderate damages, and stage III, severe lesions. The alterations of stage I include hepatocytes with irregular shaped cellular hypertrophy, irregular shaped nucleus, nuclear hypertrophy, cytoplasmic vacuolation, hepatocytes with lateral nucleus, infiltration of leukocytes, sinusoid dilation, dilation of Disse's space, fibrosis and cloudy swelling. Such alterations, including blood congestion, melano-macrophage aggregates, nuclear vacuolation, cytoplasmic degeneration, lipid metaplasia, and hyaline droplets degeneration, are mainly observed in stage II. The severe lesions are necrosis. Fu *et al.* (2017) reported that multiple metals exposure caused sinusoidal congestion, stasis, vacuolation of hepatocytes, necrosis, leukocyte infiltration, and presence of granuloma. Van Dyk *et al.* (2007) showed a histological response in Cd and Zn exposed *Oreochromis mossambicus* with the most prevalent histological characteristics identified being hyalinization of hepatocytes, increased vacuolation associated with lipid accumulation, congestion of blood vessels, and cellular swelling. Such histological changes identified within the hepatocytes may have been the result of various biochemical lesions. Vacuolation of hepatocytes is associated with the inhibition of protein synthesis, energy depletion, disaggregation of microtubules, or shifts in substrate utilization (Hinton and Lauren, 1990). Hyalinization is said to be the result of disturbances of protein synthesis (Van Dyk *et al.*, 2007). Cellular swelling occurs either directly by denaturation of volume-regulating ATPases or indirectly by disruption of the cellular energy transfer processes required for ionic regulation (Hinton and Lauren, 1990). Alanine transaminase (ALT) and aspartate transaminase (AST), the most abundant liver enzymes released into the blood from damaged or dead hepatocytes, are involved in the metabolism of pyruvate and oxaloacetate (Omar *et al.*, 2014). Previous studies showed that AST and ALT increased in polluted areas, which could be considered as biomarkers of severe liver damage. The plasma levels of AST and ALT may increase due to the protein breakdown to provide more energy against the metal pollution stress (De Smet and Blust, 2001). Damage of the hepatocyte membrane may lead to the release of transferase enzymes into the blood and synthesis of these enzymes in the liver (Vutukuru *et al.*, 2007).

Gastrointestinal Tract

Heavy metals will enter the digestive tract of fish *via* food and water that they consumed, causing a deterioration of structures and functions along the gut. Pedla *et al.* (2002a, b) observed the mucosal sloughing and increased mucosal production in the mucosal lining of the gastrointestinal tract of lake whitefish exposed to dietary As. Prominent histopathological changes, including edema, atrophy in the mucosal layer, hemorrhage between blood vessels, blood congestion and aggregations of inflammatory cells were observed in the intestinal tissue of *Labeo rohita* inhabiting industrial waste contaminated water (Sultana *et al.*, 2016). The fundamental gastrointestinal lesions appear to be increased permeability of the small blood vessels, leading to fluid loss and hypotension (Haque and Roy, 2012). The intestine of European sea bass *Dicentrarchus labrax* (L.) treated with Cd, exhibited focally degeneration of villi with swollen tips, myelinoid bodies observed in the cytoplasm of almost all enterocytes, vacuoles of different sizes and degenerated mitochondria (cristolysis and deformation of the external membrane); at the higher concentrations acute cell swelling occurred, moreover, many enterocytes were full of autophagolysosomes (Giari *et al.*, 2007). Low concentration of As exposure (7 mg/L) to stinging catfish *Heteropneustes fossilis* (Bloch, 1794) induced the histopathological changes in intestine, including partial intactness of serosa but more or less organized mucosa and disorganized villi at 15 d, and partially damage of muscles but disorganized, slightly swollen and shorten of villi at 60 d. High concentration of As exposure (20 mg/L) showed damaged serosa disorganized and consequent fusion of mucosa, degeneration and edema between the intestinal submucosa and lamina propria at 15 d, and exhibited the increases in number of goblet (mucosal) cells, width of the lamina propria and degeneration of villi at 60 d (Choudhury, 2014).

Kidney

The kidney is involved in the removal of wastes from the blood (Fänge, 1986) and is severely affected by different toxic chemicals (Kumar and Pant, 1981), thus renal histopathological changes are widely studied in fish when acutely and chronically exposed to heavy metals. Fatima *et al.* (2015) found the kidney of *Channa striatus* and *H. fossilis*, inhabiting in seriously heavy metal polluted Kali River Estuary of Northern India, showed degenerated renal tubules, pyknosis, hemorrhages and glomerular degeneration in *C. striatus*, and necrosis, tissue vacuolization, congested renal tubules and hemorrhages in *H. fossilis*, respectively. Similar changes have been reported by Mishra and Mohanty (2008) where acute toxicity impacts of Cr(VI) on the kidney of *Channa punctatus*, showing hypertrophy of epithelial cells of renal tubules, glomerular contraction in the Bowman's capsules, and necrosis of hematopoietic tissues. Roy and

Bhattacharya (2006) found that 1/20 (3.8 mg/L) and 1/10 LC_{50} (7.6 mg/L) doses of As_2O_3 resulted in shrinkage in the glomerulus with a resultant increase in Bowman's space on the first day followed by enlargement of the glomerulus and normalization of Bowman's space during the last phase. Increase in Bowman's space suggests an increase in the filtration rate and consequently in urine volume. Similar results were also observed in rainbow trout (*Oncorhynchus mykiss*) exposed to 1 mg/L and 2 mg/L Ni for 21 d (Topal *et al.*, 2017), *L. rohita* exposed to 1/10-1/3 LC_{50} of Pb (3.42-11.4 mg/L) for 15-60 d (Brraich and Kaur, 2017), and common carp (*Cyprinus carpio*) exposed to 1/10 LC_{50} of organic selenium (Selemax) (0.054 mg/L) for 28 d (Yeganeh *et al.*, 2016).

Brain

Brain is considerably vulnerable to heavy metals due to its neurological functions crucial for survival (Baatrup, 1991, Green and Planchart, 2017). In fact, the structural and functional effects of heavy metals on the nervous system especially the brain have been largely studied. Patnaik *et al.* (2011) showed that both Cd and Pb at sublethal concentration affected neuronal cell degeneration, swelling of pyramidal cells, loss of nissl substances, vacuolization and dystrophic changes after 28 d of exposure. Vacuolization in brain tissue may be the result of glycolysis leading to microsomal and mitochondrial dysfunctions. Severe necrosis of neuronal cells in the cerebrum indicating loss of nissl substances were also supported by the studies of Loganathan *et al.* (2006) due to 10 mg/L Zn exposure. Fatima *et al.* (2015) reported that the brain of *C. striatus* from seriously heavy metal contaminated estuary showed spongiosis, neuronal degeneration, macrophages, and inflammation. Such changes were also evident from the study conducted by Berntssen *et al.* (2003) in Atlantic salmon (*Salmo salar*) where the fish were exposed to Hg. Furthermore, Berntssen *et al.* (2003) demonstrated that Hg exposure could alter the major biochemical constituents, such as lipids, proteins and nucleic of the brain tissues of the *S. salar*, significantly reduce neural enzyme activity (5-fold reduced monoamine oxidase activity), and reduce overall post-feeding activity behavior. In the brain, the optic tectum (OT) receives information from optic nerves and its large size is related to the importance of the vision in fish. Several studies highlighted the sensitivity of this organ in fish following metallic exposures (Azizishirazi *et al.*, 2015, Cambier *et al.*, 2012, Low and Higgs, 2015). Naïja *et al.* (2018) indicated that peacock blennies (*Salaria pavo*) exhibited several damages in the optic tectum and the cerebellum and 3 reaction patterns were identified for each organ.

Gonad

Heavy metals can inhibit reproduction of teleost fish by interfering with endocrine systems (Sumpter, 2005). Several effects, such as testicular/ovarian degeneration, inhibition of spermatogenesis and steroidogenesis in testis, induction of oocyte in testis (testis-ova) and synthesis of vitellogenin, which is a precursor of yolk protein, have been detected in gonad of male and female fish (Ferreira *et al.*, 2004, Hansen *et al.*, 1998, Hassanin *et al.*, 2002, Kavanagh *et al.*, 2004, Matthiessen *et al.*, 2002, Sepulveda *et al.*, 2003). A previous monitoring research in the Mekong Delta of Vietnam showed that some of the male catfish *Pangasianodon hypophthalmus* displayed inhibition of spermatogenesis, in particular, those animals showing markedly low gonadosomatic index (GSI) values. In small testis, spermatogenesis did not proceed beyond spermatogonia, and oncotic necroses (karyolysis, swelling and, loss of structure of cell) were observed among spermatogonia. Furthermore, Sertoli cells hypertrophied and contained vacuoles (Yamaguchi *et al.*, 2007). The 11-ketotestosterone (KT) is the main androgen of teleost and regulates spermatogenesis. Previous studies using Japanese eel testicular organ culture system showed that treatment with 11-KT induced proliferation of spermatogonia and progression of spermatogenesis *via* the synthesis of various factors in Sertoli cells (Miura and Miura, 2001, Miura and Miura, 2003). Fritzie *et al.* (2009) demonstrated that a low dose (0.1 µg/L) of As may inhibit 11-KT synthesis *via* suppression of steroidogenic enzyme activities, such as 3-β-hydroxysteroid dehydrogenase, which might continuously inhibit spermatogenesis. Furthermore, exposure of testis to the high concentration of As (100 µg/L) produced reactive oxygen species (ROS) in testis, which consequently caused apoptosis of germ cells, especially after induction of spermatogenesis by human chorionic gonadotropin (hCG) *via* 11-KT synthesis. These findings suggest that a low dose of As exposure inhibits spermatogenesis *via* suppression of steroidogenic enzyme activity and expression, while a high dose of this compound induces oxidative stress-mediated germ cell apoptosis.

Biochemical and Physiological Changes

Carbohydrate

Carbohydrates serve as the primary and immediate energy source and play an important role in maintaining homeostasis for fish exposed to stress condition (Umminger, 1970). The effects of heavy metal stressors on carbohydrates metabolism in fish are primarily focused on glucose, glycogen, and lactic contents (Das *et al.*, 2001, David *et al.*, 2005), partly on the vital intermediate products and key enzymes in glycometabolism processes, such as glycolysis, gluconeogenesis, tricarboxylic acid cycle, and pentose phosphate pathway (Sastry and Subhadra, 1982). Among these, the blood glucose and hepatic glycogen level are used as indicators of environmental stress and reflect the changes in carbohydrates metabolism under hypoxia and stress conditions (Al-Asgah *et al.*, 2015, Mohamed and Gad, 2008, Srivastava and Srivastava, 2008). Blood glucose level is usually increased due to glycogenolysis during acute condition (Garg *et al.*, 2009), which is considered as a general secondary response to the stress in fish (Sepici-Dinçel *et al.*, 2009). This hyperglycemic (increase in blood glucose) condition were also detected in *H. fossilis* and *Saccobranchus fossilis* exposed to Ni and Cr (Nath and Kumar, 1988, Radhakrishnaiah *et al.*, 1992), *C. batrachus* Linn subjected to As (Kumari and Ahsan, 2011b, Kumari and Ahsan, 2011a, Kumari *et al.*, 2012), *Mastacembelus armatus* and *C. punctatus* exposed to the mixture of heavy metal salts (Javed and Usmani, 2013, Javed and Usmani, 2015). However, glucose and glycogen are excessively utilized in fish to fulfill the energy requirement for detoxification of toxic substances. It has been reported by most investigators that the blood glucose level firstly elevates and then declines until it attains a depleted level under prolonged sublethal exposure to heavy metals (David *et al.*, 2005, Srivastava and Srivastava, 2008). Consistently, the glycogen level in the liver and muscle is commonly reduced under the stress of heavy metals. Depletion of the glycogen content in the liver and muscle was observed by many workers in *Mystus cavasius* exposed to electroplating industrial effluent (Palanisamy *et al.*, 2011), *C. punctatus* exposed to distillery effluent (Maruthi and Rao, 2000). In the liver, glycogen mobilized to glucose whereas in muscle glycogen/glucose served as the readily available source of energy, thus hypoglycemia was observed. Low glucose and glycogen content in chronically exposed fish observed in polluted waters could be due to improper gluconeogenesis.

Zutshi *et al.* (2010) reported the serum glucose showed high levels initially and then low concentration in *L. rohita*. Javed and Usmani (2013) assessed the effect of heavy metal (Cu, Ni, Fe, Co, Mn, Cr, Zn) pollution on glycogen metabolism of *M. armatus* and observed a significant elevation in blood glucose but a reduction of glycogen in the liver and muscle. Metal stress-induced release of catecholamines and glucocorticoids from adrenal tissues of fish (Wedemeyer and Yasutake, 1977) are known to cause elevated blood glucose levels in fish. According to Ramakritinan *et al.* (2005), glycogenolysis resulting from chronic exposure to distillery effluent may be due to a stress-induced increase in circulating catecholamines in *C. carpio*. Khanna and Gill (1975) stated the possibility of imbalance in pancreatic hormones involved in carbohydrate metabolism, due to damage of pancreas caused by toxic substances. Administration of cobalt chloride and cobalt nitrate in *C. punctatus* was shown to induce hyperglycemia accompanied with degranulation and vacuolization of pancreatic tissue in the initial stages and damage of β-cells in later stages. The altered insulin-secreting capacity of pancreatic β-cells may explain the hyperglycemia reported in fish. Ana *et al.* (2006) reviewed that As interfered with transcriptional factors involved in insulin-related gene expression resulting in a decline in insulin production, leading to hyperglycemia. Some investigation also showed that heavy metals could decrease the glycogen reserve in fish (Levesque *et al.*, 2002) by affecting the activities of enzymes that played a role in the carbohydrate metabolism. Cd decreased the glycogen reserves in *H. fossilis* by stimulating glycolytic enzymes like lactate dehydrogenase, pyruvate dehydrogenase and succinate dehydrogenase (Sastry and Subhadra, 1982).

Lipid

Lipids act as a major energy source, and lipid stores support various physiological, developmental and reproductive processes (Polakof *et al.*, 2010, Tocher, 2003). At present, a limited number of studies have demonstrated that heavy metal exposure could affect lipid metabolism in fish by investigating the change of lipid content and enzymatic activities of lipogenesis and lipolysis (Chen *et al.*, 2013c, Chen *et al.*, 2013a, Chen *et al.*, 2015, Chen *et al.*, 2013b, Song *et al.*, 2013, Song *et al.*, 2014, Zheng *et al.*, 2013a, Zheng *et al.*, 2013b, Zheng *et al.*, 2015, Zheng *et al.*, 2013c). Reduction of lipid content has been observed in many studies due to the utilization of lipids for energy demand under stress condition (Das *et al.*, 2001, Garg *et al.*, 2009, Shukla *et al.*, 2002). However, increased lipid content has also been observed in javelin goby *Synechogobius hasta* (Chen *et al.*, 2013c, Liu *et al.*, 2011, Song *et al.*, 2013, Song *et al.*, 2014) and yellow catfish *Pelteobagrus fulvidraco* (Chen *et al.*, 2013a, Chen *et al.*, 2015, Chen *et al.*, 2013b, Zheng *et al.*, 2013a, Zheng *et al.*, 2015). Zheng *et al.* (2013a) found that chronic Zn exposure increased hepatic lipid content, whereas the opposite result was observed in yellow catfish subjected to chronic Cu exposure (Chen *et al.*, 2013c). Thus, it seems that different metal elements can differentially influence lipid metabolism in fish. On the other hand, for the same metal, the response of different tissues appears to be different. Lipid accumulation results from the balance between synthesis of fatty acids (lipogenesis) and fat catabolism *via* β-oxidation (lipolysis), and several key enzymes and transcriptional factors are involved in these metabolic processes (Pierron *et al.*, 2007). 6-phosphogluconate dehydrogenase (6PGD) and glucose-6-phosphate dehydrogenase (G6PD) are the key regulatory enzymes involved in NADPH production, essential for fatty acid biosynthesis (Carvalho and Fernandes, 2008). Fatty acid synthetase (FAS) is the main lipogenic enzyme which produces fatty acid (Richard *et al.*, 2006). Carnitine palmitoyltransferase 1 (CPT1) is considered to be the main regulatory enzyme in long-chain fatty acid oxidation because it catalyzes the conversion of fatty acid-CoAs into fatty acid-carnitines for entry into the mitochondrial matrix (Kerner and Hoppel, 2000). Lipoprotein lipase (LPL) hydrolyzes triacylglycerols present in plasma lipoproteins and supplies free fatty acids for storage in adipose tissue, or for oxidation in other tissues (Nilssonehle *et al.*, 1980) and plays a pivotal role in regulating lipid content in fish (Albalat *et al.*, 2007). On the other hand, transcription factors, such as peroxisome proliferator-activated receptor (PPAR) α and γ, play an intermediary role in lipid homeostasis by orchestrating the gene transcription of the enzymes involved in lipid metabolism (Guyenet and Schwartz, 2012). Chen *et al.* (2013a) investigated the effect of waterborne Cd exposure on lipid metabolism in the liver and muscle of *P. fulvidraco* and indicated that Cd triggered hepatic lipid accumulation through the improvement of lipogenesis, and that lipid homeostasis in muscle was probably conducted by the down-regulation of both lipogenesis and lipolysis. In the liver, the lipid content, the activities and mRNA expression of lipogenic enzymes (6PGD, G6PD, FAS) and LPL activity increased with increasing waterborne Cd concentrations. However, the mRNA expressions of LPL and PPARα were down-regulated by Cd exposure. CPT1 activity, as well as the mRNA expressions of CPT1 and PPARγ showed no significant differences among the treatments. In muscle, lipid contents showed no significant differences among the treatments. The mRNA expression of 6PGD, FAS, CPT1, LPL, PPARα and PPARγ were down-regulated by Cd exposure. Similarly, Zheng *et al.* (2013a) demonstrated that chronic (8 weeks) exposure to low Zn levels apparently increased LPL activity and reduced hepatic CPT1 activity, leading to lipid accumulation through the inhibition of lipolysis and the improvement of lipogenesis. In contrast, the acute (96 h) exposure to high Zn level reduced hepatosomatic index (HSI) and LPL activity, and increased CPT1 activity, inducing lipid depletion mainly *via* up-regulated lipolysis and reduced import of lipids into liver. Furthermore, lipid peroxidation is another indicator of abnormal lipids metabolism due to oxidative stress induced by heavy metal (McRae *et al.*, 2016). Lipid peroxidation has been reported frequently for metal-exposed fish and the level shows large differences with exposure concentration and time (Altikat *et al.*, 2015, Bagnyukova *et al.*, 2007).

Protein

Proteins participate in virtually every biological process and are key targets of heavy metals, which perturb its function and activity (Beyersmann and Hartwig, 2008). Heavy metals interfere with the physiological activity of specific, particularly susceptible proteins through diverse modes of interaction; they may: (1) bind to free thiols or other functional groups in proteins; (2) displace essential metal ions in metalloproteins; or (3) catalyze oxidation of amino acid side chains (Tamás *et al.*, 2014). For example, Hg binds strongly to ligands, especially those with R-SH or R-S-S-R groups, including cysteinyl and histidyl side chains of protein, amino, and carboxyl groups on enzymes, and disrupts the enzymes and substrates involved in both aerobic and anaerobic metabolism (Vallee and Ulmer, 1972). Cd inhibits thiol transferases (glutathione reductase, thioredoxin reductase, thioredoxin) *in vitro*, possibly by binding to cysteine residues in their active sites (Chrestensen *et al.*, 2000, Mohanty and Samanta, 2016). Inhibition of thiol transferases probably leads to increased oxidative stress and cell damage. Cd may also displace Zn and Ca ions from metalloproteins, Zn finger proteins, Ca^{2+} ATPase and Na^+/K^+-ATPase (Atli and Canli, 2007, Faller *et al.*, 2005, Hartwig, 2001). The high toxicity of As(III) results from its greater affinity with the sulfhydryl groups of biomolecules (enzymes, receptor, or coenzymes), whereas As(V) does not directly bind to the sulfhydryl group, but can replace inorganic phosphate in the enzymatic reaction of glyceraldehyde 3-phosphate dehydrogenase and phosphorylation reactions of the ATP production to exert its toxic effects (Hughes, 2002, Kumari *et al.*, 2017). Furthermore, recent studies have revealed an additional mode of metal action targeted at proteins in a non-native state; certain heavy metals and metalloids have been found to inhibit the *in vitro* refolding of chemically denatured proteins (Jacobson *et al.*, 2012, Ramadan *et al.*, 2009, Sharma *et al.*, 2008), to interfere with protein folding *in vivo* and to cause aggregation of nascent proteins in living cells (Holland *et al.*, 2007, Pan *et al.*, 2010, Rudolph *et al.*, 1978). By interfering with the folding process, heavy metal ions and metalloids profoundly affect protein homeostasis and cell viability.

"Omics" technologies, such as genomics and proteomics, have enabled simultaneous assessment of the expression profiles of hundreds and/or thousands of genes/proteins, intending to detect critical genes/proteins and pathways disrupted by exposure to harmful chemicals and environmental stressors (Merrick, 2006, Ung *et al.*, 2010). Recently omics-based approaches have been applied well to investigate the mechanisms of heavy metal toxicity in fish (Berg *et al.*, 2010, Keyvanshokooh *et al.*, 2009, Wang *et al.*, 2010, Wang *et al.*, 2011). Wang *et al.* (2013) used quantitative proteomic analysis and demonstrated that chronic Hg exposure to marine medaka *Oryzias melastigma* caused oxidative stress due to seven upregulated protein spots (Cathepsin D, Glutathione S-transferase, DJ-1 protein, Peroxiredoxin-1, Natural killer enhancing factor, *etc.*), cytoskeleton disruption due to the variations of cytoskeletal proteins (Keratin, Novel protein similar to vertebrate plectin 1, *etc.*), and altered energy metabolism due to decreased expression of the respiratory proteins (ATP synthase subunit d, mitochondrial, Electron-transferring-flavoprotein dehydrogenase, *etc.*). Costa *et al.* (2012) also revealed the hepatic proteome changes in *Solea senegalensis* exposed to contaminated estuarine sediments and found forty-one cytosolic proteins were deregulated. Among the deregulated cytosolic proteins, nineteen were able to be identified, taking part in multiple cellular processes, such as anti-oxidative defense, energy production, proteolysis, and contaminant catabolism (especially oxidoreductase enzymes).

Antioxidant Enzymes and Oxidative Stress

Heavy metals are known to result in the production of reactive oxygen species (ROS), and the induced oxidative stress is presumed to be one of the key mechanisms of toxic action (Lushchak, 2011, Regoli and Giuliani, 2014). ROS include hydrogen peroxide (H_2O_2), hydroxyl radical (•OH), singlet oxygen, hydroperoxyl radical (HO_2•), superoxide anion radical (O_2•‾), and other oxygen-derived species (Lushchak, 2011, Regoli and Giuliani, 2014). Protection against ROS, which are generated continuously during aerobic metabolism, is normally achieved by antioxidant enzymes, such as Cu/Zn superoxide dismutase (SOD), which catalyze the dismutation of superoxide radical to hydrogen peroxide, catalase (CAT) acting on hydrogen peroxide, peroxidase (POD) converting hydrogen peroxide to H_2O, glutathione peroxidase (GPx) facilitating the oxidation of glutathione (GSH), glutathione reductase (GR) reducing oxidized glutathione (GSSG) back to GSH, and Glutathione-S-transferase (GST) possessing detoxifying activities towards lipid hydroperoxides generated by organic pollutants, such as heavy metals (Ferreira *et al.*, 2008, Okamoto and Colepicolo, 1998, Padmini *et al.*, 2009, Srikanth *et al.*, 2013). In accord with ROS formation caused by Cu exposure in fish gill cells and hepatocytes (Bopp *et al.*, 2008, Krumschnabel *et al.*, 2005), CAT gene expression and enzymatic activity have been reported to increase within a few days of exposure in gills, liver, and kidney of fish (Craig *et al.*, 2007, Hansen *et al.*, 2006). However, it appears that CAT gene expression and enzymatic activity are not tightly correlated and that CAT activity is controlled in part by enzymatic activation rather than transcription (Craig *et al.*, 2007, Hansen *et al.*, 2006). It is also clear that Cu can contribute to oxidative stress by inhibiting antioxidant enzymes, as illustrated by observations of decreased CAT activity in gill, hepatic, and renal tissue in Cu-exposed fish (Ahmad *et al.*, 2005, Hansen *et al.*, 2007, Sampaio *et al.*, 2008). Similarly, diverse responses are reported for GPx and SOD, which may show increased messenger RNA (mRNA) expression or lack of expression change despite increased and decreased GPx and SOD enzymes activity levels have been reported from fish during heavy metal exposure (Ahmad *et al.*, 2005, Hansen *et al.*, 2007, Hansen *et al.*, 2006, Sampaio *et al.*, 2008, Sanchez *et al.*, 2005, Vutukuru *et al.*, 2006). One possible reason for the variable reported responses in SOD (as well as CAT and GPx) activity may be that both transcriptional and enzymatic responses appear to be transient, even during continued metal exposure (Sanchez *et al.*, 2005). This transient response can reasonably be interpreted as a complex interaction between a need to defend against the accumulation of ROS on one hand and the direct inhibitory action of metal on these antioxidant enzymes on the other hand.

In addition, reducing agents, such as GSH, ascorbate, β-carotene, and α-tocopherol, are involved in protection against oxidative stress (Valavanidis *et al.*, 2006). Appropriate levels of reduced GSH are important for homeostatic redox balance, because the first line of cellular defense against metals is chelation and detoxification as well as scavenging of oxyradicals by reduced GSH (Srikanth *et al.*, 2013). The ratio of GSH *versus* GSSG is determined by GPx and GR which can be altered by metal exposure as described above. In addition to altering the GSH/GSSG ratio, most metals have been reported to inhibit glutathione synthetase (Ventura-Lima *et al.*, 2011) and furthermore form stable complexes with GSH, hence decreasing GSH levels in the cytosol. Indeed, reduced GSH levels have been observed in gills, liver, and kidney from fish exposed to heavy metals. Glutathione reductase mRNA expression may be increased in gill and hepatic tissue during waterborne Cu exposure (Hansen *et al.*, 2006, Minghetti *et al.*, 2008). The shared function of glutathione reductase and glutathione peroxidase in glutathione turnover is illustrated by the positive correlation in expression of these enzymes in the gills and liver of Cu-exposed brown trout (Hansen *et al.*, 2006).

Metallothionein (MT) is generally accepted as a metal scavenger, with two of the four isoforms (MT-1 and MT-2) being inducible in response to metals and other stimuli (Bourdineaud *et al.*, 2006). Both free metal ions and free oxygen radicals are known to increase MT mRNA, a response mediated by a combination of antioxidant responses elements and metal responsive elements (MREs), suggesting a direct role for MT in antioxidant defense (Chiaverini and De, 2010, Kling and Olsson, 2000). Indeed, MT is highly efficient in quenching superoxide radicals when compared to SOD and GSH. Both elevated MT expression and protein levels in target organs for heavy metal accumulation have been observed in a large number of metal exposure studies. Heat shock protein (HSP70) acts as a molecular chaperone and forms an important part of the cellular response to oxidative stress by protecting the protein machinery (Rajeshkumar *et al.*, 2013). Constitutive forms of HSP70 are present in unstressed cells, whereas the inducible form is synthesized by fish in response to stressors, including heavy metal (Boone and Vijayan, 2002, Feng *et al.*, 2003). Heavy metal induced elevation of HSP70 levels have been reported from the gills, liver, and kidney in numerous studies (Hansen *et al.*, 2006, Rajeshkumar *et al.*, 2013).

Hematological Changes

Hematological parameters have been widely used to evaluate the health status of fish exposed to heavy metals (Miandare *et al.*, 2017, Hwang *et al.*, 2016). The red blood cell (RBC) count, hematocrit value, and hemoglobin concentration are usually decreased when fish exposed to metals. The RBC count, hematocrit, and hemoglobin concentration were significantly decreased in *Sebastes schlegelii* exposed to dietary Cd and Pb (Kim and Kang, 2017, Kang *et al.*, 2005). Waterborne Se exposure also decreased the RBC count of *Pagrus major* (Kim and Kang, 2014). Acute exposure to Cr(VI) could induce a significant decrease in total erythrocyte count, hemoglobin percent and absolute value mean cell hemoglobin (MCH) in *L. rohita* both at the end of 24 h and 96 h exposure

(Vutukuru, 2005). The *O. mykiss* collected at Cu contaminated sites exhibited decreased hematocrit, leukocrit, and percentage of lymphocytes in blood compared to that at reference sites (Dethloff *et al.*, 2001). But dietary Ni exposure didn't change the concentrations of glucose, hemoglobin, and hematocrit in *Coregonus clupeaformis* (Ptashynski *et al.*, 2002).

The serum glutamate oxaloacetate transminase (GOT) and glutamate pyruvate transminase (GPT) concentrations increased in *S. schlegelii* after Cu exposure (Kim and Kang, 2004), similar hematological changes were observed after Pb exposure (Kim and Kang, 2017). Dennis Lemly (1993) reported a reduction in respiratory capacity and an increase in respiratory demand and oxygen consumption in *Lepomis macrochirus* exposed to Se. After chronic exposure to Cu for 21 d, the hemoglobin, hematocrit, RBC count, mean corpuscular volume (MCV), MCH, and mean corpuscular hemoglobin concentration (MCHC) were found to be decreased in *M. cephalus*, whereas white blood cells (WBC) count increased in Cu-treated fish. Plasma aspartate aminotransferase (AST), alanine aminotransferase (ALT), and lactate dehydrogenase (LDH) activity increased in treated groups; however, chronic Cu exposure significantly decreased plasma alkaline phosphatase (ALP) activity compared to the control group (Akbary *et al.*, 2018). Dietary Cd exposure could also increase glucose and Mg concentration in serum (Kang *et al.*, 2005). Combined (Cd+Pb+Cr+Ni) metal exposure in *C. carpio* increased the level of serum iron and copper (Vinodhini and Narayanan, 2009).

RISK ASSESSMENT

Human exposure assessment of trace metals in fish are assessed by estimated daily intake (EDI) (µg/kg/d) and hazard quotient (HQ). The EDI can be calculated by the following formula:

$$EDI = C_{fish} \times IR_{fish}/BW \qquad (9)$$

where C_{fish} is the metal concentration of fish (µg/g); IR_{fish} is daily intake of fish (g/d); BW is mean weight of target people (kg). The metal concentration usually uses the value in the muscle of fish because muscle is the mainly edible part. The daily intake of fish from different regions can refer the FAOSTAT "Food Supply - Livestock and Fish Primary Equivalent".

The HQ was calculated as follows:

$$HQ=EDI/RfD \qquad (10)$$

where the reference doses (RfD) is the standard reference doses of metal. If the HQ was over 1, it means that the fish has potential risk to human.

The reference dose (RfD) of metals are shown in Table **2**. The WHO-Joint FAO expert committee on food additives (JECFA) reported that the no observed adverse effect level (NOAEL) of Al is 30 mg/kg bw/d, and the provisional tolerable weekly intake (PTWI) is 2 mg/kg bw. Environmental Protection Agency (US-EPA) gives the RfD of 0.3 µg/kg/d for inorganic As. The JECFA use the benchmark dose for a 0.5% increased incidence for lung cancer (BMDL0.5) of 3 µg/kg/d. Cu is not carcinogenic in either humans or animals, and Cu salts are not embryotoxic in rodents. On this basis, the JECFA gives a provisional value for a maximum tolerable intake (PMTDI) of 0.5 mg/kg bw/d from all sources. The PMTDI for Fe is 0.8 mg/kg bw/d. The Pb chronic dietary exposure corresponding to a decrease of 1 intelligence quotient (IQ) point was estimated to be 0.6 µg/kg bw/d, and 1.2 µg/kg bw/d for 1 mmHg increase in blood pressure.

Fuentes-Gandara *et al.* (2018) used the formula (10) to assess the risk of fish collected from the Colombian Caribean Sea, and found that there was no health risk from most of the metals. The HQ of 10 elements in wild marine fish from the coast of China Sea were all less than 1, thus indicating there was no obvious health risk from the intake of metals through these marine fish (Zhang and Wang, 2012).

Copat *et al.* (2018) use the Target Hazard Quotient (THQ) to assess the risk of developing chronic systemic effects derived from seafood consumption.

$$THQ=(EF \times ED \times IR \times C)/(RfD \times BW \times AT) \tag{11}$$

According to EPA guideline, it is assumed an exposure duration (ED) of 26 years, an exposure frequency (EF) (day/yr) of 365 days/year, a lifetime (LT) of 70 yr and an average time (AT) equal to EF×LT. By comparing the THQ between the year 2012 and 2017, the author found a decreased risk to develop chronic systemic effects derived from consumption of local seafood (Copat *et al.*, 2018). The THQ also showed that the edible fish species from the Izmir Bay can be consumed safely (Pazi *et al.*, 2017).

Table 2. Permissible Tolerable Daily Intake (PIDI) and Reference Dose (RfD) of heavy metals.

Heavy Metal	PIDI (µg/kg/d)	RfD (µg/kg/d)
Al	1000[c]	-
As	3[c]	0.3[a] (inorganic)

(Table 2) cont.....

Heavy Metal	PIDI (µg/kg/d)	RfD (µg/kg/d)
Cd	0.83[b]	1[a]
Co	10[c]	-
Cr(VI)	5[c]	3[a]
Cu	500[b]	-
Fe	800[b]	-
Mn	-	140[a]
Ni	-	20[a]
Pb	-	-
Zn	300[b, c]	300[a]

[a] Obtained from the Integrated Risk Information System, USEPA. (https:// www.epa.gov/iris)
[b] Obtained from the JECFA (http://apps.who.int/food-additives-contaminants- jecfa-database/Search.aspx)
[c] Obtained from the ATSDR. (http://www.atsdr.cdc.gov/substances/index.asp)

CONCLUSION

Heavy metals are the widespread environmental contaminant in marine environment due to natural and anthropogenic sources, and their occurrence and chemistry are complicated. Their bioaccumulation in marine fish are also complicated, tremendous geographic variance, interspecies and intraspecies differences in heavy metal concentrations and accumulation exist among different fish species. Generally, the heavy metal levels in the carnivore and demersal fish species are always higher than those found in omnivore and pelagic species due to their variations in feeding habits and habitats. Moreover, the levels of heavy metals in metabolic activity tissues, such as liver and kidney, are higher than that in uptake tissues, such as gill and gastrointestinal tract, whereas the levels of heavy metals in store and edible tissues (muscles) are the lowest. Such differences may be explained by different biokinetic parameters (k_u, AE, k_e), aquatic animals' physiology (IR, g), metal geochemistry (k_d), and environmental conditions (TSS). The high concentrations of heavy metals in fish can affect various physiological systems, such as behaviour, organ histopathology, glycolipid metabolism, enzymatic activities, oxidative stress, and hematological changes, which finally poses threat to humans. Thus, it is important to understand the bioaccumulation mechanisms and toxicological effects of heavy metals to alleviate its damage to fish and its consumer, contributing to a more scientific monitoring of water quality and management of the safety of aquatic products.

CONSENT FOR PUBLICATION

Not applicable.

CONFLICT OF INTEREST

The authors confirm that this chapter contents have no conflict of interest.

ACKNOWLEDGEMENT

None

REFERENCES

Adali, Y & Koca, YB (2016) Effects of pollution on some tissues of fish collected from different regions of Buyuk Menderes River: a histopathology study. *J Environ Prot Ecol,* 17, 477-87.

Ahmad, I, Oliveira, M, Pacheco, M & Santos, MA (2005) *Anguilla anguilla* L. oxidative stress biomarkers responses to copper exposure with or without beta-naphthoflavone pre-exposure. *Chemosphere,* 61, 267-75. [http://dx.doi.org/10.1016/j.chemosphere.2005.01.069] [PMID: 16168750]

Akbary, P, Sartipi Yarahmadi, S & Jahanbakhshi, A (2018) Hematological, hepatic enzymes' activity and oxidative stress responses of gray mullet (*Mugil cephalus*) after sub-acute exposure to copper oxide. *Environ Sci Pollut Res Int,* 25, 1800-8. [http://dx.doi.org/10.1007/s11356-017-0582-1] [PMID: 29101705]

Akter, KF, Owens, G, Davey, DE & Naidu, R (2005) Arsenic speciation and toxicity in biological systems.*Reviews of environmental contamination and toxicology* Springer, New York 97-149. [http://dx.doi.org/10.1007/0-387-27565-7_3]

Al-Asgah, NA, Abdel-Warith, A-WA, Younis, SM & Allam, HY (2015) Haematological and biochemical parameters and tissue accumulations of cadmium in *Oreochromis niloticus* exposed to various concentrations of cadmium chloride. *Saudi J Biol Sci,* 22, 543-50. [http://dx.doi.org/10.1016/j.sjbs.2015.01.002] [PMID: 26288556]

Al-Reasi, HA, Wood, CM & Smith, DS (2011) Physicochemical and spectroscopic properties of natural organic matter (NOM) from various sources and implications for ameliorative effects on metal toxicity to aquatic biota. *Aquat Toxicol,* 103, 179-90. [http://dx.doi.org/10.1016/j.aquatox.2011.02.015] [PMID: 21470554]

Albalat, A, Saera-Vila, A, Capilla, E, Gutiérrez, J, Pérez-Sánchez, J & Navarro, I (2007) Insulin regulation of lipoprotein lipase (LPL) activity and expression in gilthead sea bream (*Sparus aurata*). *Comp Biochem Physiol B Biochem Mol Biol,* 148, 151-9. [http://dx.doi.org/10.1016/j.cbpb.2007.05.004] [PMID: 17600746]

Altikat, S, Uysal, K, Kuru, HI, Kavasoglu, M, Ozturk, GN & Kucuk, A (2015) The effect of arsenic on some antioxidant enzyme activities and lipid peroxidation in various tissues of mirror carp (*Cyprinus carpio carpio*). *Environ Sci Pollut Res Int,* 22, 3212-8. [http://dx.doi.org/10.1007/s11356-014-2896-6] [PMID: 24770925]

Navas-Acien, A, Silbergeld, EK, Streeter, RA, Clark, JM, Burke, TA & Guallar, E (2006) Arsenic exposure and type 2 diabetes: a systematic review of the experimental and epidemiological evidence. *Environ Health Perspect,* 114, 641-8. [http://dx.doi.org/10.1289/ehp.8551] [PMID: 16675414]

Vallee, BL & Ulmer, DD (1972) Biochemical effects of mercury, cadmium, and lead. *Annu Rev Biochem,* 41, 91-128. [http://dx.doi.org/10.1146/annurev.bi.41.070172.000515] [PMID: 4570963]

Annabi, A, Messaoudi, I, Kerkeni, A & Said, K (2011) Cadmium accumulation and histological lesion in mosquitofish (*Gambusia affinis*) tissues following acute and chronic exposure. *Int J Environ Res,* 5, 745-56.

Arockia Vasanthi, L, Revathi, P, Mini, J & Munuswamy, N (2013) Integrated use of histological and ultrastructural biomarkers in *Mugil cephalus* for assessing heavy metal pollution in Ennore estuary, Chennai.

Chemosphere, 91, 1156-64.
[http://dx.doi.org/10.1016/j.chemosphere.2013.01.021] [PMID: 23415490]

Atchison, GJ, Henry, MG & Sandheinrich, MB (1987) Effects of metals on fish behavior-a review. *Environ Biol Fishes,* 18, 11-25.
[http://dx.doi.org/10.1007/BF00002324]

Ateş, A, Türkmen, M & Tepe, Y (2015) Assessment of heavy metals in fourteen marine fish species of four Turkish Seas. *Indian J Geo-Mar Sci,* 44, 49-55.

Atli, G & Canli, M (2007) Enzymatic responses to metal exposures in a freshwater fish *Oreochromis niloticus. Comp Biochem Physiol C Toxicol Pharmacol,* 145, 282-7.
[http://dx.doi.org/10.1016/j.cbpc.2006.12.012] [PMID: 17289437]

Atli, G & Canli, M (2013) Metals (Ag$^{(+)}$, Cd$^{(2+)}$, Cr$^{(6+)}$) affect ATPase activity in the gill, kidney, and muscle of freshwater fish *Oreochromis niloticus* following acute and chronic exposures. *Environ Toxicol,* 28, 707-17.
[http://dx.doi.org/10.1002/tox.20766] [PMID: 21901811]

Azizishirazi, A, Dew, WA, Bougas, B, Bernatchez, L & Pyle, GG (2015) Dietary sodium protects fish against copper-induced olfactory impairment. *Aquat Toxicol,* 161, 1-9.
[http://dx.doi.org/10.1016/j.aquatox.2015.01.017] [PMID: 25646894]

Baatrup, E (1991) Structural and functional effects of heavy metals on the nervous system, including sense organs, of fish. *Comp Biochem Physiol C Comp Pharmacol Toxicol,* 100, 253-7.
[http://dx.doi.org/10.1016/0742-8413(91)90163-N] [PMID: 1677859]

Bagnyukova, TV, Luzhna, LI, Pogribny, IP & Lushchak, VI (2007) Oxidative stress and antioxidant defenses in goldfish liver in response to short-term exposure to arsenite. *Environ Mol Mutagen,* 48, 658-65.
[http://dx.doi.org/10.1002/em.20328] [PMID: 17685460]

Barillet, S, Larno, V, Floriani, M, Devaux, A & Adam-Guillermin, C (2010) Ultrastructural effects on gill, muscle, and gonadal tissues induced in zebrafish (*Danio rerio*) by a waterborne uranium exposure. *Aquat Toxicol,* 100, 295-302.
[http://dx.doi.org/10.1016/j.aquatox.2010.08.002] [PMID: 20822817]

Baudin, JP & Fritsch, AF (1989) Relative contributions of food and water in the accumulation of ^{60}Co by a freshwater fish. *Water Res,* 23, 817-23.
[http://dx.doi.org/10.1016/0043-1354(89)90004-3]

Berg, K, Puntervoll, P, Valdersnes, S & Goksøyr, A (2010) Responses in the brain proteome of Atlantic cod (*Gadus morhua*) exposed to methylmercury. *Aquat Toxicol,* 100, 51-65.
[http://dx.doi.org/10.1016/j.aquatox.2010.07.008] [PMID: 20701987]

Berntssen, MHG, Aatland, A & Handy, RD (2003) Chronic dietary mercury exposure causes oxidative stress, brain lesions, and altered behaviour in Atlantic salmon (*Salmo salar*) parr. *Aquat Toxicol,* 65, 55-72.
[http://dx.doi.org/10.1016/S0166-445X(03)00104-8] [PMID: 12932701]

Besser, JM, Brumbaugh, WG, May, TW & Schmitt, CJ (2007) Biomonitoring of lead, zinc, and cadmium in streams draining lead-mining and non-mining areas, southeast Missouri, USA. *Environ Monit Assess,* 129, 227-41. a
[http://dx.doi.org/10.1007/s10661-006-9356-9] [PMID: 16957839]

Besser, JM, Mebane, CA, Mount, DR, Ivey, CD, Kunz, JL, Greer, IE, May, TW & Ingersoll, CG (2007) Sensitivity of mottled sculpins (*Cottus bairdi*) and rainbow trout (*Onchorhynchus mykiss*) to acute and chronic toxicity of cadmium, copper, and zinc. *Environ Toxicol Chem,* 26, 1657-65. b
[http://dx.doi.org/10.1897/06-571R.1] [PMID: 17702339]

Beyersmann, D & Hartwig, A (2008) Carcinogenic metal compounds: recent insight into molecular and cellular mechanisms. *Arch Toxicol,* 82, 493-512.
[http://dx.doi.org/10.1007/s00204-008-0313-y] [PMID: 18496671]

Bilandžić, N, Ðokić, M & Sedak, M (2011) Metal content determination in four fish species from the Adriatic Sea. *Food Chem,* 124, 1005-10.

[http://dx.doi.org/10.1016/j.foodchem.2010.07.060]

Boone, AN & Vijayan, MM (2002) Constitutive heat shock protein 70 (HSC70) expression in rainbow trout hepatocytes: effect of heat shock and heavy metal exposure. *Comp Biochem Physiol C Toxicol Pharmacol,* 132, 223-33.
[http://dx.doi.org/10.1016/S1532-0456(02)00066-2] [PMID: 12106899]

Bopp, SK, Abicht, HK & Knauer, K (2008) Copper-induced oxidative stress in rainbow trout gill cells. *Aquat Toxicol,* 86, 197-204.
[http://dx.doi.org/10.1016/j.aquatox.2007.10.014] [PMID: 18063143]

Bourdineaud, JP, Baudrimont, M, Gonzalez, P & Moreau, JL (2006) Challenging the model for induction of metallothionein gene expression. *Biochimie,* 88, 1787-92.
[http://dx.doi.org/10.1016/j.biochi.2006.07.021] [PMID: 16935407]

Brraich, OS & Kaur, M (2017) Histopathological alterations in the kidneys of *Labeo rohita* due to lead toxicity. *J Environ Biol,* 38, 257-62.
[http://dx.doi.org/10.22438/jeb/38/2/MS-52]

Cambier, S, Gonzalez, P, Mesmer-Dudons, N, Brèthes, D, Fujimura, M & Bourdineaud, J-P (2012) Effects of dietary methylmercury on the zebrafish brain: histological, mitochondrial, and gene transcription analyses. *Biometals,* 25, 165-80.
[http://dx.doi.org/10.1007/s10534-011-9494-6] [PMID: 21947502]

Carvalho, CDS & Fernandes, MN (2008) Effect of copper on liver key enzymes of anaerobic glucose metabolism from freshwater tropical fish Prochilodus lineatus. *Comparative Biochemistry and Physiology a-Molecular & Integrative Physiology,* 151, 437-42.

Castro-González, MI & Méndez-Armenta, M (2008) Heavy metals: Implications associated to fish consumption. *Environ Toxicol Pharmacol,* 26, 263-71.
[http://dx.doi.org/10.1016/j.etap.2008.06.001] [PMID: 21791373]

Chakoumakos, C, Russo, RC & Thurston, RV (1979) Toxicity of copper to cutthroat trout (*Salmo Clarki*) under different conditions of alkalinity, pH, and hardness. *Environ Sci Technol,* 13, 213-9.
[http://dx.doi.org/10.1021/es60150a013]

Chen, CY & Folt, CL (2000) Bioaccumulation and diminution of arsenic and lead in a freshwater food web. *Environ Sci Technol,* 34, 3878-84.
[http://dx.doi.org/10.1021/es991070c]

Chen, QL, Gong, Y, Luo, Z, Zheng, JL & Zhu, QL (2013) Differential effect of waterborne cadmium exposure on lipid metabolism in liver and muscle of yellow catfish *Pelteobagrus fulvidraco. Aquat Toxicol,* 142-143, 380-6. a
[http://dx.doi.org/10.1016/j.aquatox.2013.09.011] [PMID: 24095957]

Chen, QL, Luo, Z, Liu, CX & Zheng, JL (2015) Differential effects of dietary Cu deficiency and excess on carnitine status, kinetics and expression of CPT I in liver and muscle of yellow catfish *Pelteobagrus fulvidraco. Comp Biochem Physiol B Biochem Mol Biol,* 188, 24-30.
[http://dx.doi.org/10.1016/j.cbpb.2015.06.002] [PMID: 26086439]

Chen, QL, Luo, Z, Pan, YX, Zheng, JL, Zhu, QL, Sun, LD, Zhuo, MQ & Hu, W (2013) Differential induction of enzymes and genes involved in lipid metabolism in liver and visceral adipose tissue of juvenile yellow catfish *Pelteobagrus fulvidraco* exposed to copper. *Aquat Toxicol,* 136-137, 72-8. b
[http://dx.doi.org/10.1016/j.aquatox.2013.04.003] [PMID: 23660017]

Chen, QL, Luo, Z, Liu, X, Song, YF, Liu, CX, Zheng, JL & Zhao, YH (2013) Effects of waterborne chronic copper exposure on hepatic lipid metabolism and metal-element composition in *Synechogobius hasta. Arch Environ Contam Toxicol,* 64, 301-15. c
[http://dx.doi.org/10.1007/s00244-012-9835-7] [PMID: 23229194]

Chiaverini, N & De Ley, M (2010) Protective effect of metallothionein on oxidative stress-induced DNA damage. *Free Radic Res,* 44, 605-13.

[http://dx.doi.org/10.3109/10715761003692511] [PMID: 20380594]

Choudhury, TR (2014) Accumulation and histopathological effects of arsenic in tissues of shingi fish (stinging catfish) *Heteropneustes fossilis* (Bloch, 1794). *Journal of the Asiatic Society of Bangladesh Science,* 39, 221-30.
[http://dx.doi.org/10.3329/jasbs.v39i2.17861]

Chrestensen, CA, Starke, DW & Mieyal, JJ (2000) Acute cadmium exposure inactivates thioltransferase (Glutaredoxin), inhibits intracellular reduction of protein-glutathionyl-mixed disulfides, and initiates apoptosis. *J Biol Chem,* 275, 26556-65.
[http://dx.doi.org/10.1074/jbc.M004097200] [PMID: 10854441]

Çiçek, E, Avsar, D, Yeldan, H & Manasirli, M (2008) Heavy metal concentrations in fish (*Mullus barbatus, Pagellus erythrinus* and *Saurida undosquamis*) from Iskenderun Bay, Turkey. *Fresenius Environ Bull,* 17, 1251-6.

Coğun, H, Yüzereroğlu, TA, Kargin, F & Firat, O (2005) Seasonal variation and tissue distribution of heavy metals in shrimp and fish species from the yumurtalik coast of iskenderun gulf, mediterranean. *Bull Environ Contam Toxicol,* 75, 707-15.
[http://dx.doi.org/10.1007/s00128-005-0809-6] [PMID: 16400551]

Coğun, HY, Yüzereroğlu, TA, Firat, O, Gök, G & Kargin, F (2006) Metal concentrations in fish species from the northeast Mediterranean Sea. *Environ Monit Assess,* 121, 431-8.
[http://dx.doi.org/10.1007/s10661-005-9142-0] [PMID: 16752037]

Copat, C, Grasso, A, Fiore, M, Cristaldi, A, Zuccarello, P, Signorelli, SS, Conti, GO & Ferrante, M (2018) Trace elements in seafood from the Mediterranean sea: An exposure risk assessment. *Food Chem Toxicol,* 115, 13-9.
[http://dx.doi.org/10.1016/j.fct.2018.03.001] [PMID: 29510219]

Costa, PM, Chicano-Gálvez, E, Caeiro, S, Lobo, J, Martins, M, Ferreira, AM, Caetano, M, Vale, C, Alhama-Carmona, J, Lopez-Barea, J, DelValls, TA & Costa, MH (2012) Hepatic proteome changes in *Solea senegalensis* exposed to contaminated estuarine sediments: a laboratory and *in situ* survey. *Ecotoxicology,* 21, 1194-207.
[http://dx.doi.org/10.1007/s10646-012-0874-7] [PMID: 22362511]

Craig, PM, Wood, CM & McClelland, GB (2007) Oxidative stress response and gene expression with acute copper exposure in zebrafish (*Danio rerio*). *Am J Physiol Regul Integr Comp Physiol,* 293, R1882-92.
[http://dx.doi.org/10.1152/ajpregu.00383.2007] [PMID: 17855494]

Croteau, MN, Luoma, SN & Stewart, AR (2005) Trophic transfer of metals along freshwater food webs: Evidence of cadmium biomagnification in nature. *Limnol Oceanogr,* 50, 1511-9.
[http://dx.doi.org/10.4319/lo.2005.50.5.1511]

Cui, B, Zhang, Q, Zhang, K, Liu, X & Zhang, H (2011) Analyzing trophic transfer of heavy metals for food webs in the newly-formed wetlands of the Yellow River Delta, China. *Environ Pollut,* 159, 1297-306.
[http://dx.doi.org/10.1016/j.envpol.2011.01.024] [PMID: 21306806]

Das, S, Patro, SK & Sahu, BK (2001) Biochemical changes induced by mercury in the liver of penaeid prawns *Penaeus indicus* and *P. monodon* (Crustacea: Penaeidae) from Rushikulya estuary, east coast of India. *Indian Journal of Marineences,* 30, 246-52.

David, M, Shivakumar, R, Mushigeri, SB & Kuri, RC (2005) Blood glucose and glycogen levels as indicators of stress in the freshwater fish, *Labeo rohita* under fenvalerate intoxication. *Journal of Ecotoxicology & Environmental Monitoring,* 15, 1-5.

De Smet, H & Blust, R (2001) Stress responses and changes in protein metabolism in carp *Cyprinus carpio* during cadmium exposure. *Ecotoxicol Environ Saf,* 48, 255-62.
[http://dx.doi.org/10.1006/eesa.2000.2011] [PMID: 11222034]

DeForest, DK, Brix, KV & Adams, WJ (2007) Assessing metal bioaccumulation in aquatic environments: the inverse relationship between bioaccumulation factors, trophic transfer factors and exposure concentration.

Aquat Toxicol, 84, 236-46.
[http://dx.doi.org/10.1016/j.aquatox.2007.02.022] [PMID: 17673306]

Demayo, A, Taylor, MC, Taylor, KW & Hodson, PV (1982) Toxic effects of lead and lead compounds on human health, aquatic life, wildlife plants, and livestock. *CRC Crit Rev Environ Control,* 12, 257-305.
[http://dx.doi.org/10.1080/10643388209381698]

Lemly, AD (1993) Metabolic stress during winter increases the toxicity of selenium to fish. *Aquat Toxicol,* 27, 133-58.
[http://dx.doi.org/10.1016/0166-445X(93)90051-2]

Dethloff, GM, Bailey, HC & Maier, KJ (2001) Effects of dissolved copper on select hematological, biochemical, and immunological parameters of wild rainbow trout (*Oncorhynchus mykiss*). *Arch Environ Contam Toxicol,* 40, 371-80.
[http://dx.doi.org/10.1007/s002440010185] [PMID: 11443368]

Devi, R & Banerjee, TK (2007) Toxicopathological impact of sub-lethal concentration of lead nitrate on the aerial respiratory organs of 'murrel' *Channa Striata* (Bloch, Pisces). *Iran J Environ Health Sci Eng,* 4, 249-56.

Donat, J & Dryden, C (2001) Transition metals and heavy metal speciation *Encyclopedia of Ocean Sciences* 3027-35.

Du, S, Zhou, YY & Zhang, L (2017) [Application of stable isotopes (δ^{13}C and δ^{15}N) in studies on heavy metals bioaccumulation in Daya Bay food web]. *Ying Yong Sheng Tai Xue Bao,* 28, 2327-38.
[PMID: 29741067]

Dural, M & Bïckïcï, E (2010) Distribution of trace elements in the tissues of *Upeneus pori* and *Upeneus molucensis* from the Eastern cost of Mediterranean, Iskenderun Bay, Turkey. *J Anim Vet Adv,* 9, 1380-3.
[http://dx.doi.org/10.3923/javaa.2010.1380.1383]

Dural, M, Bickici, E & Manasirli, M (2010) Heavy metal concentrations in different tissue of *Mullus Barbatus* and *Mullus Surmuletus* from Iskenderun Bay, Eastern coast of Mediterranean, Turkey. *Rapport Commission International Mer Mediterranea, CIESM,* 39, 499.

Dural, M, Genc, E, Sangun, MK & Güner, O (2011) Accumulation of some heavy metals in *Hysterothylacium aduncum* (Nematoda) and its host sea bream, *Sparus aurata* (Sparidae) from North-Eastern Mediterranean Sea (Iskenderun Bay). *Environ Monit Assess,* 174, 147-55.
[http://dx.doi.org/10.1007/s10661-010-1445-0] [PMID: 20422284]

Dural, M, Genc, E, Yemenicioğlu, S & Kemal Sangun, M (2010) Accumulation of some heavy metals seasonally in *Hysterotylacium aduncum* (Nematoda) and its host Red Sea Bream, *Pagellus erythrinus* (Sparidae) from Gulf of Iskenderun (North-eastern Mediterranean). *Bull Environ Contam Toxicol,* 84, 125-31.
[http://dx.doi.org/10.1007/s00128-009-9904-4] [PMID: 19946663]

Dural, M, Göksu, MZ, Özak, AA & Derici, B (2006) Bioaccumulation of some heavy metals in different tissues of Dicentrarchus labrax L, 1758, Sparus aurata L, 1758 and Mugil cephalus L, 1758 from the Camlik lagoon of the eastern coast of Mediterranean (Turkey). *Environ Monit Assess,* 118, 65-74.
[http://dx.doi.org/10.1007/s10661-006-0987-7] [PMID: 16897534]

Eagles-Smith, CA, Suchanek, TH, Colwell, AE & Anderson, NL (2008) Mercury trophic transfer in a eutrophic lake: the importance of habitat-specific foraging. *Ecol Appl,* 18 (Suppl.), A196-212.
[http://dx.doi.org/10.1890/06-1476.1] [PMID: 19475925]

Ersoy, B & Çelik, M (2009) Essential elements and contaminants in tissues of commercial pelagic fish from the Eastern Mediterranean Sea. *J Sci Food Agric,* 89, 1615-21.
[http://dx.doi.org/10.1002/jsfa.3646]

Ersoy, B & Çelik, M (2010) The essential and toxic elements in tissues of six commercial demersal fish from Eastern Mediterranean Sea. *Food Chem Toxicol,* 48, 1377-82.
[http://dx.doi.org/10.1016/j.fct.2010.03.004] [PMID: 20214948]

Fänge, R (1986) Physiology of haemopoiesis.*Fish Physiology Recent Advances.* Springer, Dordrecht.
[http://dx.doi.org/10.1007/978-94-011-6558-7_1]

Faller, P, Kienzler, K & Krieger-Liszkay, A (2005) Mechanism of Cd^{2+} toxicity: Cd^{2+} inhibits photoactivation of Photosystem II by competitive binding to the essential Ca^{2+} site. *Biochim Biophys Acta,* 1706, 158-64.
[http://dx.doi.org/10.1016/j.bbabio.2004.10.005] [PMID: 15620376]

Farag, AM, Boese, CJ, Woodward, DF & Bergman, HL (1994) Physiological changes and tissue metal accumulation in rainbow trout exposed to foodborne and waterborne metals. *Environ Toxicol Chem,* 13, 2021-9.
[http://dx.doi.org/10.1002/etc.5620131215]

Farag, AM, Woodward, DF, Goldstein, JN, Brumbaugh, W & Meyer, JS (1998) Concentrations of metals associated with mining waste in sediments, biofilm, benthic macroinvertebrates, and fish from the Coeur d'Alene River basin, Idaho. *Arch Environ Contam Toxicol,* 34, 119-27.
[http://dx.doi.org/10.1007/s002449900295] [PMID: 9469853]

Fatima, M, Usmani, N, Firdaus, F, Zafeer, MF, Ahmad, S, Akhtar, K, Dawar Husain, SM, Ahmad, MH, Anis, E & Mobarak Hossain, M (2015) *In vivo* induction of antioxidant response and oxidative stress associated with genotoxicity and histopathological alteration in two commercial fish species due to heavy metals exposure in northern India (Kali) river. *Comp Biochem Physiol C Toxicol Pharmacol,* 176-177, 17-30.
[http://dx.doi.org/10.1016/j.cbpc.2015.07.004] [PMID: 26191657]

Feist, SW, Stentiford, GD, Kent, ML, Ribeiro Santos, A & Lorance, P (2015) Histopathological assessment of liver and gonad pathology in continental slope fish from the northeast Atlantic Ocean. *Mar Environ Res,* 106, 42-50.
[http://dx.doi.org/10.1016/j.marenvres.2015.02.004] [PMID: 25756900]

Feng, Q, Boone, AN & Vijayan, MM (2003) Copper impact on heat shock protein 70 expression and apoptosis in rainbow trout hepatocytes. *Comp Biochem Physiol C Toxicol Pharmacol,* 135C, 345-55.
[http://dx.doi.org/10.1016/S1532-0456(03)00137-6] [PMID: 12927909]

Ferreira, M, Antunes, P, Gil, O, Vale, C & Reis-Henriques, MA (2004) Organochlorine contaminants in flounder (*Platichthys flesus*) and mullet (*Mugil cephalus*) from Douro estuary, and their use as sentinel species for environmental monitoring. *Aquat Toxicol,* 69, 347-57.
[http://dx.doi.org/10.1016/j.aquatox.2004.06.005] [PMID: 15312718]

Ferreira, M, Caetano, M, Costa, J, Pousão-Ferreira, P, Vale, C & Reis-Henriques, MA (2008) Metal accumulation and oxidative stress responses in, cultured and wild, white seabream from Northwest Atlantic. *Sci Total Environ,* 407, 638-46.
[http://dx.doi.org/10.1016/j.scitotenv.2008.07.058] [PMID: 18783819]

Ferriss, BE & Essington, TE (2014) Does trophic structure dictate mercury concentrations in top predators? A comparative analysis of pelagic food webs in the Pacific Ocean. *Ecol Modell,* 278, 18-28.
[http://dx.doi.org/10.1016/j.ecolmodel.2014.01.029]

Fonseca, AR, Sanches Fernandes, LF, Fontainhas-Fernandes, A, Monteiro, SM & Pacheco, FAL (2017) The impact of freshwater metal concentrations on the severity of histopathological changes in fish gills: A statistical perspective. *Sci Total Environ,* 599-600, 217-26.
[http://dx.doi.org/10.1016/j.scitotenv.2017.04.196] [PMID: 28477478]

Celino, FT, Yamaguchi, S, Miura, C & Miura, T (2009) Arsenic inhibits *in vitro* spermatogenesis and induces germ cell apoptosis in Japanese eel (*Anguilla japonica*). *Reproduction,* 138, 279-87.
[http://dx.doi.org/10.1530/REP-09-0167] [PMID: 19494047]

Fu, D, Bridle, A, Leef, M, Norte Dos Santos, C & Nowak, B (2017) Hepatic expression of metal-related genes and gill histology in sand flathead (*Platycephalus bassensis*) from a metal contaminated estuary. *Mar Environ Res,* 131, 80-9.
[http://dx.doi.org/10.1016/j.marenvres.2017.09.014] [PMID: 28943068]

Fu, F & Wang, Q (2011) Removal of heavy metal ions from wastewaters: a review. *J Environ Manage,* 92,

407-18.
[http://dx.doi.org/10.1016/j.jenvman.2010.11.011] [PMID: 21138785]

Fuentes-Gandara, F, Pinedo-Hernández, J, Marrugo-Negrete, J & Díez, S (2018) Human health impacts of exposure to metals through extreme consumption of fish from the Colombian Caribbean Sea. *Environ Geochem Health*, 40, 229-42.
[http://dx.doi.org/10.1007/s10653-016-9896-z] [PMID: 27878501]

Garg, S, Gupta, RK & Jain, KL (2009) Sublethal effects of heavy metals on biochemical composition and their recovery in Indian major carps. *J Hazard Mater*, 163, 1369-84.
[http://dx.doi.org/10.1016/j.jhazmat.2008.07.118] [PMID: 18775601]

Gaworecki, KM, Chapman, RW, Neely, MG, D'Amico, AR & Bain, LJ (2012) Arsenic exposure to killifish during embryogenesis alters muscle development. *Toxicol Sci*, 125, 522-31.
[http://dx.doi.org/10.1093/toxsci/kfr302] [PMID: 22058191]

Gensemer, RW & Playle, RC (1999) The bioavailability and toxicity of aluminum in aquatic environments. *Crit Rev Environ Sci Technol*, 29, 315-450.
[http://dx.doi.org/10.1080/10643389991259245]

Giari, L, Manera, M, Simoni, E & Dezfuli, BS (2007) Cellular alterations in different organs of European sea bass *Dicentrarchus labrax* (L.) exposed to cadmium. *Chemosphere*, 67, 1171-81.
[http://dx.doi.org/10.1016/j.chemosphere.2006.10.061] [PMID: 17188326]

Goto, D & Wallace, WG (2010) Metal intracellular partitioning as a detoxification mechanism for mummichogs (*Fundulus heteroclitus*) living in metal-polluted salt marshes. *Mar Environ Res*, 69, 163-71.
[http://dx.doi.org/10.1016/j.marenvres.2009.09.008] [PMID: 19853291]

Green, AJ & Planchart, A (2018) The neurological toxicity of heavy metals: A fish perspective. *Comp Biochem Physiol C Toxicol Pharmacol*, 208, 12-9.
[http://dx.doi.org/10.1016/j.cbpc.2017.11.008] [PMID: 29199130]

Green, DWJ, Williams, KA & Pascoe, D (1986) The acute and chronic toxicity of cadmium to different life history stages of the freshwater crustacean *Asellus aquaticus* (L). *Arch Environ Contam Toxicol*, 15, 465-71.
[http://dx.doi.org/10.1007/BF01056557]

Grosell, M, Blanchard, J, Brix, KV & Gerdes, R (2007) Physiology is pivotal for interactions between salinity and acute copper toxicity to fish and invertebrates. *Aquat Toxicol*, 84, 162-72.
[http://dx.doi.org/10.1016/j.aquatox.2007.03.026] [PMID: 17643507]

Gu, YG, Lin, Q, Huang, HH, Wang, LG, Ning, JJ & Du, FY (2017) Heavy metals in fish tissues/stomach contents in four marine wild commercially valuable fish species from the western continental shelf of South China Sea. *Mar Pollut Bull*, 114, 1125-9.
[http://dx.doi.org/10.1016/j.marpolbul.2016.10.040] [PMID: 27765407]

Gu, YG, Lin, Q, Wang, XH, Du, FY, Yu, ZL & Huang, HH (2015) Heavy metal concentrations in wild fishes captured from the South China Sea and associated health risks. *Mar Pollut Bull*, 96, 508-12. a
[http://dx.doi.org/10.1016/j.marpolbul.2015.04.022] [PMID: 25913793]

Gu, YG, Lin, Q, Yu, ZL, Wang, XN, Ke, CL & Ning, JJ (2015) Speciation and risk of heavy metals in sediments and human health implications of heavy metals in edible nekton in Beibu Gulf, China: A case study of Qinzhou Bay. *Mar Pollut Bull*, 101, 852-9. b
[http://dx.doi.org/10.1016/j.marpolbul.2015.11.019] [PMID: 26578296]

Guan, R & Wang, WX (2004) Dietary assimilation and elimination of Cd, Se, and Zn by *Daphnia magna* at different metal concentrations. *Environ Toxicol Chem*, 23, 2689-98.
[http://dx.doi.org/10.1897/03-503] [PMID: 15559285]

Guyenet, SJ & Schwartz, MW (2012) Clinical review: Regulation of food intake, energy balance, and body fat mass: implications for the pathogenesis and treatment of obesity. *J Clin Endocrinol Metab*, 97, 745-55.
[http://dx.doi.org/10.1210/jc.2011-2525] [PMID: 22238401]

Hamilton, SJ & Mehrle, PM (1986) Metallothionein in fish-review of its importance in assessing stress from

metal contaminants. *Trans Am Fish Soc,* 115, 596-609.
[http://dx.doi.org/10.1577/1548-8659(1986)115<596:MIF>2.0.CO;2]

Hansen, BH, Garmo, OA, Olsvik, PA & Andersen, RA (2007) Gill metal binding and stress gene transcription in brown trout (*Salmo trutta*) exposed to metal environments: the effect of pre-exposure in natural populations. *Environ Toxicol Chem,* 26, 944-53.
[http://dx.doi.org/10.1897/06-380R.1] [PMID: 17521141]

Hansen, BH, Rømma, S, Søfteland, LIR, Olsvik, PA & Andersen, RA (2006) Induction and activity of oxidative stress-related proteins during waterborne Cu-exposure in brown trout (*Salmo trutta*). *Chemosphere,* 65, 1707-14.
[http://dx.doi.org/10.1016/j.chemosphere.2006.04.088] [PMID: 16780922]

Hansen, PD, Dizer, H, Hock, B, Marx, A, Sherry, J, Mcmaster, M & Blaise, C (1998) Vitellogenin - a biomarker for endocrine disruptors. *Trends Analyt Chem,* 17, 448-51.
[http://dx.doi.org/10.1016/S0165-9936(98)00020-X]

Haque, MS & Roy, SK (2012) Acute effects of arsenic on the regulation of metabolic activities in liver of fresh water fishes (Taki) during cold acclimation. *Jordan J Biol Sci,* 5, 91-7.

Hartwig, A (2001) Zinc finger proteins as potential targets for toxic metal ions: differential effects on structure and function. *Antioxid Redox Signal,* 3, 625-34.
[http://dx.doi.org/10.1089/15230860152542970] [PMID: 11554449]

Hassanin, A, Kuwahara, S, , Nurhidayat, Tsukamoto, Y, Ogawa, K, Hiramatsu, K & Sasaki, F (2002) Gonadosomatic index and testis morphology of common carp (*Cyprinus carpio*) in rivers contaminated with estrogenic chemicals. *J Vet Med Sci,* 64, 921-6.
[http://dx.doi.org/10.1292/jvms.64.921] [PMID: 12419869]

Hilmy, AM, Shabana, MB & Daabees, AY (1985) Effects of cadmium toxicity upon the *in vivo* and *in vitro* activity of proteins and five enzymes in blood serum and tissue homogenates of *Mugil cephalus*. *Comp Biochem Physiol C Comp Pharmacol Toxicol,* 81, 145-53.
[http://dx.doi.org/10.1016/0742-8413(85)90106-9] [PMID: 2861041]

Hinton, DE & Lauren, DJ (1990) Integrative histopathological approaches to detecting effects of environmental stressors on fishes. *Am Fish Soc Symp,* 8, 51-66.

Holland, S, Lodwig, E, Sideri, T, Reader, T, Clarke, I, Gkargkas, K, Hoyle, DC, Delneri, D, Oliver, SG & Avery, SV (2007) Application of the comprehensive set of heterozygous yeast deletion mutants to elucidate the molecular basis of cellular chromium toxicity. *Genome Biology,* 8, 1-10.
[http://dx.doi.org/10.1186/gb-2007-8-12-r268]

Hontela, A (1998) Interrenal dysfunction in fish from contaminated sites: *in vivo* and *in vitro* assessment. *Environ Toxicol Chem,* 17, 44-8.
[http://dx.doi.org/10.1002/etc.5620170107]

Huang, H, Ping, X, Li, L, Liao, Y & Shen, X (2011) Concentrations and assessment of heavy metals in seawater,sediments and aquatic organisms at Yangtze River estuary in spring and summer. *Shengtai Huanjing Xuebao,* 20, 898-903.

Hughes, MF (2002) Arsenic toxicity and potential mechanisms of action. *Toxicol Lett,* 133, 1-16.
[http://dx.doi.org/10.1016/S0378-4274(02)00084-X] [PMID: 12076506]

Hunt, JW, Anderson, BS, Phillips, BM, Tjeerdema, RS, Puckett, HM, Stephenson, M, Tucker, DW & Watson, D (2002) Acutei and chronic toxicity of nickel to marine organisms: implications for water quality criteria. *Environ Toxicol Chem,* 21, 2423-30.
[http://dx.doi.org/10.1002/etc.5620211122] [PMID: 12389922]

Hwang, IK, Kim, KW, Kim, JH & Kang, JC (2016) Toxic effects and depuration after the dietary lead(II) exposure on the bioaccumulation and hematological parameters in starry flounder (*Platichthys stellatus*). *Environ Toxicol Pharmacol,* 45, 328-33.
[http://dx.doi.org/10.1016/j.etap.2016.06.017] [PMID: 27362663]

Ikemoto, T, Tu, NPC, Okuda, N, Iwata, A, Omori, K, Tanabe, S, Tuyen, BC & Takeuchi, I (2008) Biomagnification of trace elements in the aquatic food web in the Mekong Delta, South Vietnam using stable carbon and nitrogen isotope analysis. *Arch Environ Contam Toxicol,* 54, 504-15.
[http://dx.doi.org/10.1007/s00244-007-9058-5] [PMID: 18026776]

Ip, CCM, Li, XD, Zhang, G, Wong, CSC & Zhang, WL (2005) Heavy metal and Pb isotopic compositions of aquatic organisms in the Pearl River Estuary, South China. *Environ Pollut,* 138, 494-504.
[http://dx.doi.org/10.1016/j.envpol.2005.04.016] [PMID: 15970366]

Jacobson, T, Navarrete, C, Sharma, SK, Sideri, TC, Ibstedt, S, Priya, S, Grant, CM, Christen, P, Goloubinoff, P & Tamás, MJ (2012) Arsenite interferes with protein folding and triggers formation of protein aggregates in yeast. *J Cell Sci,* 125, 5073-83.
[http://dx.doi.org/10.1242/jcs.107029] [PMID: 22946053]

Javed, M & Usmani, N (2013) Assessment of heavy metal (Cu, Ni, Fe, Co, Mn, Cr, Zn) pollution in effluent dominated rivulet water and their effect on glycogen metabolism and histology of *Mastacembelus armatus*. *Springerplus,* 2, 390.
[http://dx.doi.org/10.1186/2193-1801-2-390] [PMID: 24133639]

Javed, M & Usmani, N (2015) Stress response of biomolecules (carbohydrate, protein and lipid profiles) in fish *Channa punctatus* inhabiting river polluted by Thermal Power Plant effluent. *Saudi J Biol Sci,* 22, 237-42.
[http://dx.doi.org/10.1016/j.sjbs.2014.09.021] [PMID: 25737659]

Jiraungkoorskul, W, Upatham, ES, Kruatrachue, M, Sahaphong, S & Pokethitiyook, P (2002) Histopathological effects of roundup, a glyphosate herbicide, on Nile tilapia (*Oreochromis niloticus*). *Sci Asia,* 28, 260-7.
[http://dx.doi.org/10.2306/scienceasia1513-1874.2002.28.121]

Jureša, D & Blanuša, M (2003) Mercury, arsenic, lead and cadmium in fish and shellfish from the Adriatic Sea. *Food Addit Contam,* 20, 241-6.
[http://dx.doi.org/10.1080/0265203021000055379] [PMID: 12623648]

Kalay, M, Ay, O & Canli, M (1999) Heavy metal concentrations in fish tissues from the northeast Mediterranean Sea. *Bull Environ Contam Toxicol,* 63, 673-81.
[http://dx.doi.org/10.1007/s001289901033] [PMID: 10541689]

Kang, JC, Kim, SG & Jang, SW (2005) Growth and hematological changes of rockfish, *Sebastes schlegeli* (Hilgendorf) exposed to dietary Cu and Cd. *J World Aquacult Soc,* 36, 188-95.
[http://dx.doi.org/10.1111/j.1749-7345.2005.tb00384.x]

Kargin, F (1996) Seasonal changes in levels of heavy metals in tissues of *Mullus barbatus* and *Sparus aurata* collected from Iskenderun Gulf (Turkey). *Water Air Soil Pollut,* 90, 557-62.
[http://dx.doi.org/10.1007/BF00282669]

Karimi, R, Chen, CY & Folt, CL (2016) Comparing nearshore benthic and pelagic prey as mercury sources to lake fish: the importance of prey quality and mercury content. *Sci Total Environ,* 565, 211-21.
[http://dx.doi.org/10.1016/j.scitotenv.2016.04.162] [PMID: 27173839]

Kasumyan, AO (2001) Effects of chemical pollutants on foraging behavior and sensitivity of fish to food stimuli. *J Ichthyol,* 41, 76-87.

Kavanagh, RJ, Balch, GC, Kiparissis, Y, Niimi, AJ, Sherry, J, Tinson, C & Metcalfe, CD (2004) Endocrine disruption and altered gonadal development in white perch (*Morone americana*) from the lower Great Lakes region. *Environ Health Perspect,* 112, 898-902.
[http://dx.doi.org/10.1289/ehp.6514] [PMID: 15175179]

Kerner, J & Hoppel, C (2000) Fatty acid import into mitochondria. *Biochim Biophys Acta,* 1486, 1-17.
[http://dx.doi.org/10.1016/S1388-1981(00)00044-5] [PMID: 10856709]

Keyvanshokooh, S, Vaziri, B, Gharaei, A, Mahboudi, F, Esmaili-Sari, A & Shahriari-Moghadam, M (2009) Proteome modifications of juvenile beluga (*Huso huso*) brain as an effect of dietary methylmercury. *Comp*

Biochem Physiol Part D Genomics Proteomics, 4, 243-8.
[http://dx.doi.org/10.1016/j.cbd.2009.01.002] [PMID: 20403756]

Khanna, SS & Gill, TS (1975) Effect of cobalt salts on the glycemia and islet histology of Channa punctatus (Bloch). *Acta Anat (Basel),* 92, 194-201.
[http://dx.doi.org/10.1159/000144441] [PMID: 1098356]

Khoshnood, Z, Mokhlesi, A & Khoshnood, R (2010) Bioaccumulation of some heavy metals and histopathological alterations in liver of *Euryglossa orientalis* and *Psettodes erumei* along North Coast of the Persian Gulf. *Afr J Biotechnol,* 9, 6966-72.

Kim, JH & Kang, JC (2014) The selenium accumulation and its effect on growth, and haematological parameters in red sea bream, *Pagrus major,* exposed to waterborne selenium. *Ecotoxicol Environ Saf,* 104, 96-102.
[http://dx.doi.org/10.1016/j.ecoenv.2014.02.010] [PMID: 24636952]

Kim, JH & Kang, JC (2017) Toxic effects on bioaccumulation and hematological parameters of juvenile rockfish *Sebastes schlegelii* exposed to dietary lead (Pb) and ascorbic acid. *Chemosphere,* 176, 131-40.
[http://dx.doi.org/10.1016/j.chemosphere.2017.02.097] [PMID: 28260654]

Kim, SG & Kang, JC (2004) Effect of dietary copper exposure on accumulation, growth and hematological parameters of the juvenile rockfish, *Sebastes schlegeli. Mar Environ Res,* 58, 65-82.
[http://dx.doi.org/10.1016/j.marenvres.2003.12.004] [PMID: 15046946]

Kling, PG & Olsson, P (2000) Involvement of differential metallothionein expression in free radical sensitivity of RTG-2 and CHSE-214 cells. *Free Radic Biol Med,* 28, 1628-37.
[http://dx.doi.org/10.1016/S0891-5849(00)00277-X] [PMID: 10938459]

Koca, YB, Koca, S, Yildiz, S, Gürcü, B, Osanç, E, Tunçbaş, O & Aksoy, G (2005) Investigation of histopathological and cytogenetic effects on *Lepomis gibbosus* (Pisces: Perciformes) in the Cine stream (Aydin/Turkey) with determination of water pollution. *Environ Toxicol,* 20, 560-71.
[http://dx.doi.org/10.1002/tox.20145] [PMID: 16302173]

Krezel, A & Maret, W (2007) Dual nanomolar and picomolar Zn(II) binding properties of metallothionein. *J Am Chem Soc,* 129, 10911-21.
[http://dx.doi.org/10.1021/ja071979s] [PMID: 17696343]

Krumschnabel, G, Manzl, C, Berger, C & Hofer, B (2005) Oxidative stress, mitochondrial permeability transition, and cell death in Cu-exposed trout hepatocytes. *Toxicol Appl Pharmacol,* 209, 62-73.
[http://dx.doi.org/10.1016/j.taap.2005.03.016] [PMID: 15882883]

Kumar, S & Pant, SC (1981) Histopathologic effects of acutely toxic levels of copper & zinc on gills, liver & kidney of Puntius conchonius (Ham.). *Indian J Exp Biol,* 19, 191-4.
[PMID: 7287075]

Kumari, B & Ahsan, J (2011) Acute exposure of arsenic tri-oxide produces hyperglycemia in both sexes of an Indian teleost, *Clarias batrachus* (Linn.). *Arch Environ Contam Toxicol,* 61, 435-42. a
[http://dx.doi.org/10.1007/s00244-011-9649-z] [PMID: 21360079]

Kumari, B & Ahsan, J (2011) Study of muscle glycogen content in both sexes of an Indian teleost *Clarias batrachus* (Linn.) exposed to different concentrations of arsenic. *Fish Physiol Biochem,* 37, 161-7. b
[http://dx.doi.org/10.1007/s10695-010-9427-2] [PMID: 20730599]

Kumari, B, Ahsan, J & Kumar, V (2012) Comparative studies of liver and brain glycogen content of male and female *Clarias batrachus* (L.) after exposure of different doses of arsenic. *Toxicol Environ Chem,* 94, 1758-67.
[http://dx.doi.org/10.1080/02772248.2012.726823]

Kumari, B, Ghosh, A & Kumari, A (2013) Aresnic induced hyper and hypo-pigmentation of skin in freshwater fish *Heteropneustes fossilis. Conference proceeding of Asian Pacific Aquaculture.* Ho Chi Minh City, Vietnam

Kumari, B, Kumar, V, Sinha, AK, Ahsan, J, Ghosh, AK, Wang, HP & Deboeck, G (2017) Toxicology of

arsenic in fish and aquatic systems. *Environ Chem Lett,* 15, 43-64.
[http://dx.doi.org/10.1007/s10311-016-0588-9]

Lerebours, A, Gonzalez, P, Adam, C, Camilleri, V, Bourdineaud, JP & Garnier-Laplace, J (2009) Comparative analysis of gene expression in brain, liver, skeletal muscles, and gills of zebrafish (*Danio rerio*) exposed to environmentally relevant waterborne uranium concentrations. *Environ Toxicol Chem,* 28, 1271-8.
[http://dx.doi.org/10.1897/08-357.1] [PMID: 19146232]

Levesque, HM, Moon, TW, Campbell, PGC & Hontela, A (2002) Seasonal variation in carbohydrate and lipid metabolism of yellow perch (*Perca flavescens*) chronically exposed to metals in the field. *Aquat Toxicol,* 60, 257-67.
[http://dx.doi.org/10.1016/S0166-445X(02)00012-7] [PMID: 12200090]

Li, XD, Wang, WX & Zhu, YG (2012) Trace metal pollution in China. *Sci Total Environ,* 421-422, 1-2.
[http://dx.doi.org/10.1016/j.scitotenv.2011.09.087] [PMID: 22033360]

Liu, XJ, Luo, Z, Li, CH, Xiong, BX, Zhao, YH & Li, XD (2011) Antioxidant responses, hepatic intermediary metabolism, histology and ultrastructure in *Synechogobius hasta* exposed to waterborne cadmium. *Ecotoxicol Environ Saf,* 74, 1156-63.
[http://dx.doi.org/10.1016/j.ecoenv.2011.02.015] [PMID: 21392825]

Loganathan, K, Velmurugan, B, Hongray Howrelia, J, Selvanayagam, M & Patnaik, BB (2006) Zinc induced histological changes in brain and liver of *Labeo rohita* (Ham.). *J Environ Biol,* 27, 107-10.
[PMID: 16850886]

Low, J & Higgs, DM (2015) Sublethal effects of cadmium on auditory structure and function in fathead minnows (*Pimephales promelas*). *Fish Physiol Biochem,* 41, 357-69.
[http://dx.doi.org/10.1007/s10695-014-9988-6] [PMID: 25245458]

Luckey, TD, Venugopal, B (1977). *Physiologic and chemical basis for metal toxicity.* Springer Science & Business Media, Berlin.
[http://dx.doi.org/10.1007/978-1-4684-2952-7]

Lushchak, VI (2011) Environmentally induced oxidative stress in aquatic animals. *Aquat Toxicol,* 101, 13-30.
[http://dx.doi.org/10.1016/j.aquatox.2010.10.006] [PMID: 21074869]

Mallatt, J (1985) Fish gill structural-changes induced by toxicants and other irritants- a statistical review. *Can J Fish Aquat Sci,* 42, 630-48.
[http://dx.doi.org/10.1139/f85-083]

Manaşirli, M, Avşar, D, Yeldan, H & Mavruk, S (2016) Trace elements (Fe, Cu and Zn) accumulation in the muscle tissues of *Saurida undosquamis, Pagellus erythrinus* and *Mullus barbatus* in the Iskenderun Bay, Turkey. *Fresenius Environ Bull,* 24, 1601-6.

Marcantonio, AS, Ranzani-Paiva, MJT, Franca, FM, Dias, DC, Teixeira, PC & Ferreira, CM (2011) Toxicity of zinc sulphate for tadpoles of bullfrog (*Lithobates catesbeianus*): Acute toxicity, chronic toxicity and hematological parameters. *Bol Inst Pesca,* 37, 143-54.

Maruthi, YA & Rao, MVS (2000) Effect of distillery effluent on biochemical parameters of fish, *Channa punctatus* (Bloch). *Journal of Environment and Pollution,* 7, 111-3.

Matthiessen, P, Allen, Y, Bamber, S, Craft, J, Hurst, M, Hutchinson, T, Feist, S, Katsiadaki, I, Kirby, M, Robinson, C, Scott, S, Thain, J & Thomas, K (2002) The impact of oestrogenic and androgenic contamination on marine organisms in the United Kingdom-summary of the EDMAR programme. Endocrine Disruption in the Marine Environment. *Mar Environ Res,* 54, 645-9.
[http://dx.doi.org/10.1016/S0141-1136(02)00135-6] [PMID: 12408629]

Mcdonald, DG & Wood, CM (1993) *Branchial mechanisms of acclimation to metals in freshwater fish.* Springer, Dordrecht 297-321.
[http://dx.doi.org/10.1007/978-94-011-2304-4_12]

McRae, NK, Gaw, S & Glover, CN (2016) Mechanisms of zinc toxicity in the galaxiid fish, *Galaxias maculatus. Comp Biochem Physiol C Toxicol Pharmacol,* 179, 184-90.

[http://dx.doi.org/10.1016/j.cbpc.2015.10.010] [PMID: 26510681]

Melvin, SD & Wilson, SP (2013) The utility of behavioral studies for aquatic toxicology testing: a meta-analysis. *Chemosphere*, 93, 2217-23.
[http://dx.doi.org/10.1016/j.chemosphere.2013.07.036] [PMID: 23958442]

Merrick, BA (2006) Toxicoproteomics in liver injury and inflammation. *Ann N Y Acad Sci*, 1076, 707-17.
[http://dx.doi.org/10.1196/annals.1371.017] [PMID: 17119248]

Miandare, HK, Yarahmadi, P, Jalali, MA, Aramoon, A, Mohammadi, MJ & Orrego, R (2017) Exposure of rainbow trout to industrial wastewater is associated with immune-related gene expression and hematological changes. *Int J Environ Sci Technol*, 14, 623-30.
[http://dx.doi.org/10.1007/s13762-016-1185-y]

Minghetti, M, Leaver, MJ, Carpenè, E & George, SG (2008) Copper transporter 1, metallothionein and glutathione reductase genes are differentially expressed in tissues of sea bream (*Sparus aurata*) after exposure to dietary or waterborne copper. *Comp Biochem Physiol C Toxicol Pharmacol*, 147, 450-9.
[http://dx.doi.org/10.1016/j.cbpc.2008.01.014] [PMID: 18304880]

Mirza, RS, Green, WW, Connor, S, Weeks, ACW, Wood, CM & Pyle, GG (2009) Do you smell what I smell? Olfactory impairment in wild yellow perch from metal-contaminated waters. *Ecotoxicol Environ Saf*, 72, 677-83.
[http://dx.doi.org/10.1016/j.ecoenv.2008.10.001] [PMID: 19108892]

Mishra, AK & Mohanty, B (2008) Acute toxicity impacts of hexavalent chromium on behavior and histopathology of gill, kidney and liver of the freshwater fish, *Channa punctatus* (Bloch). *Environ Toxicol Pharmacol*, 26, 136-41.
[http://dx.doi.org/10.1016/j.etap.2008.02.010] [PMID: 21783901]

Miura, T & Miura, C (2001) Japanese eel: A model for analysis of spermatogenesis. *Zool Sci*, 18, 1055-63.
[http://dx.doi.org/10.2108/zsj.18.1055]

Miura, T & Miura, CI (2003) Molecular control mechanisms of fish spermatogenesis. *Fish Physiol Biochem*, 28, 181-6.
[http://dx.doi.org/10.1023/B:FISH.0000030522.71779.47]

Mohamed, FAS & Gad, NS (2008) Environmental pollution induced biochemical changes in tissues of *Tilapia zillii*, *Solea vulgaris* and *Mugil capito* from Lake Qarun, Egypt. *Glob Vet*, 2, 327-36.

Mohanty, D & Samanta, L (2016) Multivariate analysis of potential biomarkers of oxidative stress in *Notopterus notopterus* tissues from Mahanadi River as a function of concentration of heavy metals. *Chemosphere*, 155, 28-38.
[http://dx.doi.org/10.1016/j.chemosphere.2016.04.035] [PMID: 27105150]

Moreira-Santos, M, Donato, C, Lopes, I & Ribeiro, R (2008) Avoidance tests with small fish: determination of the median avoidance concentration and of the lowest-observed-effect gradient. *Environ Toxicol Chem*, 27, 1576-82.
[http://dx.doi.org/10.1897/07-094.1] [PMID: 18260687]

Morel, FMM & Price, NM (2003) The biogeochemical cycles of trace metals in the oceans. *Science*, 300, 944-7.
[http://dx.doi.org/10.1126/science.1083545] [PMID: 12738853]

Mughal, MS, Malkani, N & Jahan, N (2004) Histopathological changes in kidney and muscle of *Labeo rohita* due to cadmium intoxication. *Biologia*, 50, 39-45.

Mushiake, K, Muroga, K & Nakai, T (1984) Increased susceptibility of Japanese eel *Anguilla japonica* to *Edwardsiella tarda* and *Pseudomonas anguilliseptica* following exposure to copper. *Nippon Suisan Gakkaishi*, 50, 1797-801.
[http://dx.doi.org/10.2331/suisan.50.1797]

Muyssen, BTA, Brix, KV, Deforest, DK & Janssen, CR (2004) Nickel essentiality and homeostasis in aquatic organisms. *Environ Rev*, 12, 113-31.

[http://dx.doi.org/10.1139/a04-004]

Naïja, A, Kestemont, P, Chénais, B, Haouas, Z, Blust, R, Helal, AN & Marchand, J (2018) Effects of Hg sublethal exposure in the brain of peacock blennies *Salaria pavo*: Molecular, physiological and histopathological analysis. *Chemosphere,* 193, 1094-104.
[http://dx.doi.org/10.1016/j.chemosphere.2017.11.118] [PMID: 29874737]

Nath, K & Kumar, N (1988) Hexavalent chromium: toxicity and its impact on certain aspects of carbohydrate metabolism of the freshwater teleost, *Colisa fasciatus. Sci Total Environ,* 72, 175-81.
[http://dx.doi.org/10.1016/0048-9697(88)90016-2] [PMID: 3406728]

Neff, JM (1997) Ecotoxicology of arsenic in the marine environment. *Environ Toxicol Chem,* 16, 917-27.
[http://dx.doi.org/10.1002/etc.5620160511]

Neff, JM (2002). *Bioaccumulation in marine organisms: effect of contaminants from oil well produced water.* Elsevier.

Nfon, E, Cousins, IT, Järvinen, O, Mukherjee, AB, Verta, M & Broman, D (2009) Trophodynamics of mercury and other trace elements in a pelagic food chain from the Baltic Sea. *Sci Total Environ,* 407, 6267-74.
[http://dx.doi.org/10.1016/j.scitotenv.2009.08.032] [PMID: 19767059]

Ng, TYT, Rainbow, PS, Amiard-Triquet, C, Amiard, JC & Wang, WX (2007) Metallothionein turnover, cytosolic distribution and the uptake of Cd by the green mussel *Perna viridis. Aquat Toxicol,* 84, 153-61.
[http://dx.doi.org/10.1016/j.aquatox.2007.01.010] [PMID: 17640747]

Nieboer, E & Richardson, DHS (1980) The replacement of the non-descript term heavy-metals by a biologically and chemically significant classification of metal-ions. *Environ Pollut B,* 1, 3-26.
[http://dx.doi.org/10.1016/0143-148X(80)90017-8]

Nilsson-Ehle, P, Garfinkel, AS & Schotz, MC (1980) Lipolytic enzymes and plasma lipoprotein metabolism. *Annu Rev Biochem,* 49, 667-93.
[http://dx.doi.org/10.1146/annurev.bi.49.070180.003315] [PMID: 6996570]

Nowak, B (1992) Histological-changes in gills induced by residues of endosulfan. *Aquat Toxicol,* 23, 65-83.
[http://dx.doi.org/10.1016/0166-445X(92)90012-C]

Okamoto, OK & Colepicolo, P (1998) Response of superoxide dismutase to pollutant metal stress in the marine dinoflagellate Gonyaulax polyedra. *Comp Biochem Physiol C Pharmacol Toxicol Endocrinol,* 119, 67-73.
[http://dx.doi.org/10.1016/S0742-8413(97)00192-8] [PMID: 9568375]

Omar, WA, Saleh, YS & Marie, MAS (2014) Integrating multiple fish biomarkers and risk assessment as indicators of metal pollution along the Red Sea coast of Hodeida, Yemen Republic. *Ecotoxicol Environ Saf,* 110, 221-31.
[http://dx.doi.org/10.1016/j.ecoenv.2014.09.004] [PMID: 25261609]

Onsanit, S, Ke, C, Wang, X, Wang, KJ & Wang, WX (2010) Trace elements in two marine fish cultured in fish cages in Fujian province, China. *Environ Pollut,* 158, 1334-42.
[http://dx.doi.org/10.1016/j.envpol.2010.01.012] [PMID: 20149944]

Padmini, E, Usha Rani, M & Vijaya Geetha, B (2009) Studies on antioxidant status in *Mugil cephalus* in response to heavy metal pollution at Ennore estuary. *Environ Monit Assess,* 155, 215-25.
[http://dx.doi.org/10.1007/s10661-008-0430-3] [PMID: 18629443]

Palanisamy, PG, Sasikala, D, Mallikaraj, NB & Natarajan, GM (2011) Electroplating industrial effluent chromium induced changes in carbohydrates metabolism in air breathing cat fish *Mystus cavasius* (Ham). *Asian J Exp Biol Sci,* 2, 521-4.

Pan, K & Wang, WX (2008) The subcellular fate of cadmium and zinc in the scallop *Chlamys nobilis* during waterborne and dietary metal exposure. *Aquat Toxicol,* 90, 253-60. a
[http://dx.doi.org/10.1016/j.aquatox.2008.09.010] [PMID: 18992948]

Pan, K & Wang, WX (2012) Trace metal contamination in estuarine and coastal environments in China. *Sci Total Environ*, 421-422, 3-16.
[http://dx.doi.org/10.1016/j.scitotenv.2011.03.013] [PMID: 21470665]

Pan, K & Wang, WX (2008) Validation of biokinetic model of metals in the scallop *Chlamys nobilis* in complex field environments. *Environ Sci Technol*, 42, 6285-90. b
[http://dx.doi.org/10.1021/es800652u] [PMID: 18767700]

Pan, X, Reissman, S, Douglas, NR, Huang, Z, Yuan, DS, Wang, X, McCaffery, JM, Frydman, J & Boeke, JD (2010) Trivalent arsenic inhibits the functions of chaperonin complex. *Genetics*, 186, 725-34.
[http://dx.doi.org/10.1534/genetics.110.117655] [PMID: 20660648]

Pandey, S, Parvez, S, Ansari, RA, Ali, M, Kaur, M, Hayat, F, Ahmad, F & Raisuddin, S (2008) Effects of exposure to multiple trace metals on biochemical, histuological and ultrastructural features of gills of a freshwater fish, *Channa punctata* Bloch. *Chem Biol Interact*, 174, 183-92.
[http://dx.doi.org/10.1016/j.cbi.2008.05.014] [PMID: 18586230]

Papanikolaou, G & Pantopoulos, K (2005) Iron metabolism and toxicity. *Toxicol Appl Pharmacol*, 202, 199-211.
[http://dx.doi.org/10.1016/j.taap.2004.06.021] [PMID: 15629195]

Patnaik, BB, Howrelia, HJ, Mathews, T & Selvanayagam, M (2011) Histopathology of gill, liver, muscle and brain of *Cyprinus carpio* communis L. exposed to sublethal concentration of lead and cadmium. *Afr J Biotechnol*, 10, 12218-23.

Pazi, I, Gonul, LT, Kucuksezgin, F, Avaz, G, Tolun, L, Unluoglu, A, Karaaslan, Y, Gucver, SM, Koc Orhon, A, Siltu, E & Olmez, G (2017) Potential risk assessment of metals in edible fish species for human consumption from the Eastern Aegean Sea. *Mar Pollut Bull*, 120, 409-13.
[http://dx.doi.org/10.1016/j.marpolbul.2017.05.004] [PMID: 28479148]

Pedlar, RM, Ptashynski, MD, Evans, R & Klaverkamp, JF (2002) Toxicological effects of dietary arsenic exposure in lake whitefish (*Coregonus clupeaformis*). *Aquat Toxicol*, 57, 167-89. a
[http://dx.doi.org/10.1016/S0166-445X(01)00198-9] [PMID: 11891005]

Pedlar, RM, Ptashynski, MD, Wautier, KG, Evans, RE, Baron, CL & Klaverkamp, JF (2002) The accumulation, distribution, and toxicological effects of dietary arsenic exposure in lake whitefish (*Coregonus clupeaformis*) and lake trout (*Salvelinus namaycush*). *Comp Biochem Physiol C Toxicol Pharmacol*, 131, 73-91. b
[http://dx.doi.org/10.1016/S1532-0456(01)00281-2] [PMID: 11796327]

Pierron, F, Baudrimont, M, Bossy, A, Bourdineaud, JP, Brèthes, D, Elie, P & Massabuau, JC (2007) Impairment of lipid storage by cadmium in the European eel (*Anguilla anguilla*). *Aquat Toxicol*, 81, 304-11.
[http://dx.doi.org/10.1016/j.aquatox.2006.12.014] [PMID: 17276523]

Polakof, S, Médale, F, Skiba-Cassy, S, Corraze, G & Panserat, S (2010) Molecular regulation of lipid metabolism in liver and muscle of rainbow trout subjected to acute and chronic insulin treatments. *Domest Anim Endocrinol*, 39, 26-33.
[http://dx.doi.org/10.1016/j.domaniend.2010.01.003] [PMID: 20181454]

Połeć-Pawlak, K, Ruzik, R & Lipiec, E (2007) Investigation of Cd(II), Pb(II) and Cu(I) complexation by glutathione and its component amino acids by ESI-MS and size exclusion chromatography coupled to ICP-MS and ESI-MS. *Talanta*, 72, 1564-72.
[http://dx.doi.org/10.1016/j.talanta.2007.02.008] [PMID: 19071798]

Poleksic, V, Lenhardt, M, Jaric, I, Djordjevic, D, Gacic, Z, Cvijanovic, G & Raskovic, B (2010) Liver, gills, and skin histopathology and heavy metal content of the Danube sterlet (*Acipenser ruthenus* Linnaeus, 1758). *Environ Toxicol Chem*, 29, 515-21.
[http://dx.doi.org/10.1002/etc.82] [PMID: 20821473]

Ptashynski, MD, Pedlar, RM, Evans, RE, Baron, CL & Klaverkamp, JF (2002) Toxicology of dietary nickel in lake whitefish (*Coregonus clupeaformis*). *Aquat Toxicol*, 58, 229-47.

[http://dx.doi.org/10.1016/S0166-445X(01)00239-9] [PMID: 12007877]

Radhakrishnaiah, K, Venkataramana, P, Suresh, A & Sivaramakrishna, B (1992) Effects of lethal and sublethal concentrations of copper on glycolysis in liver and muscle of the freshwater teleost, *Labeo rohita* (Hamilton). *J Environ Biol,* 13, 63-8.

Rahman, MA, Hasegawa, H & Lim, RP (2012) Bioaccumulation, biotransformation and trophic transfer of arsenic in the aquatic food chain. *Environ Res,* 116, 118-35.
[http://dx.doi.org/10.1016/j.envres.2012.03.014] [PMID: 22534144]

Rajeshkumar, S, Mini, J & Munuswamy, N (2013) Effects of heavy metals on antioxidants and expression of HSP70 in different tissues of Milk fish (*Chanos chanos*) of Kaattuppalli Island, Chennai, India. *Ecotoxicol Environ Saf,* 98, 8-18.
[http://dx.doi.org/10.1016/j.ecoenv.2013.07.029] [PMID: 24021871]

Ramadan, D, Rancy, PC, Nagarkar, RP, Schneider, JP & Thorpe, C (2009) Arsenic(III) species inhibit oxidative protein folding *in vitro*. *Biochemistry,* 48, 424-32.
[http://dx.doi.org/10.1021/bi801988x] [PMID: 19102631]

Ramakritinan, CM, Kumaraguru, AK & Balasubramanian, MP (2005) Impact of distillery effluent on carbohydrate metabolism of freshwater fish, *Cyprinus carpio*. *Ecotoxicology,* 14, 693-707.
[http://dx.doi.org/10.1007/s10646-005-0019-3] [PMID: 16151610]

Regoli, F & Giuliani, ME (2014) Oxidative pathways of chemical toxicity and oxidative stress biomarkers in marine organisms. *Mar Environ Res,* 93, 106-17.
[http://dx.doi.org/10.1016/j.marenvres.2013.07.006] [PMID: 23942183]

Richard, N, Kaushik, S, Larroquet, L, Panserat, S & Corraze, G (2006) Replacing dietary fish oil by vegetable oils has little effect on lipogenesis, lipid transport and tissue lipid uptake in rainbow trout (*Oncorhynchus mykiss*). *Br J Nutr,* 96, 299-309.
[http://dx.doi.org/10.1079/BJN20061821] [PMID: 16923224]

Richards, JG, Curtis, PJ, Burnison, BK & Playle, RC (2001) Effects of natural organic matter source on reducing metal toxicity to rainbow trout (*Oncorhynchus mykiss*) and on metal binding to their gills. *Environ Toxicol Chem,* 20, 1159-66.
[http://dx.doi.org/10.1002/etc.5620200604] [PMID: 11392125]

Roy, S & Bhattacharya, S (2006) Arsenic-induced histopathology and synthesis of stress proteins in liver and kidney of *Channa punctatus*. *Ecotoxicol Environ Saf,* 65, 218-29.
[http://dx.doi.org/10.1016/j.ecoenv.2005.07.005] [PMID: 16150489]

Rudolph, R, Gerschitz, J & Jaenicke, R (1978) Effect of zinc(II) on the refolding and reactivation of liver alcohol dehydrogenase. *Eur J Biochem,* 87, 601-6.
[http://dx.doi.org/10.1111/j.1432-1033.1978.tb12412.x] [PMID: 679951]

Sabullah, MK, Ahmad, SA, Shukor, MY, Gansau, AJ, Syed, MA, Sulaiman, MR & Shamaan, NA (2015) Heavy metal biomarker: Fish behavior, cellular alteration, enzymatic reaction and proteomics approaches. *Int Food Res J,* 22, 435-54.

Salamat, N, Ardeshir, RA, Movahedinia, A & Rastgar, S (2017) Liver histophysiological alterations in pelagic and benthic fish as biomarkers for marine environmental assessment. *Int J Environ Res,* 11, 251-62.
[http://dx.doi.org/10.1007/s41742-017-0023-5]

Garcia Sampaio, F, de Lima Boijink, C, Tie Oba, E, Romagueira Bichara dos Santos, L, Lúcia Kalinin, A & Tadeu Rantin, F (2008) Antioxidant defenses and biochemical changes in pacu (*Piaractus mesopotamicus*) in response to single and combined copper and hypoxia exposure. *Comp Biochem Physiol C Toxicol Pharmacol,* 147, 43-51.
[http://dx.doi.org/10.1016/j.cbpc.2007.07.009] [PMID: 17728190]

Sanchez, W, Palluel, O, Meunier, L, Coquery, M, Porcher, JM & Aït-Aïssa, S (2005) Copper-induced oxidative stress in three-spined stickleback: relationship with hepatic metal levels. *Environ Toxicol Pharmacol,* 19, 177-83.

[http://dx.doi.org/10.1016/j.etap.2004.07.003] [PMID: 21783474]

Sastry, KV & Subhadra, K (1982) Effect of cadmium on some aspects of carbohydrate metabolism in a freshwater catfish *Heteropneustes fossilis*. *Toxicol Lett*, 14, 45-55.
[http://dx.doi.org/10.1016/0378-4274(82)90008-X] [PMID: 7157415]

Schramm, VL, Wedler, FC (1986). *Manganese in Metabolism and Enzyme Function*. Elsevier.

Scott, GR & Sloman, KA (2004) The effects of environmental pollutants on complex fish behaviour: integrating behavioural and physiological indicators of toxicity. *Aquat Toxicol*, 68, 369-92.
[http://dx.doi.org/10.1016/j.aquatox.2004.03.016] [PMID: 15177953]

Seenayya, G & Prahalad, AK (1987) *In situ* compartmentation and biomagnification of chromium and manganese in industrially polluted Husainsagar Lake, Hyderabad, India. *Water Air Soil Pollut*, 35, 233-9.
[http://dx.doi.org/10.1007/BF00290932]

Sepici-Dinçel, A, Cağlan Karasu Benli, A, Selvi, M, Sarikaya, R, Şahin, D, Ayhan Ozkul, I & Erkoç, F (2009) Sublethal cyfluthrin toxicity to carp (*Cyprinus carpio* L.) fingerlings: biochemical, hematological, histopathological alterations. *Ecotoxicol Environ Saf*, 72, 1433-9.
[http://dx.doi.org/10.1016/j.ecoenv.2009.01.008] [PMID: 19286258]

Sepúlveda, MS, Quinn, BP, Denslow, ND, Holm, SE & Gross, TS (2003) Effects of pulp and paper mill effluents on reproductive success of largemouth bass. *Environ Toxicol Chem*, 22, 205-13.
[http://dx.doi.org/10.1002/etc.5620220127] [PMID: 12503766]

Settle, DM & Patterson, CC (1980) Lead in albacore: guide to lead pollution in Americans. *Science*, 207, 1167-76.
[http://dx.doi.org/10.1126/science.6986654] [PMID: 6986654]

Sharma, SK, Goloubinoff, P & Christen, P (2008) Heavy metal ions are potent inhibitors of protein folding. *Biochem Biophys Res Commun*, 372, 341-5.
[http://dx.doi.org/10.1016/j.bbrc.2008.05.052] [PMID: 18501191]

Shukla, V, Rathi, P & Sastry, KV (2002) Effect of cadmium individually and in combination with other metals on the nutritive value of fresh water fish, *Channa punctatus*. *J Environ Biol*, 23, 105-10.
[PMID: 12602846]

Singh, AK & Banerjee, TK (2008) Toxic impact of sodium arsenate (Na_2HAsO_4. $7H_2O$) on skin epidermis of the air-breathing catfish *Clarias batrachus* (Linn). *Vet Arh*, 78, 73-88.

Skidmore, JF & Tovell, PWA (1972) Toxic effects of zinc sulphate on the gills of rainbow trout. *Water Res*, 6, 217-IN4.
[http://dx.doi.org/10.1016/0043-1354(72)90001-2]

Sloman, KA (2007) Effects of trace metals on salmonid fish: The role of social hierarchies. *Appl Anim Behav Sci*, 104, 326-45.
[http://dx.doi.org/10.1016/j.applanim.2006.09.003]

Song, YF, Luo, Z, Huang, C, Liu, X, Pan, YX & Chen, QL (2013) Effects of calcium and copper exposure on lipogenic metabolism, metal element compositions and histology in *Synechogobius hasta*. *Fish Physiol Biochem*, 39, 1641-56.
[http://dx.doi.org/10.1007/s10695-013-9816-4] [PMID: 23743594]

Song, YF, Luo, Z, Pan, YX, Liu, X, Huang, C & Chen, QL (2014) Effects of copper and cadmium on lipogenic metabolism and metal element composition in the javelin goby (*Synechogobius hasta*) after single and combined exposure. *Arch Environ Contam Toxicol*, 67, 167-80.
[http://dx.doi.org/10.1007/s00244-014-0011-0] [PMID: 24595737]

Sonne, C, Bach, L, Søndergaard, J, Rigét, FF, Dietz, R, Mosbech, A, Leifsson, PS & Gustavson, K (2014) Evaluation of the use of common sculpin (*Myoxocephalus scorpius*) organ histology as bioindicator for element exposure in the fjord of the mining area Maarmorilik, West Greenland. *Environ Res*, 133, 304-11.
[http://dx.doi.org/10.1016/j.envres.2014.05.031] [PMID: 24991745]

Spry, DJ & Wiener, JG (1991) Metal bioavailability and toxicity to fish in low-alkalinity lakes: A critical review. *Environ Pollut,* 71, 243-304.
[http://dx.doi.org/10.1016/0269-7491(91)90034-T] [PMID: 15092121]

Srikanth, K, Pereira, E, Duarte, AC & Ahmad, I (2013) Glutathione and its dependent enzymes' modulatory responses to toxic metals and metalloids in fish-a review. *Environ Sci Pollut Res Int,* 20, 2133-49.
[http://dx.doi.org/10.1007/s11356-012-1459-y] [PMID: 23334549]

Srivastava, R & Srivastava, N (2008) Changes in nutritive value of fish, *Channa punctatus* after chronic exposure to zinc. *J Environ Biol,* 29, 299-302.
[PMID: 18972681]

Suedel, BC, Boraczek, JA, Peddicord, RK, Clifford, PA & Dillon, TM (1994) Trophic transfer and biomagnification potential of contaminants in aquatic ecosystems. *Rev Environ Contam Toxicol,* 136, 21-89.
[http://dx.doi.org/10.1007/978-1-4612-2656-7_2] [PMID: 8029491]

Sultana, T, Butt, K, Sultana, S, Al-Ghanim, KA, Mubashra, R, Bashir, N, Ahmed, Z, Ashraf, A & Mahboob, S (2016) Histopathological changes in liver, gills and intestine of *Labeo rohita* inhabiting industrial waste contaminated water of River Ravi. *Pak J Zool,* 48, 1171-7.

Sumpter, JP (2005) Endocrine disrupters in the aquatic environment: An overview. *Acta Hydrochim Hydrobiol,* 33, 9-16.
[http://dx.doi.org/10.1002/aheh.200400555]

Sunita, C & Banerjee, TK (2003) Toxic impact of the inorganic salt zinc chloride on the skin (an accessory water breathing organ) of the air-breathing 'murrel' *Channa striata. Res J Chem Environ,* 7, 18-23.

Svecevičius, G (1999) Fish avoidance response to heavy metals and their mixtures. *Acta Zool Litu,* 9, 103-13.
[http://dx.doi.org/10.1080/13921657.1999.10512293]

Türkmen, A, Türkmen, M, Tepe, Y & Akyurt, İ (2005) Heavy metals in three commercially valuable fish species from İskenderun Bay, Northern East Mediterranean Sea, Turkey. *Food Chem,* 91, 167-72.
[http://dx.doi.org/10.1016/j.foodchem.2004.08.008]

Türkmen, A, Türkmen, M, Tepe, Y, Mazlum, Y & Oymael, S (2006) Metal concentrations in blue crab (*Callinectes sapidus*) and mullet (*Mugil cephalus*) in Iskenderun Bay, Northern East Mediterranean, Turkey. *Bull Environ Contam Toxicol,* 77, 186-93.
[http://dx.doi.org/10.1007/s00128-006-1049-0] [PMID: 16977519]

Türkmen, A, Tepe, Y, Türkmen, M & Mutlu, E (2009) Heavy metal contaminants in tissues of the garfish, *Belone belone* L., 1761, and the bluefish, *Pomatomus saltatrix* L., 1766, from Turkey waters. *Bull Environ Contam Toxicol,* 82, 70-4. a
[http://dx.doi.org/10.1007/s00128-008-9553-z] [PMID: 18784894]

Türkmen, M, Türkmen, A, Tepe, Y, Töre, Y & Ateş, A (2009) Determination of metals in fish species from Aegean and Mediterranean seas. *Food Chem,* 113, 233-7. b
[http://dx.doi.org/10.1016/j.foodchem.2008.06.071]

Türkmen, M, Tepe, Y, Türkmen, A, Kemal Sangün, M, Ateş, A & Genç, E (2013) Assessment of heavy metal contamination in various tissues of six ray species from İskenderun Bay, northeastern Mediterranean Sea. *Bull Environ Contam Toxicol,* 90, 702-7.
[http://dx.doi.org/10.1007/s00128-013-0978-7] [PMID: 23519497]

Tamás, MJ, Sharma, SK, Ibstedt, S, Jacobson, T & Christen, P (2014) Heavy metals and metalloids as a cause for protein misfolding and aggregation. *Biomolecules,* 4, 252-67.
[http://dx.doi.org/10.3390/biom4010252] [PMID: 24970215]

Tepe, Y (2009) Metal concentrations in eight fish species from Aegean and Mediterranean Seas. *Environ Monit Assess,* 159, 501-9.
[http://dx.doi.org/10.1007/s10661-008-0646-2] [PMID: 19067206]

Tepe, Y, Türkmen, M & Türkmen, A (2008) Assessment of heavy metals in two commercial fish species of

four Turkish seas. *Environ Monit Assess,* 146, 277-84.
[http://dx.doi.org/10.1007/s10661-007-0079-3] [PMID: 18034361]

Tierney, KB, Baldwin, DH, Hara, TJ, Ross, PS, Scholz, NL & Kennedy, CJ (2010) Olfactory toxicity in fishes. *Aquat Toxicol,* 96, 2-26.
[http://dx.doi.org/10.1016/j.aquatox.2009.09.019] [PMID: 19931199]

Tocher, DR (2003) Metabolism and functions of lipids and fatty acids in teleost fish. *Rev Fish Sci,* 11, 107-84.
[http://dx.doi.org/10.1080/713610925]

Topal, A, Atamanalp, M, Oruç, E & Erol, HS (2017) Physiological and biochemical effects of nickel on rainbow trout (Oncorhynchus mykiss) tissues: Assessment of nuclear factor kappa B activation, oxidative stress and histopathological changes. *Chemosphere,* 166, 445-52.
[http://dx.doi.org/10.1016/j.chemosphere.2016.09.106] [PMID: 27705832]

Turan, C, Dural, M, Oksuz, A & Oztürk, B (2009) Levels of heavy metals in some commercial fish species captured from the Black Sea and Mediterranean coast of Turkey. *Bull Environ Contam Toxicol,* 82, 601-4.
[http://dx.doi.org/10.1007/s00128-008-9624-1] [PMID: 19082517]

Tuzen, M (2009) Toxic and essential trace elemental contents in fish species from the Black Sea, Turkey. *Food Chem Toxicol,* 47, 1785-90.
[http://dx.doi.org/10.1016/j.fct.2009.04.029] [PMID: 19406195]

Uluozlu, OD, Tuzen, M, Mendil, D & Soylak, M (2007) Trace metal content in nine species of fish from the Black and Aegean Seas, Turkey. *Food Chem,* 104, 835-40.
[http://dx.doi.org/10.1016/j.foodchem.2007.01.003]

Umminger, BL (1970) Physiological studies on supercooled killifish (*Fundulus heteroclitus*). 3. Carbohydrate metabolism and survival at subzero temperatures. *J Exp Zool,* 173, 159-74.
[http://dx.doi.org/10.1002/jez.1401730205] [PMID: 5437468]

Ung, CY, Lam, SH, Hlaing, MM, Winata, CL, Korzh, S, Mathavan, S & Gong, Z (2010) Mercury-induced hepatotoxicity in zebrafish: *in vivo* mechanistic insights from transcriptome analysis, phenotype anchoring and targeted gene expression validation. *BMC Genomics,* 11, 212.
[http://dx.doi.org/10.1186/1471-2164-11-212] [PMID: 20353558]

Valavanidis, A, Vlahogianni, T, Dassenakis, M & Scoullos, M (2006) Molecular biomarkers of oxidative stress in aquatic organisms in relation to toxic environmental pollutants. *Ecotoxicol Environ Saf,* 64, 178-89.
[http://dx.doi.org/10.1016/j.ecoenv.2005.03.013] [PMID: 16406578]

Vallee, BL & Auld, DS (1990) Zinc coordination, function, and structure of zinc enzymes and other proteins. *Biochemistry,* 29, 5647-59.
[http://dx.doi.org/10.1021/bi00476a001] [PMID: 2200508]

van der Oost, R, Beyer, J & Vermeulen, NPE (2003) Fish bioaccumulation and biomarkers in environmental risk assessment: a review. *Environ Toxicol Pharmacol,* 13, 57-149.
[http://dx.doi.org/10.1016/S1382-6689(02)00126-6] [PMID: 21782649]

van Dyk, JC, Pieterse, GM & van Vuren, JHJ (2007) Histological changes in the liver of *Oreochromis mossambicus* (Cichlidae) after exposure to cadmium and zinc. *Ecotoxicol Environ Saf,* 66, 432-40.
[http://dx.doi.org/10.1016/j.ecoenv.2005.10.012] [PMID: 16364439]

Varjani, SJ, Agarwal, AK, Gnansounou, E, Gurunathan, B (2018). *Bioremediation: applications for environmental protection and management.* Springer.

Veltman, K, Huijbregts, MA, Van Kolck, M, Wang, WX & Hendriks, AJ (2008) Metal bioaccumulation in aquatic species: quantification of uptake and elimination rate constants using physicochemical properties of metals and physiological characteristics of species. *Environ Sci Technol,* 42, 852-8.
[http://dx.doi.org/10.1021/es071331f] [PMID: 18323112]

Ventura-Lima, J, Bogo, MR & Monserrat, JM (2011) Arsenic toxicity in mammals and aquatic animals: a comparative biochemical approach. *Ecotoxicol Environ Saf,* 74, 211-8.

[http://dx.doi.org/10.1016/j.ecoenv.2010.11.002] [PMID: 21112631]

Ventura, MT & Paperna, I (1985) Histopathology of *Ichthyophthirim multifiliis* infections in fishes. *J Fish Biol,* 27, 185-203.
[http://dx.doi.org/10.1111/j.1095-8649.1985.tb04020.x]

Vincent, JB (2000) The biochemistry of chromium. *J Nutr,* 130, 715-8.
[http://dx.doi.org/10.1093/jn/130.4.715] [PMID: 10736319]

Vinodhini, R & Narayanan, M (2009) The impact of toxic heavy metals on the hematological parameters in common carp (*Cyprinus carpio* L.). *Iran J Environ Health Sci Eng,* 6, 23-8.

Vutukuru, SS (2005) Acute effects of hexavalent chromium on survival, oxygen consumption, hematological parameters and some biochemical profiles of the Indian major carp, *Labeo rohita. Int J Environ Res Public Health,* 2, 456-62.
[http://dx.doi.org/10.3390/ijerph2005030010] [PMID: 16819101]

Vutukuru, SS, Chintada, S, Madhavi, KR, Rao, JV & Anjaneyulu, Y (2006) Acute effects of copper on superoxide dismutase, catalase and lipid peroxidation in the freshwater teleost fish, *Esomus danricus. Fish Physiol Biochem,* 32, 221-9.
[http://dx.doi.org/10.1007/s10695-006-9004-x]

Vutukuru, SS, Prabhath, NA, Raghavender, M & Yerramilli, A (2007) Effect of arsenic and chromium on the serum amino-transferases activity in Indian major carp, *Labeo rohita. Int J Environ Res Public Health,* 4, 224-7.
[http://dx.doi.org/10.3390/ijerph2007030005] [PMID: 17911661]

Wallace, WG, Lee, BG & Luoma, SN (2003) Subcellular compartmentalization of Cd and Zn in two bivalves. I. Significance of metal-sensitive fractions (MSF) and biologically detoxified metal (BDM). *Mar Ecol Prog Ser,* 249, 183-97.
[http://dx.doi.org/10.3354/meps249183]

Wang, M, Chan, LL, Si, M, Hong, H & Wang, D (2010) Proteomic analysis of hepatic tissue of zebrafish (*Danio rerio*) experimentally exposed to chronic microcystin-LR. *Toxicol Sci,* 113, 60-9.
[http://dx.doi.org/10.1093/toxsci/kfp248] [PMID: 19822598]

Wang, M, Wang, Y, Wang, J, Lin, L, Hong, H & Wang, D (2011) Proteome profiles in medaka (*Oryzias melastigma*) liver and brain experimentally exposed to acute inorganic mercury. *Aquat Toxicol,* 103, 129-39.
[http://dx.doi.org/10.1016/j.aquatox.2011.02.020] [PMID: 21458406]

Wang, M, Wang, Y, Zhang, L, Wang, J, Hong, H & Wang, D (2013) Quantitative proteomic analysis reveals the mode-of-action for chronic mercury hepatotoxicity to marine medaka (*Oryzias melastigma*). *Aquat Toxicol,* 130-131, 123-31.
[http://dx.doi.org/10.1016/j.aquatox.2013.01.012] [PMID: 23416409]

Wang, R & Wang, WX (2012) Contrasting mercury accumulation patterns in tilapia (*Oreochromis niloticus*) and implications on somatic growth dilution. *Aquat Toxicol,* 114-115, 23-30.
[http://dx.doi.org/10.1016/j.aquatox.2012.02.014] [PMID: 22417761]

Wang, WX & Rainbow, PS (2008) Comparative approaches to understand metal bioaccumulation in aquatic animals. *Comp Biochem Physiol C Toxicol Pharmacol,* 148, 315-23.
[http://dx.doi.org/10.1016/j.cbpc.2008.04.003] [PMID: 18502695]

Wang, WX (2002) Interactions of trace metals and different marine food chains. *Mar Ecol Prog Ser,* 243, 295-309.
[http://dx.doi.org/10.3354/meps243295]

Wang, WX (2016) *Bioaccumulation and biomonitoring Marine Ecotoxicology,* 99-119.

Wang, WX, Fisher, NS & Luoma, SN (1996) Kinetic determinations of trace element bioaccumulation in the mussel *Mytilus edulis. Mar Ecol Prog Ser,* 140, 91-113.
[http://dx.doi.org/10.3354/meps140091]

Wang, WX & Schrenk, D (2012) Contamination of marine molluscs with heavy metals. *Chemical Contaminants & Residues in Food,* 535-51.

Wang, WX & Wong, RSK (2003) Bioaccumulation kinetics and exposure pathways of inorganic mercury and methylmercury in a marine fish, the sweetlips *Plectorhinchus gibbosus. Mar Ecol Prog Ser,* 261, 257-68. [http://dx.doi.org/10.3354/meps261257]

Wedemeyer, GA & Yasutake, WT (1977) *Clinical methods for the assessment of the effects of environmental stress on fish health (No 89).* US Fish and Wildlife Service.

Winterbourn, MJ, McDiffett, WF & Eppley, SJ (2000) Aluminium and iron burdens of aquatic biota in New Zealand streams contaminated by acid mine drainage: effects of trophic level. *Sci Total Environ,* 254, 45-54. [http://dx.doi.org/10.1016/S0048-9697(00)00437-X] [PMID: 10845446]

Wood, CM, Farrell, AP, Brauner, CJ (2011). *Homeostasis and toxicology of essential metals Fish Physiology.* Academic Press.

Wood, CM, Farrell, AP, Brauner, CJ (2012). *Homeostasis and toxicology of non-essential metals Fish Physiology.* Academic Press.

Yamaguchi, S, Miura, C, Ito, A, Agusa, T, Iwata, H, Tanabe, S, Tuyen, BC & Miura, T (2007) Effects of lead, molybdenum, rubidium, arsenic and organochlorines on spermatogenesis in fish: monitoring at Mekong Delta area and *in vitro* experiment. *Aquat Toxicol,* 83, 43-51. [http://dx.doi.org/10.1016/j.aquatox.2007.03.010] [PMID: 17448548]

Yeganeh, S, Adel, M, Ahmadvand, S, Ahmadvand, S & Velisek, J (2016) Toxicity of organic selenium (Selemax) and its effects on haematological and biochemical parameters and histopathological changes of common carp (*Cyprinus carpio* L., 1758). *Toxin Rev,* 35, 207-13. [http://dx.doi.org/10.1080/15569543.2016.1213749]

Yilmaz, AB (2003) Levels of heavy metals (Fe, Cu, Ni, Cr, Pb, and Zn) in tissue of *Mugil cephalus* and *Trachurus mediterraneus* from Iskenderun Bay, Turkey. *Environ Res,* 92, 277-81. [http://dx.doi.org/10.1016/S0013-9351(02)00082-8] [PMID: 12804525]

Yilmaz, AB (2005) Comparison of heavy metal levels of grey mullet (*Mugil cephalus* L.) and sea bream (*Sparus aurata* L.) caught in İskenderun Bay (Turkey). *Turk J Vet Anim Sci,* 29, 257-62.

Yılmaz, AB, Sangün, MK, Yağlıoğlu, D & Turan, C (2010) Metals (major, essential to non-essential) composition of the different tissues of three demersal fish species from İskenderun Bay, Turkey. *Food Chem,* 123, 410-5. [http://dx.doi.org/10.1016/j.foodchem.2010.04.057]

Yılmaz, AB, Yanar, A & Alkan, EN (2017) Review of heavy metal accumulation on aquatic environment in Northern East Mediterrenean Sea part I: some essential metals. *Rev Environ Health,* 32, 119-63. [http://dx.doi.org/10.1515/reveh-2016-0065] [PMID: 28182578]

Yılmaz, AB, Yanar, A & Alkan, EN (2018) Review of heavy metal accumulation in aquatic environment of Northern East Mediterrenean Sea Part II: some non-essential metals. *Pollution,* 4, 143-81.

Zhang, L & Wang, WX (2007) Waterborne cadmium and zinc uptake in a euryhaline teleost *Acanthopagrus schlegeli* acclimated to different salinities. *Aquat Toxicol,* 84, 173-81. [http://dx.doi.org/10.1016/j.aquatox.2007.03.027] [PMID: 17675173]

Zhang, L & Wang, WX (2006) Significance of subcellular metal distribution in prey in influencing the trophic transfer of metals in a marine fish. *Limnol Oceanogr,* 51, 2008-17. [http://dx.doi.org/10.4319/lo.2006.51.5.2008]

Zhang, L & Wang, WX (2007) Size-dependence of the potential for metal biomagnification in early life stages of marine fish. *Environ Toxicol Chem,* 26, 787-94. b [http://dx.doi.org/10.1897/06-348R.1] [PMID: 17447565]

Zhang, W & Wang, WX (2012) Large-scale spatial and interspecies differences in trace elements and stable isotopes in marine wild fish from Chinese waters. *J Hazard Mater,* 215-216, 65-74.

[http://dx.doi.org/10.1016/j.jhazmat.2012.02.032] [PMID: 22410727]

Zheng, JL, Luo, Z, Liu, CX, Chen, QL, Tan, XY, Zhu, QL & Gong, Y (2013) Differential effects of acute and chronic zinc (Zn) exposure on hepatic lipid deposition and metabolism in yellow catfish *Pelteobagrus fulvidraco. Aquat Toxicol,* 132-133, 173-81. a
[http://dx.doi.org/10.1016/j.aquatox.2013.02.002] [PMID: 23523964]

Zheng, JL, Luo, Z, Zhu, QL, Chen, QL & Gong, Y (2013) Molecular characterization, tissue distribution and kinetic analysis of carnitine palmitoyltransferase I in juvenile yellow catfish *Pelteobagrus fulvidraco. Genomics,* 101, 195-203. b
[http://dx.doi.org/10.1016/j.ygeno.2012.12.002] [PMID: 23238057]

Zheng, JL, Luo, Z, Zhu, QL, Hu, W, Zhuo, MQ, Pan, YX, Song, YF & Chen, QL (2015) Different effect of dietborne and waterborne Zn exposure on lipid deposition and metabolism in juvenile yellow catfish *Pelteobagrus fulvidraco. Aquat Toxicol,* 159, 90-8.
[http://dx.doi.org/10.1016/j.aquatox.2014.12.003] [PMID: 25531431]

Zheng, JL, Luo, Z, Zhu, QL, Tan, XY, Chen, QL, Sun, LD & Hu, W (2013) Molecular cloning and expression pattern of 11 genes involved in lipid metabolism in yellow catfish *Pelteobagrus fulvidraco. Gene,* 531, 53-63. c
[http://dx.doi.org/10.1016/j.gene.2013.08.028] [PMID: 23988502]

Zhou, Q, Zhang, J, Fu, J, Shi, J & Jiang, G (2008) Biomonitoring: an appealing tool for assessment of metal pollution in the aquatic ecosystem. *Anal Chim Acta,* 606, 135-50.
[http://dx.doi.org/10.1016/j.aca.2007.11.018] [PMID: 18082645]

Zitko, P, Carson, WV & Carson, WG (1973) Prediction of incipient lethal levels of copper to juvenile Atlantic salmon in the presence of humic acid by cupric electrode. *Bull Environ Contam Toxicol,* 10, 265-71.
[http://dx.doi.org/10.1007/BF01684814] [PMID: 4766128]

Zitko, V & Carson, WG (1976) A mechanism of the effects of water hardness on the lethality of heavy metals to fish. *Chemosphere,* 5, 299-303.
[http://dx.doi.org/10.1016/0045-6535(76)90003-5]

Zutshi, B, Prasad, SGR & Nagaraja, R (2010) Alteration in hematology of *Labeo rohita* under stress of pollution from Lakes of Bangalore, Karnataka, India. *Environ Monit Assess,* 168, 11-9.
[http://dx.doi.org/10.1007/s10661-009-1087-2] [PMID: 19603276]

CHAPTER 9

Effects of Microplastics in Marine Ecosystem

Fenghua Jiang[*], **Chengjun Sun, Jingxi Li** and **Wei Cao**

Marine Ecology Center, the First Institute of Oceanography, Ministry of Natural Resources, Qingdao, 266061, China

Abstract: The increasing global production and widespread use of plastic have led to an accumulation of large amounts of plastic debris in the ocean. Microplastic exists in the marine environment on a global scale and can harm a great variety of marine organisms. Pollution caused by microplastic and associated pollutants, such as organic chemicals and heavy metals, threatens the survival of marine organisms. In this chapter, the following contents are summarized: (i) the distribution of microplastic in marine environment; (ii) the presence of microplastic in marine environment and organisms; (iii) the effects of microplastic on marine ecology, including the toxic effects on marine organisms, the effects on distribution of pollutants, and the combined pollution caused by microplastic and associated pollutants; and (iv) several aspects to work on the management and research of microplastic are proposed. Extensive research on microplastic pollution is going on.

Keywords: Marine Plastic Debris, Microplastics, Persistent Organic Pollutants, Heavy Metals, Ecological Effect, Combined Pollution.

INTRODUCTION

The current human history has been referred to as the plastic age. Plastic is being used in industry, agriculture and everywhere in our daily life due to its special properties, such as low density, good malleability, durability, low cost, *etc*. The global production of plastics reached 348 million tons in 2017 (Fig. **1**) (PlasticEurope, 2018). Unfortunately, a vast majority of the produced plastics are for single use. The very properties that make plastic so useful (*e.g.* low density, durability) also make them problematic in the environment. Due to the high disposal rate, low recycle percentage, and indiscriminate mismanagement, a large number of plastic litters enter into the sea and accumulate on shorelines, floating in the oceans and becoming the most numerous and ubiquitous component of marine litter on a global scale.

[*] **Corresponding author Fenghua Jiang:** Marine Ecology Center, the First Institute of Oceanography, Ministry of Natural Resources, Qingdao, 266061, China; Tel: +86-532-8896 2711; Fax: +86-532-8896 3253; E-mail: jiangfh@fio.org.cn

De-Sheng Pei & Muhammad Junaid (Eds.)
All rights reserved-© 2019 Bentham Science Publishers

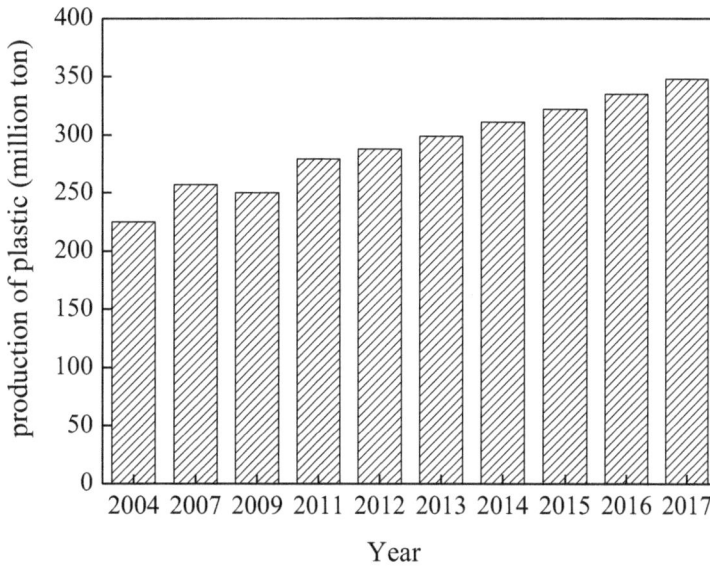

Fig. (1). The global production of plastic (PlasticEurope, 2018).

Plastic can potentially last several hundred to thousand years in the environment. There is a rising concern regarding the accumulation of floating plastic debris. United Nations Environment Programme (UNEP) began to focus on the plastic debris in the marine environment from 2011. It was reported in the first United Nations conference on the environment that the economic losses and costs caused by the generous plastic debris to the marine ecosystems might exceed US $13 billion per year. Marine plastic debris was ranked one of the ten most noteworthy urgent environmental problems in the annals of UNEP in June 2014 (UNEP, 2014). It was estimated that about 480 to 1279 tons of plastic debris was entered into the ocean in 2010 (Jambeck *et al.*, 2015).

Microplastics (MPs) are defined as small particles of plastic debris less than 5 mm in diameter by the National Oceanic and Atmospheric Administration (Wright *et al.*, 2013). They enter into the marine environment from the direct sources or primary sources, such as industrial accidental spillages during the processes of transportation and usage or the release of microbeads used in cosmetics through wastewaters (Browne *et al.*, 2015). Degradation and fragmentation of larger plastic items into small plastic fragments under the action of ultraviolet light, heat, wind, and waves represent an indirect source or "secondary source" of MPs input to the environment (Barnes *et al.*, 2009, Andrady, 2015).

According to the shape and morphotype, MPs are usually classified into the fiber, fragment, pellet or granule, and flake or film. Fiber mainly comes from clothing, disposable diapers, and fishery gears. The Fragment is usually as irregular shapes and from larger plastic items that have been broken by UV light, physical and chemical actions. Pellet or granule is mainly from preproduction plastic and daily cosmetics, such as toothpaste, shampoo, facial cleanser, *etc*. Flake or film is a thin sheet from plastic bags or other packaging materials. With respect to the polymer type, there are some common kinds of materials, including polyethylene (PE), polypropylene (PP), polyvinyl chloride (PVC), polystyrene (PS), polyethylene terephthalate (PET), polyester, nylon, polyamide, acrylic, polystyrene butadiene styrene, polyurethane, *etc*.

MPs are considered as a new emerging pollutant and concerned researchers have started to study their effects and risks in the marine environment (for reviews, see Marris, 2014; Perkins, 2014). MPs pollution has been listed as the second major scientific problems in the field of environmental and ecological science in 2015. The plastic (including MPs) pollution of the marine environment was also considered as the major global environmental problems together with ocean acidification, de-oxygenation, ocean warming (Williamson *et al.*, 2016). Fig. (**2**) lists the number of publications on MPs during the year 2004 to February, 2019, showing the rapid increasing concern for plastic pollution. Jamieson *et al* (2017) reported that the persistent organic pollutants were detected in the organism collected from Mariana Trench, and MPs were considered as carriers of these pollutants. The results indicated that the impact of MPs on marine ecosystem might be far beyond what the human expect.

MICROPLASTICS IN THE MARINE ENVIRONMENT

As a new emerging pollutant, MPs have been found in high numbers in seawater and sediments. There are numerous reports about the distribution and characteristics of MPs in the marine environment. Here we briefly summarize their distribution in seawater and sediments.

The distribution of MPs in the sea is greatly influenced by currents. They distribute widespread in everywhere of the ocean, and high abundance is observed in certain regions (Law, *et al.*, 2010). MPs are universally found in the nearshore area, bay, strait, and around the area of islands, with their abundance varies significantly (Dubaish and Liebezeit, 2013; Collignon *et al.*, 2014; Desforges *et al.*, 2014; Song *et al.*, 2014; Zhao *et al.*, 2014). The abundances of MPs were 4137 ± 2461 and 0.167 ± 0.138 items m^{-3}, respectively, in samples from the Yangtze Estuarine and East China Sea (Zhao *et al.*, 2014). More than 90% of MPs were from 0.5 to 5 mm in size by the number of items, and the most frequent geometries were fibers, followed by granules and films (Zhao *et al.*, 2014). The abundances of MPs ranged from 8 to 9200 items/m³ in subsurface seawater in the northeastern Pacific Ocean, with the lowest abundance was in offshore Pacific water and the highest was in the Queen Charlotte Sound, and about 75% of the particles were fibers (Desforges *et al.*, 2014).

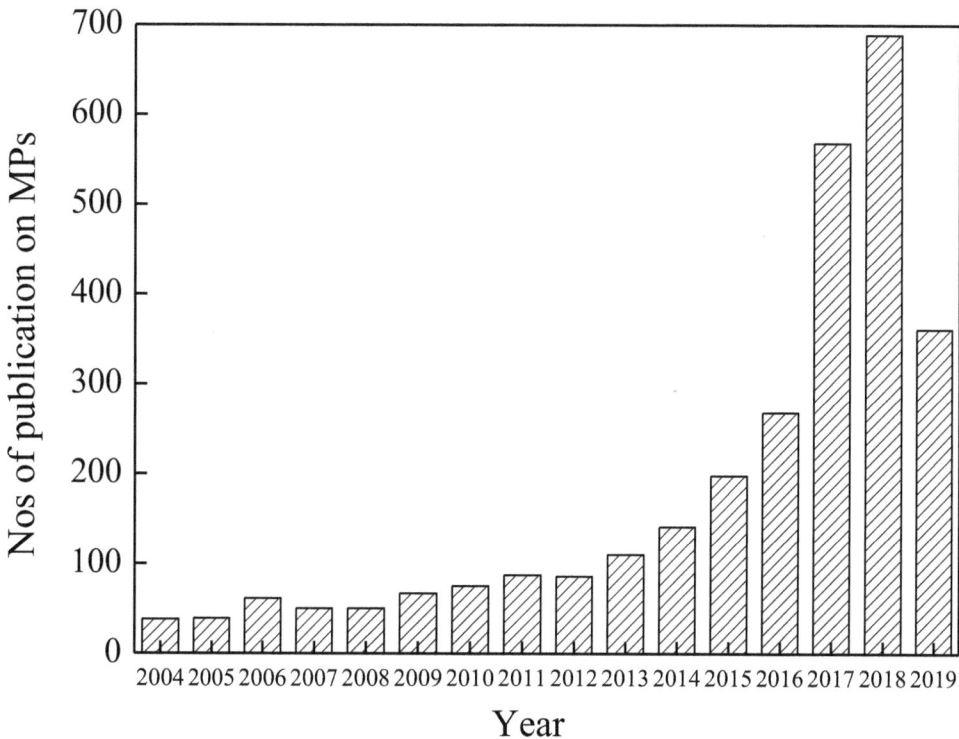

Fig. (2). Number of publication on microplastics based on Science Direct search (Up to 2019.02.10 in Science Direct with microplastic as a keyword).

The plastic on the surface of the open ocean mostly accumulates in the convergence zone of the five subtropical gyres (Fig. **3**) (Sportdiver, 2016). The abundance of MPs usually increases from the outer to the inner region of the accumulation area in the gyres, with concentrations of < 50g km^{-2}, > 500g km^{-2}, and $1500 \sim 2000$ g km^{-2} in the nonaccumulation zone, outer accumulation zone, and inner accumulation zone, respectively (Cózar *et al.*, 2014). Based on the above values, between 7000 and 35,000 tons of plastic were estimated to be floating in the open ocean (Cózar *et al.*, 2014). Eriksen et al (2014) also estimated more than five trillion pieces of plastic and more than 250,000 tons are floating in the oceans.

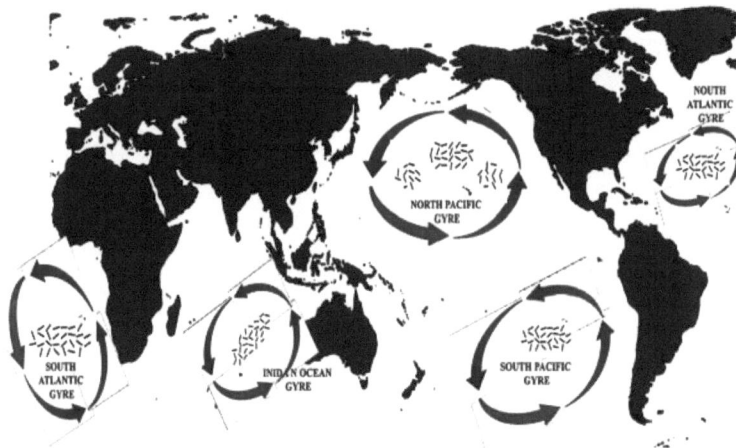

Fig. (3). The sketch map of five patches of marine debris in the open ocean (**Adapted from Sportdiver**, 2016).

The reports on the investigation of MPs mainly focus on the coastal sediments from leisure beach, estuary, sea area around island (Claessens *et al.*, 2011; Collignon *et al.*, 2012; Liebezeit and Dubaish, 2012; Imhof *et al.*, 2013; Lee *et al.*, 2013a; Baztan *et al.*, 2014; Frias *et al.*, 2016). The abundance of MPs in coastal sediments ranged from 0.21 to 7700 items m^{-2}, varied greatly in diverse regions due to the geographical position and seasonal difference (Imhof *et al.*, 2013). In the beaches near Nakdong River Estuary, South Korea, sediments collected after rainy season contained more MPs than those before the rainy season, and styrofoam was the most abundant item (Lee *et al.*, 2013a). The distribution of MPs was associated with wastewater effluent, the industrial installations, the degree of shore exposure and complex tidal flow patterns in the intertidal sediments in Scapa Flow, Orkney (Blumenröder *et al.*, 2017). The abundances of MPs ranged from 72 ± 24 to 1512 ± 187 items kg^{-1} of dry weight, showing significant variability within sampling locations in beach across Europe. The majority of MPs were fibers in shape, < 1 mm in size, blue or black in color, and identified as polyester, PE and PP (Lots *et al.*, 2017).

Sediment from deep sea contains less MPs than that from nearshore area. MPs accounted for a total of 31 particles in 27 sediment samples from Southern Portuguese shelf waters, and the majority was microfiber and PP was the major polymer type (Frias *et al.*, 2016). The abundance of MPs in coastal shallow sediments from the Mediterranean Sea was 0.90±0.10 items g^{-1} dry weight. A high proportion of fiber was found close to population areas while fragment was more common in marine protected areas (Alomar *et al.*, 2016). In sediments along the Saudi Arabian Red Sea, MPs generally associated with areas of high human activity, the abundance reached up to 1g m^{-2} and 160 items m^{-2}, and PE was the main polymer (Ruiz-Compean *et al.*, 2017).

MICROPLASTICS IN MARINE ORGANSIMS

The tiny size of MPs gives them the potential to be ingested by a wide range of biota in benthic and pelagic ecosystems (Barnes *et al.*, 2009). Ingestion of MP by marine organisms is less visible than entanglement. A wide variety of animals at different levels in the food web ingest MPs for food by mistake. The availability of MPs to animals at the base of the food chain increases with decreasing of MP's size.

In Zooplankton

Zooplankton ingests primary producers, and as the food for the higher trophic levels, they play important roles in marine ecosystems. In the North-Western Mediterranean Sea, the average ratio between MPs and mesozooplankton by weights was 0.5 and this might induce a potential confusion for zooplankton feeders (Collignon *et al.*, 2012). Natural zooplankton sampled from 16 stations in South China were found to ingest MPs, with fiber accounting for the most proportion (70%) and polyester as the main component (Sun *et al.*, 2017). The encounter rates of MP/zooplankton increase with trophic levels.

In Benthic Invertebrates

Benthic invertebrates, such as mussels or oysters, are investigated for MPs pollution, as they have been widely used in biomonitoring surveys in coastal waters (Beyer *et al.*, 2017). The exposure experiments show that benthic invertebrates, sea cucumbers (*Echinodermata, Holothuroidea*) not only ingest small nylon and PVC fragments along with sediment, but also ingest significantly more plastic fragments than predicted given the ratio of plastic to sand grains in the sediment (Graham and Thompson, 2009). The abundance and properties of MPs in benthic organisms vary among different regions. The investigation result indicated that 75% of mussels *Perna perna* from the coast of São Paulo State ingest white and irregularly-shaped MPs (Santana *et al.*, 2016). The number of total MPs varied from 0.9 to 4.6 items g^{-1} and from 1.5 to 7.6 items individual^{-1} (items ind^{-1}) in *Mytilus edulis* along the coastal waters of China, with wild groups containing more MPs than farmed groups and fiber being the most common MPs followed by the fragment. The proportion of MPs less than 250 μm in size accounted for 17% to 79% of the total MPs (Li *et al.*, 2016). Depending on different regions, there could be more MPs in farmed mussels than in wild mussels. For example, MPs was found in 80% of farmed *Mytilus galloprovincialis* and 40% in wild *Mytilus galloprovincialis*, from Qingdao, with fiber being the dominant shape (Ding *et al.*, 2018). Fig. (**4**) shows the typical images of MPs in bivalves from Qingdao, China. At the French Atlantic coasts, MPs quantity was evaluated to be 0.61 ± 0.56 items ind^{-1} and 2.1 ± 1.7 items ind^{-1} in blue mussel (*Mytilus edulis*) and the Pacific oyster (*Crassostrea gigas*), respectively. In those samples, fragment is the most common shape. Half of the MPs was grey in color, and about 50% ranged between 50 to 100 µm in size (Phuong *et al.*, 2018). The feeding type affects MPs ingestion by invertebrate organisms, with bivalves (*Mytilus trossulus* and *Macoma balthica*) containing significantly higher amounts of MPs beads. Free-swimming crustaceans ingest more beads than the benthic animals (Setälä *et al.*, 2016).

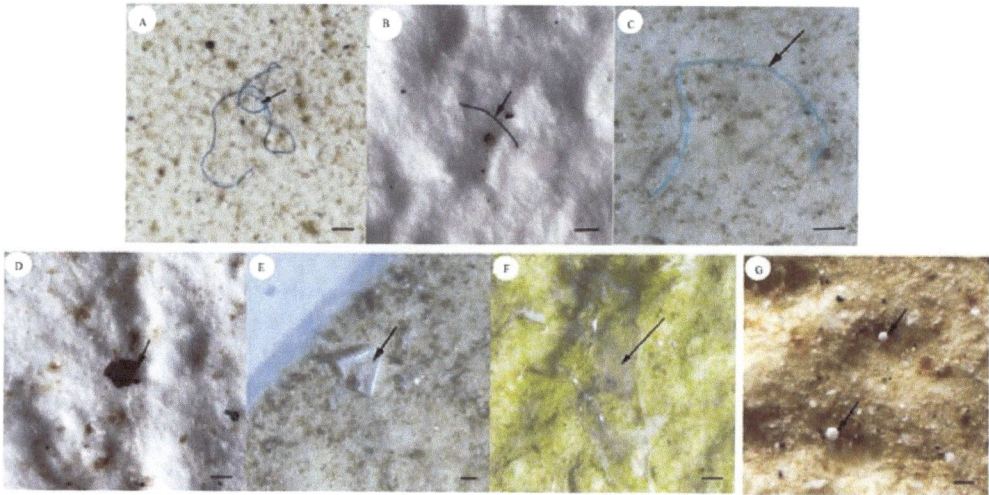

Fig. (4). Images of different shapes of microplastics in the bivalves, *Chlamys farreri* and *Mytilus galloprovincialis*, from Qingdao, China. The images showed by arrows in A, B and C are fibers, those in D, E and F are fragments, and that in G is granules. The scale bar is 100 µm (provided by **Ding**) .

In Fishes

The polymer type and occurrence of MPs detected in fish are also reported by many researchers. The ingestion of MPs is influenced by the environmental condition, foraging and feeding type and the physical condition of fish. In the Mediterranean Sea, several semipelagic and demersal fish species were found to contain microfibers in their stomach contents, including *Boops boops* (57.8%), *Mullus barbatus* (33.3%), and *Trachurus ovatus* (24.3%) (Battaglia *et al.*, 2016; Bellas *et al.*, 2016; Nadal *et al.*, 2016). In two commercial fish species in the western Mediterranean Sea, *Sardina pilchardus* and *Engraulis encrasicolus*, about 15% of them contain MPs, with fiber was the dominant shape (83%) and PET was the most common polymer type (30%) (Compa *et al.*, 2018). These results also indicated that larger individuals of *S. pilchardus* with better physical condition were less likely to ingest MPs (Compa *et al.*, 2018). While in the fishes from the North Sea, only 1 out of 400 individuals (0.25%) was found to contain plastic polymethylmethacrylate (Hermsen *et al.*, 2017). The occurrence of MPs in six marine fish species from the Texas Gulf Coast was 42.4%, with fiber dominating (86.4%). The omnivorous fish presented higher content of MPs fibers than herbivores and carnivores (Peters *et al.*, 2017; Mizraji *et al.*, 2017). Presence of MPs was 100% in fishes collected from Río de la Plata estuary, and fiber represented 96% of MPs found. The spatial differences in mean number of MPs in fishes indicated the environmental availability of MPs could be of great importance to explain the differences found among sampling sites (Pazos *et al.*, 2017). Degraded hard fragments and threads as dominant MPs, ranging from 1.1 to 4.9 (3.8 ± 2.4) mm in length, were found in only 2.1% of the fish captured along the coast of the Southeast Pacific. This low prevalence of MPs might be due to the small human population and highly dynamic oceanographic processes in this area (Ory *et al.*, 2018).

In Turtles

In juvenile oceanic-stage of loggerheads sea turtles (*Caretta caretta*) collected from North Atlantic subtropical gyre, 83% of individuals were found to have ingested marine plastic debris with PE and PP being the main components (Pham *et al.*, 2017). In the Southwestern Atlantic, 70% of the dead green turtles between 2005 and 2013 contained MPs. Fragment was the most abundant in weight, while laminar and thread MPs are the most types in occurrence (Vélez-Rubio *et al.*, 2018). A negative correlation was detected between the presence of debris and turtle's size, indicating the early juvenile stage of turtle is the most vulnerable to the threat of ingested plastic.

In Seabirds

Plastic debris was found in the digestive tracts and stomach of seabirds with varying degrees. There were higher occurrences of plastic particles similar to natural food in shape, size, color, *etc.* (Spear *et al.*, 1995; Kühn and van Franeker, 2012). A negative relationship was found between plastic ingestion and physical condition in seabirds (Spear *et al.*, 1995). The occurrence of plastic was 95% in the stomachs of northern fulmars from the North Sea during 2003 ~ 2007 (van Franeker *et al.*, 2011), while 79% of the seabird from the Labrador Sea had ingested plastic, with an average of 11.6 items ind[-1] or 0.151 g ind[-1] (Avery-Gomm *et al.*, 2018). Acampora *et al.* (2014) reported that 67% of short-tailed shearwaters (*Puffinus tenuirostris*) ingested anthropogenic debris, and juvenile birds were more likely to ingest debris than adult birds. The plastic occurrence was 41.5% in *Cassin's Auklets* stranded along the Washington and Oregon coasts in 2014 (Floren and Shugart, 2017). MPs presence was 89.5% in 143 northern fulmars from 2008 to 2013 and 64% in 25 sooty shearwaters (*Ardenna grisea*) collected in 2011 ~ 2012 on Oregon and Washington beaches, and average plastic abundance were 19.5 items ind^{-1} and 0.46 g ind^{-1} for fulmars and 13.3 items ind^{-1} and 0.335 g ind^{-1} for shearwaters (Terepocki *et al.*, 2017).

MPs mistakenly eaten by marine animals may accumulate in their stomachs and cause physical blockage of the digestive system, resulting in damage of the digestion system, decreased absorption of nutrients and even lead to starvation and death. From all the above, it can be seen that the plastic, especially the MPs, distribute so widely that they might potentially threaten the survival of more and more marine organism.

ECOLOGICAL EFFECT OF MICROPLASTICS

The special properties of MPs make them provide sites for microorganism growth, and concentrate persistent organic pollutants and heavy metals. These may cause important threats to marine ecosystems (Dantas *et al.*, 2012; Jantz *et al.*, 2013). There are many reports focusing on toxic effects of MPs on marine animals (Collignon *et al.*, 2012; Derraik, 2002), such as sea urchins (Kaposi *et al.*, 2014), mussels (Li *et al.*, 2016; Ding *et al.*, 2009), crabs (Farrell and Nelson, 2013) and fishes (Rochman *et al.*, 2014). The major toxic mechanism was ingestion of MPs and sorbent for toxic organics (Besseling *et al.*, 2013).

Effects of Plastic Debris on Marine Biota

Fragmentation of plastic debris is one source of MPs. The most visible effects of large plastic debris (such as synthetic fiber rope, plastic sheet, strapping tape, abandoned fishing gear and ropes, trawl, *etc.*) on marine wildlife are entanglement (Moore *et al.*, 2009; Allen *et al.*, 2012). The entangled organism may no longer be able to acquire food and avoid predators. Among the most studies organisms, the number of affected species has increased from 86% to 100% for marine turtles (7 of 7 species), 43 to 66% (81 of 123 species) for marine mammals and to 50% of species for seabirds, respectively, from 1997 to 2015 (Kühn *et al.*, 2015).

Effects of Microplastics on Marine Organism

The small size of MPs renders them accessible to a wide range of organism, such as zooplankton, benthic organism, and fish, with potential for physical and toxicological harm. Recent results show that the impacts of MPs on marine organism occur at many levels, including behavior, feeding, growth, and regeneration. Nanoparticles can even translocate and permeate into the lipid membranes of organisms, altering the membrane structure, protein activity and cellular function (Rossi *et al.*, 2013).

Effects on Photosynthesis

MPs has a negative effect on algal photosynthesis. For example, 20 nm PS beads at a concentration of $1.8 \sim 6.5$ mg/L inhibited photosynthesis of *Scenedesmus spp.* by being absorbed into the cellulose and causing oxidative stress (Bhattacharya *et al.*, 2010). Sjollema *et al.* (2016) found 0.05 μm polystyrene particles obviously inhibited the growth of microalgae *Dnaliella tertiolecta*, but 6 μm polystyrene particles didn't have significant effects on algae. Therefore, the size of MPs could affect their toxic effects on microalgae. Zhang *et al.*, (2017) found that PVC (diameter 1 μm) inhibited the growth of microalgae, *Skeletonema costatum* by decreasing both chlorophyll content and photosynthetic efficiency. However, PVC particles with 1 mm diameter had no effects on the growth of microalgae. Based on the results, MPs might limit the transfer of energy and substance between cells and environment and lead to the decrease of nutrition, light, CO_2, and O_2 from medium into cells.

Effects on Immunity, Reproduction and Regeneration

The filter-feeder organisms are likely to be impacted as they ingest MPs when feeding. When the mussels, *Mytilus edulis*, were exposed to the polystyrene microspheres (3.0 or 9.6 μm), the particles were accumulated in the gut of and translocated from the gut to the circulatory system, and showing significant effects of time and particle size for over 48 days. Although there were no significant biological effects were observed during the exposure period, more researches were suggested to examine the toxicological effects of longer term exposure (Browne *et al.*, 2008). The high-density polyethylene (HDPE) particles (0 ~ 80 mm) were also ingested by the mussel, *Mytilus edulis L*, and taken up into the gills and digestive gland up to 96 h (Von Moos *et al.*, 2012). The formation of granulocytomas after 3 h and a steady decrease in lysosomal stability were observed after 6 h in the digestive gland, showing notable histological changes upon uptake and a strong inflammatory response. The results provided proof of principle that MPs were taken up into cells and caused distinct adverse effects on the tissue and cellular level to the mussels (Von Moos *et al.*, 2012). Toxic effects had been observed in the egg and larva of some organisms, and toxicity degree is associated with the size of particles. In a two-generation chronic toxicity test, 12.5 mg L^{-1} and 1.25 mg L^{-1} solution of 50 nm PS particles caused mortality of *nauplii and copepodites* in the F_0 and F_2 generation, respectively, while 500 nm PS particles at a concentration of 25 mg L^{-1} induced significant decrease in survival rate compared with the control population in the F_1 generation (Lee *et al.*, 2013b). The median effect concentration of nano-PS beads modified by ammonia acid to the fertilized egg of sea urchin embryos *(Paracentrous lividus)* were 3.85 mg L^{-1} and 2.61 mg L^{-1} for 24 h and 48 h, respectively, indicating that differences in surface charges of MPs and aggregation in seawater strongly affect their embryotoxicity (Della *et al.*, 2014). When *Pacific oyster* was exposed to PS microspheres, the oocyte number, diameter, and sperm velocity decreased significantly, and the D-larval yield and larval development of offspring also decreased compared with control. This indicated that PS microsphere caused reproductive disruption and significant impacts on offspring (Sussarellu *et al.*, 2015). Micro-sized PVC particles were found to affect filtration and respiration rates as well as byssus production in Asian green mussel *Perna viridis*. Survival rate declined with increasing PVC abundance within 91 days of exposure (Rist *et al.*, 2016). By applying the molecular simulation method, Rossi *et al.* (2014) assessed the influence of nano-sized PS particles to the property of lipid membrane and found that the nano-sized PS particles may permeate into the lipid membranes of organisms, altering the membrane structure and activity of membrane protein, therefore changing cellular functions.

Plastic buried in sediments may increase the permeability of the sediments and decrease their thermal diffusivity. These changes have a variety of potential effects on the related organism process, such as those with temperature-dependent sex-determination, such as sea turtle eggs (Carson *et al.*, 2011). Green *et al.* (2015) reported that the sediments covered by plastic bags for 9 weeks became anoxic, leading to the lower primary productivity and organic matter and abundances of infaunal invertebrates.

Effects on Feeding

MPs greatly influence the benthonic filtering feeding organisms. Exposure experiments in laboratory demonstrated that benthic invertebrates may directly filter MPs (Ward and Kach, 2009; Wegner *et al.*, 2012), and the benthic sea cucumbers even selectively ingested MPs (Graham and Thompson, 2009). MPs were found at varying degree in the digestive tracts of lugworm, sea cucumber, sea squirt, and so on. In mussels and crabs, MPs were detected not only in the digestive tracts, but also in hemolymph and tissues. When mussels, *Mytilus edulis* and oysters, *Crassostrea virginica*, were exposed to aggregates of 100 nm PS beads, the uptake of 100 nm particles was significantly enhanced and nanoparticles were transported to the digestive gland, implicating the toxicological effects and transfer of nanomaterials to higher trophic levels (Ward and Kach, 2009). The blue mussel (*Mytilus edulis*) reduced its filtering activity when being exposed to 30 nm PS particles (Wegner *et al.*, 2012). Ingestion of 20 μm PS beads (75 MPs mL^{-1}) can significantly alter the feeding capacity of the pelagic copepod *Calanus helgolandicus* (Cole *et al.*, 2015).

Habitats Provider

MPs can provide habitats for invertebrates, bacteria and viruses in the deep sea, act as vectors for invasion species and affect the distribution and dispersal of marine organisms in the marine environment (Goldstein *et al.*, 2012; Schlining *et al*, 2013). In the North Pacific Subtropical Gyre (NPSG), high concentrations of MPs resulted in increasing *Halobates sericeus* egg densities because MPs provide the substrate for oviposition (Goldstein *et al.*, 2012). Kirstein *et al.* (2016) discovered potentially pathogenic *Vibrio* parahaemolyticus on a number of MP particles (PE, PP, and PS) from the North/Baltic Sea. Seventeen different fouling species were discovered on plastic from Mersin Bay, NE Levantine coast of Turkey (Gündoğdu *et al.*, 2017). The pathogenic fish bacteria *Aeromonas salmonicida* was identified on MPs in the North Adriatic (Viršek *et al.*, 2017). The plastic colonized by pathogenic microorganisms can travel a long distance and lead to invasion and disease.

Effects of Microplastics on the Distribution and Transportation of Pollutants

Plastic is carrier and transporter of organic contaminants, including components of the plastic itself (*e.g.* plasticizers), persistent organic pollutants (POPs), such as polychlorinated biphenyls (PCBs), polycyclic aromatic hydrocarbons (PAHs), and heavy metals. Pollutants can be absorbed onto or into the plastic. The concentrations of PCBs and PAHs on the MPs particles from coastal of Portuguese ranged from 273 to 307 ng g^{-1} and 100 to 300 ng g^{-1}, respectively (Mizukawa *et al.*, 2013). PE and PP were the main constitutes of MPs in the North Pacific Gyre, and the absorbed PCBs were from 1 to 233 ng g^{-1} (Rios and Jones, 2015). Because MPs are widely spread in the environment and can be transported over long distances by wind, currents and wave action before becoming temporarily or permanently stranded, they can affect the global distribution of pollutants (Mato *et al.*, 2001; Endo *et al.*, 2005; Rios *et al.*, 2007; Karapanagioti and Klontza, 2008; Zarfl and Matthies, 2010; Engler, 2012).

The pollutants associated with MPs can settle to the seafloor. The release of pollutants from sediment may be a source of contamination in the deep sea areas (Rios *et al.*, 2007). Ogata *et al.* (2009) collected PE pellets at 30 beaches from 17 countries and analyzed organochlorine compounds. They found the spatial pattern reflected regional differences in the usage of PCBs and organochlorine pesticides. Simulating experiments in the laboratory suggested that the adsorption capacities of heavy metals were greater for beached pellets than for virgin pellets, indicating that plastic may represent an important vehicle for the transport of metals in the marine environment (Holmes *et al.*, 2012). Therefore, MPs are considered as vectors for increasing source of anthropogenic contamination in the aquatic environment.

Impact Factors on Pollutants Sorption to Microplastics

The type and physical makeup of domains in polymers are important factors influencing the sorption of organic compounds to the surface of plastic (Guo *et al.*, 2012). Sorption behaviors and impact factors of pollutants on the surface of MPs are investigated by laboratory or in situ experiments (Guo *et al.*, 2012; Boucher *et al.*, 2016; Fisner *et al.*, 2017; Vedolin *et al.*, 2018). Apparent adsorption coefficients of POPs to plastic surfaces are in the order of $10^5 \sim 10^6$ (Mato *et al.*, 2001) and vary for each POP-plastic combination (Teuten *et al.*, 2009; Guo *et al.*, 2012; Rochman *et al.*, 2013a). The partition coefficient between plastic and water is the most important parameter influencing the desorption half-time (Lee *et al.*, 2018). The experiment results of Guo et al (2012) suggested that mobility and abundance of rubbery domains in polymers (PE, PS and polyphenylene oxide) regulated the sorption behaviors hydrophobic organic contaminants (phenanthrene, naphthalene, lindane, and 1-naphthol). The concentrations of PAHs and PCBs sorbed to PE and PP were much greater than those sorbed to PET and PVC (Rochman *et al.*, 2013a). The affinity of PE to phthalate was stronger than that to ammonium perfluorooctanoate (Bakir *et al.*, 2014a).

The size of MPs and other properties, such as color, also affect their affinity for pollutants. The quantity of PCBs absorbed on PS nanoparticles with 70 nm in diameter was $1 \sim 2$ orders higher than that on MPs with the diameter ranged from 10 to 180 μm (Velzeboer et al, 2014). Lighter colored PE and PP pellets contain low molecular weight PAHs, while darker pellets contained high molecular weight PAHs (Fisner *et al.*, 2017). The metal adsorption on pellets collected from the coast of São Paulo State in southeastern Brazil was greater than that on virgin pellets (Vedolin *et al.*, 2018).

The environmental conditions (such as pH and salinity) are also impact factors on the sorption behavior of chemicals to polymers. The quantity of PCBs absorbed on nano- and micro- particles increased under conditions of lower pH and higher salinity (Velzeboer *et al.*, 2014). But sorption of DDTs and phenanthrene to MPs was more related to environmental pollutant concentrations than to the salinity (Bakir *et al.*, 2014b).

Combined Pollution from Microplastics and Associated Pollutant

Plasticizer, antioxidant, fire retardants and antibacterial agents are added in the plastic for enhancing polymer properties and durability. Such chemical substances include phthalates, bisphenol A, and brominated compounds. Some phthalates, such as butyl benzyl and dibutyl phthalates, are estrogenic endocrine disruptors. The brominated compounds, such as polybrominated diphenyl ethers (PBDEs) and hexabromocyclododecane (HBCD), are carcinogens and mutagens. These additives can be released from the plastic to the surrounding environment during the use, disposal and recycling phase of plastic products (Law and Thompson, 2014; Hahladakis *et al.*, 2018; Kedzierski *et al.*, 2018). Suhrhoff *et al.* (2016) estimated the global annual release of additives from plastic into the marine environment to be between 35 and 917 tons, of which most are derived from plasticized PVC. Gandara E Silva *et al.* (2016) reported that the toxicity of the leachate from beached pellets to mussel embryo (*Perna perna*) was much higher than that of virgin pellets, suggesting contaminants adsorbed onto the surface of beached pellets were responsible for the high toxicity of leachate from beached pellets, while the toxicity of leachate from virgin pellets was mainly due to plastic additives. However, the results of Martínez-Gómez *et al.* (2017) indicated that the leachates of virgin fluorescent PS microspheres (diameter 6 μm) and virgin high-density PE fluff (diameter 0~80 μm) caused a higher toxicity on sea urchin gametes and embryos than aged materials.

MPs readily accumulates POPs and heavy metals from surroundings water during transportation process, increasing their concentrations by orders of magnitude (Law and Thompson, 2014; Ogata *et al.*, 2009; Holmes *et al.*, 2012). The associated pollutants together with plastic are accumulated in the organisms and across marine trophic levels and cause combined toxic effects on marine organisms (Wright et al, 2013). Graham and Thompson (2009) reported that PCBs attached to the plastic debris could cause adverse effects to sea cucumbers (*Echinodermata*), and could transfer to their predators through trophic chains. Japanese medaka (*Oryzias latipes*) altered gene expression and suffered liver toxicity and pathology after being exposed to marine-PE pellets (diameter <1 mm) treatment, suggesting the ingestion of plastic debris at environmentally relevant concentrations may alter endocrine system function in adult fish (Rochman *et al.*, 2013b).

When mussels are exposed to a solution of MPs with absorbed pyrene, the MPs is found in hemolymph, gills and especially digestive tissues. A marked accumulation of pyrene in the tissue indicated contaminated MPs can transfer PAH to mussels (Avio *et al.*, 2015). By analysis of biomarker gene expression, the bioavailability of phenanthrene and 17α-ethinylestradiol to larval zebrafish was found to decrease after absorbed by PVC particles (Sleight *et al.*, 2017). The toxicity of the mixtures of MPs and mercury to European seabass (*Dicentrarchus labrax*) were higher than that of MPs alone and mercury alone, indicating MPs influence the bioaccumulation of mercury by *Dicentrarchus labrax* juveniles (Barboza *et al.*, 2018).

CONCLUSION AND FURTHER TRENDS

Conclusion

In summary, the plastic debris and MPs in the marine environment are becoming a global problem and is threatening the marine ecosystems. The plastic debris from various sources will eventually become MPs entering the ocean and participate in the biogeochemical cycle process. They will affect the distribution and transportation of pollutants, impact the condition and survival of organisms, can be transferred along the food chain. The processes and effects are illustrated as in Fig. (**5**). It is important to study the effects of MPs and assess the risks to human and environmental health.

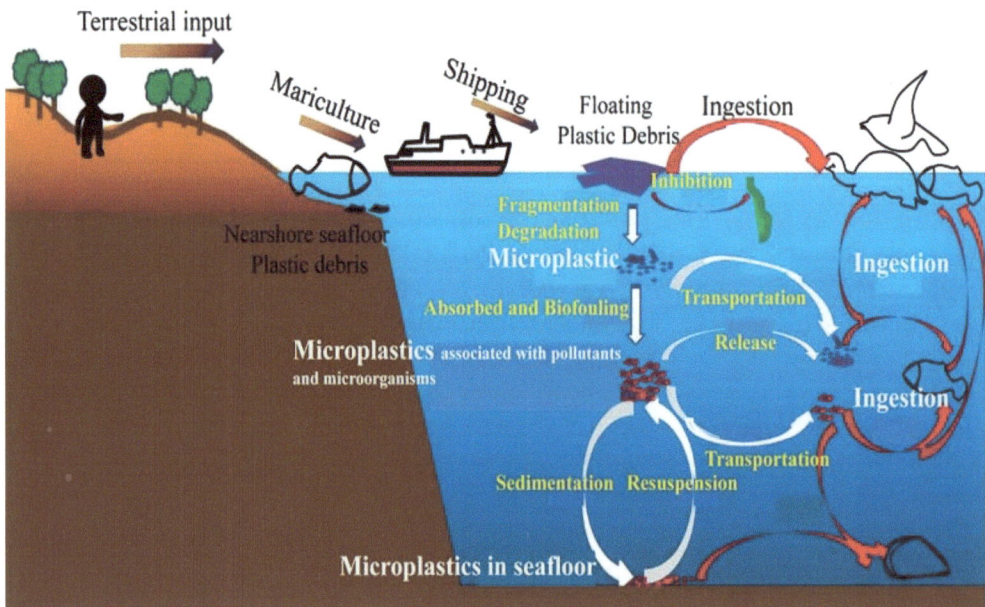

Fig. (5). The illustration of the source, distribution, and ecological and environmental effects of microplastic (Adapted from Sun *et al.*, 2016).

Future Work on the Management and Research of Microplastics

MPs pollution is currently one of the most serious problems in marine ecosystems, which has caused concern of the United Nations Environment Assembly (Haward, 2018). There is a growing consensus that we need to find solutions to the problem. In the future, there are several aspects we need to work on:

At first, it is essential to formulate an international agreement to address the marine plastic issue for sustainable development goals. We need to prevent plastic debris from entering the marine environment. The disposal of plastic by ships has been proposed to be prohibited and land-based source of pollution entering the marine environment is also suggested to be monitored and controlled by states at the United Nations Environment Assembly conference. The international convention for the prevention of pollution from ships (MARPOL) enacted in 1988 is an important international initiative to prevent ships from discarding their garbage at sea. A policy framework including law and waste management strategies, education, source identification and increased monitoring to mitigate plastic marine pollution has been developed for the Canadian context (Pettipas, *et al.*, 2016). More international collaboration on combating the plastic problem is needed in the future.

Secondly, it is very important to change the way we produce, use and dispose of plastic items. Developing a more circular economy and rethinking the using manner of plastic materials has considerable potential to bring greater resource efficiency. It is also important to improve public awareness of the detrimental effects of marine plastic pollution. Forums should be held to present innovative solutions to reduce the amount of plastic entering the marine environment.

Thirdly, scientific research work must be continued in order to understand the scale and scope of marine plastic pollution. Though there have already a lot of work done on plastic debris about the fate, degradation, and interaction, many questions remain unanswered. Future studies on micro- and nanoplastic should focus on the most urgent topics: i) a standard procedure for analysis which is essential to reveal the distribution of MPs in the global scope; ii) the biotoxicity, such as regeneration and long-term effects on marine organisms, and trophic transfer of micro- and nanoplastic combined with pollutants through the food chain; iii) the effect of hydrodynamics on the distribution and transportation of micro- and nanoplastics; iv) system and standards of ecological risk assessment on the micro- and nanoplastic pollution.

CONSENT FOR PUBLICATION

Not applicable.

CONFLICT OF INTEREST

The authors confirm that this chapter contents have no conflict of interest.

ACKNOWLEDGEMENTS

This work was supported by the Basic Scientific Fund for National Public Research Institutes of China (No. 2016Q02/ 2017Y03), the Research Cooperation and Exchange of Marine Litter and Microplastics, China (No. QY0518011), the Second Sino-German Cooperation in Marine Sciences (No. QY0518016), and the Special Operational Funding Project of the State Oceanic Administration-Marine Environmental Monitoring/Assessment and Capacity Upgrading (No. BJ1318004, JDKC0518008).

REFERENCES

Acampora, H, Schuyler, QA, Townsend, KA & Hardesty, BD (2014) Comparing plastic ingestion in juvenile and adult stranded short-tailed shearwaters (*Puffinus tenuirostris*) in eastern Australia. *Mar Pollut Bull,* 78, 63-8.
[http://dx.doi.org/10.1016/j.marpolbul.2013.11.009] [PMID: 24295596]

Allen, R, Jarvis, D, Sayer, S & Mills, C (2012) Entanglement of grey seals *Halichoerus grypus* at a haul out site in Cornwall, UK. *Mar Pollut Bull,* 64, 2815-9.
[http://dx.doi.org/10.1016/j.marpolbul.2012.09.005] [PMID: 23117201]

Alomar, C, Estarellas, F & Deudero, S (2016) Microplastics in the Mediterranean Sea: Deposition in coastal shallow sediments, spatial variation and preferential grain size. *Mar Environ Res,* 115, 1-10.
[http://dx.doi.org/10.1016/j.marenvres.2016.01.005] [PMID: 26803229]

Andrady, AL (2015) *Marine anthropogenic litter,* Springer International Publishing, Berlin, 57-72.
[http://dx.doi.org/10.1007/978-3-319-16510-3_3]

Avery-Gomm, S, Provencher, JF, Liboiron, M, Poon, FE & Smith, PA (2018) Plastic pollution in the Labrador Sea: An assessment using the seabird northern fulmar *Fulmarus glacialis* as a biological monitoring species. *Mar Pollut Bull,* 127, 817-22.
[http://dx.doi.org/10.1016/j.marpolbul.2017.10.001] [PMID: 29055560]

Avio, CG, Gorbi, S, Milan, M, Benedetti, M, Fattorini, D, d'Errico, G, Pauletto, M, Bargelloni, L & Regoli, F (2015) Pollutants bioavailability and toxicological risk from microplastics to marine mussels. *Environ Pollut,* 198, 211-22.
[http://dx.doi.org/10.1016/j.envpol.2014.12.021] [PMID: 25637744]

Bakir, A, Rowland, SJ & Thompson, RC (2014a) Enhanced desorption of persistent organic pollutants from microplastics under simulated physiological conditions. *Environ Pollut,* 185, 16-23.
[http://dx.doi.org/10.1016/j.envpol.2013.10.007] [PMID: 24212067]

Bakir, A, Rowland, SJ & Thompson, RC (2014b) Transport of persistent organic pollutants by microplastics in estuarine conditions. *Estuar Coast Shelf Sci,* 140, 14-21.
[http://dx.doi.org/10.1016/j.ecss.2014.01.004]

Barboza, LGA, Vieira, LR, Branco, V, Figueiredo, N, Carvalho, F, Carvalho, C & Guilhermino, L (2018) Microplastics cause neurotoxicity, oxidative damage and energy-related changes and interact with the bioaccumulation of mercury in the European seabass, *Dicentrarchus labrax* (Linnaeus, 1758). *Aquat Toxicol,* 195, 49-57.
[http://dx.doi.org/10.1016/j.aquatox.2017.12.008] [PMID: 29287173]

Barnes, DKA, Galgani, F, Thompson, RC & Barlaz, M (2009) Accumulation and fragmentation of plastic debris in global environments. *Philos Trans R Soc Lond B Biol Sci,* 364, 1985-98.
[http://dx.doi.org/10.1098/rstb.2008.0205] [PMID: 19528051]

Battaglia, P, Pedà, C, Musolino, S, Esposito, V, Andaloro, F & Romeo, T (2016) Diet and first documented data on plastic ingestion of *Trachinotus ovatus* L. 1758 (Pisces:Carangidae) from the Strait of Messina

(central Mediterranean Sea). *Ital J Zool (Modena),* 83, 121-9.
[http://dx.doi.org/10.1080/11250003.2015.1114157]

Baztan, J, Carrasco, A, Chouinard, O, Cleaud, M, Gabaldon, JE, Huck, T, Jaffrès, L, Jorgensen, B, Miguelez, A, Paillard, C & Vanderlinden, JP (2014) Protected areas in the Atlantic facing the hazards of micro-plastic pollution: first diagnosis of three islands in the Canary Current. *Mar Pollut Bull,* 80, 302-11.
[http://dx.doi.org/10.1016/j.marpolbul.2013.12.052] [PMID: 24433999]

Bellas, J, Martínez-Armental, J, Martínez-Cámara, A, Besada, V & Martínez-Gómez, C (2016) Ingestion of microplastics by demersal fish from the Spanish Atlantic and Mediterranean coasts. *Mar Pollut Bull,* 109, 55-60.
[http://dx.doi.org/10.1016/j.marpolbul.2016.06.026] [PMID: 27289284]

Besseling, E, Wegner, A, Foekema, EM, van den Heuvel-Greve, MJ & Koelmans, AA (2013) Effects of microplastic on fitness and PCB bioaccumulation by the lugworm *Arenicola marina* (L.). *Environ Sci Technol,* 47, 593-600.
[http://dx.doi.org/10.1021/es302763x] [PMID: 23181424]

Beyer, J, Green, NW, Brooks, S, Allan, IJ, Ruus, A, Gomes, T, Bråte, ILN & Schøyen, M (2017) Blue mussels (*Mytilus edulis* spp.) as sentinel organisms in coastal pollution monitoring: A review. *Mar Environ Res,* 130, 338-65.
[http://dx.doi.org/10.1016/j.marenvres.2017.07.024] [PMID: 28802590]

Bhattacharya, P, Turner, JP & Ke, PC (2010) Physical adsorption of charged plastic nanoparticles affects algal photosynthesis. *J Phys Chem C,* 114, 16556-61.
[http://dx.doi.org/10.1021/jp1054759]

Blumenröder, J, Sechet, P, Kakkonen, JE & Hartl, MGJ (2017) Microplastic contamination of intertidal sediments of Scapa Flow, Orkney: A first assessment. *Mar Pollut Bull,* 124, 112-20.
[http://dx.doi.org/10.1016/j.marpolbul.2017.07.009] [PMID: 28709522]

Boucher, C, Morin, M & Bendell, LI (2016) The influence of cosmetic microbeads on the sorptive behavior of cadmium and lead within intertidal sediments: A laboratory study. *Reg Stud Mar Sci,* 3, 1-7.
[http://dx.doi.org/10.1016/j.rsma.2015.11.009]

Browne, MA, Dissanayake, A, Galloway, TS, Lowe, DM & Thompson, RC (2008) Ingested microscopic plastic translocates to the circulatory system of the mussel, *Mytilus edulis* (L). *Environ Sci Technol,* 42, 5026-31.
[http://dx.doi.org/10.1021/es800249a] [PMID: 18678044]

Browne, MA, Crump, P, Niven, SJ, Teuten, E, Tonkin, A, Galloway, TS & Thompson, R (2015) Accumulation of microplastic on shorelines worldwide: Sources and sinks. *Environ Sci Technol,* 49, 9175-9.

Carson, HS, Colbert, SL, Kaylor, MJ & McDermid, KJ (2011) Small plastic debris changes water movement and heat transfer through beach sediments. *Mar Pollut Bull,* 62, 1708-13.
[http://dx.doi.org/10.1016/j.marpolbul.2011.05.032] [PMID: 21700298]

Claessens, M, De Meester, S, Van Landuyt, L, De Clerck, K & Janssen, CR (2011) Occurrence and distribution of microplastics in marine sediments along the Belgian coast. *Mar Pollut Bull,* 62, 2199-204.
[http://dx.doi.org/10.1016/j.marpolbul.2011.06.030] [PMID: 21802098]

Cole, M, Lindeque, P, Fileman, E, Halsband, C & Galloway, TS (2015) The impact of polystyrene microplastics on feeding, function and fecundity in the marine copepod *Calanus helgolandicus. Environ Sci Technol,* 49, 1130-7.
[http://dx.doi.org/10.1021/es504525u] [PMID: 25563688]

Collignon, A, Hecq, JH, Glagani, F, Voisin, P, Collard, F & Goffart, A (2012) Neustonic microplastic and zooplankton in the North Western Mediterranean Sea. *Mar Pollut Bull,* 64, 861-4.
[http://dx.doi.org/10.1016/j.marpolbul.2012.01.011] [PMID: 22325448]

Collignon, A, Hecq, JH, Galgani, F, Collard, F & Goffart, A (2014) Annual variation in neustonic micro- and meso-plastic particles and zooplankton in the Bay of Calvi (Mediterranean-Corsica). *Mar Pollut Bull,* 79,

293-8.
[http://dx.doi.org/10.1016/j.marpolbul.2013.11.023] [PMID: 24360334]

Compa, M, Ventero, A, Iglesias, M & Deudero, S (2018) Ingestion of microplastics and natural fibres in *Sardina pilchardus* (Walbaum, 1792) and *Engraulis encrasicolus* (Linnaeus, 1758) along the Spanish Mediterranean coast. *Mar Pollut Bull,* 128, 89-96.
[http://dx.doi.org/10.1016/j.marpolbul.2018.01.009] [PMID: 29571417]

Cózar, A, Echevarría, F, González-Gordillo, JI, Irigoien, X, Úbeda, B, Hernández-León, S, Palma, ÁT, Navarro, S, García-de-Lomas, J, Ruiz, A, Fernández-de-Puelles, ML & Duarte, CM (2014) Plastic debris in the open ocean. *Proc Natl Acad Sci USA,* 111, 10239-44.
[http://dx.doi.org/10.1073/pnas.1314705111] [PMID: 24982135]

Dantas, DV, Barletta, M & da Costa, MF (2012) The seasonal and spatial patterns of ingestion of polyfilament nylon fragments by estuarine drums (*Sciaenidae*). *Environ Sci Pollut Res Int,* 19, 600-6.
[http://dx.doi.org/10.1007/s11356-011-0579-0] [PMID: 21845453]

Della Torre, C, Bergami, E, Salvati, A, Faleri, C, Cirino, P, Dawson, KA & Corsi, I (2014) Accumulation and embryotoxicity of polystyrene nanoparticles at early stage of development of sea urchin embryos *Paracentrotus lividus*. *Environ Sci Technol,* 48, 12302-11.
[http://dx.doi.org/10.1021/es502569w] [PMID: 25260196]

Derraik, JGB (2002) The pollution of the marine environment by plastic debris: a review. *Mar Pollut Bull,* 44, 842-52.
[http://dx.doi.org/10.1016/S0025-326X(02)00220-5] [PMID: 12405208]

Desforges, JPW, Galbraith, M, Dangerfield, N & Ross, PS (2014) Widespread distribution of microplastics in subsurface seawater in the NE Pacific Ocean. *Mar Pollut Bull,* 79, 94-9.
[http://dx.doi.org/10.1016/j.marpolbul.2013.12.035] [PMID: 24398418]

Ding, JF, Li, JX, Sun, CJ, He, CF, Jiang, FH, Gao, FL & Zheng, L (2018) Separation and identification of microplastics in digestive system of bivalves. *Chin J Anal Chem,* 2018, 690-7.
[http://dx.doi.org/10.1016/S1872-2040(18)61086-2]

Ding, JF, Li, JX, Sun, CF, Jiang, FH, Ju, P, Qu, LY, Zheng, YF & He, CF (2019) Detection of microplastics in local marine organisms using a muti-technology system. *Analytical Methods,* 11, 78-87.
[http://dx.doi.org/10.1007/s11270-012-1352-9]

Dubaish, F & Liebezeit, G (2013) Suspended microplastics and black carbon particles in the Jade system, southern North Sea. *Water Air Soil Pollut,* 224, 1-8.
[http://dx.doi.org/10.1007/s11270-012-1352-9]

Endo, S, Takizawa, R, Okuda, K, Takada, H, Chiba, K, Kanehiro, H, Ogi, H, Yamashita, R & Date, T (2005) Concentration of polychlorinated biphenyls (PCBs) in beached resin pellets: variability among individual particles and regional differences. *Mar Pollut Bull,* 50, 1103-14.
[http://dx.doi.org/10.1016/j.marpolbul.2005.04.030] [PMID: 15896813]

Engler, RE (2012) The complex interaction between marine debris and toxic chemicals in the ocean. *Environ Sci Technol,* 46, 12302-15.
[http://dx.doi.org/10.1021/es3027105] [PMID: 23088563]

Eriksen, M, Lebreton, LCM, Carson, HS, Thiel, M, Moore, CJ, Borerro, JC, Galgani, F, Ryan, PG & Reisser, J (2014) Plastic pollution in the world's oceans: More than 5 trillion plastic pieces weighing over 250,000 tons afoat at sea. *PLoS One,* 9e111913
[http://dx.doi.org/10.1371/journal.pone.0111913] [PMID: 25494041]

Farrell, P & Nelson, K (2013) Trophic level transfer of microplasic: Mytilus edulis (L.) to Carcinus maenas (L.). *Environ Pollut,* 177, 1-3.
[http://dx.doi.org/10.1016/j.marpolbul.2017.06.072] [PMID: 28679482]

Fisner, M, Majer, A, Taniguchi, S, Bícego, M, Turra, A & Gorman, D (2017) Colour spectrum and resin-type determine the concentration and composition of Polycyclic Aromatic Hydrocarbons (PAHs) in plastic pellets.

Mar Pollut Bull, 122, 323-30.
[http://dx.doi.org/10.1016/j.marpolbul.2017.06.072] [PMID: 28679482]

Floren, HP & Shugart, GW (2017) Plastic in Cassin's Auklets (*Ptychoramphus aleuticus*) from the 2014 stranding on the Northeast Pacific Coast. *Mar Pollut Bull,* 117, 496-8.
[http://dx.doi.org/10.1016/j.marpolbul.2017.01.076] [PMID: 28160979]

Frias, JPGL, Gago, J, Otero, V & Sobral, P (2016) Microplastics in coastal sediments from Southern Portuguese shelf waters. *Mar Environ Res,* 114, 24-30.
[http://dx.doi.org/10.1016/j.marenvres.2015.12.006] [PMID: 26748246]

Gandara E Silva, PP, Nobre, CR, Resaffe, P, Pereira, CDS, Gusmão, F & Gusmão, F (2016) Leachate from microplastics impairs larval development in brown mussels. *Water Res,* 106, 364-70.
[http://dx.doi.org/10.1016/j.watres.2016.10.016] [PMID: 27750125]

Goldstein, MC, Rosenberg, M & Cheng, L (2012) Increased oceanic microplastic debris enhances oviposition in an endemic pelagic insect. *Biol Lett,* 8, 817-20.
[http://dx.doi.org/10.1098/rsbl.2012.0298] [PMID: 22573831]

Graham, ER & Thompson, JT (2009) Deposit- and suspension-feeding sea cucumbers (*Echinodermata*) ingest plastic fragments. *J Exp Mar Biol Ecol,* 368, 22-9.
[http://dx.doi.org/10.1016/j.jembe.2008.09.007]

Green, DS, Boots, B, Blockley, DJ, Rocha, C & Thompson, R (2015) Impacts of discarded plastic bags on marine assemblages and ecosystem functioning. *Environ Sci Technol,* 49, 5380-9.
[http://dx.doi.org/10.1021/acs.est.5b00277] [PMID: 25822754]

Guo, XY, Wang, XL, Zhou, XZ, Kong, XZ, Tao, S & Xing, BS (2012) Sorption of four hydrophobic organic compounds by three chemically distinct polymers: role of chemical and physical composition. *Environ Sci Technol,* 46, 7252-9.
[http://dx.doi.org/10.1021/es301386z] [PMID: 22676433]

Gündoğdu, S, Çevik, C & Karaca, S (2017) Fouling assemblage of benthic plastic debris collected from Mersin Bay, NE Levantine coast of Turkey. *Mar Pollut Bull,* 124, 147-54.
[http://dx.doi.org/10.1016/j.marpolbul.2017.07.023] [PMID: 28716475]

Hahladakis, JN, Velis, CA, Weber, R, Iacovidou, E & Purnell, P (2018) An overview of chemical additives present in plastics: Migration, release, fate and environmental impact during their use, disposal and recycling. *J Hazard Mater,* 344, 179-99.
[http://dx.doi.org/10.1016/j.jhazmat.2017.10.014] [PMID: 29035713]

Haward, M (2018) Plastic pollution of the world's seas and oceans as a contemporary challenge in ocean governance. *Nat Commun,* 9, 667.
[http://dx.doi.org/10.1038/s41467-018-03104-3] [PMID: 29445166]

Hermsen, E, Pompe, R, Besseling, E & Koelmans, AA (2017) Detection of low numbers of microplastics in North Sea fish using strict quality assurance criteria. *Mar Pollut Bull,* 122, 253-8.
[http://dx.doi.org/10.1016/j.marpolbul.2017.06.051] [PMID: 28655459]

Holmes, LA, Turner, A & Thompson, RC (2012) Adsorption of trace metals to plastic resin pellets in the marine environment. *Environ Pollut,* 160, 42-8.
[http://dx.doi.org/10.1016/j.envpol.2011.08.052] [PMID: 22035924]

Imhof, HK, Ivleva, NP, Schmid, J, Niessner, R & Laforsch, C (2013) Contamination of beach sediments of a subalpine lake with microplastic particles. *Curr Biol,* 23, 867-8.
[http://dx.doi.org/10.1016/j.cub.2013.09.001] [PMID: 24112978]

Jambeck, JR, Geyer, R, Wilcox, C, Siegler, TR, Perryman, M, Andrady, A, Narayan, R & Law, KL (2015) Plastic waste inputs from land into the ocean. *Science,* 347, 768-71.
[http://dx.doi.org/10.1126/science.1260352] [PMID: 25678662]

Jamieson, AJ, Malkocs, T, Piertney, SB, Fujii, T & Zhang, Z (2017) Bioaccumulation of persistent organic pollutants in the deepest ocean fauna. *Nature Ecology and evolution,* 1, 0051.

Jantz, LA, Morishige, CL, Bruland, GL & Lepczyk, CA (2013) Ingestion of plastic marine debris by longnose lancetfish (*Alepisaurus ferox*) in the North Pacific Ocean. *Mar Pollut Bull,* 69, 97-104.
[http://dx.doi.org/10.1016/j.marpolbul.2013.01.019] [PMID: 23465573]

Kaposi, KL, Mos, B, Kelaher, BP & Dworjanyn, SA (2014) Ingestion of microplastic has limited impact on a marine larva. *Environ Sci Technol,* 48, 1638-45.
[http://dx.doi.org/10.1021/es404295e] [PMID: 24341789]

Karapanagioti, HK & Klontza, I (2008) Testing phenanthrene distribution properties of virgin plastic pellets and plastic eroded pellets found on Lesvos island beaches (Greece). *Mar Environ Res,* 65, 283-90.
[http://dx.doi.org/10.1016/j.marenvres.2007.11.005] [PMID: 18164383]

Kedzierski, M, D'Almeida, M, Magueresse, A, Le Grand, A, Duval, H, César, G, Sire, O, Bruzaud, S & Le Tilly, V (2018) Threat of plastic ageing in marine environment. Adsorption/desorption of micropollutants. *Mar Pollut Bull,* 127, 684-94.
[http://dx.doi.org/10.1016/j.marpolbul.2017.12.059] [PMID: 29475712]

Kirstein, IV, Kirmizi, S, Wichels, A, Garin-Fernandez, A, Erler, R, Löder, M & Gerdts, G (2016) Dangerous hitchhikers? Evidence for potentially *pathogenic Vibrio* spp. on microplastic particles. *Mar Environ Res,* 120, 1-8.
[http://dx.doi.org/10.1016/j.marenvres.2016.07.004] [PMID: 27411093]

Kühn, S & van Franeker, JA (2012) Plastic ingestion by the northern fulmar (*Fulmarus glacialis*) in Iceland. *Mar Pollut Bull,* 64, 1252-4.
[http://dx.doi.org/10.1016/j.marpolbul.2012.02.027] [PMID: 22455662]

Kühn, S, Rebolledo, ELB & van Franeker, JA (2015) *Marine Anthropogenic Litter,* Springer International Publishing, Berlin 75-116.
[http://dx.doi.org/10.1007/978-3-319-16510-3_4]

Law, KL, Morét-Ferguson, S, Maximenko, NA, Proskurowski, G, Peacock, EE, Hafner, J & Reddy, CM (2010) Plastic accumulation in the North Atlantic subtropical gyre. *Science,* 329, 1185-8.
[http://dx.doi.org/10.1126/science.1192321] [PMID: 20724586]

Law, KL & Thompson, RC (2014) Oceans. Microplastics in the seas. *Science,* 345, 144-5.
[http://dx.doi.org/10.1126/science.1254065] [PMID: 25013051]

Lee, J, Hong, S, Song, YK, Hong, SH, Jang, YC, Jang, M, Heo, NW, Han, GM, Lee, MJ, Kang, D & Shim, WJ (2013a) Relationships among the abundances of plastic debris in different size classes on beaches in South Korea. *Mar Pollut Bull,* 77, 349-54.
[http://dx.doi.org/10.1016/j.marpolbul.2013.08.013] [PMID: 24054782]

Lee, KW, Shim, WJ, Kwon, OY & Kang, JH (2013b) Size-dependent effects of micro polystyrene particles in the marine copepod *Tigriopus japonicus. Environ Sci Technol,* 47, 11278-83.
[http://dx.doi.org/10.1021/es401932b] [PMID: 23988225]

Lee, H, Byun, DE, Kim, JM & Kwon, JH (2018) Desorption modeling of hydrophobic organic chemicals from plastic sheets using experimentally determined diffusion coefficients in plastics. *Mar Pollut Bull,* 126, 312-7.
[http://dx.doi.org/10.1016/j.marpolbul.2017.11.032] [PMID: 29421104]

Li, JN, Qu, XY, Su, L, Zhang, WW, Yang, DQ, Kolandhasamy, P, Li, DJ & Shi, HH (2016) Microplastics in mussels along the coastal waters of China. *Environ Pollut,* 214, 177-84.
[http://dx.doi.org/10.1016/j.envpol.2016.04.012] [PMID: 27086073]

Liebezeit, G & Dubaish, F (2012) Microplastics in beaches of the East Frisian islands Spiekeroog and Kachelotplate. *Bull Environ Contam Toxicol,* 89, 213-7.
[http://dx.doi.org/10.1007/s00128-012-0642-7] [PMID: 22526995]

Lots, FAE, Behrens, P, Vijver, MG, Horton, AA & Bosker, T (2017) A large-scale investigation of microplastic contamination: Abundance and characteristics of microplastics in European beach sediment. *Mar Pollut Bull,* 123, 219-26.

[http://dx.doi.org/10.1016/j.marpolbul.2017.08.057] [PMID: 28893402]

Marris, E (2014) Fate of ocean plastic remains a mystery. *Nature News.*
[http://dx.doi.org/10.1038/nature.2014.16508]

Martínez-Gómez, C, León, VM, Calles, S, Gomáriz-Olcina, M & Vethaak, AD (2017) The adverse effects of virgin microplastics on the fertilization and larval development of sea urchins. *Mar Environ Res*, 130, 69-76.
[http://dx.doi.org/10.1016/j.marenvres.2017.06.016] [PMID: 28716299]

Mato, Y, Isobe, T, Takada, H, Kanehiro, H, Ohtake, C & Kaminuma, T (2001) Plastic resin pellets as a transport medium for toxic chemicals in the marine environment. *Environ Sci Technol*, 35, 318-24.
[http://dx.doi.org/10.1021/es0010498] [PMID: 11347604]

Mizraji, R, Ahrendt, C, Perez-Venegas, D, Vargas, J, Pulgar, J, Aldana, M, Patricio Ojeda, F, Duarte, C & Galbán-Malagón, C (2017) Is the feeding type related with the content of microplastics in intertidal fish gut? *Mar Pollut Bull*, 116, 498-500.
[http://dx.doi.org/10.1016/j.marpolbul.2017.01.008] [PMID: 28063703]

Mizukawa, K, Takada, H, Ito, M, Geok, YB, Hosoda, J, Yamashita, R, Saha, M, Suzuki, S, Miguez, C, Frias, J, Antunes, JC, Sobral, P, Santos, I, Micaelo, C & Ferreira, AM (2013) Monitoring of a wide range of organic micropollutants on the Portuguese coast using plastic resin pellets. *Mar Pollut Bull*, 70, 296-302.
[http://dx.doi.org/10.1016/j.marpolbul.2013.02.008] [PMID: 23499535]

Moore, E, Lyday, S, Roletto, J, Litle, K, Parrish, JK, Nevins, H, Harvey, J, Mortenson, J, Greig, D, Piazza, M, Hermance, A, Lee, D, Adams, D, Allen, S & Kell, S (2009) Entanglements of marine mammals and seabirds in central California and the north-west coast of the United States 2001-2005. *Mar Pollut Bull*, 58, 1045-51.
[http://dx.doi.org/10.1016/j.marpolbul.2009.02.006] [PMID: 19344921]

Nadal, MA, Alomar, C & Deudero, S (2016) High levels of microplastic ingestion by the semipelagic fish bogue *Boops boops* (L.) around the Balearic Islands. *Environ Pollut*, 214, 517-23.
[http://dx.doi.org/10.1016/j.envpol.2016.04.054] [PMID: 27131810]

Ogata, Y, Takada, H, Mizukawa, K, Hirai, H, Iwasa, S, Endo, S, Mato, Y, Saha, M, Okuda, K, Nakashima, A, Murakami, M, Zurcher, N, Booyatumanondo, R, Zakaria, MP, Dung, LQ, Gordon, M, Miguez, C, Suzuki, S, Moore, C, Karapanagioti, HK, Weerts, S, McClurg, T, Burres, E, Smith, W, Van Velkenburg, M, Lang, JS, Lang, RC, Laursen, D, Danner, B, Stewardson, N & Thompson, RC (2009) International Pellet Watch: global monitoring of persistent organic pollutants (POPs) in coastal waters. 1. Initial phase data on PCBs, DDTs, and HCHs. *Mar Pollut Bull*, 58, 1437-46.
[http://dx.doi.org/10.1016/j.marpolbul.2009.06.014] [PMID: 19635625]

Ory, N, Chagnon, C, Felix, F, Fernández, C, Ferreira, JL, Gallardo, C, Ordóñez, OG, Henostroza, A, Laaz, E, Mizraji, R, Mojica, H, Haro, VM, Medina, LQ, Preciado, M, Sobral, P, Urbina, MA & Thiel, M (2018) Low prevalence of microplastic contamination in planktivorous fish species from the southeast Pacific Ocean. *Mar Pollut Bull*, 127, 211-6.
[http://dx.doi.org/10.1016/j.marpolbul.2017.12.016] [PMID: 29475656]

Pazos, RS, Maiztegui, T, Colautti, DC, Paracampo, AH & Gómez, N (2017) Microplastics in gut contents of coastal freshwater fish from Río de la Plata estuary. *Mar Pollut Bull*, 122, 85-90.
[http://dx.doi.org/10.1016/j.marpolbul.2017.06.007] [PMID: 28633946]

Perkins, S (2014) Plastic waste taints the ocean floors. *Nature News.*
[http://dx.doi.org/10.1038/nature.2014.16581]

Peters, CA, Thomas, PA, Rieper, KB & Bratton, SP (2017) Foraging preferences influence microplastic ingestion by six marine fish species from the Texas Gulf Coast. *Mar Pollut Bull*, 124, 82-8.
[http://dx.doi.org/10.1016/j.marpolbul.2017.06.080] [PMID: 28705629]

Pettipas, S, Bernier, M & Walker, TR (2016) A Canadian policy framework to mitigate plastic marine pollution. *Mar Policy*, 68, 117-22.
[http://dx.doi.org/10.1016/j.marpol.2016.02.025]

Pham, CK, Rodríguez, Y, Dauphin, A, Carriço, R, Frias, JPGL, Vandeperre, F, Otero, V, Santos, MR, Martins, HR, Bolten, AB & Bjorndal, KA (2017) Plastic ingestion in oceanic-stage loggerhead sea turtles (*Caretta caretta*) off the North Atlantic subtropical gyre. *Mar Pollut Bull,* 121, 222-9.
[http://dx.doi.org/10.1016/j.marpolbul.2017.06.008] [PMID: 28606614]

Phuong, NN, Poirier, L, Pham, QT, Lagarde, F & Zalouk-Vergnoux, A (2018) *PlasticEurope.*https://dio.org/10.1016/j.marpolbul.2017.10.054

PlasticEurope (2018) https://www.plasticseurope.org/en/resources/market-data/

Rios, LM, Moore, C & Jones, PR (2007) Persistent organic pollutants carried by synthetic polymers in the ocean environment. *Mar Pollut Bull,* 54, 1230-7.
[http://dx.doi.org/10.1016/j.marpolbul.2007.03.022] [PMID: 17532349]

Rios, LM & Jones, PR (2015) Characterisation of microplastics and toxic chemicals extracted from microplastic samples from the North Pacific Gyre. *Environ Chem,* 12, 611-7.
[http://dx.doi.org/10.1071/EN14236]

Rist, SE, Assidqi, K, Zamani, NP, Appel, D, Perschke, M, Huhn, M & Lenz, M (2016) Suspended micro-sized PVC particles impair the performance and decrease survival in the Asian green mussel *Perna viridis. Mar Pollut Bull,* 111, 213-20.
[http://dx.doi.org/10.1016/j.marpolbul.2016.07.006] [PMID: 27491368]

Rochman, CM, Hoh, E, Hentschel, BT & Kaye, S (2013) Long-term field measurement of sorption of organic contaminants to five types of plastic pellets: implications for plastic marine debris. *Environ Sci Technol,* 47, 1646-54. a
[http://dx.doi.org/10.1021/es303700s] [PMID: 23270427]

Rochman, CM, Hoh, E, Kurobe, T & Teh, SJ (2013b) Ingested plastic transfers hazardous chemicals to fish and induces hepatic stress. *Sci Rep,* 3, 3263.
[http://dx.doi.org/10.1038/srep03263] [PMID: 24263561]

Rochman, CM, Kurobe, T, Flores, I & Teh, SJ (2014) Early warning signs of endocrine disruption in adult fish from the ingestion of polyethylene with and without sorbed chemical pollutants from the marine environment. *Sci Total Environ,* 493, 656-61.
[http://dx.doi.org/10.1016/j.scitotenv.2014.06.051] [PMID: 24995635]

Rossi, G, Barnoud, J & Monticelli, L (2014) Polystyrene nanoparticles perturb lipid membranes. *J Phys Chem Lett,* 5, 241-6.
[http://dx.doi.org/10.1021/jz402234c] [PMID: 26276207]

Ruiz-Compean, P, Ellis, J, Cúrdia, J, Payumo, R, Langner, U, Jones, B & Carvalho, S (2017) Baseline evaluation of sediment contamination in the shallow coastal areas of Saudi Arabian Red Sea. *Mar Pollut Bull,* 123, 205-18.
[http://dx.doi.org/10.1016/j.marpolbul.2017.08.059] [PMID: 28916352]

Santana, MF, Ascer, LG, Custódio, MR, Moreira, FT & Turra, A (2016) Microplastic contamination in natural mussel beds from a Brazilian urbanized coastal region: Rapid evaluation through bioassessment. *Mar Pollut Bull,* 106, 183-9.
[http://dx.doi.org/10.1016/j.marpolbul.2016.02.074] [PMID: 26980138]

Schlining, K, Von Thun, S, Kuhnz, L, Schlining, B, Lundsten, L, Stout, NJ, Chaney, L & Connor, J (2013) Debris in the deep: Using a 22-year vido annotation database to survey marine litter in Monterey Canyon, Central California, USA. *Deep Sea Res Part I Oceanogr Res Pap,* 79, 96-105.
[http://dx.doi.org/10.1016/j.dsr.2013.05.006]

Setälä, O, Norkko, J & Lehtiniemi, M (2016) Feeding type affects microplastic ingestion in a coastal invertebrate community. *Mar Pollut Bull,* 102, 95-101.
[http://dx.doi.org/10.1016/j.marpolbul.2015.11.053] [PMID: 26700887]

Sjollema, SB, Redondo-Hasselerharm, P, Leslie, HA, Kraak, MHS & Vethaak, AD (2016) Do plastic particles affect microalgal photosynthesis and growth? *Aquat Toxicol,* 170, 259-61.

[http://dx.doi.org/10.1016/j.aquatox.2015.12.002] [PMID: 26675372]

Sleight, VA, Bakir, A, Thompson, RC & Henry, TB (2017) Assessment of microplastic-sorbed contaminant bioavailability through analysis of biomarker gene expression in larval zebrafish. *Mar Pollut Bull,* 116, 291-7.
[http://dx.doi.org/10.1016/j.marpolbul.2016.12.055] [PMID: 28089550]

Song, YK, Hong, SH, Jang, M, Kang, JH, Kwon, OY, Han, GM & Shim, WJ (2014) Large accumulation of micro-sized synthetic polymer particles in the sea surface microlayer. *Environ Sci Technol,* 48, 9014-21.
[http://dx.doi.org/10.1021/es501757s] [PMID: 25059595]

Spear, LB, Ainley, DG & Ribic, CA (1995) Incidence of plastic in seabirds from the tropical pacific, 1984-1991: Relation with distribution of species, sex, age, season, year and body weight. *Mar Environ Res,* 40, 123-46.
[http://dx.doi.org/10.1016/0141-1136(94)00140-K]

Sportdiver (2016) *What Is a Garbage Patch?.*http://www.sportdiver.com/ great-pacific-garbage-patch-f-cts-map

Suhrhoff, TJ & Scholz-Böttcher, BM (2016) Qualitative impact of salinity, UV radiation and turbulence on leaching of organic plastic additives from four common plastics - A lab experiment. *Mar Pollut Bull,* 102, 84-94.
[http://dx.doi.org/10.1016/j.marpolbul.2015.11.054] [PMID: 26696590]

Sun, CJ, Jiang, FH, Li, JX & Zheng, L (2016) The research progress in source, distribution, ecological and environmental effects of marine microplastics. *Advances in Marine Science,* 34, 449-61.

Sun, XX, Li, QJ, Zhu, ML, Liang, JH, Zheng, S & Zhao, YF (2017) Ingestion of microplastics by natural zooplankton groups in the northern South China Sea. *Mar Pollut Bull,* 115, 217-24.
[http://dx.doi.org/10.1016/j.marpolbul.2016.12.004] [PMID: 27964856]

Sussarellu, R, Suquet, M, Thomas, Y, Lambert, C, Fabioux, C, Pernet, MEJ, Le Goïc, N, Quillien, V, Mingant, C, Epelboin, Y, Corporeau, C, Guyomarch, J, Robbens, J, Paul-Pont, I, Soudant, P & Huvet, A (2016) Oyster reproduction is affected by exposure to polystyrene microplastics. *Proc Natl Acad Sci USA,* 113, 2430-5.
[http://dx.doi.org/10.1073/pnas.1519019113] [PMID: 26831072]

Terepocki, AK, Brush, AT, Kleine, LU, Shugart, GW & Hodum, P (2017) Size and dynamics of microplastic in gastrointestinal tracts of Northern Fulmars (*Fulmarus glacialis*) and Sooty Shearwaters (*Ardenna grisea*). *Mar Pollut Bull,* 116, 143-50.
[http://dx.doi.org/10.1016/j.marpolbul.2016.12.064] [PMID: 28063702]

Teuten, EL, Saquing, JM, Knappe, DRU, Barlaz, MA, Jonsson, S, Björn, A, Rowland, SJ, Thompson, RC, Galloway, TS, Yamashita, R, Ochi, D, Watanuki, Y, Moore, C, Viet, PH, Tana, TS, Prudente, M, Boonyatumanond, R, Zakaria, MP, Akkhavong, K, Ogata, Y, Hirai, H, Iwasa, S, Mizukawa, K, Hagino, Y, Imamura, A, Saha, M & Takada, H (2009) Transport and release of chemicals from plastics to the environment and to wildlife. *Philos Trans R Soc Lond B Biol Sci,* 364, 2027-45.
[http://dx.doi.org/10.1098/rstb.2008.0284] [PMID: 19528054]

UNEP (2014) In the United Nations Environment Programme year book and Valuing Plastic: The Business Case for Measuring, Managing and Plastic Use in the Consumer Goods Industry. *UNEP.*

van Franeker, JA, Blaize, C, Danielsen, J, Fairclough, K, Gollan, J, Guse, N, Hansen, PL, Heubeck, M, Jensen, JK, Le Guillou, G, Olsen, B, Olsen, KO, Pedersen, J, Stienen, EWM & Turner, DM (2011) Monitoring plastic ingestion by the northern fulmar *Fulmarus glacialis* in the North Sea. *Environ Pollut,* 159, 2609-15.
[http://dx.doi.org/10.1016/j.envpol.2011.06.008] [PMID: 21737191]

Vedolin, MC, Teophilo, CYS, Turra, A & Figueira, RCL (2018) Spatial variability in the concentrations of metals in beached microplastics. *Mar Pollut Bull,* 129, 487-93.
[http://dx.doi.org/10.1016/j.marpolbul.2017.10.019] [PMID: 29033167]

Velzeboer, I, Kwadijk, CJAF & Koelmans, AA (2014) Strong sorption of PCBs to nanoplastics, microplastics, carbon nanotubes, and fullerenes. *Environ Sci Technol,* 48, 4869-76.
[http://dx.doi.org/10.1021/es405721v] [PMID: 24689832]

Vélez-Rubio, GM, Teryda, N, Asaroff, PE, Estrades, A, Rodriguez, D & Tomás, J (2018) Differential impact of marine debris ingestion during ontogenetic dietary shift of green turtles in Uruguayan waters. *Mar Pollut Bull,* 127, 603-11.
[http://dx.doi.org/10.1016/j.marpolbul.2017.12.053] [PMID: 29475703]

Viršek, MK, Lovšin, MN, Koren, Š, Kržan, A & Peterlin, M (2017) Microplastics as a vector for the transport of the bacterial fish pathogen species *Aeromonas salmonicida. Mar Pollut Bull,* 125, 301-9.
[http://dx.doi.org/10.1016/j.marpolbul.2017.08.024] [PMID: 28889914]

Von Moos, N, Burkhardt-Holm, P & Köhler, A (2012) Uptake and effects of microplastics on cells and tissue of the blue mussel *Mytilus edulis* L. after an experimental exposure. *Environ Sci Technol,* 46, 11327-35.
[http://dx.doi.org/10.1021/es302332w] [PMID: 22963286]

Ward, JE & Kach, DJ (2009) Marine aggregates facilitate ingestion of nanoparticles by suspension-feeding bivalves. *Mar Environ Res,* 68, 137-42.
[http://dx.doi.org/10.1016/j.marenvres.2009.05.002] [PMID: 19525006]

Wegner, A, Besseling, E, Foekema, EM, Kamermans, P & Koelmans, AA (2012) Effects of nanopolystyrene on the feeding behavior of the blue mussel (*Mytilus edulis L.*). *Environ Toxicol Chem,* 31, 2490-7.
[http://dx.doi.org/10.1002/etc.1984] [PMID: 22893562]

Williamson, P, Smythe-Wright, D, Burkill, P (2016). Future of the Ocean and its Seas: a non-governmental scientific perspective on seven marine research issues of G7 interest.ICSU-IAPSO-IUGG-SCOR, Paris.

Wright, SL, Thompson, RC & Galloway, TS (2013) The physical impacts of microplastics on marine organisms: a review. *Environ Pollut,* 178, 483-92.
[http://dx.doi.org/10.1016/j.envpol.2013.02.031] [PMID: 23545014]

Zarfl, C & Matthies, M (2010) Are marine plastic particles transport vectors for organic pollutants to the Arctic? *Mar Pollut Bull,* 60, 1810-4.
[http://dx.doi.org/10.1016/j.marpolbul.2010.05.026] [PMID: 20579675]

Zhang, C, Chen, XH, Wang, JT & Tan, LA (2017) Toxic effects of microplastic on marine microalgae *Skeletonema costatum*: Interactions between microplastic and algae. *Environ Pollut,* 220, 1282-8.
[http://dx.doi.org/10.1016/j.envpol.2016.11.005] [PMID: 27876228]

Zhao, SY, Zhu, LX, Wang, T & Li, DJ (2014) Suspended microplastics in the surface water of the Yangtze Estuary System, China: First observations on occurrence, distribution. *Mar Pollut Bull,* 86, 562-8.

CHAPTER 10

Toxicity Evaluation in Flora and Fauna Exposed to Marine Pollution

Mazhar Iqbal Zafar[1,*], Mehtabidah Ali[1], Abida Farooqi[1], Riffat Naseem Malik[1], Zahid Iqbal[2] and Shahbaz Ahmad[3]

[1] *Department of Environmental Sciences, Faculty of Biological Sciences, Quaid-i-Azam University, Islamabad 45320, Pakistan*

[2] *Department of Pharmacology, Al-Nafees Medical College & Hospital, Isra University, Islamabad Campus, Islamabad, Pakistan*

[3] *Institute of Agricultural Sciences, University of the Punjab, Lahore 54590, Pakistan*

Abstract: Evaluating the toxicity in flora and fauna due to marine pollutants has attracted immense scientific, regulatory and public attention over the past years. In recent years, types and levels of contaminants in the marine environment have increased as a result of anthropogenic activities worldwide. These chemical substances are accumulated in the tissues of marine organisms and exerting harmful impacts on marine flora and fauna. Published literature on the biological effects of marine pollution revealed that the effects and distribution of marine pollutants have been increased significantly. This chapter focuses on better understanding of the toxicity evaluation of marine biota and has been divided into four main sections: (i) categories of marine pollutants affecting marine flora and fauna (ii) pollutant sources, routes of exposure and toxicological impacts on marine organisms (iii) impacts of pollutants specifically on marine flora (iv) bioassay studies at the organism level discussing marine toxicity.

Keywords: Marine Pollution, Toxicological Impacts, Plastics, Oil Spills, Bioassays, Bioaccumulation, Biomarkers.

INTRODUCTION

According to the United Nations Convention on the Law of the Sea, pollution is defined as the introduction of substances or energy directly or indirectly into the marine environment, which is likely to result in toxic impacts, for instance, harm to living resources and marine life, hindrance to marine activities, including fishing and other legitimate uses of the sea, impairment of quality for use of the

* **Corresponding author Mazhar Iqbal Zafar:** Department of Environmental Sciences, Faculty of Biological Sciences, Quaid-i-Azam University, Islamabad 45320, Pakistan; Tel/Fax: +92-51-90644182; E-mail: mzafar@qau.edu.pk

De-Sheng Pei & Muhammad Junaid (Eds.)
All rights reserved-© 2019 Bentham Science Publishers

sea water. Williams (1996) declared that there is only one type of pollution exists that is marine pollution, because every pollutant, whether in the air or on land ultimately sinks up in the ocean. The main sources of pollutant emissions in marine are human settlements and resource use, such as industrial development, urbanization, agricultural activities, tourism and infrastructural development and construction, Contaminants that pose major threats to marine flora and fauna are anticipated as oil spills and plastic debris (Islam and Tanaka, 2004).

Marine environment is considered as the dynamic and diverse network of habitat for a number of species, consequently many complex physical and ecological processes take place that interact with humans and their activities at many levels (Islam and Tanaka, 2004). Marine habitats with associated communities are classified into diverse ecosystems, for instance, salt marshes, open ocean, coral reefs, deep sea and shores, *etc.* Albeit, they are all linked with each other and the impacts on one ecosystem can affect others. Evaluating the impacts of pollutants in the marine environment, the function and structure of ecosystem are considered as important components. The profit human gains from different ecosystems are known as ecosystem services that include, fish, shellfish and other seafood's we consume. The other ecosystem services include recreational, economic and aesthetic benefits we derive from the sea (Barbier *et al.*, 2011).

The marine planktons of open oceans contribute a lot in the preservation of environment by transferring carbon to the deep sea, thus help in the maintenance of our atmosphere. Apart from this, open oceans and deep seas are also habitat to many fish that are being caught for food. Furthermore, marine planktons are the main source of food for young fish and many other marine species and capture sediments and organic waste that runs off the land (Raven *et al.*, 2005). A few decades back, anthropogenic activities have severely affected marine life, for instance, mining activities *i.e.* copper and gold mining. These all pollutants interact with the life cycles of many marine organisms and severely affect their life cycle (Harley *et al.*, 2006).

The list of flora and fauna exposed to marine pollution and severely affected are outlined in Table **1**.

Table 1. Fauna and Flora of marine biome exposed to marine pollutants.

Fauna		Flora
Green Sea Turtles	Tiger Shark	Dead Man's Fingers
Manatees	Fish	Green Feather
Parrotfish	Sailfish	Halimeda
Hermit crabs	Mahi-mahi	Leafy Flat-Blade

CATEGORIES OF MAJOR MARINE POLLUTANTS

Over the past few decades, marine pollution has become a major concern around the globe (Griffit *et al.*, 2008). Contaminants, such as oil-based products, pesticides, fertilizers, heavy metals, accidental oil spills, and plastic materials, have made survival of marine organisms difficult. Table **2** summarize a range of marine pollutants along with their sources and effects on marine life. Additionally, the percentage of pollutants entering the oceans annually is illustrated in Fig. (**1**).

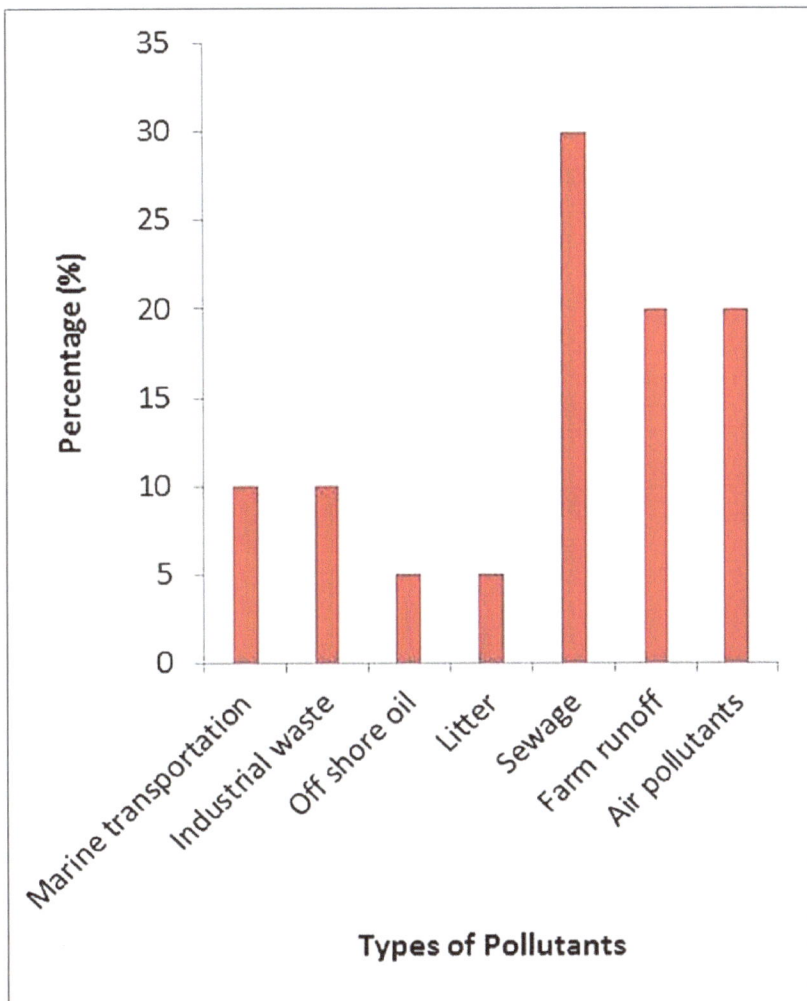

Fig. (1). Percentage of the total number of pollutants entering the oceans annually. This figure is developed by extracting information from the international report of the coastal cleanup (The Ring Leaders Programme, International Coastal Cleanup 2015).

Table 2. Sources of marine pollutants and their effects on fauna and flora.

Type	Sources	Adverse Effects
Plastics	Fishing nets, beach waste, cargo ships, waste from plastics industry and landfills	Marine life trapped in plastic debris or is consumed mistakenly as food, could remain in the marine environment for 200 to 400 years
Oil spills	46% from cars and heavy machinery, 32% from industries, 13% from accidents at sea	Causes larval dearth and other diseases in marine life, Especially near the coast, oil slicks destroy marine life
Nutrients	50% from runoff and sewage, 50% contribution from crops, livestock, and other land use.	Cause micro- and macro algal blooms in marine water, algal blooms decompose algae and this leads to a shortage of oxygen in water, toxins released can eliminate marine life
Water from Sediments	Mining effluents, forestry, farming, and coastal dredging	Turbidity slows down the photosynthesis, It also blocks fish gills, suffocates and engulf coastal ecosystems
Pathogens	Livestock and Sewage	Coastal swimming areas and seafood are contaminated by pathogens. Diseases such as cholera and typhoid are commonly spread by pathogens
Persistent Toxins, such as PCBs	Main sources are wastewater discharge from cities, pesticides used for agricultural purposes and seepage from landfills	Cause diseases in marine life, contaminate food-webs and seafood, fat-soluble toxin accumulation causes reproductive failure in predators

Table 2 is adapted from the following Source: World Watch Institute (http://www.gdrc.org/oceans/marinepollution.html)

MAJOR MARINE POLLUTANTS

Based upon the source, chemistry and effects following are the two major marine pollutants (Halpern *et al.*, 2008).

1. Plastics
2. Oil Spills

Plastics

Plastic particles are found in large quantities in the aquatic environment. However, they are normally observed along beaches, sediments and at the water surface. The durability and longevity, two main characteristics of plastics, allow them to stick with in the aquatic surroundings for longer time. Most plastics are found in the marine environment as micro plastics. The size of micro-plastics is equal to the size of a grain of sand. Although many methods have been used to isolate and identify micro plastics existing in the marine environment, for the identification of micro-plastics, the plastic particles are standardized. Majority of research studies reported in literature shed light on how plastics are taken in by marine organisms. Either, they are ingested directly or with the help of prey (Gregory and Andrady, 2003).

Accumulation of plastics by aquatic organisms is a burning issue because plastics readily accumulate in freshwater environments. The accumulation of plastics has hazardous health impacts in aquatic organisms. Presence of plastics in the aquatic environment is generally classified into two groups (macroplastics and microplastics). Plastic items which have a diameter greater than 5 mm are referred to macro plastics, i.e. disposable cups. The substances which have a diameter less than 5 mm are called as micro plastics such as, micro beads. Furthermore, on the basis of sources microplastics are divided into two types, *i.e.* primary microplastics and secondary microplastics. The manufactured products are primary sources and the breakdown of macroplastics in the environment is secondary sources. Nano-plastics are also considered as a sub-category of microplastics (<100 nm size range). Literature depicts that, the number of micro plastics are increasing day by day as compared to macroplastics. Furthermore, majority of the plastic debris (90%) present in open sea are normally micro plastics (Gregory and Andrady, 2003).

Plastics also have the potential to accumulate the chemical contaminants from the surrounding aquatic environment. These chemicals persist in the oceans for a longer period of times and they resist degradation. The persistence of plastics in the marine surroundings for longer time is directly related to enhanced toxicity. For instance, if the plastic particles remain in the aquatic environment for a longer time than they build up on the particle surface over time and the contaminant becomes more concentrated and toxic (Griffit *et al.*, 2008).

Key Sources of Plastics

Land-based Sources

Land-based activities, such as intentional and accidental disposal, are the primary sources of plastic debris. Plastics, including the littering of bags, bottles and other plastic bits and pieces mainly at large public gatherings, can enter the marine environment through it. Sewage treatment services generally do not need equipment for sufficient screening of wastewater discharges for micro plastic debris, such as plastic fragments in cleaning agents. Therefore, wastewater seepage discharges are one of the main sources to bring plastics to surface waters. A recent study revealed that the wash of single piece of clothing produce more than 1,900 plastic fibers. It is difficult to remove these plastic fibers completely (Sheavly and Register, 2007).

Sea-based Sources

The main sources of plastics in the sea are as follow:

1. Commercial and recreational fishing vessels
2. Recreational boaters
3. Military activities
4. Oil platforms
5. Aquaculture farms

The enormous amount of research indicates about marine fishing vessels which contributed an estimate of 23,000 tons of plastics during the 1970s to an estimated 6.5 million tons during the 1990s (Sheavly and Register, 2007).

Exposure Routes

Various studies are conducted on the contact between marine debris and aquatic life and found that about 693 species are impacted by marine debris. Later, debris has been accumulated in aquatic organisms due to plastic ingestion (Andrady, 2011). According to (Andrady, 2011) the most commonly used plastic items might have the aggressive effects on marine organisms, such as (i) nets and ropes (24%) (ii) Garbage (20%) (iii) Wrapping (17%) i.e. plastic bags and (iv) micro plastics (11%). Research demonstrated that the impacts of plastics entangled are observed to be more severe than ingestion impacts. Typically, the detection of ingestion is less obvious. However, it is reported that since 1997 the number of species ingesting marine plastic debris has been increased and organisms may be exposed to plastics and linked pollutants *via* direct ingestion as well as indirect ingestion and dermal exposure (Andrady, 2011).

Bioaccumulation of Plastics in Marine Biota

Several aquatic organisms ingesting plastics can act as a source and sink for chemicals. In addition to that, they also act as a vector for chemicals. If chemical exposure level increases than chemicals cross the threshold limit in the organism's body, as a result, toxicity occurs in marine flora and fauna (Todd *et al.*, 2010).

Numerous studies provide evidence from the field that ingested plastic densities positively correlates with chemical concentrations. The greater the ingestion of plastics, greater is the concentration of chemicals in organisms. For instance, the presence of chemicals, such as organochlorines in basking sharks and fin whales, is the evidence of micro plastic ingestion (Todd *et al.*, 2010). Many studies on marine birds have been reported, *i.e.* dead great shearwaters (*Puffinus gravis*) in Australia contain polychlorinated biphenyls (PCBs) in their abdominal adipose tissues (Todd *et al.*, 2010). The presence of PCBs was positively correlated with plastic loads of ingested particles. In addition to this, plastic intake in the North Pacific Ocean was investigated in 12 short-tailed shearwaters (*Puffinus tenuirostris*) (Todd *et al.*, 2010).

Toxicological Impacts of Plastics related Chemicals on Marine Biota

Plastic debris enters the organisms *via* different entry routes, such as ingestion, and entanglement. Ingestion of plastic is less observable than the entanglement of plastics. Ingestion of plastics leads to direct death or indirect death of flora and fauna due to poor nutrition or dehydration. However, the exposure pathways between the chemicals sorption on plastics and the organisms ingesting plastics are generated by ingestion of plastics (Wright *et al.*, 2013). Plastics ingested by marine flora and fauna cause severe toxicity. For instance, higher concentrations of polystyrene microplastics within sediment are ingested by *A. marina*. The ingestion causes weight loss in *A. marina* (Wright *et al.*, 2013).

Besides, microplastics which are mixed with fluoranthene do not alter the bioaccumulation of fluoranthene in marine mussels, but micro plastic exposure increases hemolytic mortality in marine mussels. These microplastics also bring changes in oxidative and energetic processes of marine mussels. The combined impacts of both microplastics and fluoranthene can be more severe and led to the maximum tissue alteration and levels of anti-oxidant markers. Furthermore, in order to evaluate the toxicological impacts, the fish exposed to PCBs and PBDEs has been used. Research carried on female fish gene expression fed with the virgin plastics suggested that the changed gene expression in fish was due to the chemicals within the plastics they were fed with (Wright *et al.*, 2013). They induced endocrine-disrupting effects in fish. Similarly, after ingestion of chemicals in combination with plastics, the physiological signs of stress have been observed in test organisms *i.e.* growth, survival, and reproduction of tested marine organisms were affected. Plastics alone induce adverse effects, but the effects can be greater when marine flora and fauna are exposed to both the plastics and the chemicals sorbed with plastics (Wright *et al.*, 2013). The proportions of plastics among marine debris are shown in Table **3**.

Table 3. The worldwide ratio of plastics among the marine debris.

Location	Debris Type	Percentage (%)	Source
St. Lucia, Caribbean	Beach	51	(Corbin & Singh, 1993)
Bay of Biscay	Atlantic Seabed	92	(Galgani *et al.*, 1995)
North West Mediterranean	Seabed	77	(Galgani *et al.*, 1995)
North Pacific Ocean	Surface waters	86	(Laist, 1987)

(Table 3) cont.....

Location	Debris Type	Percentage (%)	Source
Mediterranean	Sea surface	60–70	(Morris, 1980)
Halifax Harbor	Beach	54	(Derraik, 2002)
Southern Ocean	Beach	71	(Slip & Burton, 1991)
Gulfs in W. Greece	Seabed	79–83	Stefatos *et al.*, 1999)

Oil Spills

Another main pollutant causing toxicity in marine flora and fauna is reflected as oil spills. The most destructive effect of ship pollution appears in the form of oil spills. Crude oil is extremely toxic to marine life. It suffocates marine animals to death once it entraps them. Crude oil lasts for years in the sea. It is very difficult to clean it, especially when it is split in the water. Once it is split it causes toxicity in marine flora and fauna.

Whenever an oil spill enters the marine environment, it undergoes different weathering processes, such as (i) evaporation (ii) dispersion (iii) dissolution, (iv) emulsification and (v) sinking. Due to theses weathering processes, oil spills change its character and redistribute into other parts of the environment. The impact and fate of oil in marine environment depend upon environmental conditions, *i.e.* the area where the spill occurs and the physical and chemical properties of the oil (Board, 2003).

Processes of Oil Weathering

Evaporation

Most of the fresh oil contains high proportion hydrocarbons having a low molecular weight and low boiling point, such as alkanes (<12 carbon atoms), benzene, toluene, ethylbenzene, and xylene (BTEX) compounds. Evaporation begins immediately in the atmosphere when these hydrocarbons are released in the marine environment or in sea shore due to air movement and ambient temperature. The process of evaporation gradually increases the viscosity of spilled oil in the marine environment. Evaporation is a beneficial weathering process because it lessens the extent and acute toxicity of hydrocarbons, *i.e.* if the oil leftovers at the surface of marine environment for maximum period *i.e.* hours or days, then sticky residues are left, which are relatively low toxic. The amount of oil remaining in the environment varies. For instance, if 10 tons of gasoline will be spilled into a tropical sea on a calm summer day (25°C) then it would dissolve completely in a time period of less than three hours. While, if the same weight of gasoline will be spilled in Arctic sea on a calm winter (5°C) then it would take six hours to evaporate (Helm *et al.*, 2015).

Dissolution

Most of the hydrocarbons in the marine environment are defined as insoluble because they have low solubility in seawater, while, few smaller aromatic hydrocarbons are relatively soluble, such as, benzene and toluene. Thus, when oil is thrown into the sea, some of the relatively soluble hydrocarbons are dissolved. The pace of dissolution depends upon two factors, for instance, oil composition and the viscosity of oil. Top-heavy impacts on marine organisms are mainly the result of water-soluble fraction. The reasons behind their greater impacts are due to more bioavailability fraction than the other hydrocarbons and are, therefore, more acutely toxic (Faksness, 2007).

Dispersion

The oil droplets in water are formed due to stirring of the oil in water. They mix into the water as a result of agitation mainly. The greater will be the shakeup greater will be the mixing of oil with water. The maximum oil is eventually dispersed, either they are released in subsea surface, deposited onto the shore line or spilled onto the sea surface. The larger oil droplets resurface because they mix into the water column quickly. While, the smaller oil droplets do not resurface because they are less floating. The smaller droplets mix in the column vertically and horizontally.

This process of dispersion exposes the subsurface marine life to pollution (Helm *et al.*, 2015). However, when hydrocarbons are dissolved, the concentration of dispersed oil increases in the immediate area of the release, either it is a surface slick or subsurface rising column. In the case of surface slicks, the vertical mixing is slower in deep water in comparison with lateral mixing. Finally, the oil droplets undergo biodegradation by microbes because they have a large surface area. The large surface area facilitates the microbes to biodegrade the dispersed oil droplets (Faksness, 2007).

Emulsification

Larger droplets of oil quickly mix with sea water and traps within the surface slick than water in oil emulsion forms (Helm *et al.*, 2015). The volume of the emulsion increases when water incorporates into it, *i.e.* when water incorporates into the emulsion the volume of the emulsion increases 5 times (Faksness, 2007).

The emulsions show different physical characteristics in comparison with their parent oil. Such as, they may be stable or unstable. The emulsions having high water content, *i.e.* greater than 70% are called as stable emulsions. Stable emulsions are highly viscous. They are reddish brown in color and they are also known as 'chocolate mousse'. Furthermore, they can remain stable for several weeks. With the formation of stable mousse, the rate of dispersion reduces because in warm conditions a mousse breaks down into water and constituents of oil respectively. For instance, after landing on a beach the emulsion breaks down into its constituent oil and water. However, some emulsions are extremely persistent, *i.e.* the decomposition of an unstable emulsion takes place in several days (Faksness, 2007).

Sinking

Mostly the process of sinking and sedimentation are discussed combine but from an ecological perspective, both of them are very different because the process of sinking is not responsible for producing flocks of oil. The process of sinking occurs in case the spilled oil is denser than seawater. Sinking lead to persistent accumulations on the seabed and sometimes become buried (Faksness, 2007).

Factors Influencing Oil Impacts

Effects of oil spills largely count on the conditions of marine environment. For instance, spill volume is one of the factors for determining the impacts, but not that much important. The main factors or conditions led to the impacts of spills on marine life are (i) seasonality and (ii) life style factors (Neff, 2002).

Seasonality

Most species show seasonal stages in their behavior and biology. For instance, some of them are following as migrating, breeding, and spawning, particularly in temperate and Polar Regions (Neff, 2002).

Life Style Factors

There are number of biological traits considered for the recovery of species quickly from oil spill. Few of them are longevity or lifespan, reproductive strategy and capacity, particularly numbers of offspring, mobility or dispersal potential, growth rate, feeding method and geographic distribution (Neff, 2002).

Health and Condition

Cumulative impact is defined as, "The Individual organisms, populations, communities, and ecosystems that are already stressed from another cause may be impacted by an oil spill as well". For instance, if migrating birds have not recovered from harsh weather during their journey and at the same time they are exposed to oil spills, then the migrating seabirds will be more sensitive to the effects of oiling. Multiple oil spills in the same location results in long term impacts (Neff, 2002). In relatively smaller areas long term impacts of oil spills are expected. The primary cause of long term impacts is persistent oil. The persistent oil spills are usually observed in the form of heavy residues or incorporated into muddy sediments. They are also found in habitats that are sheltered from water movement.

Impacts of Oil Spills on Marine Life

Many habitats, especially marine habitat is impacted by Spilled oil. The spilled oil is distributed in habitats in different forms. Persistent heavy deposits on sea bed and shoreline lead to loss of natural habitat. The mechanism of how oil spills impact organisms depends upon where and how organism lives. For instance, the animals and plants living or spending time on the surface of the sea or shorelines are likely to have greater impacts because of physical smothering (Saadoun, 2015). The dissolved hydrocarbons can be absorbed through gills or other exposed tissues in the water column, while, filter feeding animals capture and swallowed the dispersed droplets of oil.

When hydrocarbons are ingested or inhaled than they have impacts on organism's internal tissues. The damage and disruption of cell walls and cellular functions in marine flora and fauna is caused by the chemical toxicity of hydrocarbons. In some cases, if the dose and exposure period of hydrocarbons increases than the organisms die (Saadoun, 2015).

Recovery from Oil Spills

The natural recovery of the impacts caused by oil spills takes place, whatever the magnitude or scale of impact is caused (Bannat, 2002). Recognizing the early stages of recovery is normally useful because the latter stages are often difficult to describe precisely. Literature revealed that logic behind the recovery simply depends upon the biomass levels for each species. However, it is recognized that biological resources and many environmental factors that characterize biological habitats are in a continuous and largely unpredictable state of flux. For this reason, a damaged resource cannot necessarily be expected to go back to exactly what it was before the spill, equally, it is not possible to predict exactly what the resource in question would have been like if it had not been damaged by the spill (Bannat, 2002). Marine organisms are mostly sensitive to the toxic effects of hydrocarbons such as, oil droplets and water soluble fractions.

Most studies of natural plankton communities in the sea have found a rapid return to normal densities and community composition once the oil in water concentrations has returned to background levels. The ability of species for quick recovery is due to following factors, *i.e.* (i) short generation times (ii) production of large numbers of eggs and juveniles (iii) distribution over large areas and (iv) rapid water exchange.

Few studies have described the effects on densities of planktonic species lasting more than a few days or weeks. For example, studies following the groundwork of the Soviet tanker *Tsesis* in 1977, during which 1,000 tons of medium grade fuel oil released into the Baltic Sea showed that zooplankton biomass declined substantially close to the wreck during the first few days after the spill but was re-established within five days (Bannat, 2002). Oil contamination of zooplankton population was reported for more than three weeks and as a result short-term increase in phytoplankton biomass and primary production in the impacted area was due to decreased zooplankton grazing rates.

IMPACTS OF POLLUTANTS ON MARINE FLORA

Sea grass grows in the sediment is well-known for its ability to trap and bind with pollutants, thus the pollutants have adverse impacts on its growth (Gacia *et al.*, 2003) and (Cabac *et al.*, 2008). It is rational to admit that most benthic organisms undergoing a larval settlement phase and photosynthetic micro aquatic plants suffer a lot due to water turbidity (Kirk, 1977; Falkowski *et al.*, 1990).

Host plants are not only affected by turbidity, but it also has adverse impacts on their related epiphytes, such as microphytobenthos and microalgae (Erftemeijer and Lewis, 2006). It results in reduced biomass, nutrients, chlorophyll a content, and other photosynthetic growth parameters (Coles and McKenzie, 2004).

Research revealed that among the seagrass species, the Indopacific species (*Halophila ovalis*) are the most sensitive species to reduced light (Coles and McKenzie, 2004).

The concept of eutrophication can be defined as "The process by which water becomes enriched in dissolved nutrients (such as phosphates) and the dissolved nutrients stimulate the growth of aquatic plant life". Eutrophication always leads to depletion of dissolved oxygen. The depletion of dissolved oxygen leads to the death of phytoplankton. It may occur naturally, but most of the time it occurs due to run-off from the land, which causes the dense growth of plant life (Jorgensen and Richardson, 1996). In recent times, numerous kinds of pollutants, such as (*i.e.* sewage, detergents, and agricultural runoff) after entering the ocean, lead to eutrophication process (Kenchington, 1985) and (Edinger *et al.*, 1998).

Following a phytoplankton bloom, the decomposition of massive amounts of algae can deplete oxygen. The depletion of dissolved oxygen in water leads to anoxia and hypoxia in marine organisms. Consequently, it also causes mortality and reduced growth rates of marine flora (Breitburg, 2002).

A brief description of pollutants and their effects on marine flora are summarized in Table **4** (Todd *et al.*, 2010). Furthermore, Fig. (**2**) illustrates the percentage of pollutants which has been entangled by marine organisms in the last 25 years.

Table 4. Summary of pollutants, their sources/causes, and effects on marine flora.

Type of Pollutants	Sources	Effects on Flora
Terrigenous	Agricultural runoff, construction sites	Reduces light and thus reduces photosynthesis in seagrasses
Nitrogen and phosphorus untreated human and animal waste	Industrial discharge Jellyfish blooms	Phytoplankton blooms leading to hypoxia, Macro algae blooms that can out-compete seagrasses
Waste, litter, and plastics	Marine dredging and disposal of waste, Turbidity	Reduced biomass, nutrients, Chl-a, other photosynthetic growth parameters

This table is taken after a slight modification from the study of Todd *et al.* (2010).

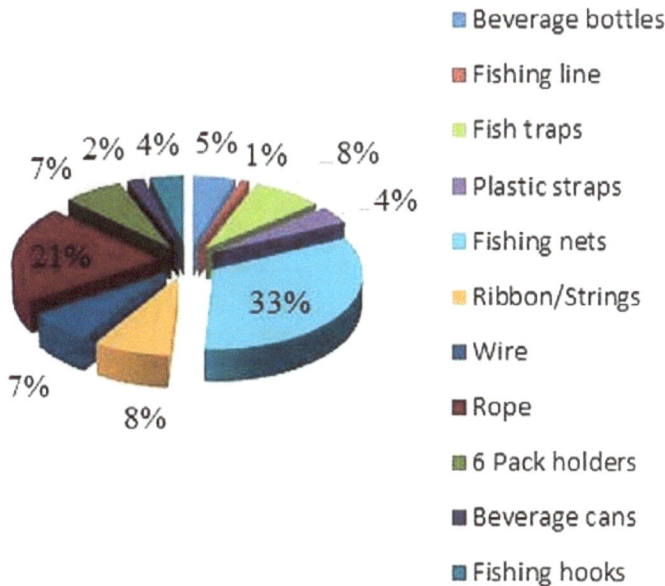

Fig. (2). Entrapment of pollutants by marine organisms for last 25 years. This figure is shaped following the information from the report of international coastal cleanup. The Ring Leaders Programme, International Coastal Cleanup (2015).

BIOMARKERS AS WARNING SIGNALS FOR TOXICITY EVALUATION

According to McCarthy and Shugart Sures (2001), biomarkers, such as body fluids, cells, and tissues, indicate the presence of contaminants (exposure biomarkers) or the extent of the host response (effect biomarkers) in organisms. Biomarkers respond in a very short period of time, for which they can act as initial warning indicators of biological effects for predicting the toxicity of environmental contaminants. CYP1A1, a specific form of biomarker for organic aromatic xenobiotics exposure, can be estimated by the measurement of ethoxyresorufin-O-deethylase (EROD) activity in fish liver (Sures, 2001). Meanwhile, the ability to metabolize organic pollutants in some marine invertebrates, especially mussels are reduced. This is partly because of the absence of efficient isoforms of P450 that are generally found among terrestrial organisms. Therefore, this biotransformation system is absent in one of the important species (mussels), suggesting that EROD stimulation in pollution biomonitoring programs using mussels as sentinels may not be effective. However, the latest research suggests the assessment of peroxisome proliferation and Multi Xenobiotics Resistance (MXR) can be an effective biomarker test for organic xenobiotics compounds in mussels (Sures, 2001).

Fishes as Bio-Indicator in Marine Life

On account of water as their habitat, fishes are particularly vulnerable and heavily exposed to pollution because they cannot run away from the harmful effects of pollutants. As compared to other invertebrates, fishes are more sensitive to many toxicants. Thus, they are used for bio-indicators to test the indication of toxicity in marine flora and fauna. Most of the trace elements are emitted from coal combustion (Chovanec, 2003).

Role of Bioassay Studies to Assess Marine Pollution

One of the important characteristic of pollutants in the marine environment is their ability to concentrate in higher magnitude in the bottom sediments compared to surface water (Ingersoll, 1995). Hence, the study of bottom sediments in coastal systems becomes important environmental studies. Previously, such characterization was majorly concerned with physio-chemical analysis (Ingersoll 1995). However, chemical analyses are not sufficient to determine the biological effect of pollutants on organisms; therefore, they cannot be used to determine their effect on ecosystems (Connor and Paul, 2000) and (Chapman *et al.*, 2003). Thus, in order to measure the potential effects of pollutants on organisms and biological communities, toxicity tests on marine systems can serve as an important supplementary tool along with the physicochemical analyses (Ingersoll,

1995). Bioassays can be used to estimate the effects caused by pollution on the marine environment (Beasley, 1993).

By measuring the toxicity in natural samples, they are used to assess the environmental quality and predict the ecological risk of contamination. Such type tests have several benefits (i) firstly, the organisms under study responds only to the bioavailable fraction of a pollutant (ii) secondly, since chemical analyses can sense only previously known compounds, bioassays can be useful in identifying new toxic elements and (iii) they can also be used to identify contaminated sites from non-contaminated sites using quantitative information on sediment toxicity. Since these tests typically use resident species, the results become also relevant ecologically. In addition, bioassay tests are carried out to compare the sensitivities of different organisms. The purpose of conducting such toxicity tests under laboratory conditions is to establish relationship between pollutants exposure and the effects caused on individual organisms. Dose-response relationships can then be further estimated for determining the relationship between the concentration or dose of the toxin and the toxic effect on an organism. Toxicity tests are considered an ideal method to assess marine pollution (Connor and Paul, 2000). They support the results of the earlier physicochemical measures by providing a direct measure of toxic adverse effects (Saco-Álvarez, 2008), (Beiras and Saco-Álvarez, 2006) and (Stronkhorst *et al.*, 2003). These bioassays are performed under highly defined, controlled and reproducible conditions. The bioassay tests are carried to evaluate or measure the variables, such as acute toxicity, long-term toxicity, bioaccumulation, endocrinal toxicity, and reproductive toxicities. The carcinogenic and mutagenic properties are also determined through bioassay tests. Marine organism used in bioassays can belong from various trophic levels (bacteria, algae, mollusks, echinoderms, annelids, fishes, *etc.*) and can be in different development phases. This enables us to investigate a wide range of biological possibilities (McPherson and Chapman, 2000) and (Nendza, 2002).

The choice of organism for bioassay test depends upon following criteria (i) their sensitivity (ii) their ecological, commercial or recreational significance (iii) their high availability and profusion (iv) their ease of culture or preservation in laboratory (v) and the simplicity of the analysis of results (Kobayashi and Okamura, 2002). A bioassay test evaluates the response of a marine organism when exposed to a toxic substance and relates it to the toxic concentration in water/sediment and the time of exposure to the pollutant. The time of exposure is kept constant in the dose-response relationship and the biological responses of individuals are determined over a series of different concentrations (Kobayashi and Okamura, 2002).

CONCLUDING REMARKS

The determination of the real toxicity of flora and fauna in marine ecosystems is challenging, so it is very important to evaluate the appropriate risk assessment. Food, bottles, nets, resin pellets, cigarette filters, and other debris items have severe destructive influences on marine flora and fauna. Successful management of toxicity in flora and fauna requires the comprehensive understanding of both marine pollutants and human behavior. Knowledge plays a key role in making appropriate choices for the costumer, when it comes to using and disposing of waste items. Proper implementation of policies, education and outreach programs can be the building blocks as initiatives for successful marine pollution prevention. The industrial sector could have a prominent role in educating its workers, and searching for technological mitigation strategies.

CONSENT FOR PUBLICATION

Not applicable.

CONFLICT OF INTEREST

The authors confirm that this chapter contents have no conflict of interest.

ACKNOWLEDGEMENT

Declared none.

REFERENCES

Andrady, AL (2011) Microplastics in the marine environment. *Mar Pollut Bull,* 62, 1596-605.
[PMID: 21742351]

Ayoub, AT (1999) Fertilizers and the environment. *Nutr Cycl Agroecosyst,* 55, 117-21.

Bannat, IM (2002) Biosurfactants production and possible use in microbial pollution remediation: A review. *Bio Resource Technology, Brebbia, CA, Oil and hydrocarbon spills iii: modeling, analysis and control,* 5, 1-12.

Barbier, EB, Hacker, SD & Kennedy, C (2011) The value of estuarine and coastal ecosystem services. *Ecological Society of America,* 81, 169-93.
[http://dx.doi.org/10.1890/10-1510.1]

Beasley, V (1993) Ecotoxicology and ecosystem health: roles for veterinarians; goals of the Envirovet program. *J Am Vet Med Assoc,* 203, 617-28.
[PMID: 8407524]

Beiras, R & Saco-Álvarez, L (2006) Toxicity of seawater and sand affected by the Prestige fuel-oil spill using bivalve and sea urchin embryogenesis bioassays. *Water Air Soil Pollut,* 177, 457-66.

Beveridge, MCM, Ross, LG & Kelly, LA (1994) Aquaculture and biodiversity. *A Journal of the Human Environment,* 23, 497-502.

Board, M & Board, OS (2003) Oil in the sea iii: Inputs, fates and effects. *National Research Council.*

Boyd, CE & Clay, JW (1998) *Shrimp aquaculture and the environment: an adviser to shrimp producers and an environmentalist present a prescription for raising shrimp responsibly.* Scientific American Magazine, New York 59-65.

Breitburg, D (2002) Effects of hypoxia and the balance between hypoxia and enrichment, on coastal fishes and fisheries. *Estuaries,* 25, 767-81.

Cabac,, OS, Santos, R & Duarte, CM (2008) The impact of sediment burial and erosion on seagrasses: a review. *Estuar Coast Shelf Sci,* 79, 354-66.

Cardoso, PG, Pardal, MA & Lillebo, AI (2004) Dynamic changes in seagrass assemblages under eutrophication and implications for recovery. *J Exp Mar Biol Ecol,* 302, 233-48.

Chapman, PM, Wang, F, Janssen, CR & Goulet, RR (2003) Conducting ecological risk assessments of inorganic metals and metalloids: current status. *Hum Ecol Risk Assess,* 9, 641-97.

Chou, LM (1996) Response of Singapore reefs to land reclamation. *Galaxea,* 85-92.

Chou, LM (2009) *The East Asian seas state of the marine environment report.* United Nations Environment Programm, Bangkok.

Chovanec, A, Hofer, R & Schiemer, F (2003) Fish as bioindicators. *Trace metals and other contaminants in the environment,* 6, 639-76.

Coles, R & McKenzie, L (2004) Trigger points and achieving targets for managers. *Seagrass Conference Townsville.* 2004

Corbin, CJ & Singh, JG (1993) Marine debris contamination of beaches in St-Lucia and Dominica. *Mar Pollut Bull,* 26, 325-8.

Corrales, RA & Maclean, JL (1995) Impacts of harmful algae on seafarming in the Asia-pacific areas. *J Appl Phycol,* 7, 151-62.

Derraik, JGB (2002) The pollution of the marine environment by plastic debris: a review. *Mar Pollut Bull,* 44, 842-52.
[PMID: 12405208]

Duarte, CM (1995) Submerged aquatic vegetation in relation to different nutrient regimes. *Ophelia,* 41, 87-112.

Edinger, EN, Jompa, J, Limmon, GV & Widjatmoko, W (1998) Reef degradation and coral biodiversity in Indonesia: effects of land-based pollution, destructive fishing practices and changes over time. *Mar Pollut Bull,* 36, 617-30.

Erftemeijer, PLA & Lewis, RRR, III (2006) Environmental impacts of dredging on seagrasses: a review. *Mar Pollut Bull,* 52, 1553-72.
[PMID: 17078974]

Faksness, LG (2007) Weathering processes in Arctic oil spills: Meso scale experiment with different ice conditions. *Cold Reg Sci Technol,* 55, 160-6.

Falkowski, PG, Jokiel, PL & Kinzie, RA (1990) *Irradiance and corals reefs* Elsevier, Amsterdam 89-107.

Gacia, E, Duarte, CM & Marba, N (2003) Sediment deposition and production in SE-Asia seagrass meadows. *Estuar Coast Mar Sci,* 909-19.

Galgani, F, Burgeot, G & Vincent, F (1995) Distribution and abundance of debris in the continental shelf of the bay of Biscay and in Seine bay. *Mar Pollut Bull,* 30, 58-62.

Gomez, ED (1990) State of the marine environment in the East Asian Seas Region. *United Nations Environment Programme Regional Seas Reports and Studies.*

Gregory, MR & Andrady, AL (2003) Plastics in the marine environment. In: Andrady, A.L., (Ed.), *Plastics and the environment, Marine Pollution Bulletin, Wiley online library* 379.

Griffitt, RJ, Luo, J, Gao, J, Bonzongo, JC & Barber, DS (2008) Effects of particle composition and species on toxicity of metallic nanomaterials in aquatic organisms. *Environ Toxicol Chem,* 27, 1972-8.
[PMID: 18690762]

Halpern, BS, Walbridge, S, Selkoe, KA, Kappel, CV, Micheli, F, D'Agrosa, C, Bruno, JF, Casey, KS, Ebert, C, Fox, HE, Fujita, R, Heinemann, D, Lenihan, HS, Madin, EM, Perry, MT, Selig, ER, Spalding, M, Steneck, R & Watson, R (2008) A global map of human impact on marine ecosystems. *Science,* 319, 948-52.
[http://dx.doi.org/10.1126/science.1149345] [PMID: 18276889]

Harley, CDG, Randall Hughes, A, Hultgren, KM, Miner, BG, Sorte, CJ, Thornber, CS, Rodriguez, LF, Tomanek, L & Williams, SL (2006) The impacts of climate change in coastal marine systems. *Ecol Lett,* 9, 228-41.
[http://dx.doi.org/10.1111/j.1461-0248.2005.00871.x] [PMID: 16958887]

Helm, RC, Costa, DP & Debruyn, TD (2015) Overview of effects of oil spills on marine mammals. Fingas, M (2015) *Hand book of oil spill science and technology* 455-71.

Ingersoll, CG (1995) Sediment tests.*Fundamentals of Aquatic Toxicology Effects, Environmental Fate and Risk Assessment* 231-55.

Islam, MS & Tanaka, M (2004) Impacts of pollution on coastal and marine ecosystems including coastal and marine fisheries and approach or management: a review and synthesis. *Marine Pollution Bulliten,* 48, 624-49.

Jorgensen, BB & Richardson, K (1996) *Eutrophication in coastal marine ecosystem.* American Geophysical Union, Washington, DC.

Kawaguti, S & Sakumoto, D (1948) The effect of light on the calcium deposition of corals. *Bull Oceanogr Inst Taiwan,* 4, 65-70.

Kenchington, R (1985) Coral reef ecosystems: a sustainable resource. *Nat Resour,* 2, 18-27.

Kirk, JTO (1977) Attenuation of light in natural waters. *Aust J Mar Freshwater Res,* 28, 497-508.

Kobayashi, N & Okamura, H (2002) Effects of new antifouling compounds on the development of sea urchin. *Mar Pollut Bull,* 44, 748-51.
[PMID: 12269477]

Laist, DW (1987) Overview of the biological effects of lost and discarded plastic debris in the marine environment. *Mar Pollut Bull,* 18, 319-26.

McPherson, C & Chapman, PM (2000) Copper effects on potential sediment test organisms: the importance of appropriate sensitivity. *Mar Pollut Bull,* 40, 656-65.

Morris, RJ (1980)) Floating plastic debris in the Mediterranean. *Marine Pollution Bulletin,* 11, 125.

Neff, JM (2002) Effects of contaminants from oil well produced water. *Bioaccumulation in Marine Organisms.*

Nendza, M (2002) Inventory of marine biotest methods for the evaluation of dredged material and sediments. *Chemosphere,* 48, 865-83.
[PMID: 12222781]

O'Connor, TP & Paul, JF (2000) Misfit between sediment toxicity and chemistry. *Mar Pollut Bull,* 40, 59-64.

Paerl, HW, Dennis, RL & Whitall, DR (2002) Atmospheric deposition of nitrogen: implications for nutrient over-enrichment of coastal waters. *Estuaries,* 25, 677-93.

Raven, J, Caldeira, K & Elderfield, H (2005) Ocean acidification due to increasing atmospheric carbon dioxide Contents. *The Royal Society,* 68.

Saadoun, IMK (2015) Impacts of oil spills on marine life. Larramendy, ML & Soloneski, S (2015) *Emerging pollutants in the environment- current and further implications* 38.

Saco, Á (2009) Advances en la estandarización del bioassay de la embryogenesis del erizo de mar (P. lividus) Para la evaluation de la contaminación marina. PhD Thesis University of Vigo, Spain .

San Diego-McGlone, ML, Azanza, RV, Villanoy, CL & Jacinto, GS (2008) Eutrophic waters, algal bloom and fish kill in fish farming areas in Bolinao, Pangasinan, Philippines. *Mar Pollut Bull,* 57, 295-301. [PMID: 18456288]

Sheavly, SB & Register, KM (2007) Marine debris and plastics: Environmental concerns, sources, impacts and solutions. *Polymer and Environment,* 15, 301-5.

Slip, DJ & Burton, HR (1991) Accumulation of fishing debris, plastic litter, and other artefacts, on Heard and Macquarie islands in the Southern ocean. *Environ Conserv,* 18, 249-54.

Stefatos, A, Charalampakis, M & Papatheodorou, G (1999) Marine debris on the sea floor of the Mediterranean Sea: Examples from two enclosed Gulfs in Western Greece. *Mar Pollut Bull,* 38, 389-93.

Stronkhorst, J, Schipper, C, Brils, J, Dubbeldam, M, Postma, J & van de Hoeven, N (2003) Using marine bioassays to classify the toxicity of Dutch harbor sediments. *Environ Toxicol Chem,* 22, 1535-47. [PMID: 12836979]

Sures, B (2001) The use of fish parasites as bioindicators of heavy metals in aquatic eco systems: A review. *Aquat Ecol,* 35, 245-55.

The Ring Leaders Programme (2015) *International Coastal Clean up.* Ocean Conservancy.

Todd, PA, Ong, X & Chou, LM (2010) Impacts of pollution on marine life in South East Asia. *Biodivers Conserv,* 19, 1063-82.

Todd, PA, Ong, X & Chou, LM (2010) Impacts of pollution on marine life in South East Asia. *Biodivers Conserv,* 19, 1063-82.

United Nations Environment Programme/Global Programme of action (2006) *The state of the marine environment: Trends and processes.*

Williams, C (1996) Combatting marine pollution from land-based activities: Australian initiatives. *Ocean Coast Manage,* 33, 1-3, 87-112. [http://dx.doi.org/10.1016/S0964-5691(96)00046-4]

Wright, SL, Thompson, RC & Galloway, TS (2013) The physical impacts of microplastics on marine organisms: a review. *Environ Pollut,* 178, 483-92. [PMID: 23545014]

Yap, LG, Azanza, RV & Talaue-McManus, L (2004) The community composition and production of phytoplankton in fish pens of Cape Bolinao, Pangasinan: a field study. *Mar Pollut Bull,* 49, 819-32. [PMID: 15530526]

CHAPTER 11

Marine Medaka (*Oryzias melastigma*) as a Model System to Study Marine Toxicology

Yan-Ling Chen[1], Naima Hamid[1,2], Muhammad Junaid[1,2] and De-Sheng Pei[1,2,*]

[1] *Research Center for Environment and Health, Chongqing Institute of Green and Intelligent Technology, Chinese Academy of Sciences, Chongqing 400714, China*

[2] *University of Chinese Academy of Sciences, Beijing 100049, China*

Abstract: Marine medaka (*Oryzias melastigma*) has been recognized as an ideal marine model fish widely used in the estuary and marine toxicological studies because of multiple favorable attributes, such as small size, short generation cycle, transparent embryos, sexual dimorphism, ease of maintenance, and wide range of salinity and temperature adaptations. Many studies have been conducted on both wild-type and transgenic fish *O. melastigma* model to evaluate the adverse effects by selecting specific biomarkers of the estuary and marine environmental pollutants. This review provides a recent research progress of the physiological effects and responsive biomarker of *O. melastigma* caused by various marine pollutants, including heavy metals, endocrine disruptors, and organic pollutants. Of note, this chapter summarizes the progress on whole-genome sequencing of *O. melastigma*, and promotes novel insights into the use of *O. melastigma* in future toxicity screening studies, targeting genetic biomarkers that highly activated by marine chemical pollutants using cutting-edge gene editing technique and bioinformatics system.

Keywords: Endocrine Disrupting Cmpounds (EDCs), Environmental Xenobiotics, Heavy Metals, Organic Pollutants, Transgenic Fish, Toxicology.

INTRODUCTION

In the past few decades, due to the rapidly increasing pollution in the marine ecosystem, several aquatic organisms have been chosen as suitable environmentally relevant models for ecotoxicity research. Zebrafish (Danio rerio) and Japanese medaka (*Oryzias latipes*) have been commonly used as fish models for eco-toxicological studies in the freshwater environment (Dodd, Curtis, Williams, & Love, 2000; Wittbrodt, Shima, & Schartl, 2002). However, a scarcity is existing related to the fish models that can be used for marine toxicological ass-

---

[*] **Corresponding author De-Sheng Pei:** Research Center for Environment and Health, Chongqing Institute of Green and Intelligent Technology, Chinese Academy of Sciences, Chongqing 400714, China; Tel/Fax: +86-23-65935812; E-mail: peids@cigit.ac.cn

De-Sheng Pei & Muhammad Junaid (Eds.)
All rights reserved-© 2019 Bentham Science Publishers

essment, and marine medaka (*Oryzias melastigma*) is one of the best available options. In classification, marine medaka (*O. melastigma*) and Japanese medaka (*O. latipes*) belong to the Beloniformes order, Adrianichthyidae family, and *Oryzias* genus. *O. melastigma*, also known as *O. dancena* or Indian medaka, generally live on the coasts of China, Korea, Japan, and India. They provide many advantages: 1) small size of adults (2-3 cm); 2) short generation cycle (3-4 months); 3) transparent embryos; 4) distinct sexual dimorphism; 5) ability of spawn daily; 6) a wide range of salinity and temperature adaptation. These advantages make them easy to breed in the laboratory and sensitively respond to diverse chemicals. Therefore, *O. melastigma* is considered a promising model organism for marine and estuarine ecotoxicological study (Kim *et al.*, 2016).

Recently, many ecotoxicological studies have been conducted using *O. melastigma* as a fish model due to their bioavailability and sensitive toxic response, such as neurotoxicity, embryotoxicity, cardiac toxicity, immunotoxicity, endocrine disruptive effect, and metabolism alteration after exposure to contaminants and other environmental stressors. It is well known that the first transgenic fish were produced in China in 1985 (Zhu, He, & Chen, 1985). For the ecotoxicological risk assessment, transgenic fish have been applied to screen and monitor aquatic contaminants, because they can offer more advanced and integrated systems for the study of toxic mechanisms (Lele & Krone, 1996; Nerbert, 2002). Similarly, fluorescent protein reporter (*e.g.* GFP, RFP) are able to monitor pollutants by emitting the real-time fluorescence signal in live embryos and organisms. Furthermore, the expression levels of heat-shock protein, *cyp1a*, and vitellogenin (Vtg) can be used to monitor the exposure risk of heavy metal, persistent organic pollutants, and estrogen or estrogen-like pollutants, respectively. Taken together, this review summarizes the finding of the wild-type and transgenic marine medaka (*O. melastigma*) that used for marine toxicological studies, and highlights the health effects after exposure to various marine pollutants, such as heavy metals, endocrine disruptors, and organic pollutants.

Embryonic Development of *O. Melastigma*

Commonly, the developmental stage of the embryos is determined by hour post fertilization (hpf) or days post fertilization (dpf). The embryonic development of medaka *O. latipes* was divided into 45 stages as described by Iwamatsu (Iwamatsu, 2004). Chen *et al.* divided the embryonic development of *O. melastigma* into 8 stages and 33 substages on the basis of the development of medaka *O. latipes* (Bo, Cai, Xu, Wang, & Au, 2011). Darve *et al.* described 24 developmental stages of *O. melastigma* according to the morphological differences (Darve, Wani, Indulkar, & Sawant, 2013). Chen *et al.* investigated the embryonic development of *O. melastigma,* and particularly observed the brain, eye, heart, pectoral fin, and the trunk muscle by using *in situ* hybridization and immunostaining technique (Chen X *et al.*, 2011). As shown in Fig. (**1**), at the initial egg stage, the egg is surrounded by a thick chorion. There is a narrow space between chorion and vitellus. Short villi are seen over the whole surface of the chorion. Oil droplets are embedded at random in the cortical cytoplasm. At the early morula stage (4 hpf), blastomeres are appeared at the animal pole of the egg. At the late gastrula stage (21 hpf), the enveloping layer entirely covers the yolk sphere. At 6 somite stage (35 hpf), the small otic vesicles can be observed, and three regions of the brain (fore-brain, mid-brain, and hindbrain) are well defined. At 34 somite stage (80 hpf), about 3/4 of the yolk sphere is encircled by the embryonic body, and the pigmentation of the melanophores preliminarily appears in the eyes. At heart development stage (145 hpf), melanophores are distributed on the whole eyes, and the heart is developed.

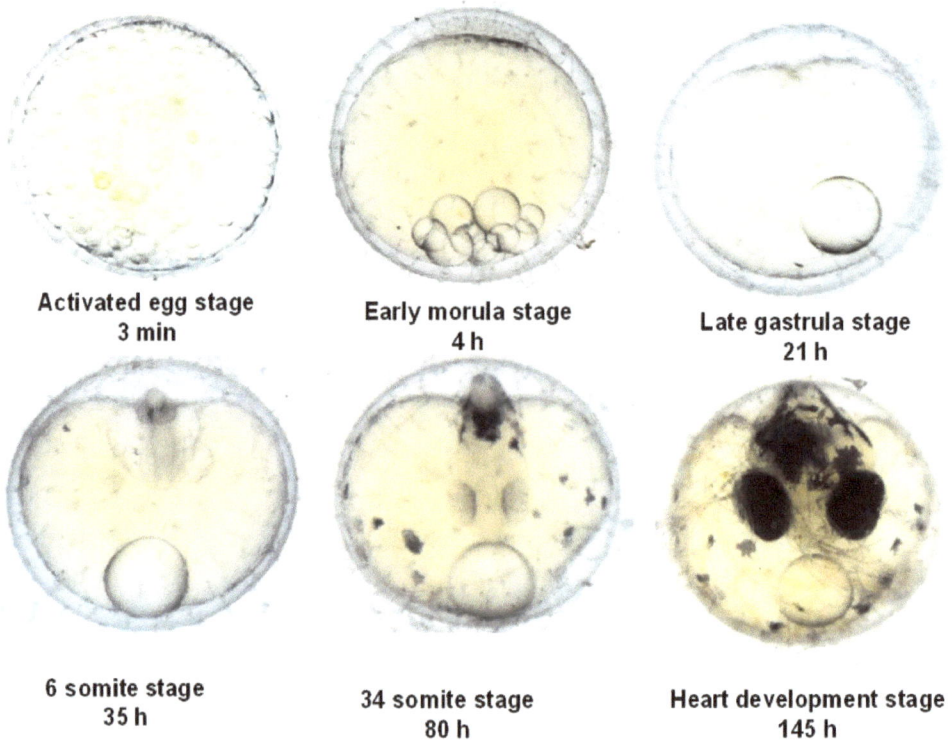

Activated egg stage
3 min

Early morula stage
4 h

Late gastrula stage
21 h

6 somite stage
35 h

34 somite stage
80 h

Heart development stage
145 h

Fig. (1). Embryonic development of marine medaka *O. melastigma.*

Whole Genome Sequencing of *O. Melastigma*

Since the embryos of medaka *O. latipes* at early development stage are extremely sensitive to pollutants or stressors of seawater (Tian *et al.*, 2014; Tseng *et al.*,

2013), medaka *O. latipes* has been adopted and proposed as a powerful tool for toxicology research (Dong S *et al.*, 2014). The whole genome information of Japanese medaka *O. latipes* is available (Kasahara *et al.*, 2007; Kong *et al.*, 2008), and most of the researchers isolated genes of *O. melastigma* in reference of *O. latipes*. Currently, the genetic chip information of *O. melastigma* can be acquired (Chen X *et al.*, 2009; S. Dong S *et al.*, 2014).

Our group (Pei lab, Chongqing Institute of Green and Intelligent Technology, Chinese Academy of Sciences) had cultured *O. melastigma* for many years (Fig. 2), and has begun *de novo* assembly and genome sequencing of medaka *O. melastigma* using the third generation sequencing technique since 2011. The third generation single-molecule sequencing technologies can generate over 10,000 bp reads or map over 100,000 bp molecules, which is unlike the second-generation sequencing (or named next-generation sequencing, NGS) that produces short reads of a few hundred base-pairs long (Lee H *et al.*, 2016). In the past decade, several groups also worked on high-throughput sequencing of the marine medaka genome and transcriptome using the second-generation sequencing technology (Huang QS *et al.*, 2012) but only generated preliminary raw genome information (Kim RO *et al.*, 2013). Recently, Kim *et al.* reported the draft genome of *O. melastigma* using the second-generation sequencing technology by developing sequencing libraries of paired ends (PEs) and mate pairs (MPs) based on the Illumina platform. They described that a total genome length of *O. melastigma* is 779.4 Mb, and the genome assembly consists of 8,602 scaffolds (N50 = 23.737 Mb). However, they did not perform the analysis of teleost genome evolution, the genetic and physical map, and the mechanism of high salinity tolerance, *etc*. The genome data of *O. melastigma* in our hands is different from the result described by Kim *et al.*, because we use the third generation sequencing technique. Definitely, we will publish a precise genome of *O. melastigma* soon.

Fig. (2). Lateral view of adult male and female *O. melastigma* cultured in our lab, Chongqing Institute of Green and Intelligent Technology, Chinese Academy of Sciences. The blue boxes indicate the different caudal fins of male and female *O. melastigma*.

Wild-Type O. Melastigma Monitoring Different Pollutants

A suitable model organism is very vital for investigating the impacts of contaminants on marine and estuarine ecosystems, which can help to understand the toxic mechanism of acute and chronic exposure. For marine ecotoxicological studies, wild-type *O. melastigma* has been suggested and applied as a fish model. The following text illustrates the physiological impacts of wild-type *O. melastigma* after exposure to heavy metals,endocrine disrupting compounds (EDCs), and organic pollutants, as listed in Table **1**.

Heavy Metals Toxicity in Wild-type O. melastigma

Metal contamination is becoming a global environmental hazard because of the increasing utilization of electronic products and nanomaterials. Moreover, metals are not chemically or biologically decomposable. Wild-type *O. melastigma* has recently developed as a new fish model for the studies of hepatotoxicity, neurotoxicity, and embryotoxicity when exposed to heavy metals. Guo *et al.* investigated the biokinetics of copper (Cu) and predicted Cu bioaccumulation during the development of marine medaka (Guo *et al.*, 2016). The accumulation of inorganic mercury in the liver and brain of *O. melastigma* can induce hepatotoxicity or neurotoxicity through the oxidative stress effects, cytoskeletal assembly dysfunction, and metabolic disorders (Wang M *et al.*, 2011; Wang M *et al.*, 2013). Moreover, with the rapid development and application of nanotechnology, the nanoparticle has become one of the most diverse metal contaminations in the environment. The potential risks of metal oxide nanoparticles on aquatic environment and organisms have caused many concerns (Joo & Zhao, 2017). Dietary exposure to silver nanoparticles (AgNPs) in wild-type *O. melastigma* caused the inhibition of Na+ /K+-ATPase and superoxide dismutase (SOD) activity (Wang J & Wang W, 2014). Exposure to AgNPs can

Table 1. Summary of the environmental pollutants and their health effects in *O. melastigma*.

Classes	Pollutants	Toxicological Research Related	Reference
Heavy metals	copper	Metal biokinetics	(Guo *et al.*, 2016)
	AgNPs	Disruption of steroidogenesis	(Degger *et al.*, 2015)
	AgNPs	Trophic toxicity	(J. Wang & Wang, 2014)
	inorganic mercury	Neurotoxicity; hepatotoxicity	(M. Wang *et al.*, 2011; M. Wang *et al.*, 2013; Y. Wang, Wang, Lin, & Wang, 2015)
	ZnONPs	Molecular biomarker HSP70 expression	(Wong, Leung, Djurisic, & Leung, 2010)
	ZnONPs	Bioavailability and toxicity	(J. Wang *et al.*, 2017)
	ZnONPs	Embryotoxicity	(Cong *et al.*, 2017)

(Table 1) cont.....

Classes	Pollutants	Toxicological Research Related	Reference
Endocrine disruptors	PBDE-47	Gender-specific modulatory and HSP regulation and expression	(Deane *et al.*, 2014)
	PBDE-47	Maternal transfer and lipid mobilization	(van de Merwe *et al.*, 2011)
	BDE-47	Immunotoxicity	(Ye RR, *et al.*, 2011)
	BDE-47	Gender-specific gene expression in livers	(Yu WK, *et al.*, 2013)
	BDE-47	Alteration of the profile of metabolites in the brain	(Lei *et al.*, 2017)
	BPA	Toxicity of cardiac development	(Huang Q, *et al.*, 2011)
	NPs and BPA	Occurrence and ecological risks of EDCs	(Xu EGB, *et al.* 2015)
	DFZ	Expression of CYP3A38 and CYP27A1	(Dong Z *et al.*, 2017)
	DFZ	Change lipid metabolism	(X. Dong *et al.*, 2016)
	DFZ	Reproductive effects	(X. Dong *et al.*, 2017)
	DCOIT	Alterations of the HPGL axis	(L. Chen *et al.*, 2016)
	DCOIT	Egg production	(L. G. Chen *et al.*, 2014)
	DCOIT	VTG expression related to estrogenic activity	(L. Chen *et al.*, 2015)
	Antiandrogens, Vinclozolin, cyproterone acetate	Gonodal development	(Kiparissis, Metcalfe, Balch, & Metcalfe, 2003)
	EE2	Reproductive behavior	(P. Y. Lee, Lin, & Chen, 2014)
	Ortho-para- DDT	Gonadal development and endocrine responses	(Metcalfe *et al.*, 2000)
	17β-estradiol	VTG response	(Marsh, Paterson, Foran, & Bennett, 2010)
	17β-estradiol	Reproductive response	(Oshima *et al.*, 2003)
	17β-estradiol	Reproductive response	(Kang *et al.*, 2002)
	17alpha-Ethinylestradiol (EE)	Transgenerational reproductive and developmental exposure	(Foran, Peterson, & Benson, 2002)
	EE2	Sex-differences in immunity	(X. Dong *et al.*, 2017)
	Benzotriazole	Reproductive effects	(Tangtian, Bo, Wenhua, Shin, & Wu, 2012)
	DEHP and PFOS	Immunotoxic effects	(Huang *et al.*, 2015)
	DEHP	Accumulation, elimination, and endocrine-disruptive effects	(T. Ye *et al.*, 2016)
	DEHP and MEHP	Endocrine disruption with sex-specific effects	(T. Ye *et al.*, 2014)
	DIM	Toxicological effects of HPGL axis	(L. Chen *et al.*, 2016)
	DIM	Development and function of gonad	(L. Chen *et al.*, 2017)

(Table 1) cont.....

Classes	Pollutants	Toxicological Research Related	Reference
Organic pollutants	PFOS	Mitochondrial dysfunction	(Q. Huang *et al.*, 2012)
	PFOS	Precocious hatching	(Wu *et al.*, 2012)
	PFOS	Endocrine disruptive effect	(Fang *et al.*, 2012)
	PFOS	Cardiac toxicity	(Huang, Fang, Wu, *et al.*, 2011)
	PFOS	Immunotoxicity	(Fang *et al.*, 2013; Huang *et al.*, 2015)
	Phe, Py, and BaP	Developmental toxicities and CYP1A induction in the embryotoxicity	(J. L. Mu *et al.*, 2012)
	PAHs and alkyl PAHs	Bioavailability and toxicity	(Liu *et al.*, 2016)
	Alkylating compounds (MNU, ENU,MMS, EMS)	Developmental toxicities	(Solomon & Faustman, 1987)
	Fungicides(Acrobat MZ, Tattoo C)	Development and adult male behavior	(Teather, Harris, Boswell, & Gray, 2001)
	alkyl-phenanthrene	Metabolism alteration	(J. Mu *et al.*, 2016)
	Toluene	Acute Toxicity	(Stoss & Haines, 1979)
	P-CTX-1	stress/immune responses, cardiac/bone development, and apoptosis	(Yan *et al.*, 2017)
	P-CTX-1	Physiological and behavioral effects	(Mak *et al.*, 2017)
	TPT	Toxicity to life-cycle	(Yi & Leung, 2017)
	WAFs	CYP-involved detoxification and endogenous steroidogenic metabolism	(Rhee *et al.*, 2013)
	WAFs	CYP1A mRNA level	(R. O. Kim *et al.*, 2013)
	WAFs	Bioaccumulation and toxicity	(J. Mu, Jin, Ma, Lin, & Wang, 2014)

Abbreviation: Estradiol (E2), 17a-ethinylestradiol (EE2), 4-nonylphenol (NP), Bisphenol A (BPA), Triphenyltin (TPT), Nonylphenols (NPs), difenoconazole (DFZ), 4,5-dichloro-2-n-octyl-4-isothiazolin-3-one (DCOIT), catalase (CAT), glutathione peroxidase (GPx), differentiation 3 (CD3), butyl benzyl phthalate (BBP), di(n-butyl) phthalate (DBP), diisodecyl phthalate (DIDP), diisononyl phthalate (DINP), di-n-octyl phthalate (DNOP), tributyltin (TBT), MNU (methylnitrosourea); ENU (ethylnitrosourea), MMS (methyl methanesulfonate), EMS (ethyl methanesulfonate).

also disrupt the regulation of steroidogenesis (Degger N *et al.*, 2015). While ZnO nanoparticles exposure can cause embryo-toxicity by increasing mortality and heart rate, reducing the hatching ratio, and boosting the malformation of newly-hatched larvae (Cong Y *et al.*, 2017). Interestingly, the antioxidative defense was found to be enhanced after exposure to high dietary ZnO nanoparticles in wild-type *O. melastigma* (Wang J *et al.*, 2017).

Endocrine Disrupting Compounds Toxicity in Wild-type O. Melastigma

Endocrine Disrupting Compounds (EDCs) can disrupt the endocrine system and function, because estrogen or androgen mimics can bind to the estrogen receptor (ER) or androgen receptor (AR) for the activation of responsive genes. Many studies have demonstrated that EDCs can induce abnormal sexual development and reproductive capabilities of fish. Polybrominated Diphenyl Ethers (PBDEs) have been widely used as flame retardants, which are released into the environment from a range of consumer products, including furniture, plastics, textiles, and electronics (Siddiqi M *et al.*, 2003). Due to their lipophilic, persistent, and bioaccumulation properties, PBDEs have a series of harmful influences on humans and wildlife (Rahman F *et al.*, 2001; Tagliaferri *et al.*, 2010). PBDEs are still found predominantly in marine environments (Rahman *et al.*, 2001; Yu M *et al*, 2009), especially at the highest concentration in marine fish (Brown FR *et al*, 2006; Meng XY *et al*, 2008). Some researchers have reported that PBDE-47 exposure can induce the bioaccumulation in the wild-type *O. melastigma* along with the maternal transfer after dietary exposure (van de Merwe *et al.*, 2011). PBDE-47 may cause gender specific modulatory related to the apoptotic cascade (Deane, van de Merwe, Hui, Wu, & Woo, 2014). Ye *et al.* also found that BDE-47 has gender-specific modulation effects on some gene expression related to immunomodulatory (Ye RR *et al.*, 2011). In addition, the transcriptional profile of liver and the metabolite profiling after exposure to BDE-47 in wild-type *O. melastigma* have been reported (Lei *et al.*, 2017; Yu WK *et al.*, 2013). 3,3'-Diindolylmethane (DIM), a compound derived from the brassica food plants, is known as a promising antifouling and chemopreventive agent (Stresser, Williams, Griffin, & Bailey, 1995). Recently, the environmental risks of DIM are gradually raising attention. Chen *et al.* investigated the toxicological effects and underlying mechanism of DIM, and found that it could disrupt the normal function of the endocrine system in a gender-specific manner *via* the hypothalamus-pituitary-gonadal-liver (HPGL) axis in *O. melastigma* (Chen L *et al.*, 2016). They also found the reproductive impairment in gonads at the transcript, protein and histological levels induced by chronic DIM exposure to F_0 *O. melastigma* (Chen L *et al.*, 2017). Di-(2-ethylhexyl)-phthalate (DEHP) has been widely used as a plastic softener in the manufacturing of various products (Koo & Lee, 2004), which is also ubiquitously distributed in the marine environments. Exposure to DEHP can induce endocrine disruption (Chen XP *et al.*, 2014; Ye T *et al.*, 2016; Ye T *et al.*, 2014), sex-specific effects (Ye T *et al.*, 2014), and immunotoxicity (Huang *et al.*, 2015) in *O. melastigma*. Other endocrine disruptors, such as BPA, DFZ, DCOIT, EE2 and benzotriazole, can cause endocrine disruption effect on *O. melastigma* as listed in Table **1**.

Organic Pollutants Toxicity in Wild-type O. Melastigma

Organic pollutants have been raised the global concerns because of their ubiquitous, persistent and carcinogenic nature. Polycyclic aromatic hydrocarbons

(PAHs), mainly derived from crude oil and petroleum products, are widely distributed in aquatic systems (Garcia KL *et al.*, 1993). PAHs can up-regulate the expression level of CYP1A by activating the aryl hydrocarbon receptor (AhR) pathway. Exposure to 3-5 ringed PAHs and ANF (a CYP1A inhibitor) showed a potentially synergistic effect in *O. melastigma* embryos (Mu JL *et al.*, 2012). The combined effect of ANF and alkyl-PAHs (a predominant form of PAHs in crude oils) caused significant developmental toxicity, genotoxicity, embryotoxicity, and alteration in metabolism (Mu J *et al.*, 2016). Interestingly, humic acid (HA) can reduce the bioaccumulation and toxicity of 3 ringed PAHs in marine medaka (Liu *et al.*, 2016). Perfluorinated compounds (PFCs) are an emerging class of environmental contaminants, which are widely used for protective coatings. Perfluorooctane sulfonate (PFOS) has been recognized as an emerging persistent organic contaminant because of their persistence, bioaccumulative property, and toxicological nature (Houde M *et al.*, 2006; Lau *et al.*, 2007). PFOS can be bioaccumulated through the food chain (Han & Fang, 2010; Shaw *et al.*, 2009), and exposure to PFOS can cause mitochondrial dysfunction (Huang Q *et al.*, 2012), precocious hatching (Wu *et al.*, 2012), endocrine disruptive effect (Fang *et al.*, 2012), cardiac toxicity (Huang Q *et al.*, 2011), and immunotoxicity (Fang *et al.*, 2013; Huang *et al.*, 2015) in *O. melastigma*. Exposure to the water-accommodated fraction (WAF) of crude oil can significantly enhance the transcripts of CYP1A (R. O. Kim *et al.*, 2013), trigger the CYP-involved detoxification mechanism, and suppress the CYP steroidogenic metabolism in the *O. melastigma* (Rhee *et al.*, 2013).

Transgenic *O. melastigma* Monitoring Different Pollutants

Transgenic fish can be used as the biosensors in aquatic ecotoxicology (Cho YS *et al.*, 2013), because transgenic fish could be genetically modified to respond specific chemicals with an easily detectable reporter. For example, the Gal4/Vp16-UAS system fused with a fluorescent marker can be applied for establishing inducible transgenic fish lines (Grabher & Wittbrodt, 2004). Cho *et al.* established several stable transgenic marine medakas (*O. dancena* and also named as *O. melastigma*) carrying the RFP reporter gene, which is the ubiquitous expression driven by b-actin promoter (Cho YS *et al.*, 2011) and the estradiol-17b (E2) inducible expression driven by endogenous choriogenin H promoter. (Cho *et al.*, 2013). Nevertheless, the transgenic strains of *O. melastigma* is relatively less available than that of the transgenic *O. latipes* at present.

Heavy Metal Toxicity in Transgenic O. melastigma

Transgenic fish have been developed driven by heat-shock protein (hsp) promoter to monitor different environmental stressors including increased temperature and heavy metals contaminants in the aquatic environment. Previous studies demonstrated that cadmium and arsenic exposure could induce the up-regulation of hsp70 in various aquatic organisms (Ireland *et al.*, 2004; Olsvik PA *et al.*, 2011; Roy & Bhattacharya, 2006; Schill RO *et al.*, 2003). Ng *et al.* developed a transgenic medaka using eGFP as a reporter driven by hsp70 promoter, named as *Tg(hsp70:eGFP)*. Exposure to mercury, arsenic, lead, and cadmium caused GFP expression in different multiple organs (Ng GH *et al.*, 2015). Currently, there are no reports of transgenic *O. melastigma* for monitoring heavy metal. A molecular element specifically responding to heavy metal was cloned in our lab and the transgenic *O. melastigma* will be established for the toxicity study of heavy metal soon.

Endocrine Disrupting Compounds Toxicity in Transgenic O. melastigma

ChgH, ChgL, and Vtg are highly sensitive biomarkers for monitoring estrogen and estrogen-like pollutants in the marine environment. Cho *et al.* established a stable transgenic marine medaka *O. dancena* driven by endogenous choriogenin H gene (chgH) promoter for monitoring E1, E3, EE2, DES, BPA, and E2 (Cho *et al.*, 2013). Exposure to 6 phthalates (BBP, DBP,DEHP, DIDP, DINP, DNOP) and their mixtures in ChgH-EGFP transgenic *O. melastigma* embryos apparently enhanced the estrogenic activity (Chen X.P. *et al.*, 2014).

Organic Pollutants Toxicity in Transgenic O. melastigma

Cytochrome P450 1A (Cyp1a) is commonly used as a biomarker for monitoring persistent organic pollutants through the activation of AhR pathway. Recently, two novel zebrafish lines *Tg(cyp1a:mCherry)* and *Tg(T-cyp1a:mCherry)* were established to sensitively monitor PAHs and TCDD in our lab (Luo *et al.*, 2018; Xie *et al.*, 2018). In *O. latipes*, transgenic medaka *Tg(cyp1a:gfp)* driven by *cyp1a* promoter was reported for detecting the presence of TCDD and other persistent organic chemicals (Boon Ng & Gong, 2011; Ng & Gong, 2013). To date, there has been no report of transgenic *O. melastigma* as a sensitive biomonitoring tool for organic pollutants.

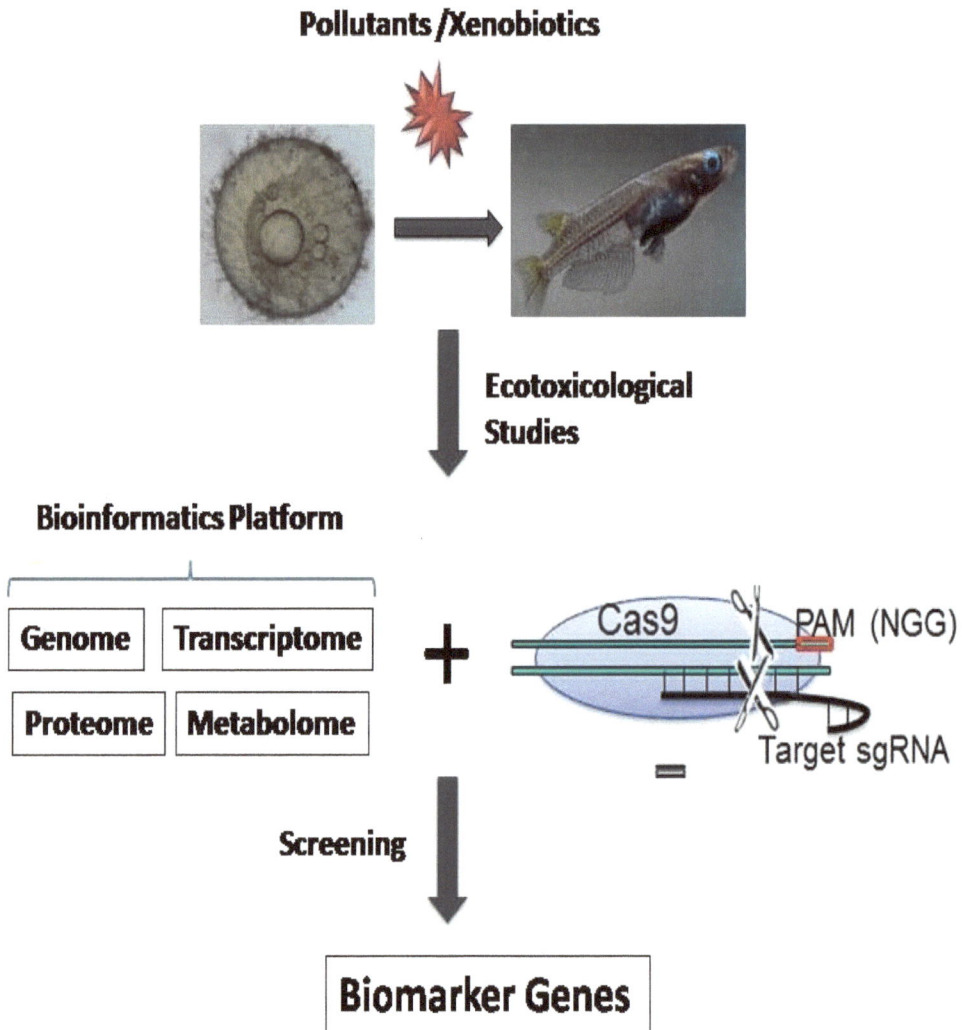

Fig. (3). The schematic diagram of screening biomarker genes induced by pollutants or xenobiotic using bioinformatic platform and the CRISPR/Cas9 system in *O. melastigma.*

CONCLUDING REMARKS

Marine medaka (*O. melastigma*) has been proposed as a feasible model for marine and estuarine ecotoxicological study. Both wild-type and transgenic marine medaka have been applied for monitoring of heavy metals, endocrine disruptors, and organic pollutants. *O. melastigma* could be a useful platform for screening of biomarker of contaminants. However, there are certain limitations and drawbacks for using *O. melastigma*. First, the whole precise genome information of *O. mel-*

astigma has not been accomplished, albeit the draft genome of *O. melastigma* is reported. Second, the approach of gene knockout using TALENs or CRISPR/Cas9 system has been well established in *O. latipes*, but there have been no relevant reports in *O. melastigma* to date (Ansai & Kinoshita, 2014; Ansai *et al.*, 2013; Chiang *et al.*, 2016). Fig. (**3**) shows the potential omics application for screening molecular biomarkers in response to environmental stressors and chemical contaminant. In conclusion, the combination of high-throughput bioinformatics systems with advanced gene editing techniques will lead to a better understanding of the complex biological processes linked to the ecotoxicological responses in *O. melastigma*.

CONSENT FOR PUBLICATION

Not applicable.

CONFLICT OF INTEREST

The authors confirm that this chapter contents have no conflict of interest.

ACKNOWLEDGEMENTS

The editors are grateful for the support from the CAS Team Project of the Belt and Road (to D.S.P), the Three Hundred Leading Talents in Scientific and Technological Innovation Program of Chongqing (No. CSTCCXLJRC201714 to D.S.P), the Program of China–Sri Lanka Joint Research and Demonstration Center for Water Technology and China–Sri Lanka Joint Center for Education and Research by Chinese Academy of Sciences, China(to D.S.P), and the University of Chinese Academy of Sciences (UCAS) for CAS-TWAS Scholarship (No. 2017A8018537001 to N.H).

REFERENCES

Ansai, S & Kinoshita, M (2014) Targeted mutagenesis using CRISPR/Cas system in medaka. *Biol Open,* 3, 362-71.
[http://dx.doi.org/10.1242/bio.20148177] [PMID: 24728957]

Ansai, S, Sakuma, T, Yamamoto, T, Ariga, H, Uemura, N, Takahashi, R & Kinoshita, M (2013) Efficient targeted mutagenesis in medaka using custom-designed transcription activator-like effector nucleases. *Genetics,* 193, 739.
[http://dx.doi.org/10.1534/genetics.112.147645]

Bo, J, Cai, L, Xu, J-H, Wang, K-J & Au, D W (2011) The marine medaka *Oryzias melastigma*–a potential marine fish model for innate immune study. *Marine pollution bulletin,* 63, 267-76.

Boon Ng, GH & Gong, Z (2011) Maize Ac/Ds transposon system leads to highly efficient germline transmission of transgenes in medaka (Oryzias latipes). *Biochimie,* 93, 1858-64.
[http://dx.doi.org/10.1016/j.biochi.2011.07.006] [PMID: 21777650]

Brown, FR, Winkler, J, Visita, P, Dhaliwal, J & Petreas, M (2006) Levels of PBDEs, PCDDs, PCDFs, and coplanar PCBs in edible fish from California coastal waters. *Chemosphere,* 64, 276-86.

[http://dx.doi.org/10.1016/j.chemosphere.2005.12.012] [PMID: 16455130]

Chen, L, Au, DWT, Hu, C, Zhang, W, Zhou, B, Cai, L, Giesy, JP & Qian, PY (2017) Linking genomic responses of gonads with reproductive impairment in marine medaka (*Oryzias melastigma*) exposed chronically to the chemopreventive and antifouling agent, 3,3'-diindolylmethane (DIM). *Aquat Toxicol,* 183, 135-43.
[http://dx.doi.org/10.1016/j.aquatox.2016.12.021] [PMID: 28063342]

Chen, L, Sun, J, Zhang, H, Au, DW, Lam, PK, Zhang, W, Bajic, VB, Qiu, JW & Qian, PY (2015) Hepatic proteomic responses in marine medaka (*Oryzias melastigma*) chronically exposed to antifouling compound butenolide [5-octylfuran-2(5H)-one] or 4,5-dichloro-2-N-octyl-4-isothiazolin-3-one (DCOIT). *Environ Sci Technol,* 49, 1851-9.
[http://dx.doi.org/10.1021/es5046748] [PMID: 25555223]

Chen, L, Ye, R, Zhang, W, Hu, C, Zhou, B, Peterson, DR, Au, DW, Lam, PK & Qian, PY (2016) Endocrine Disruption throughout the Hypothalamus-Pituitary-Gonadal-Liver (HPGL) Axis in Marine Medaka (*Oryzias melastigma*) Chronically Exposed to the Antifouling and Chemopreventive Agent, 3,3'-Diindolylmethane (DIM). *Chem Res Toxicol,* 29, 1020-8.
[http://dx.doi.org/10.1021/acs.chemrestox.6b00074] [PMID: 27092574]

Chen, L, Ye, R, Xu, Y, Gao, Z, Au, DWT & Qian, PY (2014) Comparative safety of the antifouling compound butenolide and 4,5-dichloro-2-n-octyl-4-isothiazolin-3-one (DCOIT) to the marine medaka (*Oryzias melastigma*). *Aquat Toxicol,* 149, 116-25.
[http://dx.doi.org/10.1016/j.aquatox.2014.01.023] [PMID: 24583292]

Chen, X, Li, L, Cheng, J, Chan, L L, Wang, D-Z, Wang, K-J, Baker, M E, Hardiman, G, Schlenk, D & Cheng, S H (2011) Molecular staging of marine medaka: a model organism for marine ecotoxicity study. *Marine pollution bulletin,* 63, 309-17.

Chen, X, Li, L, Wong, CKC & Cheng, SH (2009) Rapid adaptation of molecular resources from zebrafish and medaka to develop an estuarine/marine model. *Comp Biochem Physiol C Toxicol Pharmacol,* 149, 647-55.
[http://dx.doi.org/10.1016/j.cbpc.2009.01.009] [PMID: 19302835]

Chen, X, Xu, S, Tan, T, Lee, ST, Cheng, SH, Lee, FWF, Xu, SJL & Ho, KC (2014) Toxicity and estrogenic endocrine disrupting activity of phthalates and their mixtures. *Int J Environ Res Public Health,* 11, 3156-68.
[http://dx.doi.org/10.3390/ijerph110303156] [PMID: 24637910]

Chiang, YA, Kinoshita, M, Maekawa, S, Kulkarni, A, Lo, CF, Yoshiura, Y, Wang, HC & Aoki, T (2016) TALENs-mediated gene disruption of myostatin produces a larger phenotype of medaka with an apparently compromised immune system. *Fish Shellfish Immunol,* 48, 212-20.
[http://dx.doi.org/10.1016/j.fsi.2015.11.016] [PMID: 26578247]

Cho, YS, Kim, DS & Nam, YK (2013) Characterization of estrogen-responsive transgenic marine medaka Oryzias dancena germlines harboring red fluorescent protein gene under the control by endogenous choriogenin H promoter. *Transgenic Res,* 22, 501-17.
[http://dx.doi.org/10.1007/s11248-012-9650-y] [PMID: 22972478]

Cho, YS, Lee, SY, Kim, YK, Kim, DS & Nam, YK (2011) Functional ability of cytoskeletal β-actin regulator to drive constitutive and ubiquitous expression of a fluorescent reporter throughout the life cycle of transgenic marine medaka Oryzias dancena. *Transgenic Res,* 20, 1333-55.
[http://dx.doi.org/10.1007/s11248-011-9501-2] [PMID: 21437716]

Cong, Y, Jin, F, Wang, J & Mu, J (2017) The embryotoxicity of ZnO nanoparticles to marine medaka, *Oryzias melastigma. Aquat Toxicol,* 185, 11-8.
[http://dx.doi.org/10.1016/j.aquatox.2017.01.006] [PMID: 28157544]

Darve, SI, Wani, GB, Indulkar, ST & Sawant, MS (2013) Embryonic development in Indian medaka, *Oryzias melastigma* (McClelland, 1839). *Indian J Anim Res,* 47, 301-8.

Deane, EE, van de Merwe, JP, Hui, JH, Wu, RS & Woo, NY (2014) PBDE-47 exposure causes gender specific effects on apoptosis and heat shock protein expression in marine medaka, *Oryzias melastigma. Aquat*

Toxicol, 147, 57-67.
[http://dx.doi.org/10.1016/j.aquatox.2013.12.009] [PMID: 24374848]

Degger, N, Tse, AC & Wu, RS (2015) Silver nanoparticles disrupt regulation of steroidogenesis in fish ovarian cells. *Aquat Toxicol,* 169, 143-51.
[http://dx.doi.org/10.1016/j.aquatox.2015.10.015] [PMID: 26546908]

Dodd, A, Curtis, PM, Williams, LC & Love, DR (2000) Zebrafish: bridging the gap between development and disease. *Hum Mol Genet,* 9, 2443-9.
[http://dx.doi.org/10.1093/hmg/9.16.2443] [PMID: 11005800]

Dong, S, Kang, M, Wu, X & Ye, T (2014) Development of a promising fish model (*Oryzias melastigma*) for assessing multiple responses to stresses in the marine environment. *BioMed Res Int,* 2014563131
[http://dx.doi.org/10.1155/2014/563131] [PMID: 24724087]

Dong, X, Li, Y, Zhang, L, Zuo, Z, Wang, C & Chen, M (2016) Influence of difenoconazole on lipid metabolism in marine medaka (*Oryzias melastigma*). *Ecotoxicology,* 25, 982-90.
[http://dx.doi.org/10.1007/s10646-016-1655-5] [PMID: 27112457]

Dong, X, Zuo, Z, Guo, J, Li, H, Zhang, L, Chen, M, Yang, Z & Wang, C (2017) Reproductive effects of life-cycle exposure to difenoconazole on female marine medaka (*Oryzias melastigma*). *Ecotoxicology,* 26, 772-81.
[http://dx.doi.org/10.1007/s10646-017-1808-1] [PMID: 28432496]

Fang, C, Huang, Q, Ye, T, Chen, Y, Liu, L, Kang, M, Lin, Y, Shen, H & Dong, S (2013) Embryonic exposure to PFOS induces immunosuppression in the fish larvae of marine medaka. *Ecotoxicol Environ Saf,* 92, 104-11.
[http://dx.doi.org/10.1016/j.ecoenv.2013.03.005] [PMID: 23545396]

Fang, C, Wu, X, Huang, Q, Liao, Y, Liu, L, Qiu, L, Shen, H & Dong, S (2012) PFOS elicits transcriptional responses of the ER, AHR and PPAR pathways in *Oryzias melastigma* in a stage-specific manner. *Aquat Toxicol,* 106-107, 9-19.
[http://dx.doi.org/10.1016/j.aquatox.2011.10.009] [PMID: 22057250]

Foran, CM, Peterson, BN & Benson, WH (2002) Transgenerational and developmental exposure of Japanese medaka (Oryzias latipes) to ethinylestradiol results in endocrine and reproductive differences in the response to ethinylestradiol as adults. *Toxicol Sci,* 68, 389-402.
[http://dx.doi.org/10.1093/toxsci/68.2.389] [PMID: 12151635]

Garcia, KL, Delfino, JJ & Powell, DH (1993) Non-regulated organic-compounds in florida sediments. *Water Res,* 27, 1601-13.
[http://dx.doi.org/10.1016/0043-1354(93)90124-Z]

Grabher, C & Wittbrodt, J (2004) Efficient activation of gene expression using a heat-shock inducible Gal4/Vp16-UAS system in medaka. *BMC Biotechnol,* 4, 26.
[http://dx.doi.org/10.1186/1472-6750-4-26] [PMID: 15507134]

Guo, Z, Zhang, W, Du, S, Green, I, Tan, Q & Zhang, L (2016) Developmental patterns of copper bioaccumulation in a marine fish model *Oryzias melastigma*. *Aquat Toxicol,* 170, 216-22.
[http://dx.doi.org/10.1016/j.aquatox.2015.11.026] [PMID: 26675367]

Han, J & Fang, Z (2010) Estrogenic effects, reproductive impairment and developmental toxicity in ovoviparous swordtail fish (*Xiphophorus helleri*) exposed to perfluorooctane sulfonate (PFOS). *Aquat Toxicol,* 99, 281-90.
[http://dx.doi.org/10.1016/j.aquatox.2010.05.010] [PMID: 20570370]

Houde, M, Martin, JW, Letcher, RJ, Solomon, KR & Muir, DC (2006) Biological monitoring of polyfluoroalkyl substances: A review. *Environ Sci Technol,* 40, 3463-73.
[http://dx.doi.org/10.1021/es052580b] [PMID: 16786681]

Huang, Q, Chen, Y, Chi, Y, Lin, Y, Zhang, H, Fang, C & Dong, S (2015) Immunotoxic effects of perfluorooctane sulfonate and di(2-ethylhexyl) phthalate on the marine fish *Oryzias melastigma*. *Fish*

Shellfish Immunol, 44, 302-6.
[http://dx.doi.org/10.1016/j.fsi.2015.02.005] [PMID: 25687394]

Huang, Q, Dong, S, Fang, C, Wu, X, Ye, T & Lin, Y (2012) Deep sequencing-based transcriptome profiling analysis of *Oryzias melastigma* exposed to PFOS. *Aquat Toxicol,* 120-121, 54-8.
[http://dx.doi.org/10.1016/j.aquatox.2012.04.013] [PMID: 22613580]

Huang, Q, Fang, C, Chen, Y, Wu, X, Ye, T, Lin, Y & Dong, S (2011) Embryonic exposure to low concentration of bisphenol A affects the development of *Oryzias melastigma* larvae. *Environ Sci Pollut Res Int,* 19, 2506-14.
[http://dx.doi.org/10.1007/s11356-012-1034-6] [PMID: 22718145]

Huang, Q, Fang, C, Wu, X, Fan, J & Dong, S (2011) Perfluorooctane sulfonate impairs the cardiac development of a marine medaka (*Oryzias melastigma*). *Aquat Toxicol,* 105, 71-7.
[http://dx.doi.org/10.1016/j.aquatox.2011.05.012] [PMID: 21684243]

Elyse Ireland, H, Harding, SJ, Bonwick, GA, Jones, M, Smith, CJ & Williams, JHH (2004) Evaluation of heat shock protein 70 as a biomarker of environmental stress in Fucus serratus and Lemna minor. *Biomarkers,* 9, 139-55.
[http://dx.doi.org/10.1080/13547500410001732610] [PMID: 15370872]

Iwamatsu, T (2004) Stages of normal development in the medaka Oryzias latipes. *Mech Dev,* 121, 605-18.
[http://dx.doi.org/10.1016/j.mod.2004.03.012] [PMID: 15210170]

Joo, S H & Zhao, D (2017) Environmental dynamics of metal oxide nanoparticles in heterogeneous systems: A review. *J Hazard Mater,* . doi: 322, 29-47.
[http://dx.doi.org/10.1016/j.jhazmat.2016.02.068]

Kang, IJ, Yokota, H, Oshima, Y, Tsuruda, Y, Yamaguchi, T, Maeda, M, Imada, N, Tadokoro, H & Honjo, T (2002) Effect of 17β-estradiol on the reproduction of Japanese medaka (Oryzias latipes). *Chemosphere,* 47, 71-80.
[http://dx.doi.org/10.1016/S0045-6535(01)00205-3] [PMID: 11996138]

Kasahara, M, Naruse, K, Sasaki, S, Nakatani, Y, Qu, W, Ahsan, B, Yamada, T, Nagayasu, Y, Doi, K, Kasai, Y, Jindo, T, Kobayashi, D, Shimada, A, Toyoda, A, Kuroki, Y, Fujiyama, A, Sasaki, T, Shimizu, A, Asakawa, S, Shimizu, N, Hashimoto, S, Yang, J, Lee, Y, Matsushima, K, Sugano, S, Sakaizumi, M, Narita, T, Ohishi, K, Haga, S, Ohta, F, Nomoto, H, Nogata, K, Morishita, T, Endo, T, Shin-I, T, Takeda, H, Morishita, S & Kohara, Y (2007) The medaka draft genome and insights into vertebrate genome evolution. *Nature,* 447, 714-9.
[http://dx.doi.org/10.1038/nature05846] [PMID: 17554307]

Kim, BM, Kim, J, Choi, IY, Raisuddin, S, Au, DW, Leung, KM, Wu, RS, Rhee, JS & Lee, JS (2016) Omics of the marine medaka (*Oryzias melastigma*) and its relevance to marine environmental research. *Mar Environ Res,* 113, 141-52.
[http://dx.doi.org/10.1016/j.marenvres.2015.12.004] [PMID: 26716363]

Kim, R-O, Kim, B-M, Hwang, D-S, Au, DW, Jung, J-H, Shim, WJ, Leung, KM, Wu, RS, Rhee, J-S & Lee, J-S (2013) Evaluation of biomarker potential of cytochrome P450 1A (CYP1A) gene in the marine medaka, *Oryzias melastigma* exposed to water-accommodated fractions (WAFs) of Iranian crude oil. *Comp Biochem Physiol C Toxicol Pharmacol,* 157, 172-82.
[http://dx.doi.org/10.1016/j.cbpc.2012.11.003] [PMID: 23178197]

Kiparissis, Y, Metcalfe, TL, Balch, GC & Metcalfe, CD (2003) Effects of the antiandrogens, vinclozolin and cyproterone acetate on gonadal development in the Japanese medaka (Oryzias latipes). *Aquat Toxicol,* 63, 391-403.
[http://dx.doi.org/10.1016/S0166-445X(02)00189-3] [PMID: 12758004]

Kong, RY, Giesy, JP, Wu, RS, Chen, EX, Chiang, MW, Lim, PL, Yuen, BB, Yip, BW, Mok, HO & Au, DW (2008) Development of a marine fish model for studying *in vivo* molecular responses in ecotoxicology. *Aquat Toxicol,* 86, 131-41.
[http://dx.doi.org/10.1016/j.aquatox.2007.10.011] [PMID: 18055030]

Koo, HJ & Lee, BM (2004) Estimated exposure to phthalates in cosmetics and risk assessment. *J Toxicol Environ Health A,* 67, 1901-14.
[http://dx.doi.org/10.1080/15287390490513300] [PMID: 15513891]

Lau, C, Anitole, K, Hodes, C, Lai, D, Pfahles-Hutchens, A & Seed, J (2007) Perfluoroalkyl acids: a review of monitoring and toxicological findings. *Toxicol Sci,* 99, 366-94.
[http://dx.doi.org/10.1093/toxsci/kfm128] [PMID: 17519394]

Lee, H, Gurtowski, J, Yoo, S, Nattestad, M, Marcus, S, Goodwin, S, McCombie, WR & Schatz, M (2016) Third-generation sequencing and the future of genomics. *bioRxiv.*048603

Lee, PY, Lin, CY & Chen, TH (2014) Environmentally relevant exposure of 17α-ethinylestradiol impairs spawning and reproductive behavior in the brackish medaka *Oryzias melastigma. Mar Pollut Bull,* 85, 338-43.
[http://dx.doi.org/10.1016/j.marpolbul.2014.04.013] [PMID: 24775065]

Lei, EN, Yau, MS, Yeung, CC, Murphy, MB, Wong, KL & Lam, MH (2017) Profiling of Selected Functional Metabolites in the Central Nervous System of Marine Medaka (*Oryzias melastigma*) for Environmental Neurotoxicological Assessments. *Arch Environ Contam Toxicol,* 72, 269-80.
[http://dx.doi.org/10.1007/s00244-016-0342-0] [PMID: 27990605]

Lele, Z & Krone, PH (1996) The zebrafish as a model system in developmental, toxicological and transgenic research. *Biotechnol Adv,* 14, 57-72.
[http://dx.doi.org/10.1016/0734-9750(96)00004-3] [PMID: 14536924]

Liu, Y, Yang, C, Cheng, P, He, X, Zhu, Y & Zhang, Y (2016) Influences of humic acid on the bioavailability of phenanthrene and alkyl phenanthrenes to early life stages of marine medaka (*Oryzias melastigma*). *Environ Pollut,* 210, 211-6.
[http://dx.doi.org/10.1016/j.envpol.2015.12.011] [PMID: 26735166]

Luo, JJ, Su, DS, Xie, SL, Liu, Y, Liu, P, Yang, XJ & Pei, DS (2018) Hypersensitive assessment of aryl hydrocarbon receptor transcriptional activity using a novel truncated cyp1a promoter in zebrafish. *FASEB J,* 32, 2814-26.
[http://dx.doi.org/10.1096/fj.201701171R] [PMID: 29298861]

Mak, YL, Li, J, Liu, CN, Cheng, SH, Lam, PKS, Cheng, J & Chan, LL (2017) Physiological and behavioural impacts of Pacific ciguatoxin-1 (P-CTX-1) on marine medaka (*Oryzias melastigma*). *J Hazard Mater,* 321, 782-90.
[http://dx.doi.org/10.1016/j.jhazmat.2016.09.066] [PMID: 27720471]

Marsh, KE, Paterson, G, Foran, CM & Bennett, ER (2010) Variable vitellogenin response of Japanese medaka (Oryzias latipes) to weekly estrogen exposure. *Arch Environ Contam Toxicol,* 58, 793-9.
[http://dx.doi.org/10.1007/s00244-010-9468-7] [PMID: 20162267]

Meng, XZ, Yu, L, Guo, Y, Mai, BX & Zeng, EY (2008) Congener-specific distribution of polybrominated diphenyl ethers in fish of China: implication for input sources. *Environ Toxicol Chem,* 27, 67-72.
[http://dx.doi.org/10.1897/07-138.1] [PMID: 18092876]

Metcalfe, TL, Metcalfe, CD, Kiparissis, Y, Niimi, AJ, Foran, CM & Benson, WH (2000) Gonadal development and endocrine responses in Japanese medaka (Oryzias latipes) exposed to o,p′-DDT in water or through maternal transfer. *Environ Toxicol Chem,* 19, 1893-900.
[http://dx.doi.org/10.1002/etc.5620190725]

Mu, J, Jin, F, Ma, X, Lin, Z & Wang, J (2014) Comparative effects of biological and chemical dispersants on the bioavailability and toxicity of crude oil to early life stages of marine medaka (*Oryzias melastigma*). *Environ Toxicol Chem,* 33, 2576-83.
[http://dx.doi.org/10.1002/etc.2721] [PMID: 25113786]

Mu, J, Jin, F, Wang, J, Wang, Y & Cong, Y (2016) The effects of CYP1A inhibition on alkyl-phenanthrene metabolism and embryotoxicity in marine medaka (*Oryzias melastigma*). *Environ Sci Pollut Res Int,* 23, 11289-97.

[http://dx.doi.org/10.1007/s11356-016-6098-2] [PMID: 26924701]

Mu, JL, Wang, XH, Jin, F, Wang, JY & Hong, HS (2012) The role of cytochrome P4501A activity inhibition in three- to five-ringed polycyclic aromatic hydrocarbons embryotoxicity of marine medaka (*Oryzias melastigma*). *Mar Pollut Bull,* 64, 1445-51.
[http://dx.doi.org/10.1016/j.marpolbul.2012.04.007] [PMID: 22633069]

Nebert, DW, Stuart, GW, Solis, WA & Carvan, MJ, III (2002) Use of reporter genes and vertebrate DNA motifs in transgenic zebrafish as sentinels for assessing aquatic pollution. *Environ Health Perspect,* 110, A15-5.
[http://dx.doi.org/10.1289/ehp.110-a15] [PMID: 11813700]

Ng, GH & Gong, Z (2013) GFP transgenic medaka (Oryzias latipes) under the inducible cyp1a promoter provide a sensitive and convenient biological indicator for the presence of TCDD and other persistent organic chemicals. *PLoS One,* 8e64334
[http://dx.doi.org/10.1371/journal.pone.0064334] [PMID: 23700472]

Ng, GH, Xu, H, Pi, N, Kelly, BC & Gong, Z (2015) Differential GFP expression patterns induced by different heavy metals in Tg(hsp70:gfp) transgenic medaka (Oryzias latipes). *Mar Biotechnol (NY),* 17, 317-27.
[http://dx.doi.org/10.1007/s10126-015-9620-5] [PMID: 25652692]

Olsvik, PA, Brattås, M, Lie, KK & Goksøyr, A (2011) Transcriptional responses in juvenile Atlantic cod (Gadus morhua) after exposure to mercury-contaminated sediments obtained near the wreck of the German WW2 submarine U-864, and from Bergen Harbor, Western Norway. *Chemosphere,* 83, 552-63.
[http://dx.doi.org/10.1016/j.chemosphere.2010.12.019] [PMID: 21195448]

Oshima, Y, Kang, IJ, Kobayashi, M, Nakayama, K, Imada, N & Honjo, T (2003) Suppression of sexual behavior in male Japanese medaka (Oryzias latipes) exposed to 17β-estradiol. *Chemosphere,* 50, 429-36.
[http://dx.doi.org/10.1016/S0045-6535(02)00494-0] [PMID: 12656264]

Rahman, F, Langford, KH, Scrimshaw, MD & Lester, JN (2001) Polybrominated diphenyl ether (PBDE) flame retardants. *Sci Total Environ,* 275, 1-17.
[http://dx.doi.org/10.1016/S0048-9697(01)00852-X] [PMID: 11482396]

Rhee, JS, Kim, BM, Choi, BS, Choi, IY, Wu, RS, Nelson, DR & Lee, JS (2013) Whole spectrum of cytochrome P450 genes and molecular responses to water-accommodated fractions exposure in the marine medaka. *Environ Sci Technol,* 47, 4804-12.
[http://dx.doi.org/10.1021/es400186r] [PMID: 23573833]

Roy, S & Bhattacharya, S (2006) Arsenic-induced histopathology and synthesis of stress proteins in liver and kidney of Channa punctatus. *Ecotoxicol Environ Saf,* 65, 218-29.
[http://dx.doi.org/10.1016/j.ecoenv.2005.07.005] [PMID: 16150489]

Schill, RO, Görlitz, H & Köhler, HR (2003) Laboratory simulation of a mining accident: acute toxicity, hsc/hsp70 response, and recovery from stress in *Gammarus fossarum* (Crustacea, Amphipoda) exposed to a pulse of cadmium. *Biometals,* 16, 391-401.
[http://dx.doi.org/10.1023/A:1022534326034] [PMID: 12680701]

Shaw, S, Berger, ML, Brenner, D, Tao, L, Wu, Q & Kannan, K (2009) Specific accumulation of perfluorochemicals in harbor seals (Phoca vitulina concolor) from the northwest Atlantic. *Chemosphere,* 74, 1037-43.
[http://dx.doi.org/10.1016/j.chemosphere.2008.10.063] [PMID: 19101009]

Siddiqi, MA, Laessig, RH & Reed, KD (2003) Polybrominated diphenyl ethers (PBDEs): new pollutants-old diseases. *Clin Med Res,* 1, 281-90.
[http://dx.doi.org/10.3121/cmr.1.4.281] [PMID: 15931321]

Solomon, FP & Faustman, EM (1987) Developmental toxicity of four model alkylating agents on japanese medaka fish (Oryzias latipes) embryos. *Environ Toxicol Chem,* 6, 747-53.
[http://dx.doi.org/10.1002/etc.5620061004]

Stoss, F W & Haines, T A (1979)

Stresser, DM, Williams, DE, Griffin, DA & Bailey, GS (1995) Mechanisms of tumor modulation by indole-3-carbinol. Disposition and excretion in male Fischer 344 rats. *Drug Metab Dispos,* 23, 965-75. [PMID: 8565787]

Tagliaferri, S, Caglieri, A, Goldoni, M, Pinelli, S, Alinovi, R, Poli, D, Pellacani, C, Giordano, G, Mutti, A & Costa, LG (2010) Low concentrations of the brominated flame retardants BDE-47 and BDE-99 induce synergistic oxidative stress-mediated neurotoxicity in human neuroblastoma cells. *Toxicol In Vitro,* 24, 116-22. [http://dx.doi.org/10.1016/j.tiv.2009.08.020] [PMID: 19720130]

Tangtian, H, Bo, L, Wenhua, L, Shin, PK & Wu, RS (2012) Estrogenic potential of benzotriazole on marine medaka (*Oryzias melastigma*). *Ecotoxicol Environ Saf,* 80, 327-32. [http://dx.doi.org/10.1016/j.ecoenv.2012.03.020] [PMID: 22521813]

Teather, K, Harris, M, Boswell, J & Gray, M (2001) Effects of Acrobat MZ and Tattoo C on Japanese medaka (Oryzias latipes) development and adult male behavior. *Aquat Toxicol,* 51, 419-30. [http://dx.doi.org/10.1016/S0166-445X(00)00124-7] [PMID: 11090900]

Tian, L, Cheng, J, Chen, X, Cheng, SH, Mak, YL, Lam, PKS, Chan, LL & Wang, M (2014) Early developmental toxicity of saxitoxin on medaka (*Oryzias melastigma*) embryos. *Toxicon,* 77, 16-25. [http://dx.doi.org/10.1016/j.toxicon.2013.10.022] [PMID: 24184516]

Tseng, YC, Hu, MY, Stumpp, M, Lin, LY, Melzner, F & Hwang, PP (2013) CO(2)-driven seawater acidification differentially affects development and molecular plasticity along life history of fish (Oryzias latipes). *Comp Biochem Physiol A Mol Integr Physiol,* 165, 119-30. [http://dx.doi.org/10.1016/j.cbpa.2013.02.005] [PMID: 23416137]

van de Merwe, JP, Chan, AK, Lei, EN, Yau, MS, Lam, MH & Wu, RS (2011) Bioaccumulation and maternal transfer of PBDE 47 in the marine medaka (*Oryzias melastigma*) following dietary exposure. *Aquat Toxicol,* 103, 199-204. [http://dx.doi.org/10.1016/j.aquatox.2011.02.021] [PMID: 21481818]

Wang, J, Wang, A & Wang, WX (2017) Evaluation of nano-ZnOs as a novel Zn source for marine fish: importance of digestive physiology. *Nanotoxicology,* 11, 1026-39. [http://dx.doi.org/10.1080/17435390.2017.1388865] [PMID: 29050525]

Wang, J & Wang, WX (2014) Low bioavailability of silver nanoparticles presents trophic toxicity to marine medaka (*Oryzias melastigma*). *Environ Sci Technol,* 48, 8152-61. [http://dx.doi.org/10.1021/es500655z] [PMID: 24937273]

Wang, M, Wang, Y, Wang, J, Lin, L, Hong, H & Wang, D (2011) Proteome profiles in medaka (*Oryzias melastigma*) liver and brain experimentally exposed to acute inorganic mercury. *Aquat Toxicol,* 103, 129-39. [http://dx.doi.org/10.1016/j.aquatox.2011.02.020] [PMID: 21458406]

Wang, M, Wang, Y, Zhang, L, Wang, J, Hong, H & Wang, D (2013) Quantitative proteomic analysis reveals the mode-of-action for chronic mercury hepatotoxicity to marine medaka (*Oryzias melastigma*). *Aquat Toxicol,* 130-131, 123-31. [http://dx.doi.org/10.1016/j.aquatox.2013.01.012] [PMID: 23416409]

Wang, Y, Wang, D, Lin, L & Wang, M (2015) Quantitative proteomic analysis reveals proteins involved in the neurotoxicity of marine medaka *Oryzias melastigma* chronically exposed to inorganic mercury. *Chemosphere,* 119, 1126-33. [http://dx.doi.org/10.1016/j.chemosphere.2014.09.053] [PMID: 25460752]

Wittbrodt, J, Shima, A & Schartl, M (2002) Medaka-a model organism from the far East. *Nat Rev Genet,* 3, 53-64. [http://dx.doi.org/10.1038/nrg704] [PMID: 11823791]

Wong, SW, Leung, PT, Djurisić, AB & Leung, KM (2010) Toxicities of nano zinc oxide to five marine organisms: influences of aggregate size and ion solubility. *Anal Bioanal Chem,* 396, 609-18. [http://dx.doi.org/10.1007/s00216-009-3249-z] [PMID: 19902187]

Wu, X, Huang, Q, Fang, C, Ye, T, Qiu, L & Dong, S (2012) PFOS induced precocious hatching of *Oryzias melastigma*-from molecular level to individual level. *Chemosphere,* 87, 703-8.
[http://dx.doi.org/10.1016/j.chemosphere.2011.12.060] [PMID: 22273185]

Xie, SL, Junaid, M, Bian, WP, Luo, JJ, Syed, JH, Wang, C, Xiong, WX, Ma, YB, Niu, A, Yang, XJ, Zou, JX & Pei, DS (2018) Generation and application of a novel transgenic zebrafish line Tg(cyp1a:mCherry) as an *in vivo* assay to sensitively monitor PAHs and TCDD in the environment. *J Hazard Mater,* 344, 723-32.
[http://dx.doi.org/10.1016/j.jhazmat.2017.11.021] [PMID: 29154098]

Xu, EGB, Morton, B, Lee, JHW & Leung, KMY (2015) Environmental fate and ecological risks of nonylphenols and bisphenol A in the Cape D'Aguilar Marine Reserve, Hong Kong. *Mar Pollut Bull,* 91, 128-38.
[http://dx.doi.org/10.1016/j.marpolbul.2014.12.017] [PMID: 25561005]

Yan, M, Leung, PTY, Ip, JCH, Cheng, JP, Wu, JJ, Gu, JR & Lam, PKS (2017) Developmental toxicity and molecular responses of marine medaka (*Oryzias melastigma*) embryos to ciguatoxin P-CTX-1 exposure. *Aquat Toxicol,* 185, 149-59.
[http://dx.doi.org/10.1016/j.aquatox.2017.02.006] [PMID: 28214734]

Ye, RR, Lei, EN, Lam, MH, Chan, AK, Bo, J, van de Merwe, JP, Fong, AC, Yang, MM, Lee, JS, Segner, HE, Wong, CK, Wu, RS & Au, DW (2011) Gender-specific modulation of immune system complement gene expression in marine medaka *Oryzias melastigma* following dietary exposure of BDE-47. *Environ Sci Pollut Res Int,* 19, 2477-87.
[http://dx.doi.org/10.1007/s11356-012-0887-z] [PMID: 22828878]

Ye, T, Kang, M, Huang, Q, Fang, C, Chen, Y, Liu, L & Dong, S (2016) Accumulation of di(2-ethylhexyl) phthalate causes endocrine-disruptive effects in marine medaka (*Oryzias melastigma*) embryos. *Environ Toxicol,* 31, 116-27.
[http://dx.doi.org/10.1002/tox.22028] [PMID: 25066029]

Ye, T, Kang, M, Huang, Q, Fang, C, Chen, Y, Shen, H & Dong, S (2014) Exposure to DEHP and MEHP from hatching to adulthood causes reproductive dysfunction and endocrine disruption in marine medaka (*Oryzias melastigma*). *Aquat Toxicol,* 146, 115-26.
[http://dx.doi.org/10.1016/j.aquatox.2013.10.025] [PMID: 24292025]

Yi, X & Leung, KMY (2017) Assessing the toxicity of triphenyltin to different life stages of the marine medaka *Oryzias melastigma* through a series of life-cycle based experiments. *Mar Pollut Bull,* 124, 847-55.
[http://dx.doi.org/10.1016/j.marpolbul.2017.02.030] [PMID: 28242277]

Yu, M, Luo, XJ, Wu, JP, Chen, SJ & Mai, BX (2009) Bioaccumulation and trophic transfer of polybrominated diphenyl ethers (PBDEs) in biota from the Pearl River Estuary, South China. *Environ Int,* 35, 1090-5.
[http://dx.doi.org/10.1016/j.envint.2009.06.007] [PMID: 19616300]

Yu, WK, Shi, YF, Fong, CC, Chen, Y, van de Merwe, JP, Chan, AKY, Wei, F, Bo, J, Ye, R, Au, DWT, Wu, RSS & Yang, MS (2013) Gender-specific transcriptional profiling of marine medaka (*Oryzias melastigma*) liver upon BDE-47 exposure. *Comp Biochem Physiol Part D Genomics Proteomics,* 8, 255-62.
[http://dx.doi.org/10.1016/j.cbd.2013.06.004] [PMID: 23962555]

Zhang, L, Dong, X, Wang, C, Zuo, Z & Chen, M (2017) Bioaccumulation and the expression of hepatic cytochrome P450 genes in marine medaka (*Oryzias melastigma*) exposed to difenoconazole. *J Environ Sci (China),* 52, 98-104.
[http://dx.doi.org/10.1016/j.jes.2016.03.011] [PMID: 28254063]

Zhu, Z, He, L & Chen, S (1985) Novel gene transfer into the fertilized eggs of gold fish (Carassius auratus L. 1758). *J Appl Ichthyology,* 1, 31-4.
[http://dx.doi.org/10.1111/j.1439-0426.1985.tb00408.x]

CHAPTER 12

Problems of Invasive Species: A Case Study from Andaman and Nicobar Islands, Andaman Sea, India

P.M. Mohan* and **Barbilina Pam**

Department of Ocean Studies and Marine Biology, Pondicherry University Off Campus, Brookshabad, Port Blair – 744112, Andaman and Nicobar Islands, India

Abstract: A fauna or flora that was not native to the particular environment and caused harm by its proliferation was considered an invasive species. This is one among the major concern for protecting the biodiversity as well as economic loss. However, in the marine environment, this problem is further complicated due to less barrier and other common factors, such as movement of the vessels, the release of ballast waters by the tankers, and water currents. The marine Island environment concern has gain significance due to the larger distribution of benthic faunal community and its biodiversity. A study was carried out to understand the status of this problem with reference to Andaman island environment and a probable mechanism to be implemented and their status was discussed in this article.

Keywords: Invasive Species, Marine Environment, Andaman and Nicobar Islands, Andaman Sea, India.

INTRODUCTION

An invasive species has been defined as a life form, which is not native to the particular environment, but proliferates and causes harm to the native species. This problem has been considered one of the major threats to any ecosystem with reference to the environment, economy, or to human health or combination of above. Colautti *et al.*, (2004) very widely discussed the terminology "Invasive" species and its use of different organisms. However, in general terms, it is mostly accepted that an invasive species is a species that is introduced intentionally or unintentionally in an environment and proliferates causing damage to the introduced environment in all the aspects. The intentional introduction may happen through the imported live species in fisheries or aquaculture practices or

* **Corresponding author P.M. Mohan:** Department of Ocean Studies and Marine Biology, Pondicherry University Off Campus, Brookshabad, Port Blair – 744112, Andaman and Nicobar Islands, India; Tel: +91-3192-262317; +91-319--262342; 262366; 262362; 262320; Fax +91-3192-262323; E-mails: pmmtu@yahoo.com; pmmnpu@rediffmail.com

De-Sheng Pei & Muhammad Junaid (Eds.)
All rights reserved-© 2019 Bentham Science Publishers

aquarium that have been introduced into the natural system due to a certain purpose or improper management or deliberate actions. However, in the case of unintentional introduction, it is concerned with the ships or boat hull carry as biofoulers, ballast waters release or connecting major water masses by artificial means leading to the species migration from their original location to new locations. The United States estimated that out of 750,000 species around 50,000 species of plants, animals and microbes have been introduced as reported by Pimentel *et al.,* (2000). The estimated cost of loss due to this invading species was about USD 120 billion per year (Pimentel, 2005). The invasive alien species study in India was far behind with reference to its intensity, scale, and scope, due to its high taxonomic diversity and large distribution area (Khuroo *et al.,* 2007; Peh, 2010; Adhikari *et al.,* 2015). The invasive or alien species are introduced into the coastal or marine area by many different ways, such as shipping, mariculture, oil and gas exploitation, tourism, and aquarium trade. The present article deals with the status of invasive species in the tropical island coastal or marine area, specific to off Andaman and Nicobar Islands.

STATUS OF INVASIVE SPECIES IN ANDAMAN AND NICOBAR ISLANDS

Marine environment has the potential to invade by different species by the process of movement of current, migration of strong swimmers, external hull fouling, holes or crevices in the wooden hull of a ship, ballast tank waters, catastrophic events such as tsunamis, storm surges, *etc.*, translocate the macro-fauna larvae, cysts, eggs, including fauna or flora itself to the alien environment. Even though the introductory environment is alien, the introduced species survive and thrive to become a threat to the native species. This review work identified different species as an invasive species and tabulated (Table 1). Overall 182 species have the potential to invade into the alien environment by different means and survived to cause a threat to the native species. The marine algae *Monostroma oxysperma* from the Phylum Chlorophyta is the most potential species in this group of flora. The dominant alien species reported from Phylum Cnidaria has around 65 species from the reported one. The Class Scyphozoa and Hydrozoa reported as highest influenced among these phyla and these two Classes alone represented 60 species as invasive in nature. Over and above two more classes also had this trait, *i.e.* Class Cubozoa and Staurozoa. Between these Classes, 4 Orders once again represent the highest species showing invasive nature. They are Order Rhizos-tomeae (21 Nos.), Semaeostomea (10 Nos.) belonging to the Class Scyphozoa and the Order Leptothecata (15 Nos.) and Order Anthoethecata (7 Nos.) belonging to the Class Hydrozoa were found dominant in the total species identified in this Phylum. The Class Anthozoa has three Orders: Alcyonacea (1 No.), Scleractinia (1 No.), and Actiniaria (1 No.) with a representation of one species each under this trait.

Next to Phylum Cnidaria, the Phylum Chordata was identified having maximum invasive characteristic among the reported invasive species. Under the Phylum Chordata, the Class Ascidicea represented 39 species belonging to different Families. Out of these Families the Family Styclidae (9 Nos.) and Didemnidae (8 Nos.) were found dominant and remaining 4 Families Ascididae, Pyuridae, Perophoridae, and Polycitoridae have their own contribution to this activity.

The Phylum Arthropoda comes next to Phylum Chordata, which show 36 species having the invasive characters. This Phylum is represented by two Classes Malacostraca and Hexanauplia with three Orders each for their species distribution with invasive characters. The Orders are as follows: Isopoda (6 Nos.), Amphipoda (10 Nos.) and Decapoda (6 Nos.) from the Class Malacostraca and Sessilia (7 Nos.), Calanoida (6 Nos.), and Harpacticoida (1 No.) from the Class Hexanauplia.

Other than the above Phyla, there are other five Phyla, which also provide their share in the category of invasive species. They are Phylum Annelida (16 Nos.), Mollusca (8 Nos.), Ctenophora (4 Nos.), Bryozoa (6 Nos.) and Entoprocta (1 No.). Phylum Annelida represented five Orders: Sabellida (3 Nos.), Phyllodocida (4 Nos.), Terebellida (1 No.), Eunice (4Nos.), and Spionida (4 Nos.). In Phylum Mollusca, species belong to two Classes and three Orders. They are Class Gastropoda with Order Nudibranchia (1 No.) and Class Bivalvia with Orders Myida (6 Nos.) and Mytilidae (1 No.). The species identified under Phylum Bryozoa consists of one Class Gymnolaematidae and two Orders: Ctenostomatidae (1 No.) and Cheilostomatidae (5 Nos.). The Phylum Entoprocta is represented by only one Family, *i.e.* Barentisiidae with *Barentia ramosa* as the only invasive species.

INVASIVE SPECIES THREAT

Invasive species are the second largest threat to biodiversity next to environmental degradation. The anthropogenic trans-national and domestic movement leads to introduce inadvertently as well as deliberately in an environment for vanity or profit. Pimentel (2007) reported that 480,000 alien species were introduced all over the world, and these countries, such as UK, USA, South Africa, Australia, India, and Brazil, have three-fourths of these species.

The invasive species science is limited, so understanding of their impact on the environment is also limited. How a new species invades into a new habitat? It is well known that a species introduced in an environment should cross the barriers such as survival, establish its base population, reproduction and explosion then only it will act like an invasive species. The introduction normally occurred in three ways, *i.e.* transported to a new habitat by accident (a foreigner visited in the country with his own country's soil), imported for a purpose and introduced (control or propagate a species) and a regular introduction through a known methodology (ballistic waters). The invasive species have a capacity to intrinsically better competitors and have less or absence of enemies (Callaway et al., 2005). There are three models identified for the invasive species proliferation. They are the fluctuating resource availability model, the enemy release hypothesis and niche opportunity model. The fluctuating availability model is defined as the natural resources that are available in plenty, catering to the development of invasive species and the fluctuating resource leads to the dominance of the same at a point of time. The enemy release hypothesis suggesting that the invasive species are also providing an opportunity not observed by their enemy available in the environment due to less in concentration or over sightedness than it may proliferate during the right opportunity. Sometimes, it may also provide co-species, which can be harmful to the other biota existing in that invasive environment and the same favorable for the invasive species. The niche opportunity model represents, as sometimes when all the above parameters are available at a point in time, to an environment, which is conducive for the proliferation of the invasive species.

The invasive species alter the environment significantly and affect the biota in large scale by the ways of reducing or enhancing soil nutrients, their microbes, agricultural pests or even pathogens which are harmful to any faunal community. The water hyacinth introduced in India proliferated in all the water bodies destructing the natural habitat. Similarly, the zebra mussel invaded from Russia into U.S.A and destroyed the native species in large scale as well as clogging the pipeline destroying the economy of the concerned environment in an unexpected manner.

"Prevention is better than cure", is good for the invasive species because once the species invades and proliferates, it is impossible to control. Instead of that, it is possible to control through the routes of invades by the way of different quarantine methods that are considered more cost-effective than controlling after having invaded the same. The Andaman Islands, the spotted deer introduced for domestication led to killing or disappearing of more than a few tens of plant species which could not be able to reappear even after the deer was controlled at one stage. However, it is interesting to note that the Government of India keeps this species in a protected animal list and thus, cannot be killed and eliminated from the environment. Similarly, chemical spraying and biological means of control may be considered as a control measure for the invasive species proliferation. But these two methods also have its own problem by the way of health impacts to native faunal species, which develop a multifaceted problem to the particular environment or nearby environments. The biological control developed to control the alien African Giant Snail in the Hawaiian Islands by the way of introduction of Euglandina Rosea snail, a predator which ended up by removing 41 more indigenous snail species in this island environment.

To Control or monitor the invasive species is very difficult with reference to a country like India, where has a larger area, different environments, and entries leading a more complicated process for the above activities. New Zealand or Australia has a way to seal off all the entries by the way of strict quarantine procedure, while the constant monitoring was not possible in India. However, that prevention is a paramount importance for this problem.

Is it possible for Indian environment to solve this problem? Who has a lack of taxonomist as well as bit and pieces of research work carried out in different corners of this country without any coordination? Thus, this is leading to a very pathetic situation for this problem concerned. The database on this aspect is yet to be evolved, and even the available database cannot be accessed easily whenever and wherever the necessity arises.

MAJOR IMPACT OF THE MARINE INVASIVE SPECIES AND ITS STATUS

The existing report on the Andaman Sea, the Arabian Sea and the Bay of Bengal was very limited. Most of the cases were observed as a newspaper report or a passing remark on the scientific publications. The real impact as an invasive species was not studied or monitored in the above said marine waters. The existing literature suggested that the Andaman Sea has thirteen invasive marine species, such as one soft coral (*Carijoa riisei*), one hard coral (*Tubastrea coccinenea*), one ascidian (*Phallusia nigra*) and ten Jellyfishes (*Cassiopea Andromeda, Ceohea* sp., *Crabonella annandalei, Aurelia aurita, Atolla wyvillei, Eirene ceylonensis, Eirene hexanemalis, Helgicirrha malayensis, Octophialucium indicum,* and *Liriope tetraphylla*). Out of these thirteen species, only five species were identified and reported as an invasive species and the remaining species were reported only for its availability.

However, among these species, when compared with the Arabian Sea and Bay of Bengal waters, all the remaining reported 169 are invasive species. Even though these invasive species were reported as a new species or problematic species in an environment, further report on proliferation, repeated occurrence, and its impact was not studied seriously. Moreover, the impact of these invasive species was reported as newspaper information along with the opinions of academics. A few examples have been added herewith.

During the year 2013 October, Nandakumar (2013) reported that the jellyfish invaded a large number and resultant mass mortality and fouling of beaches was observed (Fig. **1**). Further, this report also stated that fishing trawlers were facing a very difficult time, especially during the post-monsoon period due to the clogging of the fishing net with jellyfish and a resultant reduction in by a catch. Over and above, it was also found that these jellyfish blooms also affected the nuclear power plants due to the clogging the water intake filters. Vikram (2014) reported that the change of currents and winds moved these open ocean creatures (Jellyfish) towards the beach and created problems for fishermen by the way of clogging the fishing net, sting on their body and reducing the catch in Bombay coastal areas.

Fig. (1). The dead jellyfish in Kerala Coast (Nandakumar, 2013).

The Bay Bengal region, near off Visakhapatnam, around 60 nautical miles, there was a bloom of jellyfish, reported by the MFV Matsya Shikari, attached to Fishery Survey of India (FSI), Visakhapatnam. They captured a 500 kg of jellyfish in a single haul. The catch was a single species *Crambionella stuhlmanni*. This bloom caused interferences to regular fisheries. The senior scientific assistant at FSI, Visakhapatnam, mentioned that the cause of this bloom may be due to the impact of Cyclone Hudhud. The Tribune India (2017) reported that Betalbatim and Velsao beaches of Goa, Arabian Sea, the western part of India, mushroomed with the jellyfish and cautioned the tourist not to venture into the sea for swimming. Johan (2017) reported about the death of different sea creatures such as jellyfish, *Balanus amphitrite*, puffer fish and *Physalia physalis* during the end of SW monsoon, in Alappuzha beach, Kerala, India. Dr. Hatha, Marine Biology Professor from Cochin University of Science and Technology opined that this incidence due to the climatic change, variation among the sea salinity, or change in the monsoon pattern. Kripa *et al.,* (2016) reported that the decline of the sardine fishery along the Kerala coast due to the jellyfish blooms during the year 2013 and 2014 which disturb the juveniles and spawning activities.

Jyeabaskaran *et al.,* (2016) reported that the jellyfish *Lychnorhiza malayensis* was in Malayan waters, which extended to the Arabian Sea from Trivandrum to Goa waters. Earlier, Nair (1951) reported that during the year 1942 and 1943 it extended up to Trivandrum for this abundance. The above information suggested that within the seventy-five years the proliferation has extended from Trivandrum to Goa. The Hindu (2018) reported that during the month of March 2018, the Kanyakumari Coast, Tamil Nadu, had suffered bloom of the jellyfish formation and the climatic factors are considered as a major factor for this variation.

ERADICATION OF INVASIVE SPECIES

A constant monitoring of the environment with reference to the basic database on their biodiversity will provide an idea of invasive species. However, for that, it is essential to have a basic database of biodiversity for an environment, which is still lacking in the marine realm of the country, even in the world ocean concern. A system should evolve to use the manpower available in the coastal academic institute to develop a database of their regular academic program with a simple monitoring and accountability process.

Unfortunately, In the Indian scenario, the Government System has a major role for marine academic and management process while the private ownership with a very minimal role in this aspect. The government system is not encouraging the long time process which in turn the basic date base creation on these aspects are not being able to succeed. Until and unless, no basic date base is available, the present day scenario with climatic variability cannot be tackled with reference to invasive species as the migration from one environment to another environment may expect increased environmental stress in a significant level. Thus, the climatic factors affecting the biodiversity and the effect of the invasive species also lead to havoc on the marine biodiversity, which cannot be predicted based on today's poor database.

The management practices are considered by the ways of preventing or minimizing the introduction on their route, prevention through early detection and rapid removal from the environment and containment. The marine environment concern, prevention is the most cost-effective method than eradication. The prevention can be applied in the areas of mariculture and aquarium trade, as well as the pretreatment of ballast water and monitoring of probable species and its eradication, when it is introduced into the system without providing time for the establishment in the new environment. Convention on Biological Diversity has clearly mentioned in the Article 8 that the necessity of production or sustainable measures to be developed for the preservation of biological diversity in an environment from the invasive species (CBD, 1993).

There are several regulations already introduced, which can be implemented at the most care and efficiency. The regulations are as follows:

• The Convention on Biological Diversity (CBD, 1992).
• The United Nations Convention on the Law of the Sea (UNCLOS, 1982).
• The International Convention for the Control and Management of Ship's Ballast Water and Sediments (BWM Convention, 2004).
• The International Convention on the Control of Harmful Anti-Fouling Systems on Ships (AFS Convention, 2001).
• The IMO Voluntary Guidelines on Biofouling (2011).
• Code of Conduct for Responsible Fisheries (1995).
• CBD Strategic Plan for Biodiversity 2011-2020.

It is high time to implement the serious efforts to identify and assess the invasive species and its impact on the marine environment.

Table 1. Invasive species list till date identified in Indian waters.

S. No.	Species	Native	Invasive	Harmful Impact	Reference
	Marine Algae				
1.	Kingdom- PLANTAE Phylum- CHLOROPHYTA Class- ULVOPHYCEAE Order- ULVALES Family- GAYRALIACE Genus- *Monostroma* Species- *Monostroma oxyspermum Gayralia oxysperma* (Kutzing), Vinogradova ex Scagel *et al.*, (1989)	Northeast Atlantic and Northwest Pacific	West coast of India: Intertidal region of Gujarat, Goa, and Maharashtra		Untawale *et al.*, 1980; Anil *et al.*, 2002; Chandra and Raghunathan, 2018 (PC); WoRMS (2018)
	Phylum- CNIDARIA (Jellyfish)				
	Class- SCYPHOZOA				
2.	Order- RHIZOSTOMEAE Family- MASTIGIIDAE Genus- *Phyllorhiza* Species- *Phyllorhiza punctata* (Australian spotted Jellyfish)	Australia and Philippines	Gulf of Mannar and Palk Bay, India	Due to its large size, during the swarming period, it causes entanglement in the shore- seine leading to the economic loss to fisher folks.	Saravanan *et al.*, 2016; Fofonoff *et al.*,2017; Chandra and Raghunathan, 2018 (PC); WoRMS (2018)

(Table 1) cont.....

S. No.	Species	Native	Invasive	Harmful Impact	Reference
3.	Order- RHIZOSTOMEAE Family- CASSIOPEIDAE Genus- *Cassiopea* Species- *Cassiopea andromeda* (Upside- Down Jellyfish)	Indo West Pacific	East Coast of India Jalda khoti, Midnapur district, West Bengal (present record); Sand head, Ganga estuary; Gulf of Mannar; Madras, Tamil Nadu; West coast of India; Maldive Islands and Andamans Island		Rao,1931; Menon, 1930 and 1936; Kramp, 1961; Ramakrishna and Sarkar, 2003; Venkataraman *et al.*, 2012; Chandra and Raghunathan, 2018 (PC); WoRMS (2018)
4.	Order- RHIZOSTOMEAE Suborder- KOLPHORAE Family- MASTIGIIDAE Genus- *Mastigias* Species- *Mastigias albipunctatus*	Red Sea, South Africa, South Pacific Ocean	East Coast of India: Madras, Tamil Nadu	No poisonous effects are reported	Ramakrishna and Sarkar, 2003; WoRMS (2018)
5.	Order- RHIZOSTOMEAE Suborder- KOLPHORAE Family- MASTIGIIDAE Genus- *Mastigias* Species- *Mastigias papua*		East Coast of India: Gulf of Mannar, Tamilnadu	No poisonous effects are reported	Ramakrishna and Sarkar, 2003; WoRMS (2018)
6.	Order- RHIZOSTOMEAE Suborder- KOLPHORA Family- MASTIGIIDAE Genus- *Mastigietta* Species- *palmipes*	South Pacific Ocean	East Coast of India: Pamban beach, Tamil Nadu (present record)		Ramakrishna and Sarkar, 2003; WoRMS (2018)
7.	Family- THYSANOSTOMATIDAE Genus- *Thysanostoma* Species- *Thysanostoma thysanura*	Indo- Pacific, Japan	East Coast of India: Puri, Orissa		Ramakrishna and Sarkar, 2003; WoRMS (2018)

(Table 1) cont.....

S. No.	Species	Native	Invasive	Harmful Impact	Reference
8.	Order- RHIZOSTOMEAE Family- MASTIGIIDAE Genus- *Thysanostoma* Species- *Thysanostoma loriferum*	Red Sea	East Coast of India: Madras coast, Tamil Nadu (Menon, 1930) as: *Lorilera Lorilera*		Ramakrishna and Sarkar, 2003; WoRMS (2018)
9.	Order- RHIZOSTOMEAE Family- RHIZOSTOMATIDAE Genus- *Rhopilema* Species- *Rhopilema hispidum*	South Pacific Ocean	Gulf of Mannar and Palk Bay, India	During May- June intensity of swarm is very high causing hindrance in operating the shore-seine for fishing activity leading to the economic loss to fisher folks who are actually operating the shore seine for shoaling fishes but invariably get jellyfishes in their net during these swarming month	Saravanan *et al.*, 2016; WoRMS (2018)

(Table 1) cont....

S. No.	Species	Native	Invasive	Harmful Impact	Reference
10.	Order- RHIZOSTOMEAE Family- CEPHEIDAE Genus- *Netrostoma* Species- *Netrostoma coerulescens* (Crown Jellyfish)	South Pacific Ocean	Gulf of Mannar and Palk Bay, India	During May- June intensity of swarm is very high causing hindrance in operating the shore-seine for fishing activity leading to the economic loss to fisher folks who are actually operating the shore seine for shoaling fishes but invariably get jellyfishes in their net during these swarming month.	Saravanan *et al.*, 2016; Ramakrishna and Sarkar, 2003; WoRMS (2018)
11.	Order- RHIZOSTOMEAE Suborder- KOLOPHORAE Family- CEPHEIDAE Genus- *Netrostoma* Species- *Netrostoma setouchianum*	Pacific Ocean	East coast of India: Madras	No poisonous effects reported	Ramakrishna and Sarkar, 2003; WoRMS (2018)
12.	Order- RHIZOSTOMEAE Family- CEPHEIDAE Genus- *Marivagia* Species- *Marivagia stellata*	Indian Ocean	Vizhinjam, Kerala, India	Whose stings resulted in severe burn like injuries	Galil *et al.*,2013; Galil *et al.*, 2015; Chandra and Raghunathan, 2018 (PC); WoRMS (2018)
13.	Order- RHIZOSTOMEAE Suborder- KOLOPHORAE Family- CEPHEIDAE Genus- *Cephea*		Madras Coast in 1930, Nicobar Islands.	No poisonous effects reported	Ramakrishna and Sarkar, 2003; WoRMS (2018)

(Table 1) cont.....

S. No.	Species	Native	Invasive	Harmful Impact	Reference
14.	Order- RHIZOSTOMEAE Suborder- KOLPHORA Family- VERSURIGIDAE Genus- *Versuriga* Species- *Mastigietta anadyomene*	South Pacific Ocean	East Coast of India: Off Sandheads, Mouth of Ganga; Bay of Bengal, West Bengal		Ramakrishna and Sarkar, 2003; WoRMS (2018)
15.	Order- RHIZOSTOMEAE Suborder- DAKTYLIOPHORAE Family- LYCHNORHIZIDAE Genus- *Lychnorhiza* Species- *Lychnorhiza malayensis*		East Coast of India: Madras coast, Tamil Nadu; Trivandrum, West coast of India		Ramakrishna and Sarkar, 2003; WoRMS (2018)
16.	Order- RHIZOSTOMEAE Suborder- DAKTYLIOPHORAE Family: CATOSTYLIDAE Genus- *Acromitus* Species- *Acromitus flagellates* (In 1910 Mayer described as *Lorifera flagellata*)		East Coast of India: Digha coast, West Bengal (Present record); Kakdwip, West Bengal Chandipore, Orissa (Present record); Puri coast, Orissa ;Madras, Tamil Nadu; Krusadai Island; Tuticorin, Tamilnadu; Gulf of Mannar or Palk Bay; Cochin backwater, Kerala; Karwar coast, Karnataka	It is observed that irritation and swelling occur if contact with the hairy surface of the skin, especially on the upper palm (Present Observation).	Ramakrishna and Sarkar, 2003; WoRMS (2018)

(Table 1) cont.....

S. No.	Species	Native	Invasive	Harmful Impact	Reference
17.	Order- RHIZOSTOMEAE Suborder- DAKTYLIOPHORAE Family- CATOSTYLIDAE Genus- *Acromitus* Species- *Acromitus rabanchatu*		East Coast of India: Digha coast, West Bengal (Present record Puri coast, Orissa (Present record); Chilka lake.	It is observed that redness of skin, irritation and slight swelling occur if contact with the hairy surface of the skin, especially on the upper palm (Present observation).	Ramakrishna and Sarkar, 2003; WoRMS (2018)
18.	Order- RHIZOSTOMEAE Suborder- DAKTYLIOPHORAE Family: CATOSTYLIDAE Genus- *Crambionella* Species- *Crambionella annandalei*	India Ocean	East Coast of India: Digha, West Bengal; Puri, Orissa (Present record); Vizag, Andhra Pradesh; Madras, Tamil Nadu (Present Record); Carbine's cove, South Andaman (Present record), Andaman	No poisonous effects reported	Ramakrishna and Sarkar, 2003; WoRMS (2018)
19.	Order- RHIZOSTOMEAE Suborder- DAKTYLIOPHORAE Family- CATOSTYLIDAE Genus- *Crambionella* Species- *Crambionella orsini*	Indian Ocean	East Coast of India: Madras, Tamil Nadu; Krusadai Island, Tamilnadu; Pondichery; Trivandrum, Kerala; India; Arabian Sea; Red Sea; Iranian Gulf.	No poisonous effects reported	Ramakrishna and Sarkar, 2003; WoRMS (2018)
20.	Order- RHIZOSTOMEAE Family- LOBONEMATIDAE Genus- *Lobonema* Species- *Lobonema smithii*	West Pacific, Off the Coast of Southern Japan	East Coast of India: Madras, Tamil Nadu	No poisonous effects reported	Ramakrishna and Sarkar, 2003; WoRMS (2018)

(Table 1) cont......

S. No.	Species	Native	Invasive	Harmful Impact	Reference
21.	Order- RHIZOSTOMEAE Family- LOBONEMATIDAE Family- LOBONEMATIDAE Genus- *Lobonemoides* Species- *Lobonemoides robustus*	Central Indo-Pacific	East coast of India: Madras, Tamil Nadu	No poisonous effects reported	Ramakrishna and Sarkar, 2003; WoRMS (2018)
22.	Order- RHIZOSTOMEAE Family- CATOSTYLIDAE Genus- *Crambionella* Species- *Crambionella stuhlmanni*	South Africa	Gulf of Mannar and Palk Bay, India	Due to its large size, during the swarming period, it causes entanglement in the shore- seine leading to the economic loss to fisher folks.	Saravanan *et al.*, 2016; WoRMS (2018)
23.	Order- SEMAEOSTOMEA Family- ULMARIDAE Subfamily- AURELIINAE Genus- *Aurelia* Species- *Aurelia aurita*	East Pacific, Western Atlantic Ocean	Gulf of Mannar and Palk Bay, India; East Coast of India; Andaman Islands, Indian Ocean	Due to its large size, during the swarming period, it causes entanglement in the shore- seine leading to the economic loss to fisher folks.	Ramakrishna and Sarkar, 2003; Saravanan *et al.*, 2016; WoRMS (2018)
24.	Order- SEMAEOSTOMEA Family- ULMARIDAE Subfamily- AURELIINAE Genus- *Aurelia* Species- *Aurelia solida*		East Coast of India: Madras, Tamilnadu; Trivandrum, Kerala, India; Maldive Islands.	No poisonous effect is reported	Ramakrishna and Sarkar, 2003; WoRMS (2018)

(Table 1) cont....

S. No.	Species	Native	Invasive	Harmful Impact	Reference
25.	Order- SEMAEOSTOMEA Family- PELAGIIDAE Genus- *Pelagia* Species- *Pelagia noctiluca*	Europe	East Coast of India: Madras, Tamil Nadu; Calicut coast, Trivandrum coast, Kerala, India	This species is well known for bioluminescence. The species is commonly known as "Sea Blubber" and inflict their injurious effects on man by their nematocyst apparatus	Ramakrishna and Sarkar, 2003; Chandra and Raghunathan, 2018 (PC); WoRMS (2018)
26.	Order- SEMAEOSTOMEAE Family- PELAGIIDAE Genus- *Chrysaora* Species- *Chrysaora caliparea*	Northern Pacific Ocean	Gulf of Mannar and Palk Bay, India	During May- June intensity of swarm is very high causing hindrance in operating the shore-seine for fishing activity leading to the economic loss to fisher folks who are actually operating the shore seine for shoaling fishes but invariably get jellyfishes in their net during these swarming month.	Saravanan *et al.*, 2016; WoRMS (2018)

(Table 1) cont....

S. No.	Species	Native	Invasive	Harmful Impact	Reference
27.	Order- SEMAEOSTOMEAE Family- CYANEIDAE Genus- *Cyanea* Species- *Cyanea nozakii* (Ghost Jellyfish)		Gulf of Mannar and Palk Bay, India	During May- June intensity of swarm is very high causing hindrance in operating the shore-seine for fishing activity leading to the economic loss to fisher folks who are actually operating the shore seine for shoaling fishes but invariably get jelly fishes in their net during these swarming month.	Saravanan *et al.*, 2016; WoRMS (2018)
28.	Order- SEMAEOSTOMEAE Family- CYANEIDAE Genus- *Cyanea* Species- *capillata* (Lion's Mane jellyfish/ Gaint Jellyfish/ Hair Jellyfish)	North Sea, Danish Straits		Commonly known as "Sea Blubbers" or "Sea Nettle" and it is known for their poisonous effect.	Ramakrishna and Sarkar, 2003; WoRMS (2018)
29.	Order- SEMAEOSTOMEAE Family- CYANEIDAE Genus- *Cyanea* Species- *Cyanea purpurea*		East Coast of India: Digha coast, West Bengal & Ekakula, Orissa (Present record); Madras, Tamil Nadu; Trivandrum, Kerala, India; Japan.		Ramakrishna and Sarkar, 2003; WoRMS (2018)

(Table 1) cont....

S. No.	Species	Native	Invasive	Harmful Impact	Reference
30.	Order- SEMAEOSTOMEAE Family- PELAGIIDAE Genus- *Chrysaora* Species- *Chrysaora helvola*		Puri coast, Orissa; Madras, Tamil Nadu		Ramakrishna and Sarkar, 2003; WoRMS (2018)
31.	Order- SEMAEOSTOMEAE Family- PELAGIIDAE Genus- *Chrysaora* Species- *Chrysaora melanaster*		East Coast of India: Talsari, Orissa (Present record); Madras coast, Tamil Nadu, Trivandrum, Kerala, India	Poisonous effects not known.	Ramakrishna and Sarkar, 2003; WoRMS (2018)
32.	Order- SEMAEOSTOMEAE Family- PELAGIIDAE Genus- *Chrysaora* Species- *Chrysaora quinquecirrha*		East Coast of India Puri, Orissa; Madras, Tamil Nadu; Trivandrum coast, India	These jelly fishes are commonly known as 'Sea Nettle' and inflict injurious wounds on human by their nematocysts	Ramakrishna and Sarkar, 2003; WoRMS (2018)
33.	Order- CORONATAE Family- ATOLLIDAE Genus- *Atolla* Species- *Atolla wyvillei* (Atolla Jellyfish or Coronate Medusa)	Gulf of Mexico	East coast of India: Off Krishna delta, Andhra Pradesh. Off Gangetic delta; Andaman sea, Lakshadweep sea		Ramakrishna and Sarkar, 2003; WoRMS (2018)
34.	Order- CORONATAE Family- NAUSITHOIDAE Genus- *Nausithoe* Species- *Nausithoe punctata*	Gulf of Mexico, East Pacific, Western Atlantic Ocean	East Coast of India: Madras, Tamil Nadu, Trivandram (Travancore) Coast and Calicut, Kerala, India, Maldive Island		Ramakrishna and Sarkar, 2003; WoRMS (2018)
35.	Order- CORONATAE Family- PERIPHYLLIDAE Genus- *Periphylla* Species- *Periphylla periphylla* (Helmet Jellyfish)		East Coast of India: Bay of Bengal; Laccadive sea and almost all seas.		Ramakrishna and J. Sarkar, (2003); WoRMS (2018)

(Table 1) cont.....

S. No.	Species	Native	Invasive	Harmful Impact	Reference
	Class- HYDROZOA				
36.	Order- LEPTOTHECATA Family- BLACKFORDIIDAE Genus- *Blackfordia* Species- *Blackfordia virginica* (Black sea Jellyfish)	Black Sea	Mumbai Bassein Creek estuarine complex in India	A higher potential predation impact on the copepod population along the estuary	Santhakumari *et al.*, 1997; Jason *et al.*, 2017; Chandra and Raghunathan, 2018 (PC); WoRMS (2018)
37.	Order- LEPTOTHECATA Family- EIRENIDAE Genus- *Eugymnanthea*		East coast of India		Anil *et al.*, 2002; Chandra and Raghunathan, 2018 (PC); WoRMS (2018)
38.	Order- LEPTOTHECATA Family- EURENIDAE Genus- *Eutima* Species- *Eutima commensalism* (Considered endemic in Cochin backwaters)		Mumbai Harbour	Dense a warm causing problem to planktons	Santhakumari *et al.*, (1997); WoRMS (2018)
39.	Order- LEPTOTHECATA Family- EIRENIDAE Genus- *Eirene* Species- *Eirene ceylonensis*	Indian Ocean and New Zealand	Mumbai Harbour Andaman Sea – Port Blair		Santhakumari *et al.*, (1997); Saneen & Padmavati, 2017; WoRMS (2018)
40.	Order- LEPTOTHECATA Family- EIRENIDAE Genus- *Eirene* Species- *Eirene hexanemalis*	Indian Ocean and New Zealand	Andaman Sea – Port Blair		Saneen & Padmavati, 2017; WoRMS (2018);
41.	Order- LEPTOTHECATA Family- EIRENIDAE Genus- *Eirene* Species- *Eirene menoni*		Thana and Bassein Creeks, Bombay		Santhakumari *et al.*, 1997; WoRMS (2018)

(Table 1) cont....

S. No.	Species	Native	Invasive	Harmful Impact	Reference
42.	Order- LEPTOTHECATA Family- EIRENIDAE Genus- *Helgicirrha* Species- *Helgicirrha malayensis*		Thana and Bassein Creeks, Bombay, Andaman Sea – Port Blair		Santhakumari *et al.,* 1997; Saneen & Padmavati, 2017; WoRMS (2018)
43.	Order- LEPTOTHECATA Family- EIRENIDAE Genus- *Eucheilota* Species- *Eucheilota menoni*		Thana and Bassein Creeks, Bombay		Santhakumari *et al.,* 1997; WoRMS (2018)
44.	Order- LEPTOTHECATA Family- PHIALUCIUM Genus- *Malagazzi* Species- *Malagazzi multitentaculatum*		Thana and Bassein Creeks, Bombay		Santhakumari *et al.,* 1997; WoRMS (2018)
45.	Order- LEPTOTHECATA Family- PHIALUCIUM Genus- *Malagazzi* Species- *Malagazzi condensum*		Thana and Bassein Creeks, Bombay		Santhakumari *et al.,* 1997; WoRMS (2018)
46.	Order- LEPTOTHECATA Family- TIAROPSIDAE Genus- *Tiaropsidium* Species- *Tiaropsidium japonicum*		Thana and Bassein Creeks, Bombay		Santhakumari *et al.,* 1997; WoRMS (2018)
47.	Order- LEPTOTHECATA Family- AEQUOREIDAE Genus- *Aequorea* Species- *Aequorea tenuis*		Thana and Bassein Creeks, Bombay		Santhakumari *et al.,* 1997; WoRMS (2018)
48.	Order- LEPTOTHECATA Family- CAMPANULARIIDAE Genua- *Clytia* Species- *Clytia brunescens*		Mumbai Harbour		Santhakumari *et al.,* (1997); WoRMS (2018)

(Table 1) cont....

S. No.	Species	Native	Invasive	Harmful Impact	Reference
49.	Order- LEPTOTHECATA Family- CAMPANULARIIDAE Genus- *Clytia* Species- *Clytia hemisphaerica* (previous name- *Phialidium bicophorum*)		Thana and Bassein Creeks, Bombay		Santhakumari *et al.*, 1997; WoRMS (2018)
50.	Order- LEPTOTHECATA Family- MALAGAZZIIDAE Genus- *Octophialucium* Species- *Octophialucium indicum*		Thana and Bassein Creeks, Bombay, Andaman Sea – Port Blair		Santhakumari *et al.*, 1997; Saneen & Padmavati, 2017; WoRMS (2018)
51.	Order- ANTHOATHECATA Family- TUBULARIIDA Genus- *Ectopleura* Species- *Ectopleura crocea* (Pink mouth hydroid)	The Atlantic coast of North America	East and West coast of India		Mammen, 1963; Gravely, 1927; Nagale and Apte, 2013; Chandra and Raghunathan, 2018 (PC); WoRMS (2018)
52.	Order- ANTHOATHECATA Family- TUBULARIIDA Genus- *Ectopleura* Species- *Ectopleura sacculifera*		Thana and Bassein Creeks, Bombay		Santhakumari *et al.*, 1997; Mammen, 1963; Gravely, 1927; Nagale and Apte, 2013; WoRMS (2018)
53.	Order- ANTHOATHECATA Family- TUBULARIIDA Genus- *Ectopleura* Species- *Ectopleura dumortieri*		Thana and Bassein Creeks, Bombay		Santhakumari *et al.*, 1997; WoRMS (2018)
54.	Order- ANTHOATHECATAE Family- MOERISIIDAE Genus- *Moerisia* Species- *Moerisia lyonsi*	Black Sea	Mumbai Harbour	Not yet Assessed	Jason *et al.*, 2017; WoRMS (2018)

(Table 1) cont....

S. No.	Species	Native	Invasive	Harmful Impact	Reference
55.	Order- ANTHOATHECATA Family- BOUGAINVILLIIDAE Genus- *Bimeria* Species- *Bimeria vestita*	Scotland	East and West coast of India		Annandale, 1907; Mammen, 1963; Fofonoff *et al.*, 2003 and 2009; Chandra and Raghunathan, 2018 (PC); WoRMS (2018)
56.	Order- ANTHOATHECATA Suborder- CAPITATA Family- MOERISIIDAE Genus- *Moerisia* Species- *Moerisia inkermanika*		Arabian Sea, Arabian sea, Vishakapatnam, West Bengal		Santhakumari *et al.*, 1997; WoRMS (2018)
57.	Order- ANTHOATHECATA Family- CAPITATA Genus- *Zanclea* Species- *Zanclea costata*		Thana and Bassein Creeks, Bombay		Santhakumari *et al.*, 1997; WoRMS (2018)
58.	Order- TRACHYMEDUSAE Family- GERYONIIDAE Genus- *Liriope* Species- *Liriope tetraphylla*	North Atlantic Ocean and Gulf of Mexico	Thana and Bassein Creeks, Bombay, Andaman Sea – Port Blair		Santhakumari *et al.*, 1997; Saneen & Padmavati, 2017; WoRMS (2018)
59.	Order- TRACHYMEDUSAE Family- RHOPALONEMA Genus- *Agluara* Species- *Agluara hemistoma*	Gulf of Mexico	Thana and Bassein Creeks, Bombay		Santhakumari *et al.*, 1997; WoRMS (2018)
60.	Order- NARCOMEDUSAE Family-SOLMUNDAEGINIDAE Genus- *Solmundella* Species- *Solmundella bitanteculata*	Arctic Ocean and Gulf of Mexico	Thana and Bassein Creeks, Bombay		Santhakumari *et al.*, 1997; WoRMS (2018)

(Table 1) cont.....

S. No.	Species	Native	Invasive	Harmful Impact	Reference
61.	Order- NARCOMEDUSAE Family- CUNINIDAE Genus- *Cunina* Species- *Cunina duplicata*	Gulf of Mexico	Thana and Bassein Creeks, Bombay		Santhakumari *et al.*, 1997; WoRMS (2018)
	Class- CUBOZOA				
62.	Order- CARYBDEIDA Family- ALATINIDAE Genus- *Alantina* Species- *Alantina alata*, Reynaud, 1830 recorded as *Charybdea madraspatana*, Menon 1930. (Box Jellyfish)	Pacific	East coast of India: Madras, Tamil Nadu; Southern Atlantic, West Indies, South Africa, West Africa, South & Tropical Pacific, Hawaiian Islands, Philippines, Sumatra, Florida, Japan, Ceylon, Bermudas, Red sea, Indo-China, Carribean sea.	Commonly known as "Sea-wasp" and have powerful stings, which inflict their strong injurious effects upon bathers by nematocysts	Ramakrishna and Sarkar, 2003; WoRMS (2018)
63.	Order- CARYBDEIDA Family- TAMOYIDAE Genus- *Tamoya* Species- *Tamoya gargantuan* (Warty Sea Wasp)		Vishakapatnam coast, Andhra Pradesh, (present record), Sand head, West Bengal, Puri coast, Orissa, Ennur, Tamil Nadu, Mergui Archipelago, Arabian sea, North America, Malay Archipelago, New South Wales, Australia, Iranian Gulf, Japan, Samoa Islands, New Guinea.	Venoumous Jellyfish	Ramakrishna and Sarkar, 2003; WoRMS (2018)

(Table 1) cont....

S. No.	Species	Native	Invasive	Harmful Impact	Reference
64.	Order- CHIRODROPIDA Family- CHIRODROPIDAE Genus- *Chironex* Species- *Chironex fleckeri*		North Queensland and Australia	It is claimed to be the most venomous marine animal known to mankind and its sting is often fatal	Ramakrishna and Sarkar, 2003; Worms (2018)
65.	Order- CHIRODROPIDA Family- CHIROPSALMIDAE Genus- *Chiropsalmus* Species- *Chiropsalmus buitendijki*		East Coast of India: Marina Beach, Chennai, Tamil Nadu; Krusadai Island, Java, Malay Archipelago, Australia.	Venoumous	Ramakrishna and J. Sarkar, (2003); WoRMS (2018)
66.	Order- CHIRODROPIDA Family- CHIROPSALMIDAE Genus- *Chiropsalmus* Species- *Chiropsalmus quadrigatus*	Malaysia, Indo-West Pacific	East coast of India: Krusadai Island, Tamil Nadu	Commonly known as 'Sea wasp' and inflict their painful injurious effects upon bathers by their nematocysts	Ramakrishna and Sarkar, 2003; WoRMS (2018)
	Class- STAUROZOA				
67.	Order- STAUROMEDUSAE Family- KISHINOUYEIDAE Genus- *Clavadosia* (Stalked Jellyfish)		Krusadai Island (Gulf of Mannar), India		Ramakrishna and Sarkar, 2003; WoRMS (2018)
	Class- ANTHOZOA Subclass- OCTOCORALLIA (SOFT CORAL)				
68.	Order- ALCYONACEA Family- CLAVULARIID Genus- *Carijoa* Species- *Carijoa riisei* (Softflake coral)	Western Atlantic Ocean and Carolina to Brazil	Andaman and Nicobar Islands, Gulf of Kachchh, Gulf of Mannar, Thiruvananthapuram and Kanyakumari		Divya *et al.*, 2012; Chandra and Raghunathan, 2018 (PC); WoRMS (2018)

(Table 1) cont.....

S. No.	Species	Native	Invasive	Harmful Impact	Reference
	(SCLERACTINIAN CORAL)				
69.	Order- SCLERACTINIA Family- DENDROPHYLLIIDAE Genus- *Tubastrea* Species- *Tubastrea coccinea* (Orange cup coral)	Brazil, Carribbean Sea, Gulf of Guinea and Gulf of Mexico	Andaman and Nicobar Islands, Gulf of Mannar, Gulf of Kachchh and Lakshadweep		Venkataraman, 2004 and 2012; Chandra and Raghunathan, 2018 (PC); WoRMS (2018)
	HEXACORALLY (Sea Anemone)				
70.	Order- ACTINIARIA Family- DIADUMENID Genus- *Diadumene* Species- *Diadumene lineate*	Japan and Hong Kong	West coast of India		Parulekar, 1968; Fautin *et al.*, 2009; Hancock *et al.*, 2017; Chandra and Raghunathan, 2018 (PC); WoRMS (2018)
	Kingdom- ANIMALIA Phylum- CTENOPHORA Class- TENTACULATA Order- LOBATA				
71.	Family- BOLINOPSIDAE Genus- *Mnemiopsis* Species- *Mnemiopsis leidyi* (Comb Jelly)- 1980s	United States	Black Sea	It devoured the eggs and larvae of a wide variety of fish that led to a collapse in fishing industry. The fish catch fell by 90% in six years.	Anil *et al.*, 2002; WoRMS (2018)
72.	Order- PLATYCTENIDA Family- CTENOPLANIDA Genus- *Vallicula* Species- *Vallicula multiformis*	Jamaica	Gulf of Kachchh		Prasade *et al.*, 2015; Chandra and Raghunathan, 2018 (PC); WoRMS (2018)

(Table 1) cont.....

S. No.	Species	Native	Invasive	Harmful Impact	Reference
	Phylum- CTENOPHORA Class- NUDA				
73.	Order- BEROIDA Genus- *Beroe* Species- *Beroe ovate*	Atlantic coasts of North and South America	West coast of India		Chopra, 1960; Purcell *et al.,* 2007; Chandra and Raghunathan, 2018 (PC); WoRMS (2018)
74.	Order- BEROIDA Genus- *Beroe* Species- *Beroe cucumis* (Pink slipper comb jelly)	Atlantic waters	West coast of India		Chandra and Raghunathan, 2018 (PC); WoRMS (2018)
	Phylum- MOLLUSCA Class- GASTROPODA				
75.	Order- NUDIBRANCHIA Family- POLICERIDAE Genus- *Thecacera* Species- *Thecacera pennigera*	Atlantic coast of Europe	Kerala		Chandra and Raghunathan, 2018 (PC); WoRMS (2018)
	Class- BIVALVIA				
76.	Order- MYIDA Family- TEREDINIDAE Genus- - *Lyrodus* Species- *Lyrodus medilobata*	Indo- Pacific, Hawaiian Islands, Marshall Islands, New Zealand, Australia, Virginia, Bermuda	West coast of India		Anil *et al.,* 2002; WoRMS (2018)
77.	Order- MYIDA Family- TEREDINIDAE Genus- *Nausitora* Species- *Nausitora dunlopei*	Cochin	Goa		Anil *et al.,* 2002; WoRMS (2018)

(Table 1) cont....

S. No.	Species	Native	Invasive	Harmful Impact	Reference
78.	Order- MYIDA Family- TEREDINIDAE Genus- *Teredo* Species- *Teredo fuller*	Gulf of Mannar	Okha		Anil *et al.*, 2002; WoRMS (2018)
79.	Order- MYDIA Family- DREISSENIDA Genus- *Dreissena* Species- *Dreissena polymorpha* (Zebra Mussel)	Europe	North America (Lake St. Clair)	It has now spread to infest more than 40% of the United States waterways, fouls the cooling water intakes of the industry and may have cost US $ 5 billion in control measures since 1984.	Anil *et al.*, 2002; WoRMS (2018)
80.	Order- MYDIA Family- DREISSENIDA Genus- *Mytilopsis* Species- *Mytilopsis sallei* (Black striped Mussel/ Caribbean false mussel)	Tropical and Subtropical Atlantic waters	East and West coast of India (Mumbai and Vishakhapatnam)	Now spread to Hong Kong and Threatened Australia.	Anil *et al.*, 2002; Chandra and Raghunathan, 2018 (PC); WoRMS (2018)
81.	Order- MYDIA Family- TEREDINIDAE Genus- *Lyrodus* Species- *medilobatus*	Indo- Pacific Hawaiian Island, New Zealand, Australia, Virginia, Bermuda	West coast of India		Santhakumaran, 1986; Chandra and Raghunathan, 2018 (PC); WoRMS (2018)
82.	Family- PHOLADIDAE Genus- *Martesia* Species- *striata* (Striated wood paddock)	Native range is UNKNOWN	East and West coast of India and Lakshadweep		Nair and Dharmaraj, 1983; Suryo Rao and Subba Rao, 1991; Chandra and Raghunathan, 2018 (PC); WoRMS (2018)

(Table 1) cont....

S. No.	Species	Native	Invasive	Harmful Impact	Reference
83.	Order- MYTILIDAE Family- MYTILIDAE Genus- *Perna* Species- *perna* (Brown Mussel)	The Red Sea and East coast of South Africa	West coast of India, Tamil Nadu		Parulekar *et al.*, 1964; Menon and Pillai, 1996; Mahapatro *et al.*, 2015; Chandra and Raghunathan, 2018 (PC); WoRMS (2018)
	Phylum- ANNELIDA Class- POLYCHAETA				
84.	Kingdom- METAZOA Phylum- ANNELIDA Class- POLYCHAETA Order- SABELLIDA Family- SERPULIDAE Genus- *Ficopomatus* Species- *Ficopomatus enigmata* (Australian Tube Worm)	Australia	Indian Ocean	Hull fouling	Anil *et al.*, 2002; Kupriynova *et al.*, 2006; WoRMS (2018)
85.	Order - SABELLIDA Family- SERPULIDAE Genus- *Protula* Species- *Protula tubularis*	Native range is unknown	Mumbai		Gaonkar *et al.*, 2010; Chandra and Raghunathan, 2018 (PC); WoRMS (2018)
86.	Order - SABELLIDA Family- SERPULIDAE Genus- *Hydroides* Species- *Hydroides elegans*	Indo-Pacific	East coast of India		Mahapatro *et al.*, 2015; Chandra and Raghunathan, 2018 (PC); WoRMS (2018)
87.	Order- PHYLLODOCIDA Family- GLYCERIDAE Genus *Glycera* Species- *Glycera longipinnis*	Philippines	Mumbai		Gaonkar *et al.*, 2010; Chandra and Raghunathan, 2018 (PC); WoRMS (2018)

(Table 1) cont.....

S. No.	Species	Native	Invasive	Harmful Impact	Reference
88.	Order- PHYLLODOCIDA Family- NEREIDIDAE Genus- *Neanthes* Species- *Neanthes cricognatha*	New Zealand	Mumbai		Gaonkar *et al.*, 2010; Chandra and Raghunathan, 2018 (PC); WoRMS (2018)
89.	Order- PHYLLODOCIDA Family- NEREIDIDAE Genus- *Nereis* Species- *Nereis falcaria*	Mumbai	Gaonkar*et al.*, 2010		Gaonkar *et al.*, 2010; Chandra and Raghunathan, 2018 (PC); WoRMS (2018)
90.	Order- PHYLLODOCIDA Family- NEREIDIDAE Genus- *Perinereis* Species- *Perinereis nuntia*	Gulf of Suez	Mumbai		Gaonkar *et al.*, 2010; Chandra and Raghunathan, 2018 (PC); WoRMS (2018)
91.	Order- TEREBELLIDA Family- CIRRATULIDAE Genus- *Protocirrineris* Species- *chrysoderma*		Mumbai		Gaonkar *et al.*, 2010; Chandra and Raghunathan, 2018 (PC); WoRMS (2018)
92.	Family- COSSURIDAE Genus- *Cossura* Species- *Cossura coasta*	Greece	Mumbai		Gaonkar *et al.*, 2010; Chandra and Raghunathan, 2018 (PC); WoRMS (2018)
93.	Family- MALDANIDAE Genus- *Petaloproctus* Species- *Petaloproctus terricolus*	San Sebastian, South West Africa	Mumbai		Gaonkar *et al.*, 2010; Chandra and Raghunathan, 2018 (PC); WoRMS (2018)
94.	Order- EUNICIDA Family- LUMBRINERIDAE Genus- *Lumbrineris* Species- *Lumbrineris japonica*	Japan	Mumbai		Gaonkar *et al.*, 2010; Chandra and Raghunathan, 2018 (PC); WoRMS (2018)

(Table 1) cont....

S. No.	Species	Native	Invasive	Harmful Impact	Reference
95.	Order- EUNICIDA Family-LUMBRINERIDAE Genus- *Lumbrineris* Species- *Lumbrineris bifilaris*	The Pacific Ocean	Mumbai		Gaonkar *et al.*, 2010; Chandra and Raghunathan, 2018 (PC); WoRMS (2018)
96.	Order- EUNICIDA Family- ONUPHIDAE Genus- *Onuphis* Species- *Onuphis eremita*	Atlantic Ocean	Bay of Bengal, Mumbai		Fauvel, 1953; Gaonkar *et al.*, 2010; Chandra and Raghunathan, 2018 (PC); WoRMS (2018)
97.	Order- EUNICIDA Genus- *Onuphis* Species- *Onuphis holobranchiata*	Japan	Mumbai		Gaonkar *et al.*, 2010; Chandra and Raghunathan, 2018 (PC); WoRMS (2018)
98.	Order- SPIONIDA Family- SPIONIDAE Genus- *Polydora* Species- *Polydora limicola*	Sea of Japan		Fouling organism on the bottom of ships.	Chandra and Raghunathan, 2018 (PC);WoRMS (2018)
99.	Order- SPIONIDA Family- SPIONIDAE Genus- *Pseudopolydora* Species- *Pseudo-polydora paucibranchiata*		Northeastern Pacific		Chandra and Raghunathan, 2018 (PC);WoRMS (2018)
100.	Order- SPIONIDA Family- SPIONIDAE Genus- *Scolelepis* Species- *Scolelepis squamata*	Gulf of Mexico and the Caribbean Sea	Mumbai		Gaonkar *et al.*, 2010; Chandra and Raghunathan, 2018 (PC)
101.	Order- SPIONIDA Family- SPIONIDAE Genus- *Malacoceros* Species- *Malacoceros indicus*	Australia	Mumbai		Gaonkar *et al.*, 2010; Chandra and Raghunathan, 2018 (PC); WoRMS (2018)

(Table 1) cont....

S. No.	Species	Native	Invasive	Harmful Impact	Reference
	Phylum- ARTHROPODA **Subphylum- CRUSTACEA** **Class- MALACOSTRACA** **Order- ISOPODA**				
102.	Family- CIROLANIDAE Genus- *Cilicaea* Species- *Cilicaea lateraillei*	Indonesia, Philippines, Sri Lanka, South Africa, Red Sea and Australia	Arabian Sea		Venugopal and Wagh, 1986; Anil *et al*., 2002; Roy and Nandi, 2017; Chandra and Raghunathan, 2018 (PC); WoRMS (2018)
103.	Family- CIROLANIDAE Genus- *Cirolana* Species- *Cirolana hardfordi*		Indian waters	Not suspected	Anil *et al*., 2002; Anil et al., 2003; Roy and Nandi, 2017; Chandra and Raghunathan, 2018 (PC); WoRMS (2018)
104.	Family- SPHAEROMATIDAE Genus- *Paradello* Species- *Paradello dianae*	Pacific coasts of North and Central America	Indian waters	Not suspected	Anil *et al*., 2002; Anil et al., 2003; Roy and Nandi, 2017; Chandra and Raghunathan, 2018 (PC); WoRMS (2018)
105.	Family- SPHAEROMATIDAE Genus- *Sphaeroma* Species- *Sphaeroma serratum*		Indian waters	Not suspected	Anil *et al*., 2002; Anil et al., 2003; Roy and Nandi, 2017; Chandra and Raghunathan, 2018 (PC); WoRMS (2018)
106.	Family- SPHAEROMATIDAE Genus- *Sphaeroma* Species- *Sphaeroma walkeri*	Indian waters		Not suspected	Anil *et al*., 2002; Anil et al., 2003; Roy and Nandi, 2017; Chandra and Raghunathan, 2018 (PC); WoRMS (2018)
107.	Family- *IDOTEIDAE* Genus- *Synidotea* Species- *Synidotea laevidorsalis*	Australia and California	Indian waters	Not suspected	Anil *et al*., 2002, Anil et al., 2003; Roy and Nandi, 2017; Chandra and Raghunathan, 2018 (PC); WoRMS (2018)

(Table 1) cont....

S. No.	Species	Native	Invasive	Harmful Impact	Reference
	Order- AMPHIPODA				
108.	Family- *STENOTOID* Genus- *Stenothoe* Species- *Stenothoe gallensis*	China and Sri Lanka	Indian waters	Not yet assessed	*Anil et al.*, 2002; Anil *etal.*.2003; Walker, 1904; Venugopal and Wagh, 1986; Roy and Nandi, 2017; Chandra and Raghunathan, 2018 (PC); WoRMS (2018)
109.	Family- *STENOTOID* Genus- *Stenothoe* Species- *Stenothoe valida*	Brazil and Australia	Indian waters	Not yet assessed	Shyamsundari, 1997; Roy and Nandi, 2017; Chandra and Raghunathan, 2018 (PC); WoRMS (2018)
110.	Family- *ISCHYROCERIDAE* Genus- *Jassa* Species- *Jassa falcata*	Black sea, British coast and Ireland	Indian waters	Not yet assessed	Shyamasudari, 1997 Roy and Nandi, 2017; Chandra and Raghunathan, 2018 (PC); WoRMS (2018)
111.	Family- *ISCHYROCERIDAE* Genus- *Jassa* Species- *Jassa marmorata*	Native distribution is unknown; first described from Rhode Island and the North Atlantic	Indian waters	Not yet assessed	*Anil et al.*, 2002; Anil *et al.*, 2003; Roy and Nandi, 2017; Chandra and Raghunathan, 2018 (PC); WoRMS (2018)
112.	Family- ISCHYROCERIDAE Genus- *Ericthonius* Species- *Ericthonius brasiliensis*	Atlantic Ocean at Rio de Janeiro	East coast of India		Venugopal and Wagh, 1986; Chandra and Raghunathan, 2018 (PC); WoRMS (2018)
113.	Family- *MAERIDAE* Genus- *Quadrimaera* Species- *Quadrimaera pacifica* (Schellenberg, 1938) as *Maera pacifica*	Australia, Madagascar, North Pacific Ocean, Panama and Republic of Mauritius	Indian waters	Not yet assessed	Venugopal and Wagh, 1986; *Anil et al.*, 2002; Roy and Nandi, 2017; Chandra and Raghunathan, 2018 (PC); WoRMS (2018)

(Table 1) cont.....

S. No.	Species	Native	Invasive	Harmful Impact	Reference
114.	Family- *MAERIDAE* Genus- *Elasmopus* Species- *Elasmopus rapax*	Gulf of California	Indian waters		Shyamasudari, 1997; Roy and Nandi, 2017; Chandra and Raghunathan, 2018 (PC); WoRMS (2018)
115.	Family- *PODOCERIDA* Genus- *Podocerus* Species- *Podocerus brasiliensis*	Atlantic Ocean at Rio de Janeiro and Brazil	Indian waters	Not yet assessed	Venugopal and Wagh, 1986; Anil *et al.*, 2002; Shyamsundari, 1997; Roy and Nandi, 2017; Chandra and Raghunathan, 2018 (PC); WoRMS (2018)
116.	Family- *COROPHIIDAE* Genus- *Monocorophium* Species- *Monocorophium acherusicum* (Costa, 1853) as *Corophium acherusicum*	Europe	Indian waters	Not yet assessed	Shyamasudari, 1997; Roy and Nandi, 2017; Chandra and Raghunathan, 2018 (PC); WoRMS (2018)
117.	Family- *CAPRELLIDAE* Genus- *Paracaprella* Species- *Paracaprella pusilla*	Brazil	Indian waters	Not yet assessed	Lacerda and Masunari, 2014; Alarcon- Ortega, 2015; Guerra-Garcia, 2010; Roy and Nandi, 2017; Chandra and Raghunathan, 2018 (PC); WoRMS (2018)
Order- DECAPODA					
118.	Family- PENAEIDAE Genus- *Penaeus* Species- *Penaeus monodon*	Indian waters		Not suspected	Anil *et al.*, 2002; Roy and Nandi, 2017; WoRMS (2018)
119.	Family- PENAEIDAE Genus- *Penaeus* Species- *Penaeus vannamei*	Eastern Pacific coast	Indian waters		Dev Roy, 2007; Chandra and Raghunathan, 2018 (PC); WoRMS (2018)

Marine Ecology: Current and Future Developments, Vol. 1   **327**

(Table 1) cont....

S. No.	Species	Native	Invasive	Harmful Impact	Reference
120.	Family- PENAEIDAE Genus- *Litopenaeus* Species- *Litopenaeus vannamei*		Indian waters	Not suspected	Dev Roy, 2007; Roy and Nandi, 2017; WoRMS (2018)
121.	Family- PORTUNIDAE Genus- *Charybdis* (charybdis) Species- *Charybdis feriata*	Indian waters		Suspected harmful species	Anil *et al.*, 2002; Roy and Nandi, 2017; WoRMS (2018)
122.	Family- PORTUNIDAE Genus- *Charybdis* (charybdis) Species- *Charybdis hellerii*	Indian waters		Known harmful species	Anil *et al.*, 2002 Roy and Nandi, 2017; WoRMS (2018)
123.	Family- PORTUNIDAE Genus- *Scylla* Species- *Scylla serrata*	Indian waters		Not suspected	Anil *et al.*, 2002, Roy and Nandi, 2017; WoRMS (2018)
	Class- HEXANAUPLIA Infraclass- CIRIPEDIA (Barnacle)				
124.	Order- *SESSILIA* Family- *BALANIDAE* Genus- *Amphibalanus* Species- *Amphibalanus amphitrite* *Amphibalanus reticulatus* (Utinomi, 1967)as *Balanus reticulates* (Utinomi, 1967) and *Balanus amphitrite hawaiiensis*, Broch	The West Pacific and Indian Oceans from Southern Africa to Southern China	Mumbai, India	Not suspected	Anil *et al.*, 2002; Anil *et al.*, 2003; Roy and Nandi, 2017; Fofonoff *et al.*, 2003 and 2009; Chandra and Raghunathan, 2018 (PC); WoRMS (2018)
125.	Order- *SESSILIA* Family- *BALANIDAE* Genus- *Amphibalanus* Species- *Amphibalanus eburneus* (Gould, 1841) as *Balanus amphitrite eburneus*	The Western Atlantic from the Southern Gulf of Maine to Venezuela	Indian waters	Not suspected	Anil *et al.*, 2002; Anil *et al.*, 2003; Roy and Nandi, 2017; Fofonoff *et al.*, 2017; Chandra and Raghunathan, 2018 (PC); WoRMS (2018)

(Table 1) cont.....

S. No.	Species	Native	Invasive	Harmful Impact	Reference
126.	Order- *SESSILIA* Family- *BALANIDAE* Genus- *Amphibalanus* Species- *Amphibalanus cirratus* (Darwin, 1854) as *Balanus amphitrite cirratus*	Indian waters		Not suspected	Anil *et al.*, 2002; Roy and Nandi, 2017; WoRMS (2018)
127.	Order- *SESSILIA* Family-*BALANIDAE* Genus- *Fistulobalamus* Species- *Fistulobalanus pallidus* (Darwin, 1854) -*Balanus amphitrite var. statsburi*	West coast of Africa	West coast of Indian		Anil *et al.*, 2002; Anil *et al.*, 2003; Roy and Nandi, 2017; Chandra and Raghunathan, 2018 (PC); WoRMS (2018)
128.	Order- *SESSILIA* Family- *BALANIDAE* Genus- *Megabalamus* Species- *Megabalanus tintinnabulum*	The North Sea	Indian waters	Known harmful species	Anil *et al.*, 2002; Anil *et al.*, 2003; Roy and Nandi, 2017; Chandra and Raghunathan, 2018 (PC); WoRMS (2018)
129.	Order- *SESSILIA* Family- *BALANIDAE* Genus- *Megabalamus* Species- *Megabalanus zebra*			Not suspected	Anil *et al.*, 2002; Roy and Nandi, 2017; WoRMS (2018)
130.	Order- *SESSILIA* Family- *BALANIDAE* Genus- *Balanus* Species- *Balanus trigonus*			Not suspected	Anil *et al.*, 2002; Roy and Nandi, 2017; WoRMS (2018)
	Class- *HEXANAUPLIA* Subclass- *COPEPODA*		Mumbai, India		
131.	Order- *CALANOIDA* Family- *CALANIDAE* Genus- *Nannocalamus* Species- *Nannocalanus minor*				*Gaonkar et al., 2010;*Roy and Nandi, 2017; Chandra and Raghunathan, 2018 (PC); WoRMS (2018)

(Table 1) cont....

S. No.	Species	Native	Invasive	Harmful Impact	Reference
132.	Order- *CALANOIDA* Family- *CALANIDAE* Genus- *Cosmocalanus*		Mumbai, India		*Gaonkar et al., 2010;*Roy and Nandi, 2017; Chandra and Raghunathan, 2018 (PC); WoRMS (2018)
133.	Order- *CALANOIDA* Family- *PARACALANIDAE* Genus- *Paracalanus*		Mumbai, India		*Gaonkar et al.,2010;*Roy and Nandi, 2017; Chandra and Raghunathan, 2018 (PC); WoRMS (2018)
134.	Order- *CALANOIDA* Family- *TORTANIDAE* Genus- *Tortanus*		Mumbai, India		*Gaonkar et al., 2010;*Roy and Nandi, 2017; Chandra and Raghunathan, 2018 (PC); WoRMS (2018)
135.	Order- *HARPACTICOIDA* Family- *TACHIDIIDAE* Genus- *Euterpina* Species- *Euterpina acutifrons*		Mumbai, India		*Gaonkar et al., 2010;*Roy and Nandi, 2017; Chandra and Raghunathan, 2018 (PC); WoRMS (2018)
	Phylum- BRYOZOA				
136.	Class- GYMNOLAEMATIDAE Order-CTENOSTOMATIDAE Family- VESICULARIIDAE Genus- *Amathia* Species- *Amathia verticillata*	The Mediterranean sea	Chennai, Vishakhapatnam, Cuddalore, Palk Bay, Colachal and Mumbai		*Robertson, 1921; Nair et al., 1991; Swami and Udhayakumar, 2010; Bhave and Apte, 2012; Fofonoff et al., 2003 and 2009;* Chandra and Raghunathan, 2018 (PC); WoRMS (2018)
137.	Order- CHEILOSTOMATIDA Family- BUGULIDAE Genus- *Bugula* Species- *Bugula neritina*	Mediterranean and America Seas	West and East Coast of India	Cosmopolitan; Potential to be a fouling pest	*Robertson, 1921; Menon and Nair, 1967; Subba and Rao, 2005; Ryland et al., 2011;* Chandra and Raghunathan, 2018 (PC); WoRMS (2018)

(Table 1) cont.....

S. No.	Species	Native	Invasive	Harmful Impact	Reference
138.	Order- CHEILOSTOMATIDA Family- BUGULIDAE Genus- *Bugulina* Species- *Bugulina stolonifera*	Swan Sea, Wales; North West Atlantic	West Coast of India		*Shrinivaasu et al., 2015; Fofonoff et al., 2003 and 2009;* Chandra and Raghunathan, 2018 (PC); WoRMS (2018)
139.	Order- CHEILOSTOMATIDA Family- BUGULIDAE Genus- *Bugulina* Species- *Bugulina flabellate*	The North East Atlantic, from Southern Norway to Morocco and The Mediterranean Sea	East Coast on India		*Menon and Nair, 1967; Shrinivaasu et al., 2015; Fofonoff et al., 2003 and 2009;* Chandra and Raghunathan, 2018 (PC); WoRMS (2018)
140.	Order- CHEILOSTOMATIDA Family- CRYPTOSULIDAE Genus- *Cryptosula* Species- *Cryptosula pallasiana*	The North East Atlantic, from Southern Norway to Morocco and The Mediterranean Sea and Black Sea and Nova Scotia to Florida	West Coast of India	The most competitive fouling organisms in ports	*Shrinivaasu et al., 2015;* Chandra and Raghunathan, 2018 (PC); WoRMS (2018)
141.	Order- CHEILOSTOMATIDA Family- MEMBRANIPORIDAE Genus- *Membranipora* Species- *Membranipora membranacea*	Eastern Canada, Nova- Scotia	East coast of India		*Shrinivaasu et al., 2015;* Chandra and Raghunathan, 2018 (PC); WoRMS (2018)
	Phylum- ENTOPROCTA				
142.	Family- BARENTISIIDAE Genus- *Barentsia* Species- *Barentsia ramose*	Pacific, California, Belgium	Indian Ocean		Anil *et al.,* 2002; Chandra and Raghunathan, 2018 (PC); WoRMS (2018)
	Phylum- CHORDATA Class- ASCIDICEA (Ascidian)				

(Table 1) cont....

S. No.	Species	Native	Invasive	Harmful Impact	Reference
143.	Family- ASCIDIDAE Genus- *Ascidia* Species- *Ascidia sydneiensis*	Indo- Pacific and Atlantic ocean, Sub Antarctic Region and East South America	Tuticorin Port- Gulf of Mannar		Abdul *et al.*, 2009; Tamilselv *et al.*, 2007; Chandra and Raghunathan, 2018 (PC); WoRMS (2018)
144.	Family- ASCIDIDAE Genus- *Ascidia* Species- *Ascidia gemmata*	Indo- West Pacific	Tuticorin Port- Gulf of Mannar		Tamilselv *et al.*, 2007; Chandra and Raghunathan, 2018 (PC); WoRMS (2018)
145.	Family- ASCIDIDAE Genus- *Phallusia* Species- *Phallusia nigra*	Panama, USA, Indo-Pacific, Atlantic, and The Mediterranean	Tuticorin harbor, Gulf of Mannar and Andaman and Nicobar Islands		Anil *et al.*, 2002; Tamilselvi *et al.*, 2007; Jhimli *et al.*, 2015; Chandra and Raghunathan, 2018 (PC); WoRMS (2018)
146.	Family- ASCIDIDAE Genus- *Phallusia* Species- *Phallusia Arabica*	Indo- West Pacific Region and Northeast Atlantic	Tuticorin Port- Gulf of Mannar		Abdul and Sivakumar, 2007; Tamilselv *et al.*, 2007; Chandra and Raghunathan, 2018 (PC); WoRMS (2018)
147.	Family- ASCIDIDAE Genus- *Phallusia* Species- *Phallusia polytrema*	Indo- West Pacific Region, East South America and Pan Tropical throughout the Caribbean	Tuticorin Port- Gulf of Mannar		Abdul and Sivakumar, 2007; Tamilselv *et al.*, 2007; Chandra and Raghunathan, 2018 (PC); WoRMS (2018)
148.	Family- PYURIDAE Genus- *Herdmania* Species- *Herdmania pallida*	Atlantic Ocean, Indo- West Pacific and The Mediterranean: Sub Antarctic region	Tuticorin Port- Gulf of Mannar		Abdul and Sivakumar, 2007; Tamilselv *et al.*, 2007; Chandra and Raghunathan, 2018 (PC); WoRMS (2018)

(Table 1) cont.....

S. No.	Species	Native	Invasive	Harmful Impact	Reference
149.	Family- PYURIDAE Genus- *Microcosmus* Species- *Microcosmus curvus*	Pacific Ocean	Tuticorin Port- Gulf of Mannar		Abdul and Sivakumar, 2007; Tamilselv *et al.*, 2007; Chandra and Raghunathan, 2018 (PC); WoRMS (2018)
150.	Family- PYURIDAE Genus- *Microcosmus* Species- *Microcosmus squamiger*	Indo- Pacific, Southwest Atlantic and The Mediterranean Sea: Sub Antactic region	Tuticorin Port- Gulf of Mannar		Abdul and Sivakumar, 2007; Tamilselv *et al.*, 2007; Chandra and Raghunathan, 2018 (PC); WoRMS (2018)
151.	Family- PYURIDAE Genus- *Microcosmus* Species- *Microcosmus exasperates*	Indo- Pacific, Southwest Atlantic and The Mediterranean Sea: East Africa, Sub Antactic region and Southeast America	Tuticorin Port- Gulf of Mannar		Abdul and Sivakumar, 2007; Tamilselv *et al.*, 2007; Chandra and Raghunathan, 2018 (PC); WoRMS (2018)
152.	Family- PYURIDAE Genus- *Hermania* Species- *Hermania momus*	Mediterranean Sea, North Atlantic Ocean and Federal Republic of Somali	South Western coast of India		Abdul and Sivakumar, 2007; Jaffar Ali *et al.*, 2015; Chandra and Raghunathan, 2018 (PC); WoRMS (2018)
153.	Family- PYURIDAE Genus-*Pyura* Species- *Pyura praeputiali*	Australia	Chile		Abdul and Sivakumar, 2007; WoRMS (2018)
154.	Family- PEROPHORIDEAE Genus- *Perophora* Species- *Perophora formosana*	Indo- West Pacific and Atlantic Ocean	Tuticorin Port- Gulf of Mannar		Tamilselv *et al.*, 2007; Chandra and Raghunathan, 2018 (PC); WoRMS (2018)
155.	Family- PEROPHORIDEAE Genus- *Ecteinascidia* Species- *Ecteinascidia garstangi*	Madagascar, South Pacific Ocean	South Western coast of India		Jaffar Ali *et al.*, 2015; Chandra and Raghunathan, 2018 (PC); WoRMS (2018)

(Table 1) cont.....

S. No.	Species	Native	Invasive	Harmful Impact	Reference
156.	Family- PEROPHORIDAE Genus- *Ecteinascidia* Species- *Ecteinascidia venui*	India			Abdul and Sivakumar, 2007; WoRMS (2018)
157.	Family- PEROPHORIDAE Genus- *Ecteinascidia* Species- *Ecteinascidia krishnani*	India			Abdul and Sivakumar, 2007; WoRMS (2018)
158.	Family- STYELIDAE Genus- *Eusynstyela* Species- *Eusynstyela tincta*	Atlantic, Mozambique, Red Sea, Gulf of Suez, Africa, Sri Lanka	Tuticorin harbor		Anil *et al.*, 2002; Chandra and Raghunathan, 2018 (PC); WoRMS (2018)
159.	Family- STYELIDAE Genus- *Styela* Species- *Styela canopus*	Indo- West Pacific, Atlantic Ocean and The Mediterranean: South and South East America	Tuticorin Port- Gulf of Mannar		Renganathan, 1981; Tamilselv *et al.*, 2007; Chandra and Raghunathan, 2018 (PC); WoRMS (2018)
160.	Family- STYELIDAE Genus- *Symplegma* Species- *Symplegma oceania*	Indo- West Pacific	Tuticorin Port- Gulf of Mannar		Abdul and Sivakumar, 2007; Tamilselv *et al.*, 2007; Chandra and Raghunathan, 2018 (PC); WoRMS (2018)
161.	Family- STYELIDAE Genus- *Symplegma* Species- *Symplegma viride*	Atlantic Ocean, Indo- West Pacific, Atlantic Ocean and The Mediterranean: South and South East America	Tuticorin Port- Gulf of Mannar		Abdul and Sivakumar, 2007; Tamilselv *et al.*, 2007; Chandra and Raghunathan, 2018 (PC); WoRMS (2018)

(Table 1) cont.....

S. No.	Species	Native	Invasive	Harmful Impact	Reference
162.	Family- STYELIDAE Genus- *Botrylloides* Species- *Botrylloides schlosseri*	South Africa, Alexander Bay and Durban Bay	South- Western coast of India and South- Eastern coast of India, Tamil Nadu		Jaffar Ali *et al.*, 2009; Tamilselv *et al.*, 2007; Jaffar Ali *et al.*, 2015; Chandra and Raghunathan, 2018 (PC); WoRMS (2018)
163.	Family- STYELIDAE Genus- *Botrylloides* Species- *Botrylloides magnicoecum*	Indo- West Pacific and Western Central Atlantic	Tuticorin Port- Gulf of Mannar		Abdul and Sivakumar, 2007; Tamilselv *et al.*, 2007; Chandra and Raghunathan, 2018 (PC); WoRMS (2018)
164.	Family- STYELIDAE Genus- *Botrylloides* Species- *Botrylloides leachi*	Northeast Atlantic, Indo- West Pacific and Mediterranean and Black Sea: Australia and Europe	Tuticorin Port- Gulf of Mannar		Abdul and Sivakumar, 2007; Jaffar Ali *et al.*, 2015; Chandra and Raghunathan, 2018 (PC); WoRMS (2018)
165.	Family- STYELIDAE Genus- *Botrylloides* Species- *Botrylloides chevalensis*	Eastern Indian Ocean, India	Tuticorin Port- Gulf of Mannar		Abdul and Sivakumar, 2007; Tamilselv *et al.*, 2007; Chandra and Raghunathan, 2018 (PC); WoRMS (2018)
166.	Family- STYELIDAE Genus- *Styela* Species- *Styela bicolor*	North Australia, Gulf of Siam, Java, Banda Sea, Ambonia and the Philippines	East coast of India Tuticorin harbour		Anil *et al.*, 2002; WoRMS (2018)

(Table 1) cont.....

S. No.	Species	Native	Invasive	Harmful Impact	Reference
167.	Family- DIDEMNIDAE Genus- *Didemnum* Species- *Didemnum psammathodes*	Indo West Pacific and Eastern Atlantic; Sub-Antarctic region, Malaya and West Africa	Tuticorin Port- Gulf of Mannar		Abdul and Sivakumar, 2007; Tamilselv *et al.*, 2011; Chandra and Raghunathan, 2018 (PC); WoRMS (2018)
168.	Family- DIDEMNIDAE Genus- *Didemnum* Species- *Didemnum candidum*	North Pacific Ocean	South Eastern coast of India- Tamil Nadu		Abdul and Sivakumar, 2007; Jaffar Ali *et al.*, 2009; Chandra and Raghunathan, 2018 (PC); WoRMS (2018)
169.	Family- DIDEMNIDAE Genus- *Didemnum* Species- *Didemnum fragile*	Australia	South Western coast of India		Abdul and Sivakumar, 2007; Jaffar Ali *et al.*, 2015; Chandra and Raghunathan, 2018 (PC); WoRMS (2018)
170.	Family- DIDEMNIDAE Genus- *Lissoclinum* Species- *Lissoclinum fragile*	Indo- Pacific and Western Central Atlantic	Tuticorin Port- Gulf of Mannar		Abdul and Sivakumar, 2007; Tamilselv *et al.*, 2011; Chandra and Raghunathan, 2018 (PC); WoRMS (2018)
171.	Family- DIDEMNIDAE Genus- *Diplosoma* Species- *Diplosoma listerianum*	South African- Alexander Bay, Durban Bay, Lange Baan Bay	South Western coast of India		Abdul and Sivakumar, 2007; Jaffar Ali *et al.*, 2015; Chandra and Raghunathan, 2018 (PC); WoRMS (2018)
172.	Family- DIDEMNIDAE Genus- *Trididemnum* Species- *Trididemnum clinides*	Indo- West Pacific	Tuticorin Port- Gulf of Mannar		Abdul and Sivakumar, 2007; Tamilselv *et al.*, 2007; Chandra and Raghunathan, 2018 (PC); WoRMS (2018)

(Table 1) cont.....

S. No.	Species	Native	Invasive	Harmful Impact	Reference
173.	Family- DIDEMNIDAE Genus- *Trididemnum* Species- *Trididemnum savignii*	Indo- Pacific and Western Central Atlantic	Tuticorin Port- Gulf of Mannar		Abdul and Sivakumar, 2007; Tamilselv *et al.*, 2007; Chandra and Raghunathan, 2018 (PC); WoRMS (2018)
174.	Family- POLYCITORIDAE Genus- *Eudistoma* Species- *Eudistoma viride*	Western Central Pacific and Indian Ocean	Tuticorin Port- Gulf of Mannar		Tamilselv *et al.*, 2007; Abdul and Sivakumar, 2007; Chandra and Raghunathan, 2018 (PC); WoRMS (2018)
175.	Family- POLYCITORIDAE Genus- *Eudistoma* Species- *Eudistoma malaysani*	Pacific Ocean and Indian Ocean	Tuticorin Port- Gulf of Mannar		Abdul and Sivakumar, 2007; Tamilselv *et al.*, 2007; Chandra and Raghunathan, 2018 (PC); WoRMS (2018)
176.	Family- POLYCITORIDAE Genus- *Eudistoma* Species- *Eudistoma lakshmiani*	India			Abdul and Sivakumar, 2007; Chandra and Raghunathan, 2018 (PC); WoRMS (2018)
177.	Family- POLYCITORIDAE Genus- *Polyclinum* Species- *Polyclinum glabrum*	Central Indo- Pacific	South Western coast of India		Tamilselv *et al.*, 2007; Chandra and Raghunathan, 2018 (PC); WoRMS (2018)
178.	Family- POLYCITORIDAE Genus- *Polyclinum* Species- *Polyclinum indicum*	India		Settles and grows on boat hulls below the water line or other submerged surfaces of the vessels as they move from one port to another along the coast.	Abdul and Sivakumar, 2007; WoRMS (2018)

(Table 1) cont.....

S. No.	Species	Native	Invasive	Harmful Impact	Reference
179.	Family- POLYCITORIDAE Genus- *Polyclinum* Species- *madrasensis*	India		Settles and grows on boat hulls below the water line or other submerged surfaces of the vessels as they move from one port to another along the coast.	Abdul and Sivakumar, 2007; WoRMS (2018)
180.	Family- POLYCITORIDAE Genus- *Aplidium* Species- *Aplidium multiplicatum*	Indo- West Pacific	Tuticorin Port- Gulf of Mannar		Tamilselv *et al.*, 2007; WoRMS (2018)
181.	Family- HOLOZOIDAE Genus- *Distaplia* Species- *Distaplia nathensis*	India			Abdul and Sivakumar, 2007; WoRMS (2018)
182.	Class- DIDEMNIDAE Genus- *Diplosoma* Species- *Diplosoma swamiensis*	South Africa- Alexander Bay, Durban Bay, Lange Baan Bay	South Western coast of India		Abdul and Sivakumar, 2007; Jaffer Ali *et al.*, 2015; WoRMS (2018)

CONCLUSION AND SUGGESTIONS

The Convention on Biological Diversity (CBD, 2000) reported that the knowledge on the alien species which threaten ecosystem, habitats or species to be used for any prevention, introduction and mitigation activities is to be compiled and disseminated for the benefit of the biodiversity concern. The Ecological Niche Modelling (ENM) may be developed for the different environments using the Global Biodiversity Information Facility (GBIF) and to identify any Hotspots (O'Donnel *et al.,* 2012) that already existed or emerge in due course of time. If the case of eradication of the invasive species is unrealistic, from the point of economically or biologically, then as proposed by Schlepfer *et al.* (2005), the native species are exposed sufficient pressure in order to develop the evolutionary change to sustain in the new environment or temporarily reduce the invasive species until the native species manage and adjust to the new environment. Another suggestion by Hopkins (1999), these jellyfish blooms can be used as edible items using the processing technology or extracting proteins and vitamins for the food supplement. Even in India, CIFT – Central Institute of Fishery Technology designed the processing technology which provides a palatable jellyfish. A similar kind of process may also be evolved from all the invasive species for better management practices for control or utilize the same.

Therefore, the biodiversity of a marine environment and the environmental parameters concern has to be monitored continuously, which will provide a tool for the future benefit and for further action in this regard.

CONSENT FOR PUBLICATION

Not applicable.

CONFLICT OF INTEREST

The authors confirm that this chapter contents have no conflict of interest.

ACKNOWLEDGEMENTS

The authors thank the authorities of Pondicherry University and Zoological Survey of India, Kolkata for providing the facilities to prepare this article. We also thank Dr. C. Ragunathan, Scientist, ZSI, Kolkatta to have a fruitful discussion on this subject.

REFERENCES

Abdul, JAH & Sivakumar, V (2007) Occurrence and distribution of ascidians in Vizhinjam bay (southwest coast of India). *J Exp Mar Biol Ecol,* 342, 189-90.
[http://dx.doi.org/10.1016/j.jembe.2006.10.041]

Abdul, JAH, Sivakumar, V & Tamilselvi, M (2009) Distribution of Alien and Cryptogenic Ascidians along the Southern Coasts of Indian Peninsula. *World Journal of Fish and Marine Sciences,* 1, 305-12.

Abdul, JAH, Tamilselvi, M, Akram, AS, Kaleem, AML & Sivakumar, V (2015) Comparative study on bioremediation of heavy metals by solitary ascidian, Phallusia nigra, between Thoothukudi and Vizhinjam ports of India. *Ecotoxicology and Environmental Safety,* 121, 93-9.

Abdul, JAH, Akram, AS & Kaleem, AML (2016) New Distribution data on ascidian fauna (Tunicata: Ascidiacea) from Mandapam coast, Gulf of Mannar, India. *Biodivers Data J,* 4
[http://dx.doi.org/10.3897/BDJ.4.e7855]

Adhikari, D, Tiwary, R & Barik, SK (2015) Modelling Hotspots for Invasive Alien Plants in India. *PLoS One,* 10e0134665
[http://dx.doi.org/10.1371/journal.pone.0134665] [PMID: 26230513]

Alarcon-Ortega, LC, Rodríguez-Troncoso, AP & Cupul-Magaña, AL (2015) First record of non-indigenous *Paracaprella pusilla* Mayer, 1890 (Crustacea: Amphipoda) in the Northern Tropical East Pacific. *BioInvasions Rec,* 4, 211-5.
[http://dx.doi.org/10.3391/bir.2015.4.3.10]

Alpa, PW, Mariah, HM & Moyle, PB (2013) Abundance, size, and diel feeding ecology of *Blackfordia virginica* (Mayer, 1910), a non-native hydrozoan in the lower Napa and Petaluma Rivers, California (USA). *Aquat Invasions,* 8, 147-56.
[http://dx.doi.org/10.3391/ai.2013.8.2.03]

Anil, AC, Venkat, K, Sawant, SS, Dileepkumar, M, Dhaegalkar, VK, Ramaiah, N, Harkantra, SN & And Ansari, ZA (2002) Marine Bioinvation: Concern for ecology and shipping. *Curr Sci,* 3, 214-8.

Anil, AC, Clarke, C, Hayes, T, Hiliard, R & Joshi, G (2003) Ballast water risk assessment: Ports of Mumbai and Jawaharlal Nehru, India. *Final Report: IMO GloBallast Monograph Series,* 11, 1-63.

Annandale, N (1907) Notes on the freshwater fauna of India: No.XI., Preliminary note on the occurrence of a medusa (*Irene ceylonensis,* Browne) in a brackish pool in the Geanges Delta and on the Hydroid stage of the species. *Proceeding of Asiatic Society of Bengal,* III, 79-81.

Bhatt, J, Singh, J, Singh, S, Tripathi, R & Kohli, R (2011) *Invasive Alien Plants an Ecological Appraisal for the Indian Subcontinent.* CABI, Wallingford, UK.
[http://dx.doi.org/10.1079/9781845939076.0000]

Bhave, V & Apte, D (2012) First record of *Okenia pellucida* Burn, 1967 (Mollusca: Nudibranchia) from India. *J Threat Taxa,* 4, 3362-5.
[http://dx.doi.org/10.11609/JoTT.o2929.3362-5]

CBD - Convention on Biological Diversity (CBD) implemented from 29th December 1993.

Chandra, K & Ragunathan, C Personal Communication regarding invasive species.

Chopra, S (1960) A note on the sudden outburst of ctenophores and medusa in the waters off Bombay. *Curr Sci,* 29, 392-3.

Colautti, RI & Hugh, JM (2004) A neutral terminology to define 'invasive' species. *Divers Distrib,* 10, 135-41.
[http://dx.doi.org/10.1111/j.1366-9516.2004.00061.x]

Devroy, MK (2007) Problems and prospects of White leg shrimp culture in India. *SEBA Newsletter,* 4, 15.

Dhivya, P, Sachithanandam, V & Mohan, PM (2012) New Record of *Carijoa riisei* at Wandoor – Mahatma Gandhi Marine National Park (MGMNP), Andaman and Nicobar Islands, India. *Indian J Geo-Mar Sci,* 41, 212-4.

Fauvel, P (1953) *Fauna of India including Pakistan, Ceylon, Burma and Malaya – Annelida, Polychaeta.* The Indian Press, Allahabad 1-507.

Fautin, DG, Tan, SH & Tan, R (2009) Sea anemones (Cnidaria: Actiniaria) of Singapure: Abundant and well known shallow water species. *Raffles Bull Zool,* 22, 121-43.

Fofonoff, PW, Ruiz, GM, Hines, AH, Steves, BD & Carlton, JT (2009) Four centuries of biological invasions in tidal waters of the Chesapeake Bay region. *Biological Invasions in Marine Ecosystems.* In: Rilov, G., Crooks, J.A., (Eds.), Springer- Verlag, Berlin, Heidelberg 479-505.
[http://dx.doi.org/10.1007/978-3-540-79236-9_28]

Fofonoff, PW, Ruiz, GM, Steves, BD & Carlton, JT (2003) In ships or on ships? Mechanisms of transfer and invasion for nonnative species to the coasts of North America. *Invasive Species: Vector and Management Strategies.* In: Ruiz, G.M., Carlton, J.T., (Eds.), Island Press, Washington, D.C. 152-82.

Galil, BS, Kumar, BA & Riyas, AJ (2013) Marivagia stellata Galil and Gershwin, 2010, (Scyphozoa: Rhizostomeae: Cepheidae), found off the coast of Kerala, India. *BioInvasions Rec,* 2, 317-8.
[http://dx.doi.org/10.3391/bir.2013.2.4.09]

Galil, BS, Boero, F, Campbell, ML, Carlton, JT, Cook, E & Fraschetti, S (2015) Double trouble: the expansion of the Suez Canal and marine bioinvasions in the Mediterranean Sea. *Biol Invasions,* 17, 973-6.
[http://dx.doi.org/10.1007/s10530-014-0778-y]

Ganapati, P & Rao, GC (1962) Ecology of the interstial fauna inhabiting the sandy beaches of Waltair coast. *J Mar Biol Assoc India,* 4, 44-57.

Gaonkar, CA, Sawant, SS, Anil, AC, Krishnamurthy, V & Harikantra, SN (2010) Changes in the occurrence of hard substratum fauna: A case study from Mumbai harbour India. *Indian J Geo-Mar Sci,* 39, 74-84.

Guerra-García, JM, Ganesh, T, Jaikumar, M & Raman, AV (2010) Caprellids (Crustacea: Amphopoda) from India. *Helgol Mar Res,* 64, 297-310.
[http://dx.doi.org/10.1007/s10152-009-0183-6]

Hancock, ZB, Goeke, JA & Wicksten, MK (2017) A sea anemone of many names: a review of the taxonomy and distribution of the invasive actiniarian *Diadumene lineata* (Diadumenidae), with records of its reappearance on the Texas coast. *ZooKeys,* 706, 1-15.
[http://dx.doi.org/10.3897/zookeys.706.19848] [PMID: 29118617]

Hopkins, J (1999) Extreme Cuisine *The Weird & Wonderful Foods that People Eat* 305.

Invasive Alien Species (2004) http://www.downtoearth.org.in/coverage/invasive-alien-species-10861

Jason, B, Teejay, A, O'Rear, JD, Cook, AD & Moyle, MPB (2017) Factors affecting distribution and abundance of jellyfish medusae in a temperate estuary: a multi-decadal study. *Biol Invasions,* 20, 105-19.
[http://dx.doi.org/10.1007/s10530-017-1518-x]

Jeyabaskaran, J, Mohan, KG, Abilash, S, Premav, D & Kripa, V (2016) Is the Scyphozoan jellyfish *Lychnorhiza malayensis* symbiotically associated with the crucifix crab *Charybdis feriatus? Curr Sci,* 110, 479-80.

Johan, H (2017) Locals Panic as Sea Fish, Even Toxic Ones, Wash up on Kerala Beach. https://www.thequint.com/news/environment/locals-panic-toxic-fish-kerala-beach

Khuroo, AA, Rashid, I, Reshi, Z, Dar, G & Wafai, B (2007) The alien flora of Kashmir Himalaya. *Biol Invasions,* 9, 269-92.
[http://dx.doi.org/10.1007/s10530-006-9032-6]

Kramp, PL (1961) Synopsis of the medusae of the world. *J Mar Biol Assoc U K,* 40, 7-469.
[http://dx.doi.org/10.1017/S0025315400007347]

Kripa, V (2016) *Investigation Report on Reasons for decline in Sardine Fishery along Kerala coast.* CMFRI, Kochi 1-6.

Lacerda, MB & Masunari, S (2014) A new species of *Paracaprella* Mayer, 1890 (Amphipoda: Caprellida: Caprellidae) from southern Brazil. *Zootaxa,* 3900, 437-45.
[http://dx.doi.org/10.11646/zootaxa.3900.3.7] [PMID: 25543748]

Lin, W, Zhou, G, Cheng, X & Xu, R (2007) Fast economic development accelerates biological invasions in

China. *PLoS One,* 2e1208
[http://dx.doi.org/10.1371/journal.pone.0001208] [PMID: 18030342]

Mahapatro, D, Panograhy, RC, Panda, S & Mishra, RK (2015) Checklist of intertidal benthic macrofauna of a brackish water coastal lagoon on east coast of India: The Chilika lake. *Int J Mater Sci,* 5, 1-13.

Mammen, TA (1963) On a Collection of hydroids from South India. I. Suborder Athecata. *J Mar Biol Assoc India,* 5, 27-61.

Marques, F, Angelico, MM, Costa, JL, Teodosio, MA, Presado, P, Fernandes, A, Chainho, P & Domingos, I (2017) Ecological aspects and potential impacts of the non-native hydromedusa *Blackfordia virginica* in a temperate estuary. *Estuar Coast Shelf Sci,* 197, 69-79.
[http://dx.doi.org/10.1016/j.ecss.2017.08.015]

Menon, MGK The Scyphomedusae of Madras and the neighboring coast. Bulletin of Madras Government Museum Natural History Section, 3(1). Madras, India: Printed by the superintendent, Govt. Press.

Menon, MGK (1936) Scyphomedusae of Krusadai Island. Bulletin of Madras Government Museum. Natural History Section 1. Madras, India: Printed by the superintendent, Govt. Press, 1-9.

Menon, NG & Pillai, CSG (1996) Marine biodiversity conservation and management. *Central Marine Fisheries Research Institute Cochin,* 1-50.

Menon, NR & Nair, NB (1967) Observations on the structure and ecology of *Victorella pavida,* Kent from the southwest coast of India. *International Revue Ges Hydrobiology,* 52, 237-56.
[http://dx.doi.org/10.1002/iroh.19670520205]

Mondal, J, Raghunathan, C & Mondal, T (2015) Diversity and Distribution of common ascidians of Andaman group of Islands. *Middle East J Sci Res,* 23, 2411-7.

Nagale, P & Apte, D (2013) Some hydroids (Cnidaria: Hydrozoa: Hydroidolina) from the Konkan coast, Maharashtra, India. *J Threat Taxa,* 5, 4814-8.
[http://dx.doi.org/10.11609/JoTT.o3484.4814-8]

Nair, KK (1951) Medusae of the Trivandrum Coast. Parti. Systematics. *Bulletin of Research Institute. University of Travancore, Series. C. Nat Sci,* 2, 47-75.

Nair, NB & Dharmaraj, K (1983) Marine wood-boring Molluscs of the Lakshadweep Archipelago. *Indian J Geo-Mar Sci,* 12, 96-9.

Nair, PSR (1991) Occurrence of bryozoan in Vellar estuarine region, Southeast coast of India. *Indian J Geo-Mar Sci,* 20, 277-9.

Nandakumar, T (2013) Jellyfish Proliferating along Kerala Coast. 13th November 2013
http://www.thehindu.com/news/national/kerala/jellyfish-proliferating-along-kerala-coast/article5357801.ece

O'Donnell, J, Gallagher, RV, Wilson, PD, Downey, PO, Hughes, L & Leishman, MR (2012) Invasion hotspots for non- native plants in Australia under current and future climates. *Glob Change Biol,* 18, 617-29.
[http://dx.doi.org/10.1111/j.1365-2486.2011.02537.x]

Parulekar, AH (1968) Sea anemones (Actinaria) of Bombay. *J Bombay Nat Hist Soc,* 65, 138-47.

Peh, KSH (2010) Invasive species in Southeast Asia: the knowledge so far. *Biodivers Conserv,* 19, 1083-99.
[http://dx.doi.org/10.1007/s10531-009-9755-7]

Pimentel, D (2005) Aquatic nuisance species in the New York State Canal and Hudson River systems and the Great Lakes Basin: an economic and environmental assessment. *Environ Manage,* 35, 692-702.
[http://dx.doi.org/10.1007/s00267-004-0214-7] [PMID: 15920670]

Piementel, D, Lach, L, Zuniga, R & Morrison, D (2000) Environmental and economic costs of nonindigenous species in the United States. *Bioscience,* 50, 53-65.
[http://dx.doi.org/10.1641/0006-3568(2000)050[0053:EAECON]2.3.CO;2]

Prasade, A, Nagale, P & Apte, D (2016) *Cassiopea andromeda* (Forsskal, 1775) in the Gulf of Kutch, India: initial discovery of the scyphistoma, and a record of the medusa in nearly a century. *Mar Biodivers Rec,* 9,

36.
[http://dx.doi.org/10.1186/s41200-016-0031-8]

Purcell, JE, Uye, S & Lo, WT (2007) Anthropogenic causes of jellyfish blooms and their direct consequences for humans: a Review. *Mar Ecol Prog Ser,* 350, 153-74.
[http://dx.doi.org/10.3354/meps07093]

(2003) On the Scyphozoa from East coast of India, including Andaman & Nicobar islands. *Rec Zool Surv India,* 101, 25-56.

Rao, HS (1931) Notes on Scyphomedusae in the Indian museum. *Rec Indian Mus,* 33, 25-62.

Renganathan, TK (1981) On the occurrence of colonial ascidian, *Didemnum psammathodes* (Sluiter, 1895) from India. *Curr Sci,* 50, 922.

Robertson, A (1921) Report on a collection of Bryozoa from the Bay of Bengal and other eastern seas. *Rec Indian Mus,* 22, 33-65.
[http://dx.doi.org/10.5962/bhl.part.1465]

Roy, MKD & Nandi, NC (2017) Marine Invasive Alien Crustaceans of India. *Journal of Aquaculture and Marine Biology,* 5, 00115.
[http://dx.doi.org/10.15406/jamb.2017.05.00115]

Ryland, JS, Bishop, JDD, DeBlauwe, H, ElNagar, A, Minchin, D, Wood, CA & Yunnie, ALE (2011) Alien species of Bugula (Bryozoa) along the Atlantic coasts of Europe. *Aquat Invasions,* 6, 17-31.
[http://dx.doi.org/10.3391/ai.2011.6.1.03]

Sanders, NJ, Gotelli, NJ, Heller, NE & Gordon, DM (2003) Community disassembly by an invasive species. *Proc Natl Acad Sci USA,* 100, 2474-7.
[http://dx.doi.org/10.1073/pnas.0437913100] [PMID: 12604772]

Saneen, CVF & Padmavati, G (2017) Distribution and Abundance of Gelatinous zooplankton in Coastal Waters of Port Blair, South Andaman. *Journal of Aquaculture and Marine Biology,* 5, 1-8.

Santhakumari, V, Ramaiah, N & Nair, VR (1997) Ecology of hydromedusae from Bombay Harbour (1997) - Thana and Bassein Creek estuarine complex. *Indian J Geo-Mar Sci,* 26, 162-8.

Santhakumari, V & Nair, NB (1985) Some ciliates from the marine wood-boring isopod *Sphaeroma. Indian J Fish,* 32, 215-23.

Saravanan, R, Ranjith, L, Joshi, KK, Jasmine, S, Abdul Nazar, AK, Syed Sadiq, I & Kingsly, JH (2016) Scyphozoan Jelly fish Diversity in the Gulf of Mannar and Palk Bay with an account the Invasive Jelly fish *Phyllorhiza punctate. First International Biodiversity Conference.* New DelhiNovember, 2016

Schlaepfer, MA, Sherman, PW & Blossey, B (2005) Introduced species as evolutionary traps. *Ecol Lett,* 8, 241-6.
[http://dx.doi.org/10.1111/j.1461-0248.2005.00730.x]

Shyamasundari, K (1997) Amphipoda. *Fouling Organisms of the Indian Ocean, Biology and Control Technology.* In: Nagabhushanam, R., Thompson, M.F., (Eds.), 363-90.

Shrinivassu, S, Venkatraman, C, Rajan, R & Venkataraman, K (2015) Marine bryozoans of India. *Lesser known marine animals of India.* In: Venkataraman, K., Ragunathan, C., Mandal, T., Raghuraman, R., (Eds.), Zoological Survey of India, Kolkatta 321-37.

Subba Rao, NV (2005) Fauna of Marine National Park, Gulf of Kutch (Gujarat): An overview conservation area Series, 23:1-79. Zoological Survey of India, Kolkatta.

Sudhir, SNV (2016) Jellyfish Bloom off Visakhapatnam Coast Reported. 10th January, 2016 https://www.deccanchronicle.com/150227/nation-current-affairs/article/jellyfish-bl-om-visakhapatnam-coast-reported

Suryarao, KV & Subbarao, NV (1991) Mollusca of Lakshadweep. Fauna of Lakshadweep. *State Fauna Series,* 2, 273-362. [Zoological Survey of India.].

Swami, BS & Udhyakumar, M (2010) Seasonal influence on settlement, distribution and diversity of organisms at Mumbai harbour. *Indian J Geo-Mar Sci,* 23, 170-2.

Tamilselvi, M, Sivakumar, V, Abdul, JAH & Thilaga, RD (2007) Distribution of Alien Tunicates (Ascidians) in Tuticorin Coast, India. *World Journal of Zoology,* 6, 164-72.

The Hindu (2018) Nov 5, 2017 http://www.tribuneindia.com/news/nation/visitors-cautioned-against-jellyfish-on-beaches-in-south-goa/492817.html

The Tribune India (2017) http://www.tribuneindia.com/news/nation/visitors- cautioned-against-jellyfish-on-beaches-in-south-goa/492817.html

Thuiller, W, Richardson, DM & Midgley, GF (2007) Will climate change promote alien plant invasions?*Biological invasions* Springer, Berlin, Heidelberg 197-211. [http://dx.doi.org/10.1007/978-3-540-36920-2_12]

Untawale, AG, Agadi, VV & Dhargalkar, VK (1980) Occurrence of Genus *Monostroma* (Ulvales, Chlorophyta) from Ratnagiri (Maharashtra). *Mahasagar–Bulletin of the Nationa Institute of Oceanography,* 13, 179-81.

Venkataraman, K, Raghunathan, C, Raghuraman, R & Sreeraj, CR (2012) *Marine Biodiversity in India.* Zoological Survey of India, Kolkatta 1-168. ISBN 978-81-8171-307-0

Venkataraman, K, Raghunathan, C, Raghuraman, R & Sreeraj, CR (2012) *Marine Biodiversity in India.*

Venugopalan, VP & Wagh, AB (1986) Comprehensive review of the records of the biota of the Indian Seas. *Mahasagar,* 19, 213-5.

Jellyfish Blooms and Ganapati Immersions, The Economic Times - September 6, 2014, Doctor in On My Plate. https://blogs.economictimes.indiatimes.com/onmyplate/jellyfish-blooms-and-ganapati-immersions/

Wakida-Kusnoki, AT, Angel, LEA, Alegandro, C & Brahma, CQ (2011) Presence of Pacific white shrimp Litopenaeus vannamei (Boon, 1931) in the Southern Gulf of Mexico. *Aquat Invasions,* 6, 129-32. [http://dx.doi.org/10.3391/ai.2011.6.S1.031]

WoRMS (2018) *World Register of Marine Species.* Available at: http://www.marinespecies.org

CHAPTER 13

Problems of Invasive Species

Jeyasingh Thelma[*]

Department of Zoology, Thiagarajar College, Madurai-625009, Tamilnadu, India

Abstract: As indicated by the World Conservation Union, obtrusive invasive species are the second most critical danger to biodiversity, after natural disasters. In their new environments, invasive alien species move to become predators, contenders, parasites, hybridizers, and affect native plants and creatures. Higher rates of multiplication, less normal predators and capacity to flourish in various conditions are some basic qualities, which can make them hard to control. Marine environments are among the most important ecosystems both from a monetary and ecological point of view. The complexity of marine biological communities and their area postures challenges for administration, valuation, and the foundation of sound strategy to defend them for these invaders. Different procedures, for example, aquarium exchange, aquaculture, channel development, dispatching, and live fish exchange have achieved the dispersal of creatures. These dispersal systems result generally in the modification of biodiversity and achieve monetary misfortunes on fisheries. Starting with a short prologue to intrusive species, this section investigates a couple of vital life forms that represent a genuine risk to the earth and the mode by which they spread. This section additionally clarifies the different effects caused by these species and the courses by which they could be controlled.

Keywords: Alien Species, Biodiversity, Invasive Species, Marine Ecosystem.

WHAT IS AN INVASIVE SPECIES?

A greater part of the earth is covered by water in which oceans and seas contribute 70% and coastline covers up to 1.6 million kilometers. The resources provided by oceans and coasts are essential for the survival and well-being of humankind in many ways. Many people not only utilize seafood as a food but seaweed provides livelihoods through sustainable harvesting. Some marine organisms, such as corals, kelp, mangroves and sea grasses, have the ability to reshape the marine environment that promotes further habitation for other organisms. Marine organisms have a chance of moving around the world with ocean currents or by attaching themselves to driftwoods (De Poorter, 2009). Development in trade and shipping has opened gate for speedy transport of the organisms too. A large

---

[*] **Corresponding author Jeyasingh Thelma:** Department of Zoology, Thiagarajar College, Madurai-625009, Tamilnadu, India; Tel/Fax: +91-7373979977 E-mail: thelmaj90@gmail.com

De-Sheng Pei & Muhammad Junaid (Eds.)
All rights reserved-© 2019 Bentham Science Publishers

volume of these organisms move around the world rapidly. It is estimated that 10 billion tons of ballast water carries 7,000 species every day, which are transferred worldwide every year (Ruiz *et al.*,1997). The term "invasive" means tending to spread very quickly and undesirably or harmfully. An invasive species is an introduced or alien species that is not an inhabitant of a precise location (Carlton, 1999). It can be a plant, fungus, or animal species that spreads to an extent supposed to cause damage to the environment (Pimentel *et al.*, 2001). Around the world, a total of 480,000 invasive species have been recorded. The primary characteristic of invasive species is that it is rapidly reproducible and it spreads antagonistically. In North America, nearly 500 species have been transported around the world every day (Carlton, 1999; Fofonoff *et al.*,2003).

The most destructive of these invaders can displace native species, alter the community structure, food webs, and modify nutrient cycling and sedimentation processes. Alien invasive species have harmed economies by lessening fisheries, fouling ships structures, and blocking intake pipes. A couple of animal varieties also influence human being by causing infection (Ruiz *et al.*,1997). Invasive species expand frequently and cause allelopathy, by contaminating or causing illness to native species by acting as vectors, or hybridize them. These invasive species can change entire biological communities by adjusting hydrology, supplement cycling, and other environment forms. Regularly similar species that undermine biodiversity likewise make grave harm different characteristic asset businesses. Not all invasive species are detrimental. In numerous territories, many species are introduced for food and nourishment.

Causes of the Dispersal of Marine Organisms

Multiple processes influence the dispersal of marine organisms. Aquarium trade, aquaculture, canal construction, shipping, and live seafood trade may be the primary reasons for the dispersal of organisms. These dispersal strategies result mostly in alteration of biodiversity and bring about economic losses on fisheries.

The construction of canals came into existence in the late 1800s and early 1900s to give ships easier way around land obstacles. The Panama canal and Suez canal are the two major canal that play a major role in providing pathway for the exchange of marine invasive species The Panama Canal, which connects the Atlantic and Pacific Oceans in 1914, provides a chiefly significant model system for marine invasions in the tropics (Ruiz *et al.*, 2009). The other one, the Suez canal connects the Mediterranean sea to the Red sea. This canal has paved way for the highly noxious jellyfish *Rhopilema nomadica*, to reach the Mediterranean Sea in the 1980s. Artificial canals provide a link to connect two drainages that have been blocked by natural barriers to prevent the exchange of organisms.

Construction of the Erie Canal has modified the scenery of New York State altering its social and economic stature. Within the canal and the water it connects, it has brought many species that are not indigenous. Eurasian watermill foil, a fast-growing aquatic plant forms a thick mat growth, which leads to the suffocation of the native plant species. Zebra mussels, Asian clam, Round goby, Sea lamprey, and water chestnut are few species that are found to be invasive in these areas.

Thus, they act as a key reason for the introduction of alien species through drainages. A few examples of organisms that bypassed natural barriers through the construction of canals include *Petromyzon marinus* in the upper great lakes *via* the Welland canal. The round goby, *Neogobius melanostomus* made its way through the Chicago Shipping and Sanitary canal to the Mississippi basin and spread throughout the country. "The shipjack herring, *Alosa chrysochloris* probably gained way into Lake Michigan *via* the Chicago Shipping Canal (Fago, 1993). The river darters, *Percina shumardi* (Becker, 1983), the gizzard shad, *Dorosoma cepedianum* (Miller, 1957) are also the best example that followed this route of invasion to inflate their ranges.

Since people started to cruise the oceans, different species have been going the world over with them. These are not constrained to valuable plants and creatures, nor to bugs, for example, pathogenic operators or rats, yet additionally incorporate critical quantities of marine living beings. Authentic records and archeological discovers demonstrate that the cruising boats of the early travelers were colonized by up to 150 distinctive marine life forms that lived on or in the wooden frames, or utilized the metal parts, for example, stay chains as a substrate. On the off chance that the development turned into an irritation, the life forms were scratched off while adrift.

In different cases, the creatures stayed on the decaying frame of a ship when it was rejected and could never again be repaired. It is not really astonishing then that numerous wood-exhausting species, for example, the shipworm *Teredo navalis* are found far and wide today. Be that as it may, it is not any more conceivable to decide if these species were at that point cosmopolitan before the European voyages of revelation. Expanding quantities of marine life forms are currently transported over the seas because of globalization, exchange, and tourism. It is evaluated that the water in counterweight tanks used to balance out vessels is separated from everyone else in charge of transporting a huge number of various species between topographically inaccessible locales. The vast majority of these exotics pass on amid the outing or at the goal, while just a little portion can effectively duplicate and shape another populace. In any case, an investigation of six harbors in North America, Australia, and New Zealand has demonstrated that, despite all impediments, one to two species could effectively build up themselves every year at every one of the locales researched.

Land hindrances can likewise be overwhelmed by waterways. More than 300 species have just relocated through the Suez Canal from the Indian Ocean into the Mediterranean Sea. Moreover, streams and different conduits are in charge of species trade, for example, between the Baltic Sea and the Black Sea. Another vital reason for the dispersal of marine life forms is the exchange of living marine creatures for aquaculture, aquaria and the nourishment business. Researchers, partition the beach front waters of the world into a sum of 232 ecoregions, which are either isolated from each other by topographical boundaries, for example, arrive connects, or are unmistakably not the same as each other as for certain natural attributes, for example, saltiness. As indicated by a report issued in 2008, new species have just been acquainted by people with no less than 84 for every penny of these 232 ecoregions.

Examinations in the North and Baltic Seas demonstrate that no less than 80 to 100 colorful species have possessed the capacity to build up themselves in every one of these zones. In San Francisco Bay, 212 outside species have just been recognized, and for the Hawaiian Islands, it is accepted that about a fourth of the marine living beings that can be seen without a magnifying instrument have been foreign made. Generally, little is thought about the appropriation of microorganisms or different plants and creatures that are difficult to recognize. Species records are additionally crude for some marine districts where access is troublesome. Specialists expect that in future extraordinary living beings will have better opportunities to set up themselves in a few locales because of atmosphere warming. Life forms from South-East Asia, for instance, which lean toward a warm atmosphere, could flourish in areas that were already excessively unpleasantly cold for them.

Top Five Examples of Marine Invasive Species and the Problems Caused by Them

Green Crab (Carcinus maenas)

In the middle of 1800s, the European green crab (Fig. **1**) first entered the U.S by sailing ship to the Cape Cod region. At the beginning of 1900s, they started to move towards the north and arrived Maine in the year 1950's. In a little while, they migrated to Nova Scotia. In 1989 they were discovered on the West Coast, in San Francisco Bay (Cohen *et al.*, 1995). Possible reasons for their migrations are that they may have come in the ballast water of ships or they may have been shipped over hidden in the kelp packing around live main lobsters or Atlantic bait worms. By 1993 they had reached Bodega Harbor, since these crabs have a floating larval stage where they had an opportunity to establish their population. In 1997, strong El Nino currents moved the crab into Oregon, Washington, and British Columbia Estuaries. High fecundity, omnivorous feeding and ability to tolerate high temperature and salinity conditions are possible reasons for its establishment and persistent survival (Crothers, 1967; Ropes, 1968; Cohen *et al.* 1995).

Fig. (1). The European green crab *Carcinus maenas.*

Species, such as smaller shore crab, clams, and small oysters are vulnerable to the attack of this green crab. The most unique feature of the green crab is its color that can range from reddish to a dark mottled green and the five spines or teeth on each side of the shell. There are three rounded lobes between the eyes, and the last pair of legs are somewhat flattened. The carapace is broad and its length rarely exceeds 3.5 to 4 inches. Females can lay a large number of eggs up to 185,000 eggs, and the development of larvae takes place offshore in the intertidal zone (Perry H, 2014). After they hatch the young crabs, they find shelter among seaweeds and seagrasses, such as *Posidonia oceanica*, until they mature as adults (Bedini, 2002).

Mechanisms including ballast water, ships' hulls, stuffing resources used to transport live marine organisms, bivalves moved for aquaculture, rafting, immigration of crab larvae on ocean currents, and the movement of submerged aquatic vegetation for coastal zone management initiatives are a small number of reasons by which *C. maenas* (Fig. **2**) had a chance to scatter and invade various regions (Perry H, 2014). Thresher *et al.,* 2003 found *C. maenas* dispersed in Australia mainly by rare long-distance events by anthropogenic activities. Ballast water and hull fouling were the reasons for the spread of *C. maenas* in Argentina (Hidalgo *et al.,* 2005). Other reasons, such as ruined seawater pipes, the movement of fouled frames of exploratory drilling platforms, and scientific investigations, may also lead to the dispersal of *C. maenas.* The regular uses of *C. maenas* in scientific investigations and accidental release or intentional release are all promising introduction vectors (Carlton and Cohen, 2003).

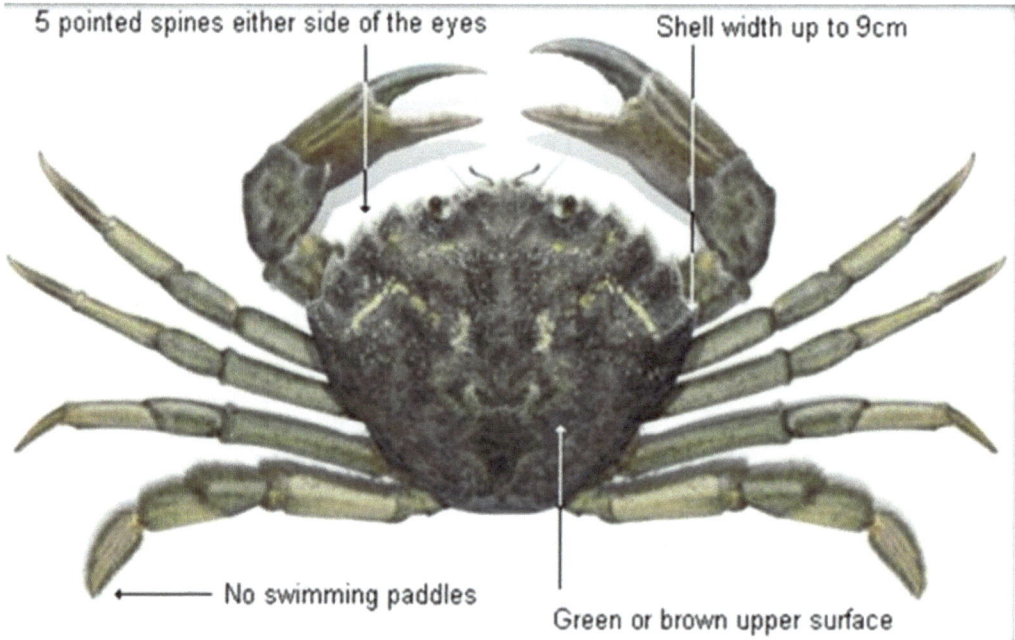

Fig. (2). Morphological features of *Carcinusmaenas* (https://www.dpi.nsw.gov.au/fishing/pests-diseases /marine-pests/found-in-nsw/european-shore-crab

The ecological impacts caused by *C. maenas* are numerous. They are primarily nocturnal, although activity also depends on the tide, and crabs can be active at any time of day (Novak, 2004). Bivalve molluscs, such as clams, oysters, and mussels, polychaetes and small crustaceans are the food for the predatory crab *C. maenas.* In addition to single–species impacts caused by *C. maenas* on single species its predation could alter community structure in intertidal sand flats through predation and prey excavation (Grosholz and Ruiz, 1995) that can have a significant reduction in flatfish population (Taylor, 2005). In California, native clams *Nutricola* spp. were preferentially predated by *C. maenas,* which in turn resulted in the reduced the population of the native clams and amplified the population of the previously introduced clam *Gemma gemma* (Grosholz, 2005).

The soft-shell clam *Mya arenaria* has been destructed by the green crab to a greater extent on the east coast of the United States and Canada, other commercially significant bivalves, such as scallops, *Argopectenir radians*, and northern quahogs. *Mercenaria mercenaria* have also faced a serious reduction in its numbers (Perry, 2014). Local commercial fisheries, such as oysters and the Dungeness crab, can be negatively impacted by *C. maenas*, by preying on the young of species, or competing with them for resources (Lafferty and Kuris, 1996).

C. maenas, an extensive marine species is tremendously hard to manage and control. Complete abolition is not possible since the adults are highly motile up to 15 km and larvae are planktonic (Gomes, 1991). The spread of *C. maenas to* satellite locations has made the removal of adults a difficult task since eradication at one will be followed by the migration of the adults from adjacent sites. The control of *C. maenas* is carried around the world due to its potentially harmful effects on the ecosystem. As a part of this control program, 10 tons of *C. maenas* were caught in Edgartown, Massachusetts in the year 1995 to protect local shellfish (Ennulty, Jones and Bax, 2002).

The predator-prey relationship observed in laboratory and field experiments by Catherine *et al.*,2005 indicates that predation by the native blue crab *Callinectes sapidus* directly affects the abundance, distribution, and mortality patterns of *Carcinus maenas* in the northwestern Atlantic. Predation by the native rock crabs *Romaleon antennarium* and *Cancer productus* in the west coast of North America competes with *C. maenas* for shelter with a native shore crab, *Hemigrapsus oregonensis* (Yamada, 2001). Studies by Jeffery *et al.*, 2005 suggests that natural enemy of the green crab, the parasitic barnacle *Sacculina carcini* acted as a potential biocontrol agent against the green crab. Out of four species (*Hemigrapsus oregonensis*, *H. nudus*, *Pachygrapsus crassipes* and *Cancer magister)* tested the specificity was 79% for green crab whereas it was 33-35% for the other crab species.

Killer Algae (Caulerpa taxifolia)

Caulerpa taxifolia Fig. (**3**) is a species of seaweed that belongs to the genus *Caulerpa*. It had its birthplace in the Indian Ocean and it is generally utilized decoratively in aquariums, as a result of its appealing shading, perfect course of action, simple foundation and support. The alga has a stem, which spreads on a level plane simply over the ocean bottom. From this stem develop vertical greenery like pinnae, whose sharp edges are level like those of the yew, henceforth the species name taxifolia. It is one of two algae on the list of the world's 100 worst invasive species compiled by the IUCN Invasive Species

Specialist Group while the other being *Undaria pinnatifida* that is native to Japan and found to be invasive in New Zealand, the United States, Western Europe, Argentina, Australia, and Mexico.

Fig. (3). *Caulerpa taxifolia.*

Caulerpa taxifolia was first identified in the Mediterranean Sea in June 2000 and then identified at 2 sites in southern California (Jousson, Pawlowski and Zaninetti, 2000). Since it was first discovered in the Mediterranean Sea in 1984, this invasive seaweed has continued to expand its range in the Mediterranean and Adriatic Seas (Meinesz, 1999). It escaped from the aquariums and has extended broadly in the Mediterranean and replaced the native plants by depleting the food sources and the habitat of marine life. At present, *Caulerpa* has inhabited thousands of hectares of the sea bottom in the Mediterranean and it is found from France to Croatia and its range in the Mediterranean will likely to continue to expand. In moist conditions, the invasive strain of *Caulerpa* can stay alive out of water for up to 10 days and can withstand low sea water. This alga can colonize most kinds of substrates including rock, sand, mud, and seagrass beds from depths ranging from less than 1 m to ~12 m. *Caulerpa* can rapidly grow and reproduce in asexual manner and dispersal occurs through fragmentation. Viable plants can arise from fragments as small as 1 cm. Ballast water discharge from transoceanic boats and illegal dumping of aquaria plants leads to the spread of the seaweed. Unintentional movement of plant material on boats, anchors, or fishing gear, or *via* algal fragments being dispersed by sea currents may be other reasons for localized dispersal of the seaweeds.

Other species of seagrasses, algae, and invertebrates that have a sedentary mode of life are being suffocated by the growth of the invasive strain of *Caulerpa* in the Mediterranean Sea. It does this by either competing with other species for food and shelter or due to the noxious effects of a compound, caulerpenyne that is present in its foliage. Native species diversity and fish habitat have been immensely reduced by *Caulerpa*. Native fish, such as Mediterranean bream, which feed on *Caulerpa*, build up the toxin caulerpenyne in their flesh. This makes the fish unfit for consumption of humankind (Meinez and Hesse, 1991). Studies by Villele and Verlaque,1995 indicated a severe threat to a seagrass species *Posidonia oceanica* by the invaded *C. taxifolia* in the northwestern Mediterranean.

The manifestations of *Caluerpa* in southern California started in 2000 by an aquarium owner who improperly dumped the unwanted contents of a marine fish tank into a storm water system that fed into Agua Hedionda Lagoon in Carlsbad where this weed was first found to be exposed. Sale or shipping of *Caulerpa taxifolia* within the state has been forbidden by California. The Noxious Weed Act proposed California in order to eradicate Caulerpa, prohibited the interstate sale and transport of the aquarium strain. More than 7 million dollars and six years were spent to eradicate the weed in Agua Hedionda Lagoon in Carlsbad near San Diego and Huntington Beach near Los Angeles. So far no other infestations of the cold water strain of *Caulerpa* have been located in the USA.

Elysia subornata, an oceanic slug found off the bank of Florida can go about as a potential predator of *C. taxifolia*. The slug sticks its proboscis into the stem and sucks out the white gooey fluid inside the stem, making the green growth end up flaccid; pale lastly prompts the demise of the alga. A compound delivered by the slug can kill the poisonous impacts of the toxin created by the alga. In any case, this slug can't make due in the cooler waters of the Mediterranean and, consequently, can't control the obtrusive alga there (Thibaut, 2001).

Sea Walnut (Mnemiopsis leidyi)

Mnemiopsis leidyi (Fig. **4**), the warty comb jelly is commonly called as the sea walnut (*Encyclopedia of Life*, 2013). It is a tentaculate ctenophore (comb jelly), a phylum noted for marine invertebrates that use cilia for swimming. They are original inhabitants of the western Atlantic coastal waters. The body is translucent and its size ranges from few millimeters to 12 cm in length (NIMPIS 2002).

Mnemiopsis leidyi is a predator that feeds on copepods and their nauplii, bivalve veligers, barnacle nauplii, larvae of fishes and eggs (Granhag, Moller and Hansson, 2011; Javidpour *et al.*, 2009a, Cowan and Houde, 1993, Purcell, Nemazie and Dorsey 1994). It is a hermaphrodite, which renders the organism to have high reproductive and growth rate that permits the organism to intensify to a greater extent (Purcell *et al.*, 2001).

Fig. (4). *Mnemiopsis leidyi.*

M. leidyi was first introduced into the Black Sea during the year 1980 (Purcell *et al.*, 2001). Transported in the ballast waters by way of the Volga Don Channel *M. leidyi* continued to invade the Caspian Sea in the middle of 1990's (Kideys 2002, Bilio and Niermann 2004). Nissum Fjord, western Jutland, Denmark had a spotting of *M. leidyi* in August 2005 (Tendal, Jensen and Riijgard, 2007), and during November 2005 in the Oslofjorden, Norway (Oliveira, 2007). Later in 2006 its presence was noted in Kiel Bay (Javidpour, Sommer and Shignova, 2006) and by the side of the west coast of Sweden (Hansson 2006) and also on the Dutch and German North Sea coast (Boersma *et al.* 2007, Kube *et al.* 2007). In 2007 it was spotted in the Danish straits (Tendal *et al.* 2007). In August, 2007 the organism was observed in the Gulf of Finland and the Bothnian Sea (Lehtiniemi *et al.*, 2007). During October 2007 it was observed in the Gulf of Gdansk, Poland (Janas and Zgrundo 2007). In the same year it was noted in Bornholm Basin bounteously (Huwer *et al.,* 2008).

The primary reason for the introduction of this alien species is the transport through ballast waters (Faasse and Bayha, 2006) since the two largest European ports, Antwerp and Rotterdam are in close proximity to the Dutch estuaries. Studies by Oliveira, 2007 suggests that the increase surface water temperature of the North Atlantic, including the North Atlantic Current and the North Sea due to climate change must be a reason for its introduction into Europe. The negative impacts of this species are numerous. It affects the habitats and indigenous organisms, mainly by feeding on copepods, fishes, larvae or eggs of fishes, and zooplanktons (Shiganova *et al* 1998, Shiganova *et al.,* 2001). *Aurelia aurita*, the native jellyfish in the Baltic, is directly competed by M. *leidyi* for food and habitat since the fecundity rate of *M. leidyi*is high when compared to *Aurelia aurita* (Mutlu *et al.,* 1994). Divers face a serious threat because of a rash called "sea bather's eruption" caused by the larvae of the parasitic sea anemone *Edwardsiella lineata* that have been found in *M. leidyi* in Sweden (Selander *et al.* 2009). The economic impacts caused by *M. leidyi*are mainly due to the organisms feeding habit since it feeds on planktonic organisms (Burrell and Van Engel, 1976). In 1989 and 1995, an estimated loss of 250 million dollars was noted in the Black Sea (NIMPIS, 2002). Reproduction of cod in the Baltic Sea is affected since *M. leidyi* be likely to get collected near the halocline where the spawning of Baltic cod occurs (Haslob *et al.*, 2007).

Veined Rapa Whelk (Rapana venosa)

Rapana venosa Fig. (**5**) is a large predatory sea snail that is native to the northwest Pacific, from Vladivostok, Russia to Hong Kong. It is a marine gastropod that belongs to the family Muricidae. *Rapana venosa* has a bulky and profound shell with a small spire. The interior of the shell has a deep orange color, which is a distinctive feature of the *R. venosa*. The outer shell color ranges from dull grey to red-brown, with prominent dark brown patches on the spiral ribs, which gives a vein-like pattern throughout the entire shell. The body whorl is large and inflated with a deep umbilicus. *Rapana venosa* has a large and elliptical aperture and a broad and smooth columella. Small elongate teeth are present at the edge of the outer lip (Mann and Harding 2000).

Rapana venosa has now become an invasive species in many different regions of the world. In 1946 it was first discovered from the Black Sea in the Novorrosick Bay soon after it invaded the Mediterranean Sea. In 1998, transported in the ballast water of ships the snail was noted in the Chesapeake Bay. The European coastal waters from Norway to Spain, and the Rio de la Plata estuary in South America have also been colonized by these *R. venosa*. The animal feeds on bivalve mollusks and also it rigorously reduced the population of shellfish in the Black Sea and the Chesapeake Bay.

R. venosa are predatory in nature. They are able to tolerate extensive environmental changes (Harding and Mann 1999, Mann and Harding 2003). They follow a reproductive strategy that facilitates the animal to mature and reproduce at a small size and juvenile period. Their lifespan has been recorded approximately 15 years with no substantiation of agedness. They are noted for their ability to lay eggs multiple times per year and thus have a high fecundity rate. The larval period is also pelagic and the extended developments of the larval period make a possible local scattering of the organism (Mann & Harding 2003).

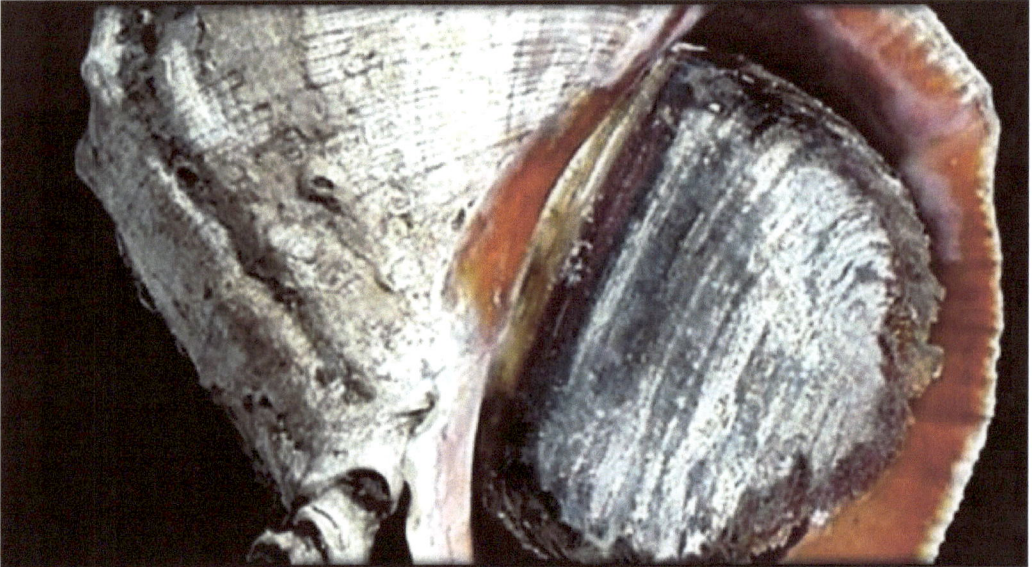

Fig. (5). *Rapana venosa.*

The ecological impacts caused by *R. venosa* are due to its predatory nature. The epifaunal bivalves are the most important organisms that are more vulnerable to the attack of *R. venosa*. They tend to proliferate in the place where cultivation of oysters and mussels are carried out. In the Black Sea, the population of edible bivalve has been declined due to the voracious predation by *R. venosa* (Zolotarev, 1996). The Gudaut oyster has reached mere extinction due to *R. venosa* (Chukhchin, 1984). Another ecological impact is that the population of the local hermit crab *Clibanarius vittatus* seems to increase due to the increased presence of the large empty *R. venosa* shells. The thick broad shell of *R. venosa* makes it less susceptible to predation by large predators like turtles (Harding and Mann 1999). ICES, 2004 has suggested that *Rapana venosa* may stay as an unchallenged predator for up to a decade, once it has reached refuge size.

Zebra Mussel (Dreissena polymorpha)

The Zebra mussel Fig. (6) is a small freshwater bivalve mollusk. It is classified under prohibited invasive species since it can deplete natural resources. The import, purchase, transport or introduction of these species is against the law except under a permit for disposal, control, research, or education. A Zebra mussel has a banded, D-shaped shell composed of two valves hinged to each other and connected by a ligament. The shells on average are one-quarter inch to one and one-half inches long. The size of the shell depends on the age of the mussel with alternating yellow and brown color bands. Zebra mussels are filter feeders and a single mussel can filter up to one quart of water per day while consuming mainly on algae. They live at the bottom of the sea and get themselves attached to any solid substrate, such as rocks, wood, plants, native mussels, and other debris. The fecundity rate is high and a female can produce 100,000 to 500,000 eggs per year. Fertilized eggs develop into young, minute, free-living larvae, called "veligers," that form shells. Later the veligers settle and attach to a stiff substrate using tiny fibers called byssal threads. Beds of zebra mussels can reach tens-o--thousands within a single square yard.

The primary manifestation of the organism in northern Italy was in Lake Garda in 1973 (Guisti and Oppi, 1973). In 1794, it was reported by Grossinger in Hungary. Beginning in about 1800, they began dispersion across western and northern Europe. Recently the inland waters in the British Isles, Spain, Portugal, and France have been colonized by zebra mussels. In 1820s Britain was rapidly colonized by Zebra mussels as explained by Kerney and Morton. Later, the existence of Zebra mussels was noted in London in 1824, and in the Union Canal near Edinburgh in 1834 (Mackie *et al.*, 1984). In 1827, zebra mussels were spotted in the Netherlands at Rotterdam. Their early dispersal was facilitated by Canals that artificially link many European waterways. Around 1920 the mussels reached Lake Mälaren in Sweden. In central Italy, they were noted in Tuscany in 2003 (Elisabetta and Simone, 2006). In 1988 they appeared in North America, and rapidly in five years, they started to spread throughout the Great Lakes and large rivers. In 1989 Zebra mussels were spotted in the Duluth Harbor. In 1993 the Mississippi River was fully occupied by Zebra mussels. As of 2015, 82 lakes and 12 rivers and streams in Minnesota have been populated by Zebra mussels.

Zebra mussels have ruined power plants, water purification facilities, ships, and spoiled beaches with decomposing mussels and spiky shells. Plankton comm-unities have been completely devoured by large populations of Zebra mussels. The availability of food for commercial and game fish have been decreased by these mollusks. It is plentiful in the fresh, tidal parts of the St. Lawrence and Hudson Rivers, and has been revealed at the head of Chesapeake Bay. They build colonies on indigenous unionid clams and therefore reduce the ability of the native clams to move, nourish them, and reproduce, ultimately leading to their deaths. In Lake St. Clair and the western basin of Lake Erie unionid clams have disappeared because of the nuisance caused by Zebra mussels (Schlosser and Nalepa, 1994).

Fig. (6). Zebra Mussel.

OTHER MAJOR PROBLEMS OF INVASIVE SPECIES

Alteration of Biodiversity

Countless alien species penetrate the native flora and fauna without dominating them. This domination increases the species association diversity. The entire species communities can face lethality and entire habitat destruction due to natural calamities. In these cases, the invasion of new species causes assemblage of a completely different species. An example of this is the evolution of *Fucus radicans* alga in the Baltic Sea, which was formed after the last Ice Age. This was the only native species that evolved in this region and all other species indigenous to this area today migrated from habitats, such as the North Sea or the White Sea.

The exchange of species between far-flung parts of the Earth has become progressively greater than before. This process initiated since Christopher Columbus traveled to America in 1492. It has therefore turned out to be more probable that species will infringe on ecoregions that are far expelled from their regular regions of starting point. At times the new species make issues. They may dislodge various local species and along these lines prompt a lessening in biodiversity. This is particularly liable to happen in the event that they have no regular adversaries in the new area.

For instance, inside just 15 years of its underlying revelation in Monaco the Australian green alga *Caulerpa taxifolia* had congested 97 percent of all the appropriate ground amongst Toulon and Genoa, and had spread into the northern Adriatic and to the extent Sicily. The alga delivers an anti-agents substance that makes it unpalatable to the herbivores. There are life forms that consume Caulerpa and have adjusted to the anti-agents, yet these species are absent in the Mediterranean Sea. The Asian green growth *Sargassum muticum* and *Gracilaria vermiculophylla* likewise shaped for all intents and purposes monospecific remain in some beach front zones after the first experience with Europe. In the middle of 1980s, the northern Pacific seastar *Asterias amurensis* established itself in south-eastern Australian waters (Buttermore, Turner, and Morice, 1994). Just two years after it was first identified in Port Philipp Bay, an extensive sound off Melbourne, in excess of hundred million examples were assessed. This starfish excessively discovered for all intents and purposes no characteristic adversaries in its new territory, empowering it to demolish loads of local starfish, mussels, crabs and snails. The biomass of the starfish, in the end, surpassed the aggregate sum of all monetarily angled marine creatures in the area.

Instances of recently invasive species uprooting local species have been recorded in 78 percent of the 232 waterfront ecoregions of the world. Numerous cases have been accounted for from the calm scopes specifically, those locales of the Earth where it is neither amazingly hot nor greatly icy. Except for Hawaii and Florida, the 20 beachfront ecoregions most firmly burdened by obtrusive marine life forms are found only in the calm North Atlantic and North Pacific or in southern Australia, and nine of these areas are in Europe. A few spots, as San Francisco Bay, are presently commanded by non-local species. There, the infringing species are regularly thought to be a risk to marine biodiversity, albeit so far not a solitary case is known in which an animal type presented from outside has caused the elimination of local living beings.

Two hundred and fifty-two presented and cryptogenic marine and estuarine species have been recognized in Australia (Hewitt, no date), in excess of 150 outsider species in Port Phillip Bay alone (Hewitt *et al.*,1999). New Zealand researchers distinguished 159 outsider marine species (Cranfield *et al.*,1998), while 212 outsider marine, estuarine and freshwater species have been accounted for in the San Francisco Bay and Delta, California (Cohen, 1998). In Hawaii,91 of the almost 400 marine species display in Pearl Harbor are outsider (Coles *et al.*,1999). In view of verifiable data, a new marine or estuarine species sets up itself each 32– 85 weeks in every one of six ports examined in the US, New Zealand and Australia, a rate that has all the earmarks of being expanding (Hewitt, no date).

Economic Impacts

Invasive marine life forms can perpetrate monetary misfortunes on fisheries. The warty comb jelly *Mnemiopsis leidyi*, local to America, achieved a crumple of the fisheries of Sea 25 years back, a territory as of now extraordinarily debilitated environmentally around then because of overfishing and eutrophication. Examples that were most likely presented with ballast water were first located there in 1982. The jellyfish spread quickly and assaulted local species, particularly, by bolstering on their eggs and hatchlings. The business fishery diminished by around 90 percent. In 1989, a check of 240 examples for each cubic meter of water was made, the best leading concentration of *M. leidyi* on the planet. Just the unintentional introduction of another organism, Beroe ovata – a predator – was viable in repulsing the populace and permitting a rebound of the fish populace. Obtrusive species are additionally causing issues on the eastern bank of North America. There the European basic shore crab *Carcinus maenas* caused a decrease in the shellfish fishery gather. Now and again invasive marine living beings can even present a danger to human wellbeing. One case of this is delineated by microalgae of the genus *Alexandrium*, which create a nerve poison. Species of *Alexandrium* have been found in numerous beachfront zones where they most likely did not exist only a couple of decades back. Such wonders can clearly have to a great degree negative impact on tourism.

Introduced species are not just transported unexpectedly in the ballast water of boats. Business visionaries regularly import marine life forms from different nations to non-indigenous living spaces for aquaculture rearing. This may give here and now business benefits, yet additionally represents the danger of imported species uprooting local life forms, prompting transitional or long haul monetary or biological harm. Studies have demonstrated that not less than 34 percent of the 269 introduced marine living beings explored were intentionally foreign for aquaculture reproducing. One case is the Japanese monster shellfish *Crassostrea gigas*, which has taken up habitation and built up itself in not less than 45 ecoregions.

In the vicinity of 1964 and 1980 specifically, a lot of young oysters, called spat, were brought into Europe. As a rule, the natural effect was wrecking. In North America and Australia, the monster oysters frame thick provinces that dislodge local species. Besides, they as often as possible reason eutrophication of the seaside waters since they discharge unpalatable particles agglutinated with bodily fluid, which cause the extra natural sullying of the water. The nearness of monster oysters in France has likewise prompted contamination of the waters. What's more, a decrease in the zooplankton and additionally bigger creatures has been watched. In the Netherlands and Germany, the giant clams tend to settle on blue mussel banks. This is debilitating the important species of the customary fishery.

It is expected that other than the giant oyster not less than 32 extra species have been unexpectedly brought into the North Sea, including the regular Atlantic slipper snail *Crepidula fornicata* and the alga *Gracilaria vermiculophylla*, both of which have turned out to be environmentally tricky. With a specific end goal to maintain a strategic distance from this sort of peril, a standard appraisal framework would be useful. This could be utilized to appraise the capability of an animal type uprooting different life forms. Moreover, it could be utilized to weigh up the points of interest and disservices of acquainting an outer species with specific natural surroundings.

By looking at dangerous and safe imported species, specialists have been striving for quite a while to recognize attributes that demonstrate a high potential for relocation of the local species. For instance, some algal species glide while others sink. Regardless of whether the species floats and would thus be able to effectively scatter depends basically on this factor. In any case, so far it has demonstrated hard to reach inferences about the relocation capability of an animal type in view of individual characteristics. Maybe it will never be conceivable to make certain expectations about the conduct of an animal category in another area, in light of the fact that various basic elements are having an effect on everything. This forecast is additionally confounded as animal groups build up itself in another territory over an expanded day and age, living through various stages. After an underlying extension stage, amid which a species flourishes, therefore, the most part takes after a decrease before the animal types have totally adjusted to the new natural surroundings. Prior to the relocation capability of an animal, varieties can be certainly assessed; it must be known which stage the species is in at a given time. In any case, that is exceptionally hard to decide.

CONCLUSION

Eradication of alien species is highly impossible. Because once they have been established, their population supersede the indigenous varieties. Hence any mechanical removal of these species would not be possible. Moreover, the marine environment is a vast area and many species dwell a dormant or free-living larval stages. During these conditions, it would be highly inefficient to follow any control measures. Introduction of any natural enemies to control the alien species population may be risky, since the natural enemies by themselves may pose a severe threat to the environment later.

Government policies must be modified and novel ecological administration procedures must be adopted with a specific end goal to control the important reasons for species introduction. It is vital this incorporates continuous monitoring and persistent checking of aquaculture and ballast water. Aquaculture has to be monitored uninterruptedly since most of the introductions are made through ballast water. Unilateral efforts at the national or neighborhood levels, nonetheless, will barely be viable. Universal methodologies rehearsed by all states circumscribing an ecoregion have more prominent odds of accomplishment

CONSENT FOR PUBLICATION

Not applicable.

CONFLICT OF INTEREST

The author confirms that this chapter contents have no conflict of interest.

ACKNOWLEDGEMENTS

Declare None.

REFERENCES

Becker, GC (1983) *Fishes of Wisconsin Univ.* Wisconsin Press, Madison.

Bedini, B (2002) Colour change and mimicry from juvenile to adult: Xanthoporessa (Olivi, 1792) (Brachyura, Xanthidae) and Carcinusmaenas (Linnaeus, 1758) (Brachyura, Portunidae). *Crustaceana,* 75, 703-10.

Bilio, M & Niermann, U (2004) Is the comb jelly really to blame for it all? Mnemiopsisleidyi and the ecological concerns about the Caspian Sea. *Mar Ecol Prog Ser,* 269, 173-83. [http://dx.doi.org/10.3354/meps269173]

Boersma, M, Malzahn, A, Greve, W & Javidpour, J (2007) The first occurrence of the ctenophore Mnemiopsisleidyi in the North Sea. *Helgol Mar Res,* 61, 153-5. [http://dx.doi.org/10.1007/s10152-006-0055-2]

Burrell, VG, Jr & Van Engel, WA (1976) Predation by and distribution of a ctenophore, Mnemiopsisleidyi A. Agassiz, in the York River estuary. *Estuar Coast Shelf Sci,* 4, 235-42. [http://dx.doi.org/10.1016/0302-3524(76)90057-8]

Buttermore, RE, Turner, E & Morice, MG (1994) The introduced northern Pacific seastar asteriasamurensis in Tasmania. 36, 21-5.

Carlton, JT & Cohen, AN (2003) Episodic global dispersal in shallow water marine organisms: The case history of the European shore crabs Carcinusmaenas and C. aestuarii. *J Biogeogr,* 30, 1809. [http://dx.doi.org/10.1111/j.1365-2699.2003.00962.x]

Carlton, JT (1999). The scale and ecological consequences of biological invasions in the world's oceans. *Sandlund OT, Schei PJ, VikenA Invasive species and biodiversity management.* Kluwer Academic Publishers, Dordrecht 95-212.

Chukchin, VD (1984) *Ecology of the gastropod molluscs of the Black Sea Academy of Sciences of the USSR.* NaukovaDumka, Kiev. (In Russian)

Cohen, AN, Carlton, JT & Fountain, MC (1995) Introduction, dispersal and potential impacts of the green crab Carcinusmaenas in San Francisco Bay, California. *Mar Biol,* 122, 225-38.

Cohen, AN & Carlton, JT (1998) Accelerating invasion rate in a highly invaded estuary. *Science,* 279, 555-8. [http://dx.doi.org/10.1126/science.279.5350.555] [PMID: 9438847]

Common Names for Sea Walnut (Mnemiopsisleidyi) *Encyclopedia of Life* Retrieved 13 December 2013.

Cowan, JH, Jr & Houde, ED (1993) Relative predation potentials of scyphomedusae, ctenophores and planktivorous fish on ichthyoplankton in Chesapeake Bay. *Mar Ecol Prog Ser,* 95, 55-65. [http://dx.doi.org/10.3354/meps095055]

Cranfield, HJ, Gordon, DP, Willan, RC, Marshall, BA, Battershill, CN, Francis, MP, Nelson, WA, Glasby, CJ & Read, GB Adventive marine species in New Zealand. National Institute of Water and Atmosphere. Technical Report 34, Wellington.

Crothers, JH (1967) The biology of the shore crab, *Carcinusmaenas* (L.). I. The background-anatomy, growth and life history. *Field Stud,* 2, 407-34.

De Poorter (2009) Marine menace: alien invasive species in the marine. IUCN Publication, IUCN-2009-011

de Villèle, X & Verlaque, M (1995) Changes and degradation in a posidoniaoceanica bed invaded by the introduced tropical alga caulerpataxifolia in the North Western Mediterranean. *Botanica Marina. Bot Mar,* 38, 79-88. [http://dx.doi.org/10.1515/botm.1995.38.1-6.79]

DeRivera, CE, Ruiz, GM, Hines, AH & Jivoff, P (2005) Biotic resistance to invasion: Native predator limits abundance and distribution of an introduced crab. *Ecology,* 86, 3367-76. [http://dx.doi.org/10.1890/05-0479]

Elisabetta, L & Simone, C (2006) New records of *Dreissenapolymorpha* (Pallas, 1771) (Mollusca: Bivalvia: Dreissenidae) from Central Italy. *Aquat Invasions,* 1, 281-3. [http://dx.doi.org/10.3391/ai.2006.1.4.11]

Faasse, MA & Bayha, KM (2006) The ctenophore Mnemiopsisleidyi A. Agassiz 1865 in coastal waters of the Netherlands: an unrecognized invasion? *Aquat Invasions,* 1, 270-7. [http://dx.doi.org/10.3391/ai.2006.1.4.9]

Fago, D (1993) Skipjack herring, alosa, expanding its range into the Great Lakes. *Can Field Nat,* 107, 352-3.

Fofonoff, PW, Ruiz, GM, Steves, B, Hines, AH & Carlton, JT (2003) *National Exotic Marine and Estuarine Species Information System.* Retrieved from: http://invasions.si.edu/nemesis/ on 29 March 2006.

Giusti, F & Oppi, E (1973) Dreissenapolymorpha (Pallas) nuovamente in Italia. (Bivalvia, Dreissenidae). *Mem Mus Civ St Nat Verona,* 20, 45-9. [in Italian].

Gomes, V (1991) First results of tagging experiments on crab *Carcinusmaenas*(L.) in the Ria de Aveiro Lagoon, Portugal. *Ciencia Biologica Ecology and Systematics,* 11, 21-9.

Granhag, L, Møller, LF & Hansson, LJ (2011) Size-specific clearance rates of the ctenophore Mnemiopsisleidyi based on in situ gut content analyses. *J Plankton Res,* 33, 1043-52. [http://dx.doi.org/10.1093/plankt/fbr010]

Grosholz, ED & Ruiz, GM (1995) Spread and potential impact of the recently introduced European green crab, *Carcinusmaenas*, in central California. *Mar Biol,* 122, 239-47.

Grosholz, ED (2005) Recent biological invasion may hasten invasional meltdown by accelerating historical introductions. *Proc Natl Acad Sci USA,* 102, 1088-91. [http://dx.doi.org/10.1073/pnas.0308547102] [PMID: 15657121]

Hansson, HG (2006) Ctenophores of the Baltic and adjacent Seas – the invader Mnemiopsis is here! *Aquat*

Invasions, 1, 295-8.
[http://dx.doi.org/10.3391/ai.2006.1.4.16]

Harding, JM & Mann, R (1999) Observations on the biology of the veined rapa whelk, *Rapanavenosa* (Valenciennes, 1846) in the Chesapeake Bay. *J Shellfish Res,* 18, 9-17.

Haslob, H, Clemmensen, C, Schaber, M, Hinrichsen, HH, Schmidt, JO, Voss, R, Kraus, G & Köster, FW (2007) Invading Mnemiopsisleidyi as a potential threat to Baltic fish. *Mar Ecol Prog Ser,* 349, 303-6.
[http://dx.doi.org/10.3354/meps07283]

Hewitt, CL In: BorgesseEM,ChircopA,McConnellML,editors. Marine biosecurity issues in the world oceans: global activities and Australian directions. Ocean Yearbook 17, University of Chicago Press, in press.

Hewitt, C, Campbell, M, Thresher, R & Martin, R (1999) Marine biological invasions of Port Phillip Bay, Victoria *CRIMP Technical Report 20, CSIRO Marine Research, Hobart, Tasmania.*

Hidalgo, FJ, Barón, PJ & Orensanz, JML (2005) A prediction come true: the green crab invades the Patagonian coast. *Biol Invasions,* 7, 547-52.
[http://dx.doi.org/10.1007/s10530-004-5452-3]

Huwer, B, Storr-Paulsen, M, Riisgård, HU & Haslob, H (2008) Abundance, horizontal and vertical distribution of the invasive ctenophore Mnemiopsisleidyi in the central Baltic Sea, November 2007. *Aquat Invasions,* 3, 113-24. [available online].
[http://dx.doi.org/10.3391/ai.2008.3.2.1]

Janas, U & Zgrundo, A (2007) First record of Mnemiopsisleidyi A. Agassiz, 1865 in the Gulf of Gdansk (southern Baltic Sea). *Aquat Invasions,* 2, 450-4.
[http://dx.doi.org/10.3391/ai.2007.2.4.18]

Javidpour, J, Molinero, JC, Lehmann, A, Hansen, T & Sommer, U (2009) Annual assessment of the predation of Mnemiopsisleidyi in a new invaded environment, the Kiel Fjord (Western Baltic Sea): a matter of concern? *J Plankton Res,* 31, 729-38.
[http://dx.doi.org/10.1093/plankt/fbp021]

Javidpour, J, Sommer, U & Shiganova, T (2006) First record of Mnemiopsisleidyi A. Agassiz 1865 in the Baltic Sea. *Aquat Invasions,* 1, 299-302.
[http://dx.doi.org/10.3391/ai.2006.1.4.17]

Jeffrey, G, Torchin, E, Mark, K & Armand, L (2005) Host specificity of Sacculinacarcini, a potential biological control agent of the introduced European green crab Carcinusmaenas in California. *Biol Invasions,* 7, 895-912.
[http://dx.doi.org/10.1007/s10530-003-2981-0]

Jousson, O, Pawlowski, J, Zaninetti, L, Zechman, FW, Dini, F, Di Guiseppe, G, Woodfield, R, Millar, A & Meinesz, A (2000) Invasive alga reaches California. *Nature,* 408, 157-8.
[http://dx.doi.org/10.1038/35041623] [PMID: 11089959]

Kideys, AE (2002) Ecology. Fall and rise of the Black Sea ecosystem. *Science,* 297, 1482-4.
[http://dx.doi.org/10.1126/science.1073002] [PMID: 12202806]

Kube, S, Postel, L, Honnef, C & Augustin, CB (2007) Mnemiopsisleidyi in the Baltic Sea – distribution and overwintering between autumn 2006 and spring 2007. *Aquat Invasions,* 2, 137-45.
[http://dx.doi.org/10.3391/ai.2007.2.2.9]

Lafferty, KD & Kuris, KD (1996) Biological control of marine pests. *Ecology. Ecological Society of America,* 77, 1989-2000.

Lehtiniemi, M, Paakkonen, JP, Flinkman, J, Gorokhova, E, Karjalainen, M, Viitasalo, S & Björk, H (2007) Distribution and abundance of the American comb jelly (Mnemiopsisleidyi) – A rapid invasion to the Northern Baltic Sea during 2007. *Aquat Invasions,* 2, 445-9.
[http://dx.doi.org/10.3391/ai.2007.2.4.17]

Mackie, G, Gibbons, W, Muncaster, B & Gray, I (1989) *The Zebra Mussel, Dreissenapolymorpha: A synthesis of European Experiences and a preview for North America.* Ontario Ministry of Environment.

Mann, R & Harding, JM (2000) Invasion of the North American Atlantic coast by a large predatory Asian mollusc. *Biol Invasions,* 2, 7-22.
[http://dx.doi.org/10.1023/A:1010038325620]

Mann, R & Harding, JM (2003) Salinity tolerance of larval Rapana venosa: implications for dispersal and establishment of an invading predatory gastropod on the North American Atlantic coast. *Biol Bull,* 204, 96-103. [Woods Hole].
[http://dx.doi.org/10.2307/1543499] [PMID: 12588748]

McEnnulty, FR, Jones, TE & Bax, NJ (2002) The Web-Based Rapid Response Toolbox. Web publication. Archived from the original on May 19, 2006. Retrieved March 9, 2006.

Meinesz, A (1999) *Killer algae.* University of Chicago Press, Chicago.

Meinesz, A & Hesse, B (1991) Introduction et invasion de l'alguetropicale Caulerpataxifolia enMéditerranéenord-occidentale. *OceanologiaActa,* 14, 415-26.

Miller, RR (1957) Origin and dispersal of the alewife, *Alosapseudoharengus*and the gizzard shad, *Drosomacepedianum,* in the Great lakes. *Trans Am Fish Soc,* 86, 97-111.
[http://dx.doi.org/10.1577/1548-8659(1956)86[97:OADOTA]2.0.CO;2]

Mutlu, E, Bingel, F, Gucu, AC, Melnikov, VV, Niermann, U, Ostrovskaya, NA & Zaika, VE (1994) Distribution of the new invader Mnemiopsis sp. and the resident Aurelia aurita and Pleurobrachia pileus populations in the Black Sea in the years 1991-1993. *ICES J Mar Sci,* 51, 407-21.
[http://dx.doi.org/10.1006/jmsc.1994.1042]

NIMPIS (2002) Mnemiopsis leidyi reproduction & life cycle. National Introduced Marine Pest Information System (Eds: Hewitt C.L., Martin R.B., Sliwa C., McEnnulty, F.R., Murphy, N.E., Jones T. & Cooper, S.). Web publication, (Accessed June 5, 2008)

Novak, M (2004) Diurnal activity in a group of Maine decapod. *Crustaceana,* 77, 603-20.
[http://dx.doi.org/10.1163/1568540041717975]

Oliveira, OMP (2007) The presence of the ctenophore Mnemiopsisleidyi in the Oslofjorden and considerations on the initial invasion pathways to the North and Baltic Seas. European Research Network on Aquatic Invasive Species. *Aquat Invasions,* 2, 185-9.
[http://dx.doi.org/10.3391/ai.2007.2.3.5]

Oliveira, OMP (2007) The presence of the ctenophore Mnemiopsisleidyi in the Oslofjorden and considerations on the initial invasion pathways to the North and Baltic Seas. European Research Network on Aquatic Invasive Species. *Aquat Invasions,* 2, 185-9.
[http://dx.doi.org/10.3391/ai.2007.2.3.5]

Perry, H (2014) *Carcinusmaenas USGS nonindigenous aquatic species database.* Google Scholar.

Pimentel, D, McNair, S, Janecka, J, Wightman, J, Simmonds, C & O'Connell, C (2001) Economic and environmental threats of alien plant, animal, and microbe invasions. *Agric Ecosyst Environ,* 84, 1-20.
[http://dx.doi.org/10.1016/S0167-8809(00)00178-X]

Purcell, JE, Nemazie, DA & Dorsey, SE (1994) Predation mortality of by anchovy Anchoamitchilli eggs and larvae due to scyphomedusae and ctenophores in Chesapeake Bay. *Mar Ecol Prog Ser,* 114, 47-58.
[http://dx.doi.org/10.3354/meps114047]

Purcell, JE, Shiganova, TA, Decker, MB & Houde, ED (2001) The ctenophore Mnemiopsis in native and exotic habitats: U.S. estuaries *versus* the Black Sea basin. *Hydrobiologia,* 451, 145-76.
[http://dx.doi.org/10.1023/A:1011826618539]

Ropes, JW (1968) The feeding habits of the green crab, *Carcinusmaenas* (L.). *Fish Bull,* 67, 183-203.

Ruiz, GM, Carlton, JT, Grosholz, ED & Hines, AH (1997) Global invasions of marine and estuarine habitats by non-indigenous species: mechanisms, extent, and consequences. *Am Zool,* 37, 621-32.

[http://dx.doi.org/10.1093/icb/37.6.621]

Ruiz, GM, Torchin, ME & Grant, K (2009) Using the Panama Canal to test predictions about tropical marine invasions. In: Lang, M.A., (Ed.), *Proceedings of the Smithsonian Marine Science Symposium,* vol. 38

Schloesser, DW & Nalepa, TF (1994) Dramatic decline of unionid bivalves in offshore waters of western Lake Erie after infestation by the zebra mussel, Dreissenapolymorpha. USGS Great Lakes Science Center. (Assessed October 20, 2014)

Selander, E, Møller, LF, Sundberg, P & Tiselius, P (2009) Parasitic anemone infects the invasive ctenophore Mnemiopsisleidyi in the North East Atlantic. *Biol Invasions,* 12, 1003-9.
[http://dx.doi.org/10.1007/s10530-009-9552-y]

Shiganova, TA, Kıdeys, AE, Gücü, AC, Niermann, U & Khoroshilov, VS (1998) Changes in species diversity and abundance of the main components of the Black Sea pelagic community during the last decade. In: Results, L. Ivanov, Oguz, T., (Eds.), *Symposium on Sci,* Vol. 1, 171-88.

Shiganova, TA, Mirzoyan, ZA, Studenikina, EA, Volovik, SP, Siokou-Frangou, I, Zervoudaki, S, Christou, ED, Skirta, AY & Dumont, HJ (2001) Population development of the invader ctenophore Mnemiopsisleidyi in the Black Sea and other seas of the Mediterranean basin. *Mar Biol,* 139, 431-45.
[http://dx.doi.org/10.1007/s002270100554]

Taylor, DL (2005) Predatory impact of the green crab (*Carcinusmaenas Linnaeus*) on post– settlement winter flounder (*Pseudopleuronectesamericanus Walbaum*) as revealed by immunological dietary analysis. *J Exp Mar Biol Ecol,* 324, 112-26.
[http://dx.doi.org/10.1016/j.jembe.2005.04.014]

Tendal, OS, Jensen, KR & Riisgård, HU (2007) Invasive ctenophore Mnemiopsisleidyi widely distributed in Danish waters. *Aquat Invasions,* 2, 455-60.
[http://dx.doi.org/10.3391/ai.2007.2.4.19]

Thibaut, T (2001) Elysiasubornata a potential control agent of the alga Caulerpataxifolia in the Mediterranean Sea, Archived 2005-10-25 at the Wayback Machine. *J. Mar. Biol. Assoc. U. K.*

Thresher, R, Proctor, C, Ruiz, G, Gurney, R, MacKinnon, C & Walton, W (2003) Invasion dynamics of the European shore crab, Carcinusmaenas, in Australia. *Mar Biol,* 142, 867-76.
[http://dx.doi.org/10.1007/s00227-003-1011-1]

Washington Department of Fish and Wildlife. Archived from the original on June 13, 2008. Retrieved February 12, 2010.

Yamada, SB (2001) *Global invader: the European Green Crab.* Oregon Sea Grant 123.

Zolotarev, V (1996) The Black Sea ecosystem changes related to the introduction of new mollusc species. P.S.Z.N. I. *Mar Ecol (Berl),* 17, 227-36.
[http://dx.doi.org/10.1111/j.1439-0485.1996.tb00504.x]

CHAPTER 14

Disturbance and Biodiversity of Marine Protected Areas

P.M. Mohan[*], **Barbilina Pam**, **D.B.K.K. Sabith** and **Akshai Raj**

Department of Ocean Studies and Marine Biology, Pondicherry University Off Campus, Brookshabad, Port Blair – 744112, Andaman and Nicobar Islands, India

Abstract: The conservation of biological diversity and the sustainable use of its components are the major objectives of the Convention on Biological Diversity (CBD). To achieve these objectives, many marine protected areas (MPAs) were identified and developed. The success of these protected areas depends upon several factors of local concern. The failure of coastal and marine biodiversity protection was mainly caused by the environmental changes influenced by anthropogenic activities, overexploitation of resources, habitat loss because of developmental activities, and natural change in climate. This chapter highlights the status of these activities in the island environment and provides potential strategies for its protection by mainstreaming biodiversity with people's participation.

Keywords: Marine Protected Areas, Mainstreaming, Biodiversity, Andaman and Nicobar Islands, India.

INTRODUCTION

Marine Protected Areas (MPAs) are regarded as one of the global thrust that has emerged towards a holistic management approach that takes the entire ecosystems into account with essential tools for implementation (Currie *et al.*, 2008). As per IUCN (2008), the definition of MPA is "A clearly defined geographical space, recognized, dedicated and managed, through legal or other effective means, to achieve the long-term conservation of nature with associated ecosystem services and cultural values". MPAs are an effective way of protecting and conserving the marine biodiversity and maintaining the productivity of the oceans with their cultural and historical heritage for today and future generations (Laxmilatha, 2015 and Brander *et al.*, 2015). MPAs provide necessary insights with several threats and consequences, of which some may be prevented or some may be unavoidable

[*] **Corresponding author P.M. Mohan:** Department of Ocean Studies and Marine Biology, Pondicherry University Off Campus, Brookshabad, Port Blair – 744112, Andaman and Nicobar Islands, India; Tel: +91-3192-262317; +91-3192-262342; 262366; 262362; 262320; Fax: +91-3192-262323; E-mails: pmmtu@yahoo.com; pmmnpu@rediffmail.com

De-Sheng Pei & Muhammad Junaid (Eds.)
All rights reserved-© 2019 Bentham Science Publishers

with concern to tourism influx and coastal urbanization (Gray, 2010). Although conservation fact is important for MPA, there are severe impacts on the coastline and marine environment with direct antagonism with traditional small-scale fishing and other recent activities like recreational fishing and aquaculture farm due to the above said factors.

The FAO (2011) technical guidelines for Responsible Fisheries defined MPA as: "any marine geographical area that is afforded greater protection than the surrounding waters for biodiversity, conservation or fisheries management purposes will be considered an MPA". The marine ecosystem is extremely diverse and attributed to the geomorphology and climatic variations along the coast, leading to recognition of the role of MPAs as one of the important step towards increasing the effectiveness of MPAs under the CBD 2020 agenda (Simard *et al.*, 2016), as Oceanic climatological changes can lead to a profound change in marine ecosystems. Molenaar and Elferink (2009) provided a short overview on the global regime with reference to the designation, identification and the regional cooperation on the protection of the marine environment of MPAs, including recent developments in the worldwide scenario. Reker (2015) discussed how best to evaluate the effectiveness of MPAs and regulate their effectiveness in protecting biodiversity across Europe's seas. Alino (2018) mentioned about marine passages that are also considered as a strategic zone, facilitating the exchange of materials and connectivity of various marine biogeographic regions.

Dorel *et al.,* (2015) worked on the PANACHE (Protected Area Network Across the Channel Ecosystem) project, which aimed for a coherent approach for a marine protected area (MPA) on both sides of the Channel, involving two nations - France and the United Kingdom, with their management judgments but also with the common desire to address MPAs in a genuine and effective way. The review work done by Jones *et al.,* (2011) outlined the policy agenda with regards to the main components of the emerging UK marine protected area (MPA) for creating a marine conservation zones (MCZs) and marine special areas of conservation (SACs), leading to increase in 27% of MPAs in English waters. Sink (2016), studied the key aspects of the initiative to explore and unlock the economic potential of South Africa's marine and coastal environment. As agreed at the Rio 'Earth' Summit in 1992, marine sites, together with terrestrial and freshwater sites, form a part of the European Natura 2000 network of protected areas. The Marine Strategy Framework Directive (2008/56/EC) requires Member States to include in their programs in the establishment of MPAs, thus contributing to one of the key objectives of the Convention on Biological Diversity (CBD 2016). The review work was done by the European Commission in 2015, where they found that MPAs covered 6% of the European Seas by 2012, with an aim to reach 10% by 2020, even the economy and other benefits of the Natura 2000 network were also discussed (Jones and Burgess, 2005).

A contrast to the merits of MPAs, Rajagopalan (2008), came up with work saying that, though MPAs have become a tool that limits, prohibits and regulates the use-pattern and human action through certain frameworks of rights and rules, where essentiality of social components is needed to be considered for a long-term benefit of coastal communities. The existing studies say that least importance was given to the social aspect than the ecological and biological factors. Many times, in the past, over-exploitation of the world's fishery resources has been observed with regards to traditional communities. UK fishers fear that MPA restrictions beyond six nautical miles might be unilaterally imposed on them, a concern with the recent banning on pair trawling by English vessels to protect cetaceans in the south-west approaches (De Santo and Jones, 2007). Most of the MPAs in the Philippines have been established for the purpose of sustaining fisheries utilization in the adjacent fishing areas hampering the livelihood of the prevailing communities and other stakeholders in that area. The first empirical analysis was studied by Leisher *et al.,* (2007) with regards to the link between biodiversity conservation initiatives and poverty reduction. The procedure and outcome of the 2014 World Parks Congresses (WPC), which emphasized on the role of people (in particular, fishery folks) in marine conservation was studied by Charles *et al.,* (2016). This article clearly mentioned that inclusion of the human dimension in any evaluation of management practices, without this factor the MPA activities affect the local livelihoods of fishing communities (Lester and Halpern, 2008), Jones (2008, 2009). Many conservation NGOs had campaigned for the implementation of the Marine Act of a statutory, practicing no-take MPA setting ban on extractive and disturbing activities involving 30% of the national marine area, in keeping with previous recommendations (RCEP - Royal Commission on Environmental Pollution, 2004). Highly protected MPAs produce results rapidly, but benefits build up for decades. A good MPA should give protection to a broad spectrum of biodiversity, not just a handful of species. Therefore, documenting and analyzing the experiences and views of local communities, particularly fishing communities, with respect to various aspects of MPA design and implementation would help in integrating the MPA programme (Sanchirico, 2000; Fraga and Jesus, 2008). The Marine Act does not include a no-take MPA target, nor does it require any no-take MPAs, but it maintains the flexibility to ensure the right level of protection in each case, based on the evidence available (Appleby and Jones, 2012).

Chuenpagdee *et al.,* (2012) did a case study stating that MPA is not a simple management tool or an easy technical fix because of complexities existing within the process, which may surpass the expectations of promoters and stakeholders. The exponential increase in MPAs and its complexity involved within the marine ecosystems with an impact leading to various ecological changes, which is associated with the declaration of MPAs varying greatly from one place to another. Edgar *et al.*, (2007) and Kelleher (2015) represented that accurate prediction of specific goal of every MPA is needed to be specified in order to allow management agencies to assess the success of the MPA and to monitor research activities. However, Claudet (2011), had put forward a multidisciplinary approach towards the establishment of MPAs for a better result with an influential framing of MPAs as a globally accepted policy with wider vision, based on an 'event ethnography' conducted at 2008's, World Conservation Congress in Barcelona, Noella, throwing light into the vital role played by NGOs in various perspective. Drastic failures in marine fisheries management led some to suggest that marine protected areas (MPAs) are the only way to achieve sustainable fisheries (Kaiser, 2005; Arias, 2015). The growing recognition in finding the 'best' or 'right' approach needs developing governance models, frameworks and approaches that combine the striking role of states, markets, and people, enabling to move beyond ideological debates, Jones *et al.*, (2011). The MPAs had larger disadvantages when it came to surveillance enforcement and monitoring of such vast coastline area involving high cost with much more expensive technologies (Wilhelm *et al.,* 2014).

The aim of the Convention on Biological Diversity (CBD) is to conserve and sustainable use of biological diversity and distribute the benefits of genetic resources equal to the concerned local bodies and benefit claimers. All the 193 countries are parties of this act and regularly revisiting their strategies and action plans (NBSAP) to implement in best practices. These action plans may be helpful to integrate poverty alleviation in the local bodies with environmental policy, plans and investment to support both the objectives of the convention, such as development as well as support and protect the biodiversity (IIED, 2017). The foundation of life on Earth is biodiversity. However, it is in great strain due to the human activity for the developmental need. All the sectors of society have its own responsibility to conserve and use natural resources in sustainable ways.

In the last two centuries, the developmental needs of human beings for food, water, energy, textile and other necessary items, through unsustainable practices significantly changed their surrounding environment. The resultant of this need affected the quality of the clean air, water and shelter from adverse weather. Overall, it has also affected the common man for their basic subsistence needs. However, now humans in larger extent are aware of the impact of environmental deterioration on their daily life and well-being. The maintenance of the clear service, *i.e.* uncontaminated level, will be provided by the ecosystem through sustainable development of biodiversity at all levels (species ecosystem and genes) by the way of conservation of biodiversity. This sustainable development of biodiversity cannot be achieved by the isolated work of the environmental community alone. It needs a collective attempt of all parts of society, governmental agencies, non-governmental agencies and the private sector.

Mainstreaming of Biodiversity means the integration of the conservation and sustainable use of biodiversity in developmental activities such as, poverty reduction, climate change adaptation or mitigation, trade and international cooperation in the cross-sectoral areas as well as agriculture, fisheries, forestry, mining, energy, tourism and transportation in sectorial activities. By all means, it is suggested that those changes are applicable to the developmental models, strategies and paradigms. Over and above, it also clearly indicates that it should be integrated into existing or new systems, but not creating any parallel or artificial process, against the existing systems. Further, the mainstreaming is not simply pushing the biodiversity into the system. Instead, it should be actively incorporated of relevant environmental concerns into the existing sector decisions. Further, the Convention on Biological Diversity (CBD) suggested that mainstreaming is forming an integral part of biodiversity, which develop national strategies, plans, programs for the conservation and sustainable use of biological diversity and integrate into the relevant sectorial or cross sectorial plans, programs and policies, without affecting or modifying their regular process, at the same time benefit for the conservation of biodiversity.

Integrating the current development activities with biodiversity is not a simple process as well as it would not happen on its own. It needs some leader or champion who is essential to motivate or trigger this process. Otherwise, a motivation itself as an attribute exists in the process, or it should exhibit a reciprocal cooperation between the sectors. Over and above, it also shows replication, long-term engagement, and sustainability. The success of mainstreaming can be characterized by the way of positive outcomes from the short, medium and long-term process, *i.e.* the inclusion of biodiversity considerations in a policy or plan or budget of another sector. The success can be measured from the result of the particular biodiversity's sustainability, the increment of capacity, knowledge or awareness, which resultantly benefit to human well-being.

ISLAND BIODIVERSITY AND MARINE PROTECTED AREAS

International Union of Conservation of Nature and Natural Resources (IUCN) promoting the establishment and protection of Marine Biodiversity under the Commission on National Parks and Protected Areas (CNPPA) from 1986 through Marine Protected Areas (MPA) programmes. The MPA description as reported by IUCN (1998) was demarcated in Indian context by Singh (2002) and Wells *et al.*,

(1995) as Category I, Category II and Category III. Category I represented MPA of IUCN definition with trawling or fishing regulations required in the sea (Venkatraman *et al.*, 2012). Category II mainly concerned with the terrestrial environment, but the marine biodiversity areas adjoining to these territories require marking boundaries on the seaward side and have substantial in size to declare PA. Category II may move into Category I, if the potential of biodiversity proved. Category III is mainly concentrated in terrestrial and important MPA area fringes in the marine sector (Fig. **1**). Accordingly, Category I MPA has 12 sites (Table **1**), Category II has 16 groups (Table **2**) and 95 sites and Category III has 11 sites (Table **3**).

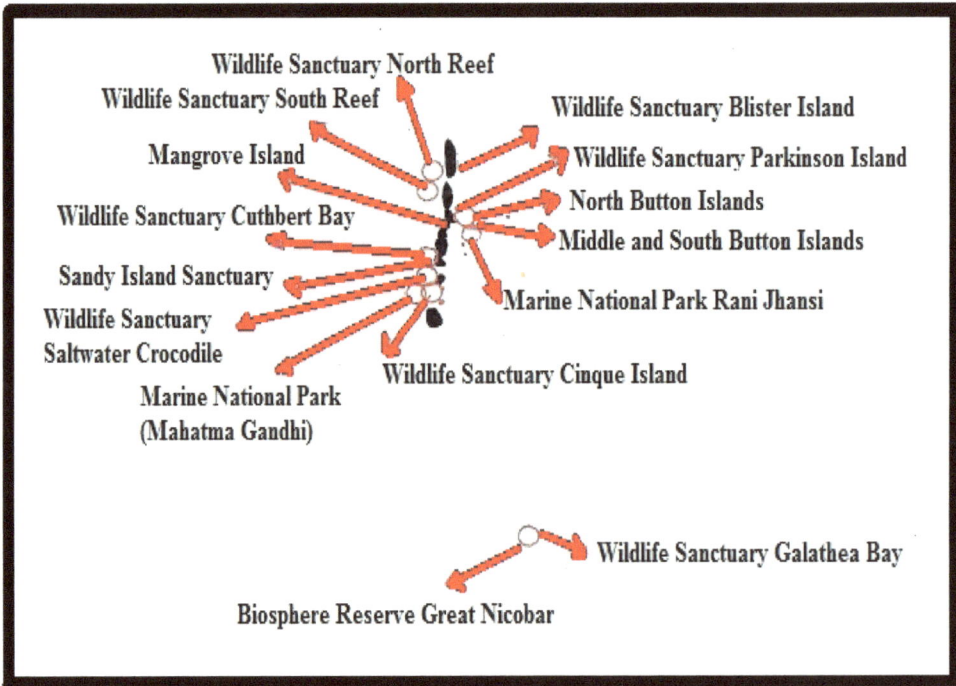

Fig. (1). Selective Marine Protected Areas of Andaman and Nicobar Islands.

Table 1. Indian Scenario on Marine Protected Areas (MPA) under Category I.

S. No.	Category I	Location	Species Under Protection
1	Mahatma Gandhi Marine National Park, Wandoor	South Andaman	Corals and associated Fishes, Sea Cucumber, Sea Anemones, *etc*. Salt water Crocodiles, Sea Turtles (Green and Olive Ridley), Dugong, Manta Ray, 300 species of birds, Spotted Deer, Flying Fox and Wild Boar.

(Table 1) cont.....

S. No.	Category I	Location	Species Under Protection
2.	Lohabarrak Sanctuary (Salt Water Crocodile), Wandoor	South Andaman	Salt water Crocodiles, Hawksbill turtle, Leather back turtle, Olive Ridley and Green turtle, Water monitor, Dolphin, Wild Boar, Spotted Deer, Himalayan palm civet, Andaman dark serpent eagle, Crested serpent eagle, White bellied eagle, Harriers, Corals and the associated fishes.
3.	Rani Jhansi Marine National Park (Ritchies Archipelago) Havelock Island	South Andaman	Corals and the associated fishes, Salt water Crocodiles, Hawksbill turtle, Leather back turtle, Olive Ridley and Green turtle, Dugong and Birds.
4.	Sundarbans National Park and Tiger Reserve, North and South 24 Pargana	West Bengal	400 Tigers adapted to saline waters, Mangroves, Fishing cats, Leopard cats, Macaque, Wild boar, Grey mongoose, Fox, Jungle cat, Pangolin, Fishes and Amphibians, Gangetic dolphins, Skipping frogs, Common toads, Snakes, Pythons, King cobra, Dog faced water snakes and common kraits.
5.	Sajnakhali Sanctuary, South 24 Pargana	West Bengal	Rich avian faunas, Migratory birds, Spotted deer, Rhesus Macaque, Wild boar, Tiger, Water monitor lizard, Otter, Fishing cat, Crocodile and Batagur terrapin.
6.	Gahirmatha Marine Sanctuary, Kendrappa	Orissa	Hawksbill turtle, Leather back turtle, Olive Ridley and Green turtle, Batagur turtle, Irrawady dolphins, Humpback porpoise, Whale shark, Sperm whale, Sea snakes and Horseshoe crab.
7.	Chilka (Nalaband) Wild Life Sanctuary, Kundara-Pur--Ganjam	Orissa	400,000 Waterfowls of different species, Greater flamingo, Gull billed tern, River tern, Little tern, White bellied sea eagles, Golden plovers, Sand pipers, Pelicons, Gulls, Herrings, Sardines, Milk fish, Ten pounder, Hilsa and mullets.
8.	Coringa Wildlife Sanctuary, East Godavari	Andhra Pradesh	Mud skippers, Crabs, Molluscs, Indian smooth otters, Fishing cat, Olive ridley turtle, Crusted serpent eagle, Black capped kingfisher and White bellied woodpecker.
9.	Gulf of Mannar Marine National Park, Tuticorin	Tamil Nadu	Corals, Sea anemones, Sea fishes, Sharks, Mammals, Reptiles- Sea turtles, Sea snakes, Water birds, Sacred shank, Pearl oysters and 11 species of sea grasses.
10.	Pulicat Lake Sanctuary (Bird), Thiruvellore	Tamil Nadu	Greater flemingos, Large billed grey pelican, Ducks, Sea gulls, Water birds and Sand pipers.
11.	Gulf of Kachchh Marine National Park and Gujrat Marine Sanctuary, Jamnagar	Gujarat	Coral reefs, Dugong, Fishes, Crabs, Prawns, Sharks, Porpoise, *Nerita textalis*, *Acanthastrea hillae*, *Barapattoia amicorum*, *Favia lacuna* and *Turbinaria frondens*.

(Table 1) cont.....

S. No.	Category I	Location	Species Under Protection
12.	Malvan Marine Sanctuary, Sindhudurg	Maharastra	Corals, Sea anemones, Molluscs, Polychaetes, Pearl oyster, Mangrove, 220 species of birds, Barking deer, Sambar deer, Wild boars, Nilgai, Leopards and 27 species of fishes.

Table 2. Indian Scenario on Marine Protected Areas (MPA) under Category II.

S. No.	Category II	Location	Species Under Protection
	Group 1	North Andaman	
1	Channel Island Sanctuary	""	Coral reefs.
2.	Landfall Island Sanctuary	""	Corals and associated marine lives.
3.	East Island Sanctuary	""	Mangroves, Corals and associated marine lives.
	Group 2	North Andaman	
4.	West Island Sanctuary		Corals and associated marine lives.
5.	White Cliff Island Sanctuary		Corals and associated marine lives.
6.	Reef Island Sanctuary		Corals and associated marine lives.
	Group 3	North Andaman	
7.	Mayo Island Sanctuary		Corals and associated marine lives.
8.	Paget Island Sanctuary		Corals and associated marine lives.
9.	Point Island Sanctuary		Corals and associated marine lives.
10.	Sheame Island Sanctuary		Mangroves, Corals and associated marine lives.
	Group 4	North Andaman	
11.	Treen Island Sanctuary		Corals and associated marine lives.
12.	Thirlby Island Sanctuary		Corals and associated marine lives.
13.	Table (Delgarno) Island Sanctuary		Mangroves, Corals and associated marine lives.
14.	Table (Excelsior) Island Sanctuary		Corals and associated marine lives.
15.	Temple Island Sanctuary		Corals and associated marine lives.
16.	Turtle Island Sanctuary		Corals and associated marine lives.
	Group 5	North Andaman	
17.	Jungle Island Sanctuary		Mangroves, Corals and associated marine lives.
18.	North Island Sanctuary		Mangroves, Corals and associated marine lives.

(Table 2) cont.....

S. No.	Category II	Location	Species Under Protection
19.	Wharf Island Sanctuary		Mangroves, Corals and associated marine lives.
20.	OX Island Sanctuary		Mangroves, Corals and associated marine lives.
21.	Ross Island Sanctuary		Corals and associated marine lives.
	Group 6	North Andaman	
22.	Shark Island Sanctuary		Corals and associated marine lives.
23.	Kwangtung Island Sanctuary		Corals and associated marine lives.
24.	Latouche Island Sanctuary		Corals and associated marine lives.
25.	North Reef Island Sanctuary		Corals and associated marine lives.
	Group 7	North Andaman	
26.	Bamboo Island Sanctuary		Mangrove habitats.
27.	Blister Island Sanctuary		Mangrove habitat crabs and fishes.
28.	Middle Curlew Island Sanctuary		Coral reefs.
29.	Dot Island Sanctuary		Corals and associated marine lives.
30.	Oliver Island Sanctuary		Mangrove associated marine lives.
31.	Goose Island Sanctuary		Corals and associated marine lives.
32.	Gander Island Sanctuary		Mangrove associates.
33.	Oyster – 1- Island Sanctuary		Mangrove and littoral forest.
34.	Orchid Island Sanctuary		Mangrove and associated marine lives.
35.	Egg Island Sanctuary		Corals and associated marine lives.
36.	Curlew B.P. Island Sanctuary		Corals and associated marine lives.
37.	Dottrel Island Sanctuary		Corals and associated marine lives.
38.	Girjan Island Sanctuary		Mangrove and intertidal habitat.
39.	Swamp Island Sanctuary		Mangrove associated marine life.
	Group 8	Middle Andaman	
40.	Sea Serpent Island Sanctuary		Mangrove associates and sea snakes.
41.	Snake – I- Island Sanctuary		Mangrove associates, Crabs, Fishes and sea snakes.
42.	Buchanan Island Sanctuary		Mangrove or marshy habitat.
43.	Bondoville Island Sanctuary		Mangrove associates.
44.	Entrance Island Sanctuary		Mangrove associates.
45.	Surat Island Sanctuary		Mangrove associates.
46.	Spike Island – I - Island Sanctuary		Mangrove associated Crabs and Fishes.

(Table 2) cont.....

S. No.	Category II	Location	Species Under Protection
47.	Bennet Island Sanctuary		Mangrove associates.
48.	Ranger Island Sanctuary		Mangrove associates.
49.	Roper Island Sanctuary		Mangrove habitats.
50.	Interview Island Sanctuary		Indian Elephant (Feral), Wild boar, Spotted deer, Dog, Salt water crocodiles, Hawks bill turtle, Water monitor, Chest nut headed bee eater, Fairy blue bird, Red vented bulbul and Indian lorikeet.
	Group 9	Middle Andaman	
51.	South Reef Island Sanctuary		Coral associates, Sea snakes and Turtles.
52.	Elat Island Sanctuary		Mangrove and Coral associated faunas.
53.	Mask Island Sanctuary		Mangrove associates.
54.	Tuf Island Sanctuary		Coral reef and other associated marine life.
55.	Hump Island Sanctuary		Evergreen forest.
	Group 10	Middle Andaman	
56.	Oyster Island Sanctuary		Marine Mangrove associates.
57.	Parkinson Island Sanctuary		Mangrove associated marine life.
58.	Cone Island Sanctuary		Mangrove associated marine life.
	Group 11	South Andaman	
59.	North Button Island National Park		Coral associated marine life, Sea snake, turtles and salt water crocodiles.
60.	Middle Button Island National Park		Coral associated marine life, Sea snake, turtles and salt water crocodiles.
61.	South Button Island National Park		Coral associated marine life, Sea snake, turtles and salt water crocodiles.
62.	East or Inglis Island Sanctuary		Corals and associated marine life.
63.	Sir Hugh Rose Island Sanctuary		Corals and associated marine life.
	Group 12	South Andaman	
66.	Stoat Island Sanctuary		Mangrove and coral associates.
67.	Talabaicha Island Sanctuary		Mangrove and coral associates.
68.	Mangroves Island Sanctuary		Mangrove associates, Crabs and fishes.
69.	Bingham Island Sanctuary		Mangrove associates.
70.	Bluff Island Sanctuary		Coral associates.
71.	Spike Island Sanctuary		Mangrove associates.

(Table 2) cont.....

S. No.	Category II	Location	Species Under Protection
	Group 13	South Andaman	
72.	Dungan Island Sanctuary		Corals and associated marine lives.
73.	Pitman Island Sanctuary		Corals and associated marine lives.
74.	Potanma Island Sanctuary		Corals and associated marine lives.
75.	Kyd Island Sanctuary		Corals and associated marine lives.
76.	James Island Sanctuary		Mangrove associates.
	Group 14	South Andaman	
77.	Patric Island Sanctuary		Corals and associated faunas.
78.	Defence Island Sanctuary		Mangrove and Coral associated marine lives.
79.	Montogomery Island Great Nicobar Biosphere Reserver Sanctuary		Coral associated marine lives.
80.	Clyde Island Sanctuary		Mangrove associated marine lives.
	Group 15	South Andaman	
81.	Cinque Island Sanctuary		Coral and associated marine lives, Sea snakes and turtles.
82.	Passage Island Sanctuary		Mangrove associates.
83.	Sisters Island Sanctuary		Coral associates.
84.	South Sentinel Island Sanctuary		Coral reefs and associated fishes.
85.	North Brother Island Sanctuary		Mangrove, Sea grass and Coral associated marine lives.
86.	South Brother Island Sanctuary		Coral associated marine lives.
	Group 16 – Outside the Clusters	Indian Waters	
87.	Peacock Island Sanctuary	North Andaman	Mangrove associates.
88.	Rowe Island Sanctuary	North Andaman	Coral associated marine lives.
89.	Arial Island Sanctuary	South Andaman	Mangrove associates.
90.	Belle Island Sanctuary	South Andaman	Mangrove associates.
91.	Sandy Island Sanctuary	South Andaman	Crabs and birds along with marine lives.
92.	Snake Island-II-Sanctuary	South Andaman	Mangrove and reef associates.
93.	Baltimalv Island Sanctuary	Nicobar	Mangrove and reef associates.
94.	Pitti Island Sanctuary	Lakshadweep	Coral associates.
95.	Halliday Island Sanctuary, Sundarbans	West Bengal	Spotted deer, Wild boar, Rhesus macaque, turtles, pythons and birds.

The loss of coastal and marine biodiversity, mainly accelerated by the environmental changes influenced by anthropogenic activities, overexploitation of resources, habitat loss due to developmental activities and natural change in climate. The estimated loss of this biodiversity is in the order of a species per day. There are three types of fauna and flora in MPA concern. They are plankton, nekton, and benthos.

Table 3. Indian Scenario on Marine Protected Areas (MPA) under Category III.

	Category III	Location	Species Under Protection
1.	Narcondam Island Sanctuary	North East of Andaman Islands	Narcondam hornbill- Endemic, 38 species of birds, reptiles, water monitor, sea snake, banded gecko, dwarf gecko and emerald gecko.
2.	Barren Island Sanctuary	East of Middle Andaman Islands	Sparse vegetation and feral goats.
3.	Saddle Peak National Park	North Andaman	Horseshoe bat, Indian wild boar, Himalayan palm civut, spotted deer, flying fox, Andaman shrew, heron, kingfishers and imperior pigeon.
4.	Cuthbert Bay Island Sanctuary	Middle Andaman	Sea turtles, reticulated python, dugong, crab eating macaque and Nicobar megapode.
5.	Mount Harriet National Park	South Andaman	Horseshoe bat, Indian wild boar, Himalayan palm civut, spotted deer, flying fox, Andaman shrew, heron, kingfishers and imperior pigeon, blue bird fairy, Andaman drango, crested serpent eagle, three toad kingfisher, Indian lorikeet and swift.
6.	Great Nicobar Biosphere Reserve	Great Nicobar	Endemic animals- 12 mammals, 35 birds, 7 reptiles, 4 amphibians and 100 insects; Nicobar crab eating macaque, Nicobar tree shrew, Nicobar wild pig, South Nicobar megapode, Great Nicobar created serpent eagle and reticulated python.
7.	Megapode Island Sanctuary	Nicobar	Megapode, water monitor, Andaman wood pigeon and white bellied sea eagle.
8.	Tillongchang Island Sanctuary	Nicobar	Megapode, water monitor, salt water crocodiles, Indian wild boar, spotted deer, crested serpent eagle, Andaman wood pigeon and white bellied sea eagle.
9.	Bhitarkanika Wild Life Sanctuary	Orissa	Sambar deer, chital, wild boar, leopards, jackals, hyenas, porcupines, otters, fishing cats, 5 species of amphibians, 9 species of lizards, 7 species of turtles, 18 species of snakes, salt water crocodiles, mud skippers, limulus crab and 170 species of birds.
10.	Point Calimere Sanctuary	Tamil Nadu	Greater flamingos, black buck antelope, spotted deer, jackal, civet, wild boar, jungle cat, bonnet macaque, black naped hare, Indian mongoose, Olive ridley turtles, dolphins and more than 100 species of birds.

(Table 3) cont.....

	Category III	Location	Species Under Protection
11.	Lothian Island Sanctuary, Sundarban	West Bengal	Gangetic dolphins, skipping frogs, common toads, tree frogs, estuarine crocodiles, chameleons, monitor lizards, Olive ridley, hawks bill, green turtles, pythons, king cobras, rat snakes, russells vipers, dog faced water snakes and common kraits.

The restricted movement of plankton always depends upon the water mass movement. The free swimmer, nekton moves its own way depending upon their own will. The benthos is defined as the living beings existing in, as well as on the bottom of the water. They cannot move around anywhere in the basin, even if their movement is present, it may be limited. Any effect on the biodiversity, the first and foremost impact is seen in the benthic community than the remaining two groups concern, because the other two can move out from the environment. However, the benthic community is always dependent upon those above two communities by the way of its availability as a straight feed or its waste or its organic matter developed by its dead remains. So, they are one among the important system with reference to trophic as well as eco-service concern. They are the immediate indicator of deterioration of the environment. The major objective of the Marine Protected Area is to protect the existing benthic community. The different concept of conservation of Marine Protected Area was not able to achieve their overall sustainable development due to failure in any one of the concepts. Even though there are different existing benthic community services, this article takes a coral reef ecosystem as an example for mainland as well as island concern. The Coral reef ecosystem is one among the benthic diversity existing in the marine ecosystem. Coral reef ecosystem can sustain on its own as well as provide the ecosystem service to more than 1800 groups of living beings, over and above the humans. The present scenario explains how to mainstream this ecosystem. Everyone is aware of the importance of coral reef ecosystem and their necessity to protect for the welfare of the other community. All the world governing systems have taken necessary steps to protect the same.

Anthropogenic activities affecting coastal habitats are a well-known phenomenon and no area of ocean is unaffected *i.e.* almost 41% impact on biodiversity is from these drivers (Halpern *et al.,* 2008). The habitat loss is the second important fact affecting the biodiversity in MPA. An example, it was observed in Goa (India) coast, the loss of sand dunes due to tourism activities are associated with loss of flora and fauna. Similarly, pollution is one among the driver for biodiversity loss in the MPA. The nutrient enrichment due to domestic sewage and agricultural runoff along with industrial effluents drives the fauna and flora community in a particular environment or kill them. The developmental activities, such as thermal power generation or nuclear power generation, also affect the coastal environment and its biodiversity (Wafar *et al.*, 2011). The next major factor that deteriorates the biodiversity is the overexploitation of resources (Qasim and Wafar, 1990), which in turn depletes the stock, locally as well as regionally, and in turn, eco-balance will be disturbed and leads to the proliferation of unwanted fauna or flora in a particular environment. This was mainly due to the lack of fishery regulations, nonadherence of mesh size for fishing gears, trawling in nearshore waters, and knowledge deprivation among the fisherfolk. Moreover, the natural calamities, such as storm surges, earthquake, tsunami, raise of temperature along the coastal regions and sea level oscillation provides cumbersome impacts on the biodiversity. The rich marine resources of Andaman and Nicobar Islands attracts foreign poachers to Indian territorial waters. Even though routinely apprehend these poachers, several of them get away undetected due to the larger amount of (500 islands) uninhabited islands and a less proportionate number of security personnel from the defense, coast guards as well as local police.

PRESENT STATUS ON THE PROTECTION OF BIODIVERSITY

There are several legal measures for the protection of the marine biodiversity of this region. The Regulation of Fishing by Foreign Vessels Act, 1981, Coastal Regulation Zone Notification, 1991 (last amended in 2011) and Wild Life (Protection) Act, 1972 are available and implemented on their status. Over and above, with more focused conservation initiative, the establishment of 9 National Parks and 96 Wildlife Sanctuaries is also implemented in this region. Overall 1619.786 sq.km area has been covered under the Protected Area Network. Overall, the Andaman and Nicobar Islands biodiversity is one of the best-protected area than any part of the world with a limited number of resource concern. However, the larger area, less manpower for the protection of these resources has also led to major confusion and gap, prevailing in the protection of these resource concerns.

ANY SOLUTIONS FOR FURTHER IMPROVEMENT

Protection of this biodiversity, as explained by CBD, the mainstreaming of biodiversity is one of the best ways for this activity. Even though it is best, this may be a challengeable task. For the betterment of this system, it is highly essential to mainstream this biodiversity within the social model. The fishing community is the prime community concerned with the coastal and marine environment, who is the first-hand beneficiary from this biodiversity. This community can be used for conserving the activities of biodiversity through their cooperative society.

On the trial basis, each designated protected area (PA) to be handed over to one co-operative society for the benefit of their fishing activities and ask them to protect this environment. This process (fisherfolk and co-operative society) will be controlled by the Union Territory Fishery Department (here Andaman and Nicobar Islands) for further advice and management of day to day fishing activities. A suitable craft and gear should be advised as well as catch also can be monitored by this Department. At the ACCF (Assistant Chief Conservator of Forest) level, the Department of Forest and Environment must practice sustainable development with a weekly or monthly advice or discussion with the community. A scientific committee consisting of members of the Coral Reef Scientist, Environment Monitoring Scientist, Coast Guard Representative, the ACCF and head of the Fishery Department responsible for this PA and a Member from the Finance section of the UT Administration.

Fisherfolk can take necessary activities daily on these PA and their product can be sold through the co-operative society concern. Their activities and product will be monitored by the Department of Fisheries on a daily basis. The judgment of their product can be inferred to the welfare of the environment. If any abnormalities are found, it can be intimated to Department of Environment of Forest immediately or on weekly meetings depending upon the situation and if immediate action is needed to be taken on this aspect. If necessary, the Scientific Advisory Committee should also be consulted for the appropriate action or mitigation. Necessary, yearly fund and the required manpower are to be allocated by the UT administration for the functioning of this system with the above component without any ambiguity as well as an outlet also should be developed for the products available from this activity. This process may be initiated for the two or three PA environment and their rigorous assessment every six months, providing the results on the progress as well as a pitfall of this system. Accordingly, yearly course correction should be made to reach a success story.

Similarly, with due course of time, other departments like tourism and servicing industry can be included depending upon the need, and its essentiality on the particular PA. Indian work environment suggests that the following shortfall may be expected and necessary remediation is to be adopted for a successful implementation of the same.

1. Interference of political personnel in the management activities.
2. Interference of religious/casts/regional group based activities in the management.
3. Who should do what or why I should do - attitude problem among the participating Institutes?
4. Non-availability of fund during the need
5. Shortage of manpower due to the official procedure for filling the vacancies, if exist.

For achieving success in mainstreaming diversity needs a good leadership (SANBI, 2017) with a full technical capacity, the identification of the entry point to influence the policymakers, as well as incorporate the sustainable development into the existing and new policies. Nevertheless, it is also essential to assess the opportunities and constraints for focusing diversity in Indian Island conditions.

CONSENT FOR PUBLICATION

Not applicable.

CONFLICT OF INTEREST

The authors confirm that this chapter contents have no conflict of interest.

ACKNOWLEDGEMENTS

The authors thank the authorities of Pondicherry University and Zoological Survey of India, Kolkatta for providing the facilities and the opportunity to prepare this article.

REFERENCES

Alino, P (2018) Marine Protected Areas in the Philippines. How much spillover do we need? innri.unuftp.is/pdf

Appleby, T & Jones, P (2012) The marine and coastal access act. A hornets' nest? *Mar Policy,* 36, 73-7.
[http://dx.doi.org/10.1016/j.marpol.2011.03.009]

Arias, A, Cinner, J, Jones, R & Pressey, R (2015) Levels and drivers of fishers' compliance with marine protected areas. *Ecol Soc,* 20, 19.
[http://dx.doi.org/10.5751/ES-07999-200419]

Brander, L, Baulcombe, C, van der Lelij, J, Eppink, F, McVittie, A, Nijsten, L & van Beukering, P (2015) WWF Scorecard 2016 – Marine Protected Areas in the Baltic Sea. *WWF report.* http://wwf.panda.org/wwf_news/?247477/

Charles, A, Westlund, L, Bartley, D, Fletcher, W, Garcia, S, Govan, H & Sanders, J (2016) Fishing livelihoods as key to marine protected areas: insights from the World Parks Congress. *Aquat Conserv,* 26, 165-84.
[http://dx.doi.org/10.1002/aqc.2648]

Chuenpagdee, R, Pascual-Ferna'ndez, J, Szelia'nszky, E, Alegret, J, Fraga, J & Jentoft, S (2013) Marine protected areas: Re-thinking their inception. *Mar Policy,* 39, 234-40.
[http://dx.doi.org/10.1016/j.marpol.2012.10.016]

Claudet, J (2011) Marine protected areas a multidisciplinary approach. *Ecology. Biodivers Conserv,* 3
[http://dx.doi.org/10.1017/CBO9781139049382.001]

CNPPA (1986) *Commission on National Parks and Protected Area, IUCN, The world's Greatest Natural Areas and Indicative inventory of natural sites of World Heritage Quality.* IUCN, Switzerland 69.

Convention on Biodiversity- CBD (2016) www.cbd.int

Currie, H, Grobler, K & Kemper, J (2008) Namibia Islands' Marine Protected Area. *WWF South Africa Report Series.* 2008/Marine/003

De Santo, E & Jones, P (2007) Offshore marine conservation policies in the North East Atlantic: emerging tensions and opportunities. *Mar Policy,* 31, 336-47.
[http://dx.doi.org/10.1016/j.marpol.2006.10.001]

IIED (2017) International Institute for Environment and Development. https://www.iied.org/national-biodiversity-strategies-action-plans-20-mainstreaming-biodiversity-development

IUCN (1998) International Union for Conservation of Natural, U.K.

IUCN (2008) *Guidelines for applying protected area management categories.* https://portals.iucn.org/library/efiles/documents/PAPS-016.pdf

Dorel, G, Mannaerts, G & Germain, L (2015) Management plan tutorial. Report prepared by Agence des aires marines protegees for the Protected Area Network Across the Channel Ecosystem (PANACHE) project. INTERREG programme France (Channel) England funded project, 40.

FAO (2011) Technical Guidelines for Responsible Fisheries No. 4, Suppl. 4 Marine Protected Areas and Fisheries, 198. http://www.fao.org/docrep/015/i2090e/i2090e.pdf

Fraga, J & Jesus, A (2008) SAMUDRA Monograph Coastal and Marine Protected Areas in Mexico. *International Collective in Support of Fishworkers,* 77.http://www.icsf.net/images/monographs/pdf/english/issue_92/92_all.pdf

Edgar, G, Russ, G & Babcock, R (2007) Marine protected areas. 534-565. https://www.researchgate.net/publication/284222610

Gray, N (2010) Sea Change: Exploring the International Effort to Promote Marine Protected Areas. *Conserv Soc,* 8, 331-8.
[http://dx.doi.org/10.4103/0972-4923.78149]

Jones, PJ & Burgess, J (2005) Building partnership capacity for the collaborative management of marine protected areas in the UK: a preliminary analysis. *J Environ Manage,* 77, 227-43.
[http://dx.doi.org/10.1016/j.jenvman.2005.04.004] [PMID: 16182438]

Jones, P (2008) Fishing industry and related perspectives on the issues raised by no-take marine protected area proposals. *Mar Policy,* 32, 749-58.
[http://dx.doi.org/10.1016/j.marpol.2007.12.009]

Jones, P (2009) Equity, justice and power issues raised by no-take marine protected area proposals. *Mar Policy,* 33, 759-65.
[http://dx.doi.org/10.1016/j.marpol.2009.02.009]

Jones, P, Qiu, W & De Santo, E (2011) Governing Marine Protected Areas - Getting the Balance Right. Technical Report, United Nations Environment Programme. MPA Governance. www.mpag.infowww.unep.org/ecosystemmanagement

Kaiser, M (2005) Are marine protected areas a red herring or fisheries panacea? *Can J Fish Aquat Sci,* 62, 1194-9.
[http://dx.doi.org/10.1139/f05-056]

Kelleher, G (2015) The importance of regional networks of Marine Protected Areas (MPAs) and how to achieve them. *An opinion piece by Graeme Kelleher AO.* http://www.oceanelders.org/elder/graeme-kelleher/

Laxmilatha, P, Sruthy, T & Varsha, M (2015) Marine Protected Area.

Leisher, C, Beukering, P & Scherl, L (2007) Nature's Investment Bank: How Marine Protected Areas Contribute to Poverty. *The Nature Conservancy* 43.

Lester, S & Halpern, B (2008) Biological responses in marine no-take reserves *versus* partially protected areas. *Mar Ecol Prog Ser,* 367, 49-56.
[http://dx.doi.org/10.3354/meps07599]

Halpern, BS, Walbridge, S, Selkoe, KA, Kappel, CV, Micheli, F, D'Agrosa, C, Bruno, JF, Casey, KS, Ebert, C, Fox, HE, Fujita, R, Heinemann, D, Lenihan, HS, Madin, EM, Perry, MT, Selig, ER, Spalding, M, Steneck,

R & Watson, R (2008) A global map of human impact on marine ecosystems. *Science,* 319, 948-52. [http://dx.doi.org/10.1126/science.1149345] [PMID: 18276889]

MoEFCC, ANI – Minstry of Environment, Forest and Climatic Change – Andaman and Nicobar Islands (2011) Committee constituted to holistically address the issue of poaching in the andaman and nicobar islands report; Government of India, Ministry of Environment and Forests.

Molenaar, J & Elferink, A (2009) Marine protected areas in areas beyond national jurisdiction. The pioneering efforts under the OSPAR convention. 5:5-20. http://www.utrechtlawreview.org/

Qasim, SZ & Wafar, MVM (1990) Marine resources. *Research and Management Optimation,* 7, 141-69.

Rajagopalan, R (2008) Marine Protected Areas in India. *Samudra monograph*, MPAs in India. https://core.ac.uk/download/pdf/11017358

Reker, J (2015) *Marine protected areas in Europe's seas An overview and perspectives for the future.* European Environment Agency.

RCEPC (2008) Report from the commission to the european parliament and the council (2008) On the progress in establishing marine protected areas (as required by Article 21 of the Marine Strategy Framework Directive 2008/56/EC). http://ec.europa.eu/environment/marine/eu-coast-and-marine olicy/implementation/pdf/marine_protected_areas.pdf

RCEP (2004) Turning the tide: addressing the impacts of fisheries on the marine environment. Twenty-fifth Report, presented to Parliament by Command of Her Majesty, Cm 6392. *Turning the tide: addressing the impacts of fisheries on the marine environment.* http://www.fcrn.org.uk/sites/default/files/Turning_the_tide_%20Report.pdf

SANBI (2017) South African National Biodiversity Institute [12th March 2017]. http://www.sanbi.org/biodiversity-science/science-policyaction/mainstreaming-biodiversity

Sanchirico, J (2000) Marine Protected Areas as Fishery Policy: A Discussion of Potential Costs and Benefits. *Resources for the Future,* 16.

Simard, F, Laffoley, D & Baxter, J (2016) Marine Protected Areas and Climate Change: Adaptation and Mitigation Synergies, Opportunity and Challenges. *International Union for Conservation of Nature and Natural Resources,* 52.

Sink, K (2016) The Marine Protected Areas debate: Implications for the proposed Phakisa Marine Protected Areas Network. *J Sci,* 1-4. [http://dx.doi.org/10.17159/sajs.2016/a0179]

Singh, HS (2002) Marine Protected Areas in India: Status of coastal wetlands and their conservation. GEER foundation, Gujarat 1-62.

Venkatraman, K, Rajan, R, Satyanarayana, CH, Ragunathan, C & Venkatraman, C (2012) Marine Ecosystems and Marine Protected Areas of India. ZSI, Kolkatta 269.

Wafar, M, Venkataraman, K, Ingole, B, Ajmal Khan, S & Lokabharathi, P (2011) State of knowledge of coastal and marine biodiversity of Indian Ocean countries. *PLoS One,* 6e14613. [http://dx.doi.org/10.1371/journal.pone.0014613] [PMID: 21297949]

Wells, S, Dwivedi, SN, Singh, S & Robso, J (1995) Marine Region 10: Central Indian Ocean In: Kelleher, G., Bleakley, C., Wells, S., (Eds.), *A Global representative systems of protected areas.*

Wilhelm, T, Sheppard, C, Sheppard, A, Gaymer, C, Parks, J, Wagnera, D & Lewis, N (2014) Large marine protected areas – advantages and challenges of going big. *Aquat Conserv,* 24, 24-30. [http://dx.doi.org/10.1002/aqc.2499]

CHAPTER 15

Monitoring of Environmental Indicators and Bacterial Pathogens in the Muthupettai Mangrove Ecosystem, Tamil Nadu, India

Rajendran Viji* and Nirmaladevi D. Shrinithivihahshini

Environmental Microbiology and Toxicology Laboratory, Department of Environmental Management, School of Environmental Sciences, Bharathidasan University, Tiruchirappalli – 620 024, Tamil Nadu, India

Abstract: Aim of this study focused on monitoring the environmental indicators and bacterial pathogens level based on the aquaculture practices that impacted on the Muthupettai mangrove ecosystem. Water samples were collected at five stations during the pre-monsoon season, and samples were analyzed by standard methods. The results of environmental parameters were shown as follows: temperature (31.4-33.2 °C), pH (7.9-8.6), EC (12-14 mS/cm), TSS (4650-5500 mg/l), TDS (36400-41650 mg/l), TS (41250-46450 mg/l) and DO (2.2-4.1 mg/l); total heterotrophic bacteria (72-294 102 cfu/mL), total coliform bacteria (9-150 101 cfu/mL), fecal coliform bacteria (4-135 101 cfu/mL), total *Enterococcus bacteria* (1-10 101 cfu/mL) and *E.coli* (2-46 101 cfu/mL); and the Pathogens: total *Vibrio* species (1-6 101 cfu/mL), total *Salmonella* species (1-3 101 cfu/mL), total *Shigella* species (1-2 101 cfu/mL), and total *Klebsiella* species (1-39 101 cfu/mL). These results were more vulnerable to the ecosystems and highly exceeded the standard permissible limits of the WHO, EU, and CPCB. Continuously discharges of untreated aquaculture effluents deteriorated the mangrove ecosystem qualities. Therefore, there is a need for a regular monitoring and systematic waste management from aquaculture, which can develop sustainable aquaculture and strictly follow the recommended management rules and regulations of aquaculture practices at national or regional level. Further research needs to improve the ecosystems qualities and maintain the rich biological diversity in the Muthupettai mangrove ecosystem.

Keywords: Physico-Chemical, Indicators, Pathogens, Shrimp Culture, Mangrove Ecosystem.

INTRODUCTION

Mangrove forests are most productive ecosystems, and covered with the 15 million hectares of forests at the interface between terrestrial, estuarine and mar-

* **Corresponding author Rajendran Viji:** Environmental Microbiology and Toxicology Laboratory, Department of Environmental Management, School of Environmental Sciences, Bharathidasan University, Tiruchirappalli – 620 024, Tamil Nadu, India; Tel: +91 431 2407072; Fax: +91 431 2407045; E-mail: biovijitech@gmail.com

De-Sheng Pei & Muhammad Junaid (Eds.)
All rights reserved-© 2019 Bentham Science Publishers

ine systems in the tropical and subtropical regions of 123 countries (Food and Agriculture Organization, 2007; Queiroz, 2017). Mangrove forests are the group of vascular plants, having special morphological, physiological and non-visible adaptations and support diverse groups of the ideal nursery and breeding ground to rich biodiversity, ranging from bacteria, fungi and algae through to invertebrates, birds, and mammals (Kantharajan *et al.*, 2017). They are provided the services for the supporting coastal livelihoods of communities with raw material and foods, coastal protection, soil erosion control, water purification, fisheries maintenance, and carbon sequestration, as well as recreation, education and research in globally (Barbier et al, 2011). The mangrove ecosystems services economical cost rate was at least US $1.6 billion and carbon sequestration rate 1.15-1.39 t/ha (6.5 billion tons) every year in around the world (Nellemann and Corcoran, 2009).

Over the past few decades, unregulated human development activities of the construction of ports, marinas, housing, and shrimp farms were the rapid level of the mangrove forests clearance in around the world (Bernardino *et al.*, 2017). The recent estimation report was (26%) 3.6 million ha mangrove forests loss (Food and Agriculture Organization, 2007, 2010; Queiroz *et al.*, 2013; Guzmán *et al.*, 2003; Ahmed *et al.*, 2017). 38% of mangrove areas were degraded and transformed the coastal aquaculture practices (1.4 million ha), shrimp culture 0.49 million ha (14%) and other forms of aquaculture in worldwide (Queiroz, *et al.*, 2017; Kauffman *et al.*, 2014). The unplanned and unregulated shrimp farming is strongly widespread destruction of mangroves in coastline countries, particularly Bangladesh, Brazil, China, India, Indonesia, Malaysia, Mexico, Myanmar, Sri Lanka, the Philippines, Thailand, and Vietnam losses more significant (Joffre and Schmitt, 2010; Ahmed and Glaser, 2016; Barraza-Guardado, 2013).

The shrimp industry is one of the most important factors to degrade the mangroves, discharging of effluents and water exchanges are multifarious impacts faced in the coastal ecosystem qualities (Kauffman *et al.*, 2014; Barraza-Guardado *et al.*, 2013). The effluent enter into the ocean can increase the continuously organic and inorganic matter, suspended solids and pathogens (Barraza-Guardado *et al.*, 2013; Cardoso-Mohedano *et al.*, 2018). Shrimp cultured wastewater directly affects the oxygen depletion, reduction of transparency, and eutrophication, which alters the benthic organisms of macrofauna populations and seawater qualities (Gengmao *et al.*, 2010; Ferreira *et al.*, 2011). The seawater contamination directly affected the rich biodiversity levels of the mangrove ecosystems, particularly more concentered the bivalves, crustaceans, fish, and birds (Sara *et al.*, 2011).

The seawater microbial contamination is a huge amount of the pathogens accumulation in filter feeding organisms of the bivalves vigorously affected the several infectious diseases to humans (Almeida and Soares, 2012). As the consumption of contaminated shellfish constitutes a potential risk to public health their hygiene-sanitary control is extremely important and legislated (Almeida and Soares, 2012; World Health Organization, 2010). Adequate legislation for safeguarding consumers can minimize the probability of shellfish microbial contamination. In Europe, the Directives 2006/113/CE (Anonymous, 2006) and 2004/41/CE (Anonymous, 2004) are guidelines to control the levels of microbiological indicators for both shellfish and overlying waters (Almeida and Soares, 2012; Anonymous, 2006). In India, mangroves occur on the West Coast, on the East Coast and on Andaman and Nicobar Islands (6,749 km²), the fourth largest mangrove area in the world. According to the Government of India survey 40% (22 400 ha) was degraded conditions. Shrimp aquaculture is responsible for about 80 percent of the conversion of mangrove and this impact surrounded the seawater highly contaminated the coastal ecosystems (Mandal and Naskar, 2008). The Muthupettai mangrove ecosystem is a one of most productive environment in Tamil Nadu, but recent years rapidly increased the shrimp industries generated the untreated wastewater discharges was highly polluted the mangrove ecosystem natural behaviors and along with the seawater qualities. Here, very few studies only monitored seawater quality level in the coastal environment. Therefore, we are focused on the mangrove ecosystem near build the number of the shrimp culture pond wastewater discharges zone of the seawater and along with the towards different zone seawater physicochemical, microbiological indicators and pathogens level estimated in the Mutthpettai mangrove ecosystems.

MATERIALS AND METHODS

Study Area

Muthupettai mangrove ecosystems are situated at the southernmost end of the Cauvery delta connected to Palk Strait, which opens to the Bay of Bengal (Lat. 10° 23' 44.52"N: Long. 79° 29' 42" E) (Fig. 1). The mangrove ecosystem covered with the 6800 ha, which the water spread area covers approximately 2720 ha. The mangrove ecosystem was declared as reserve forest since 1937. The mangrove environment is receiving a large amount of the freshwater from the river of Cauvery tributaries of the Paminiyar, Koraiyar, Kilaithangiyar, Kandankurichanar and Marakkakoraiyar rivers. Anthropogenic activities of aquaculture practices of 134 ha (1.75 h shrimp culture) major environmental issues of the Muthupet mangrove ecosystems (Jayanthi, 2010).

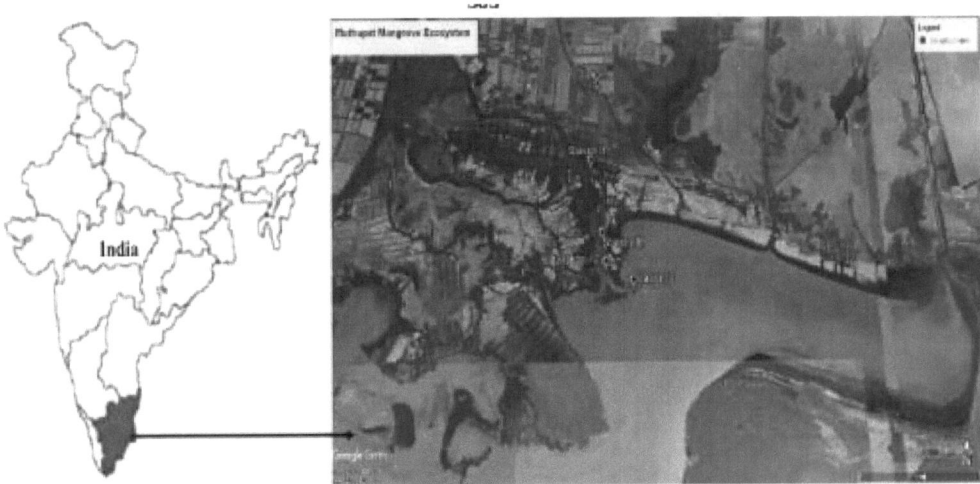

Fig. (1). Study area map and sampling station.

Selection of Sampling Station and Sample Collection

The aquaculture practices of shrimp culture effluent impact in seawater quality monitoring work conducted at the Muthupettai mangrove ecosystem during the pre-monsoon season 2017. Sampling station was selected the shrimp culture wastewater discharges of the seawater environment to approximately 1000 m distance interval of the five stations selected in towards to the offshore lagoon environment. The station-I (10°21'46.13"N; long 79°32'15.42"E), station-II (10°21'8.67"N; 79°32'15.41"E) station-III (10°20'36.82"N; 79°32'24.46"E), station-IV (10°20'24.93"N; 79°32'27.42"E) and station-V (10°20'18.07"N; 79°32'41.48"E). The water samples were collected at the 30 cm depth sample taken the sterile distilled water washed clean polyethylene bottled. Samples were stored at 4 °C in the ice box and transfer laboratory further analysis within 8 hrs.

Estimation of Physico-chemical Parameters

pH (Hanna pH meter), temperature (standard mercury-filled thermometer), and electrical conductivity (EC) were measured using (Systronics-EC-TDS meters 307). Samples were stirred gently, and stable readings were recorded. Total suspended solids (TSS) were determined by filtering seawater 1.2 μm using Millipore GF/C filter paper method; for determining total dissolved solids (TDS) and total solids (TS), hot air oven 105°C 8 h method was applied (Viji *et al.*, 2018). Dissolved oxygen (DO) was measured by the using hand DO meter (Lutron, Taiwan) by spot field analysis.

Enumeration of the Microbiological Indicator and Bacterial Pathogens

All the media were prepared with the addition of sterile distilled water and autoclaved properly. The plates were prepared 5 days prior to sampling. The microbiological pollution indicator and bacterial pathogens population densities were estimated by pour plate method. Samples were diluted by the method of serial dilution, up to 10^{-2} dilution was created by adding 1 mL of the original sample to 9 mL of sterile distilled water and each dilution 1 mL of sample was transferred in selective media. The total heterotrophic bacterial population (THB) was determined using the Zobell marine agar medium and plates were incubated at 35±2°C for 24 h. The total coliform bacterium (TCB) was determined using Macconey agar medium, and plates were incubated at 44±2°C for 24 h. Total *Enterococcus* bacteria (TEB) were enumerated by adopting the *Enterococcus* agar medium, and plates were incubated at 44±2 °C for 24 h. Total *Vibrio* species (TVS) population was counted the Thiosulfate-citrate-bile salts-sucrose agar (TCBS) medium, and plates incubated at 35±2°C for 24 h. Total *Salmonella* species (TSLS) and *Shigella* species (TSHS) colonies were enumerated by the Deoxycholate citrate agar medium and plates were incubated at 35±2°C for 24 h. Total *Klebsiella* species (TKS) was determined used to the *Klebsiella* selective agar base medium and plates were incubated at 35±2°C for 24 h. After incubation, the viable colonies were counted.

Statistical Analysis

The water samples physicochemical parameters, microbiological indicator and bacterial pathogens descriptive and Pearson correlation were analyzed using SPSS Ver.17.0 statistical software.

RESULTS AND DISCUSSION

Environmental Parameters

The environmental parameters results were highly contaminated, and exceeded the regional (CPCB) and national (EU and WHO) standards permissible limits of ecologically sensitive zones seawater quality in Muthupettai mangrove ecosystems (Table **1**).

Table 1. The environmental indicators and bacterial pathogens minimum, maximum, mean and standard deviation values in the five stations.

Parameters	Min	Max	Mean	Std
Temp (°C)	32.6	33.8	33.4	0.1
pH	7.9	8.6	8.3	0.26

(Table 1) cont.....

Parameters	Min	Max	Mean	Std
EC (mS/cm)	12.0	14.0	12.8	0.07
TSS (mg/l)	4650	5500	4910	342
TDS (mg/l)	35750	41650	39170	2132
TS (mg/l)	41250	46450	44470	2089
DO (mg/l)	2.2	4.1	3.2	0.79
THB (cfu/mL)	72	294	129	93.2
TCB (cfu/mL)	9	150	54	56.9
TFCB (cfu/mL)	4	135	44	52.8
TEB (cfu/mL)	1	10	4	3.37
E.coli (cfu/mL)	2	46	15	18.0
TVS (cfu/mL)	2	6	3	1.67
TSLS (cfu/mL)	1	3	1.6	089
TSHS (cfu/mL)	1	3	1.6	0.89
TKS (cfu/mL)	1	39	13	16.1

Temperature

Seawater temperature minimum level (32.6 °C) was observed at the station-III and maximum level (33.8 °C) observed at the station-V, respectively (Fig. **2**). There was no significant difference observed in the temperature levels among five investigated sections. The Pearson correlation result temperature no strong positive correlation with the indicators and pathogens (Table **2**). The surface seawater temperature fluctuation was not much more variation of the tropical region coastal environments, and strong fluctuation interfere the chemical and biological processes in marine ecosystems. Predominate of the coastal organism's metabolic activities, growth, feeding, and reproduction depended on the water temperature in the coastal marine ecosystems (Lawson, 2011; Babalola and Agbebi, 2013).

Table 2. Pearson correlation values of the environmental indicators and bacterial pathogens in the five stations.

	Temp	pH	EC	TSS	TDS	TS	DO	THB	TCB	TFCB	TEB	*E.coli*	TVS	TSLS	TSHS	TKS
Temp	1															
pH	.227	1														
EC	-.456	.689	1													
TSS	-.250	-.952*	-.743	1												
TDS	.088	.670	.733	-.842	1											
TS	.424	.507	.371	-.698	.903**	1										

(Table 2) cont.....

	Temp	pH	EC	TSS	TDS	TS	DO	THB	TCB	TFCB	TEB	*E.coli*	TVS	TSLS	TSHS	TKS
DO	.170	-.957*	.678	-.889*	.517	.303	1									
THB	-.300	-.920*	-.703	.991**	-.888*	-.785	-.825	1								
TCB	-.191	-.953*	-.782	.998**	-.838	-.667	-.899*	.983**	1							
TFCB	-.258	-.959*	-.736	1.00**	-.828	-.684	-.899*	.989**	.997**	1						
TEB	-.364	-.936*	-.661	.993**	-.827	-.730	-.871	.991**	.983**	.993**	1					
E.coli	-.155	-.946*	-.805	.995*	-.853	-.672	-.888*	.982**	.999**	.994**	.976**	1				
TVS	-.070	-.969*	-.831	.970**	-.753	-.522	-.946*	.933**	.982**	.972**	.939*	.982**	1			
TSLS	.054	-.896*	-.910*	.952**	-.867	-.621	-.839	.933**	.967**	.948**	.910*	.977**	.982**	1		
TSHS	.705	-.479	-.910*	-.910*	-.388	.045	-.557	.385	.520	.461	.361	.544	.977**	.688	1	
TKS	.668	-.500	-.896*	-.896*	-.358	.077	-.598	.389	.532	.475	.377	.552	.544	.685	.996**	1

** Correlation is significant at the 0.01 level (2-tailed).
* Correlation is significant at the 0.05 level (2-tailed).

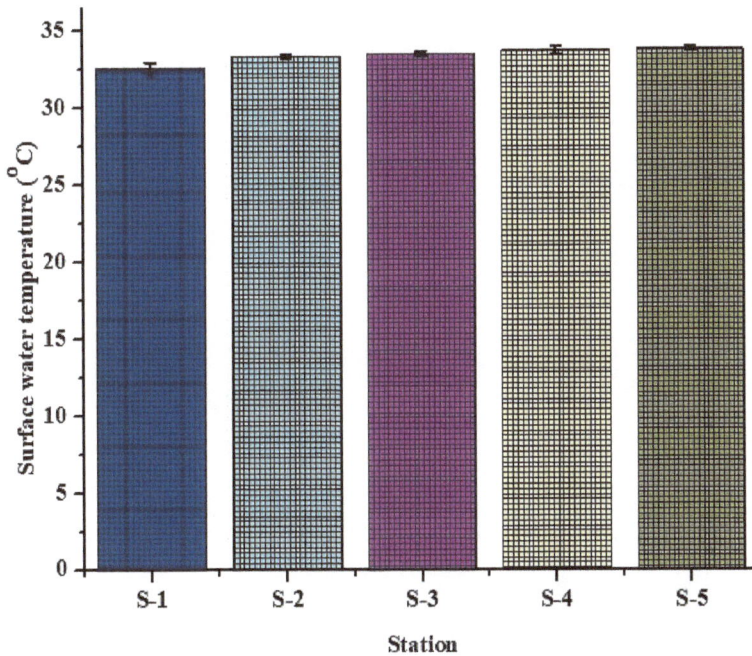

Fig. (2). The surface water temperature variation in the five stations.

Hydrogen Ion Concentration (pH)

The hydrogen ion concentration was ranged from the pH 7.9 to 8.6 in five stations. The lowest level of pH was recorded at the station-I and highest level recorded at the station-II (Fig. **3**). The Pearson correlation result was pH strong negative correlation with the TSS, THB, TCB, TFCB, TEB, *E. coli*, TVS and TSLS. The shrimp culture ponds use the acid sulfate soil, limes, and chemicals during farm operation was changed the pH level (6.0-9.0) in mangrove ecosystems (Boyd and Green, 2002). The pH level fluctuation of <6.5 and >8.5 ranges significantly reduced the biological resources in mangrove ecosystems, particularly more vulnerable condition of the benthic organisms (Ramanathan *et al.*, 2005; Shrinithivihahshini *et al.*, 2014).

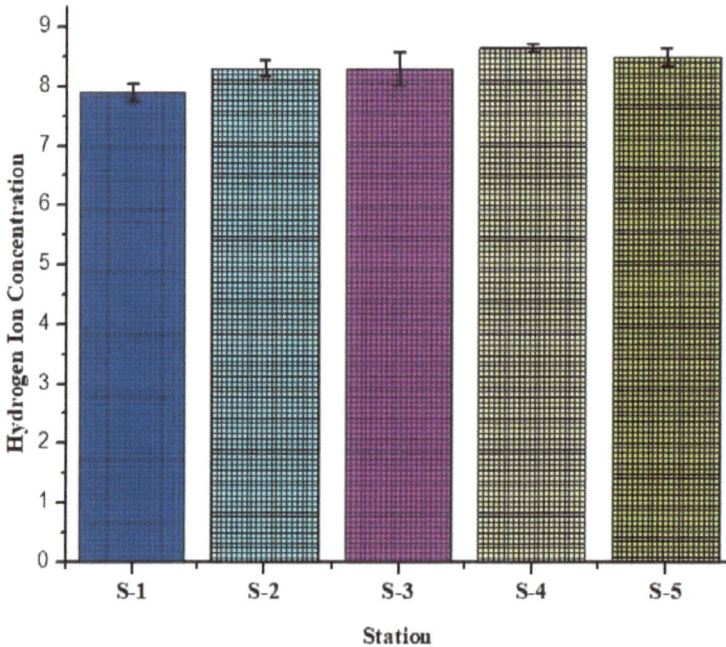

Fig. (3). The hydrogen ion concentration variation in the five stations.

Electrical Conductivity (EC)

Electrical conductivity minimal value (12.0 mS/cm) was noted at the station-I & II and maximal values (13.2 mS/cm) noted at the station-III respectively (Fig. **4**). EC strong negative correlation with the TSLS, TSHS, and TKS. The electrical conductivity values very narrow in five stations, and shrimp culture nearby seawater EC ranges was very low level registered in the study areas.

TSS, TDS, and TS

TSS minimum (4650 mg/l) maximum (5500 mg/l) in station-V and I (Fig. **5**) TSS strong positive correlation with the THB, TCB, TFCB, TEB, *E. coli*, TVS, and TSLS. The TDS low level (36400 mg/l) at the station-I and high level (41650 mg/l) at the station-III (Fig. **6**) TDS strong positive correlation with the TS. The Total solids varied from the 41250 mg/l to 46450 mg/l in five stations. A minimal level was registered at the station-I and maximum level registered at the station-III respectively (Fig. **7**). Overload of the suspended and dissolved solids was a more threatening condition of the mangrove ecosystems biodiversity. The aquaculture pond releasing wastewater, feed and shrimp culture faces, which leads to increases the TSS level and TDS (1000 mg/l) level increases in seawater significantly reduces to the spawning fishes and juveniles in mangrove ecosystems (Boyd and Green, 2002; Bui *et al.*, 2012).

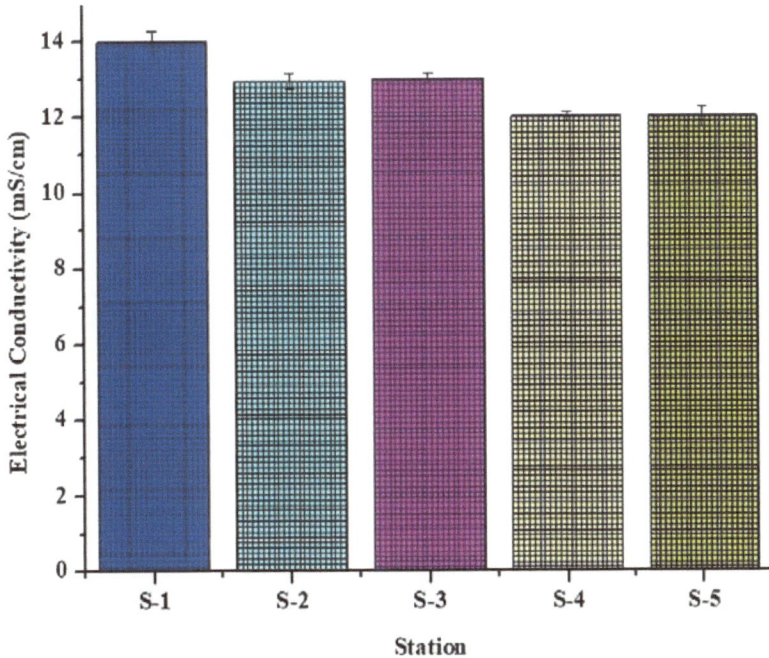

Fig. (4). The electrical conductivity variation in the five stations.

DO

The dissolved oxygen minimum level (2.2 mg/l) recorded at the station-I and maximum level was (4.1 mg/l) recorded at the station-V respectively (Fig. **8**). The Pearson correlation values of the DO strong negative correlation with the THB, TCB, and TVS. The DO level was regional and national standard permissible limit of the ecologically sensitive zone (>5.0) below ranges were recorded at the five stations (Coastal Pollution Control Board, 1995; Viji and Shrinithivihahshini, 2017). The shrimp culture discharges of the vast amount of wastewater contain enormous level of microbes, organic and fecal matter, dead animals and animal manure were completely depleted the dissolved oxygen level destroying the many aerobic organism populations, and frequently deaths of fishes from a deficit of consumed oxygen from biological decomposition of pollutants in the mangrove ecosystems (Lawson, 2011).

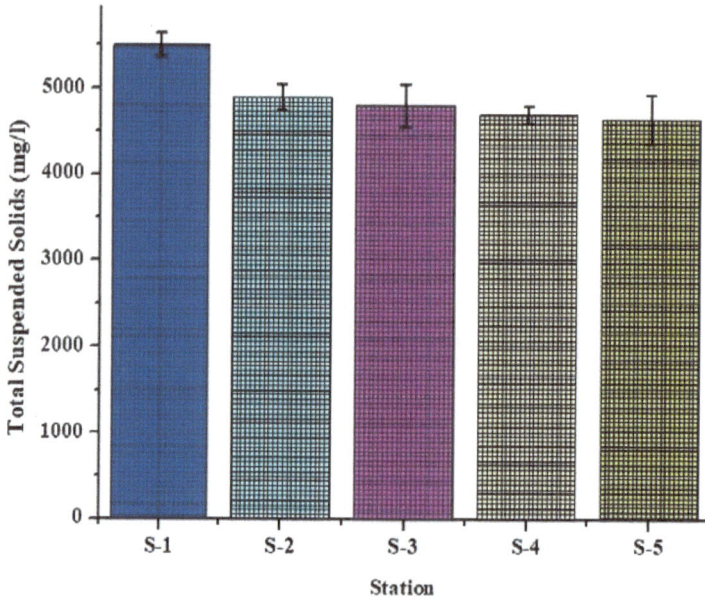

Fig. (5). The total suspended solids variation in the five stations.

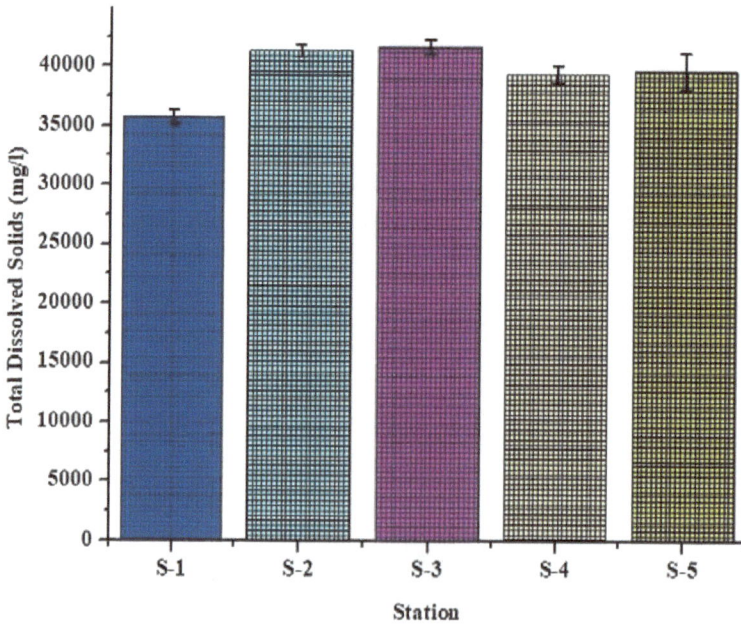

Fig. (6). The total dissolved solids variation in the five stations.

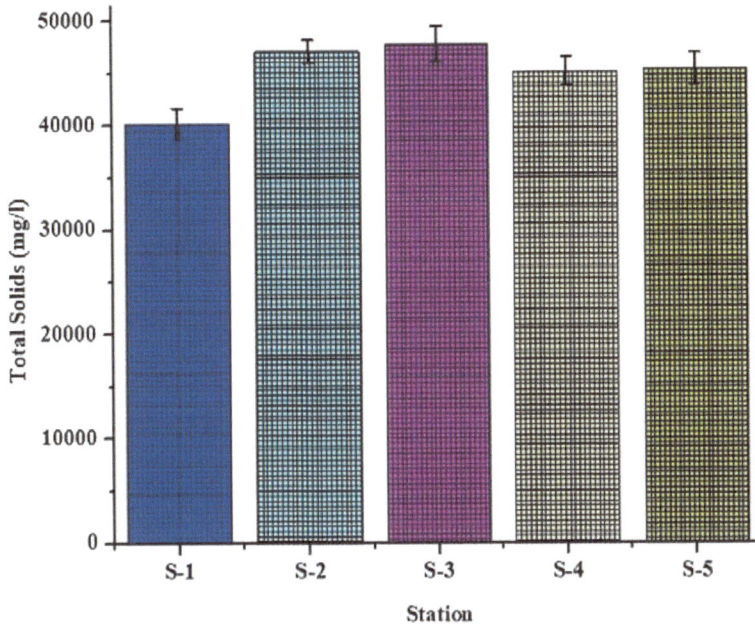

Fig. (7). The total solids variation in the five stations.

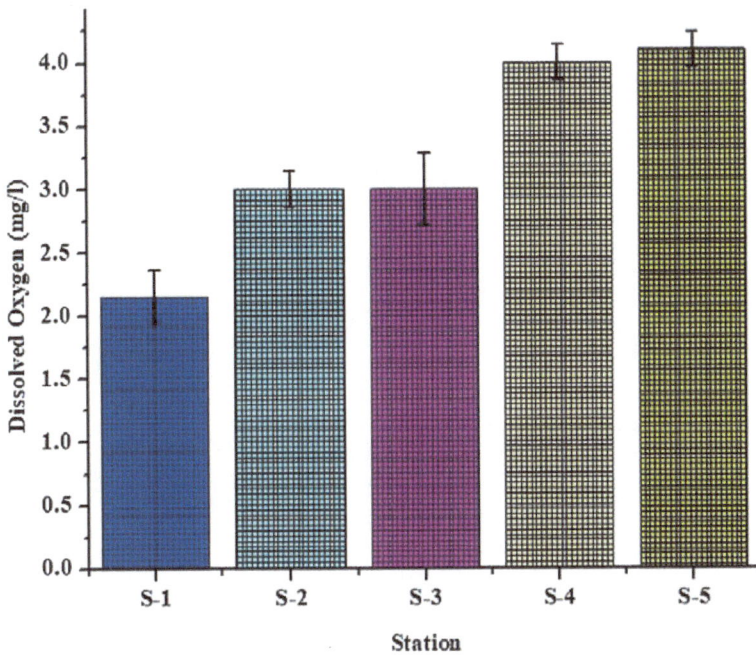

Fig. (8). The dissolved oxygen variation in the five stations.

Microbiological Pollution Indicator

The construction of unplanned aquaculture practices was an enormous level of the microbial load increased in nearby the mangrove ecosystem seawater environments (Table **1**).

Total Heterotrophic Bacteria

Total heterotrophic bacteria greatest population density (294×10^2 cfu/mL) was observed at the station-I and greatest decline value (72×10^2 cfu/mL) observed at the station-V respectively (Fig. **9**). The THB Pearson correlation result was strong positive correlation with the TSS, THB TCB, TFCB, TEB, TVS and TSLS (Table **1**). The heterotrophic bacteria result was indicative of the sanitary problems of the lagoons (Akoachere *et al.*, 2008). The shrimp culture effluent discharges were more amount of the organic load (nitrogen and phosphorus) generation enormous level of the heterotrophic bacteria load increased in the mangrove ecosystems (Rajendran *et al.*, 2018). They are contamination effect modified the physicochemical parameters and microbiological behaviors in seawater environments. The heterotrophic bacteria link between the primary producers and consumers level; and overload of bacterial pollution affects the aquatic food chains in the mangrove ecosystems (Akoachere *et al.*, 2008).

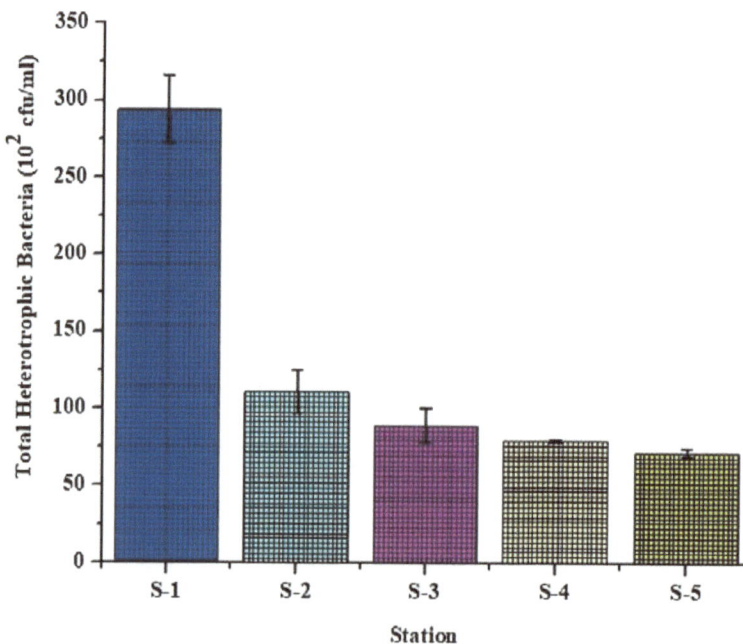

Fig. (9). The total heterotrophic bacteria variations in the five stations.

Total Coliform Bacteria

Total coliform bacteria maximum level was (150×10^1 cfu/mL) recorded at the station-I and minimum level (9×10^1 cfu/mL) was recorded at the station-V (Fig. **10**). TCB significantly correlated with the THB, TFCB, TEB, *E. coli,* and TVS. The total coliform bacteria accept the guide values up to 500/100 mL in human activities surrounded the lagoon environments (World Health Organization, 1992). Shrimp culture ponds used the cow and poultry dung and semi-liquid pig manure were high level of coliform bacteria group bacteria *Escherichia coli* O157, *Salmonella, Yersinia enterocolitica, Campylobacter, Listeria monocytogenes etc.,* contaminations risk related to the creation of unbalanced conditions in the mangrove ecosystem environments (Sung *et al.*, 2003; Heenatigala and Fernando, 2016; Guan and Holley, 2003). The coliform bacteria contamination was rapidly deceased the coastal lagoon ecological conditions and teeming with biodiversity (Hennani *et al.*, 2012).

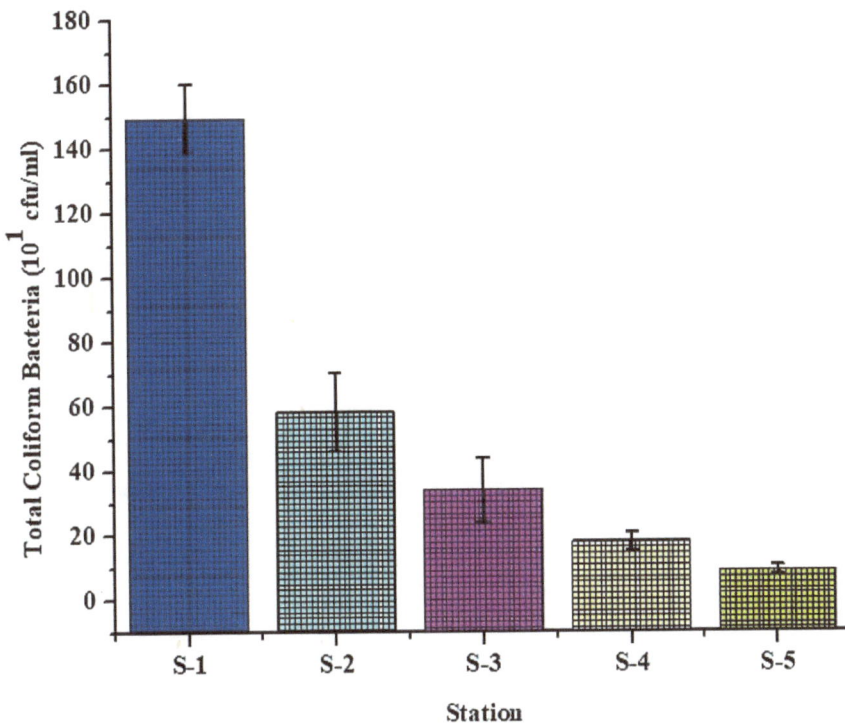

Fig. (10). The total coliform bacteria variation in the five stations.

Fecal Coliform Bacteria

Fecal coliform bacteria ranged from the 4×10^1 cfu/mL to 135×10^1 cfu/mL results in five stations (Fig. **11**) and highly exceeded with the WHO standard permissible limit 100/100 mL fecal coliforms bacteria level in lagoon environments (World Health Organization, 1992; De *et al.*, 2015). The station-I was reported with the highest manifestation of fecal coliform bacteria, while station-V was reported with the lowest number of coliform bacteria. The result of Pearson correlation showed that FCB had strong positive correlations with the THB, TCB, TEB, *E. coli*, TVS, and TSLS. The shrimp culture fecal pathogens were longtime survival for the up to 4 months, and protect and enriched the various types of the abiotic factors of manure, temperature, pH, oxygen and ammonia concentrations. The fecal pathogens major threatening factors of the biological resources in the lagoon environment, particularly filter-feeding organisms of the bivalves' oyster filter large volumes of water from their contaminated environment and are able to concentrate large numbers of pathogens accumulation affect the seafood consumers of the human (Mlejnkova and Sovova, 2013).

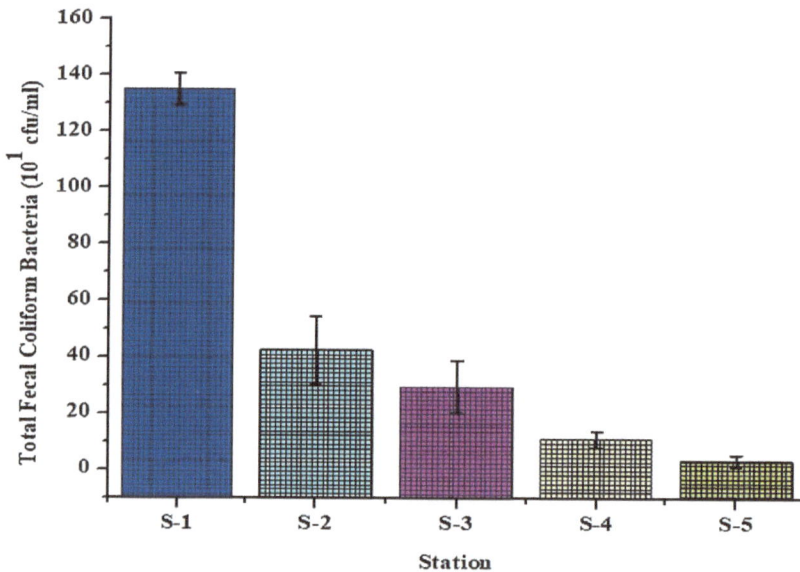

Fig. (11). The total fecal coliform bacteria variation in the five stations.

Total Enterococcus Bacteria

Total *Enterococcus* bacteria high level was (10×10^1 cfu/mL) noted at the station-I and low level (1×10^1 cfu/mL) noted at the station-5, respectively (Fig. **12**). The *Enterococcus* bacteria strong positive correlation with the TSS, THB, TCB, TFCB, TSLS and TSHS and strong negative correlation with the pH, respectively. This result higher than standard permissible limit of lagoon seawater environments (100/100 mL *Enterococcus* bacteria), and directly affect the nursery for finfish and shellfish species caught in the adjacent marine waters (World Health Organization, 1992; De *et al.*, 2015). The longtime survival of the *Enterococcus* bacteria in seawater changed the short life and reproductive cycles, specific ecological requirements of benthic communities, and sensitive changing the biodiversity level in mangrove ecosystems (Kouassi *et al.*, 1995; Debenay *et al.*, 2015).

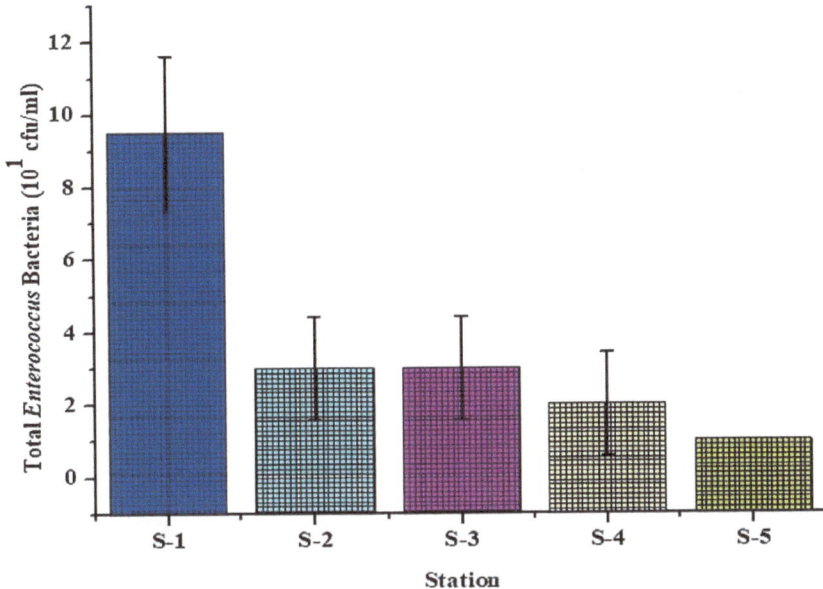

Fig. (12). The *Enterococcus* bacteria variation in the five stations.

E. coli

E. coli is the best indicator organism of fecal pollution identification in the coastal marine ecosystems. Maximal level (48×10^1 cfu/mL) registered at the station-I and minimal level (2×10^1 cfu/mL) registered at the station-V (Fig. **13**). The *E. coli* level in the five stations, current accept upper limit for *E. coli* 500 cfu/100 mL level registered in the seawater (Rogers and Haines, 2005; Venglovsky *et al.*, 2006; Elsaidy *et al.*, 2015). Shrimp pond applied animal manure can provide a favorable environment for fecal organisms *Salmonella* sp., *E. coli* O157 H7 pathogen survival and even re-growth of availability of nutrients as well as protection from UV radiation, desiccation, and temperature extremes in water (European Union, 2006; Price and Wildeboer, 2017). *E. coli* contamination may deleterious effects causes in the mangrove forest, and can adapted and access roots or leaf internal plant compartments can also be colonized (Bui *et al.*, 2012; Natvig *et al.*, 2003).

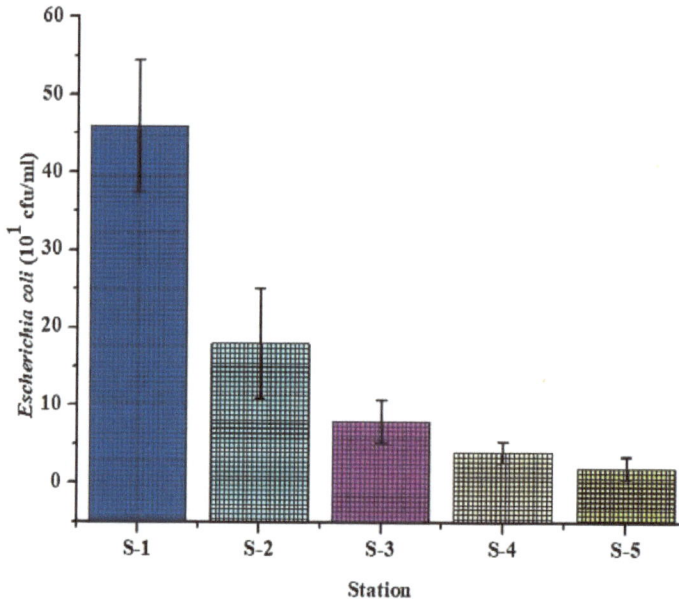

Fig. (13). The *Escherichia coli* variation in the five stations.

Bacterial Pathogens

The untreated shrimp culture effluents discharges were an enormous level of the bacterial pathogens transferred in the coastal water, and long-time survival of the bacterial pathogens vigorously affected the coastal qualities and the biological resources in the mangrove ecosystems.

Total Vibrio Species

The *Vibrio's* species results in five stations, the highest level of population density $(6 \times 10^1$ cfu/mL) observed at the station-I and lowest level $(1 \times 10^1$ cfu/mL) observed at the station-IV & V respectively (Fig. **14**). The Pearson correlation result of *Vibrio* species was strong positive correlation with the TSS, THB, TCB, TFCB, TEB, TSLS, and TSHS. Discharges from the aquaculture waste contain fecal material and uneaten food and soluble, inorganic excretory waste major sources of the *Vibrio* species contamination, and the risk of disease outbreaks increasing in lagoon environments (Menasveta, 2002; Guzman *et al.*, 2003; Barraza☐Guardado *et al.*, 2008). They are longtime survival and adapted to the high saline and alkaline conditions, high level of nutrients and suspended particles, oxygen-deficient conditions in shrimp culture wastewaters. The virulent nature of *Vibrio* species frequently affected and isolated from the marine organisms of the mollusks (oysters, clams, scallops, and octopus), other invertebrates (crabs, sea cucumbers, sea urchins, starfish, and polychaetes) or fish (turbot, gilthead seabream, spotted rose snapper, dentex, Atlantic salmon, wrasse, sand smelt, and tuna or seahorses) in around the world (Perez-Cataluna *et al.*, 2016).

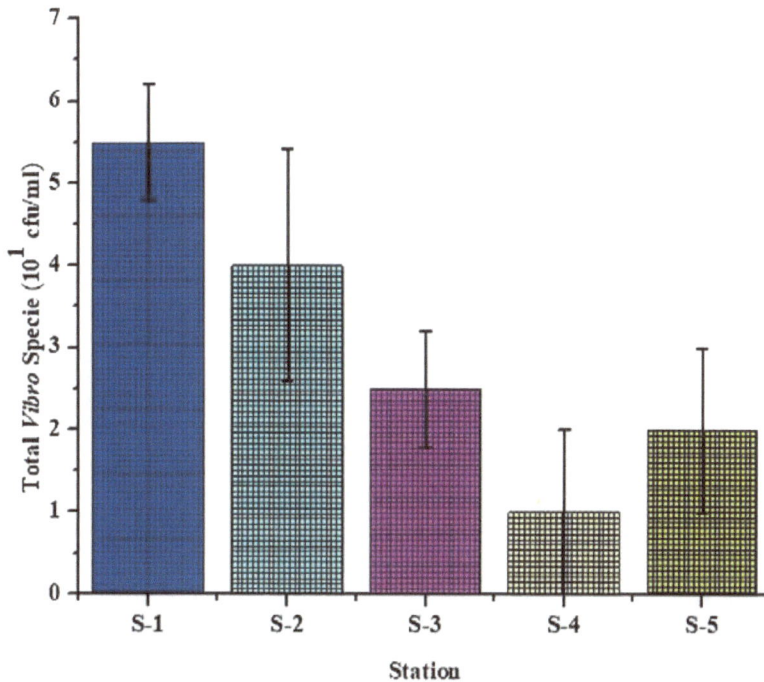

Fig. (14). The total *Vibrio* species variation in the five stations.

Total Salmonella Species

Maximum (3×10^1 cfu/mL) and minimum (1×10^1 cfu/mL) level of total *Salmonella* species population was noted at the station- I and IV& V, respectively (Fig. **15**). The *Salmonella* species strong positive correlation with the TSS, THB, TCB, TFCB, TEB, TVS, TSLS and TSHS. The presence of these *Salmonella* species primary sources to the discharge of untreated sewage into local water bodies, pond fertilization with untreated manure and also to the reuse of water resources of coastal water (Mumby *et al.*, 2004). The pathogenic *Salmonella* species contamination affect the feeding, breeding, and nursery grounds for many ecologically and commercially important species, including crabs, fish, mollusks, oysters, and shrimps, and provide habitats for many species of amphibians, birds, crustaceans, and mammals in mangrove ecosystems (Food and Agriculture Organization, 2007; Mumby *et al.*, 2004). The *Salmonella* species standard permissible limits of the 25 g of flesh and an upper limit of 230 MPN level found in the seafood organisms affect the number of foodborne diseases, such as typhoid and paratyphoid fevers, gastroenteritis, pneumonia, and bacteremia in human health (European Commission, 2015[a, b]; Wang *et al.*, 2017).

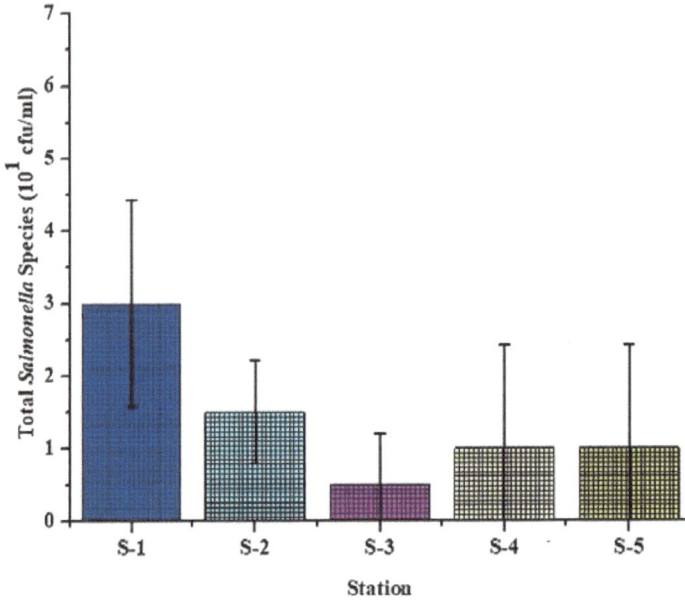

Fig. (15). The total *Salmonella* species variation in the five stations.

Total Shigella Species

Total *Shigella* species was varied from the 1×10^1 cfu/mL to 2×10^1 cfu/mL in five stations. High level was recorded at the station-I and low level recorded at the station- III, IV, and V (Fig. **16**). The *Shigella* species was strong positive correlation with the TVS. The sewage overflows, sewage systems that are not working properly, animal manure runoff, and polluted urban stormwater runoff derived from the *Shigella* species contaminated in the mangrove ecosystems. The virulent nature of *Shigella* species were deterioration of the biological diversity in mangroves environments and affect the numbers of gastroenteric diseases, such as diarrhoea (traveller's disease), dysentery, vomiting, fever, colitis, and hemolytic uremic syndrome with renal failure in coastal users (Food and Agriculture Organization, 2007; Hale, 1991; Danba *et al.*, 2015).

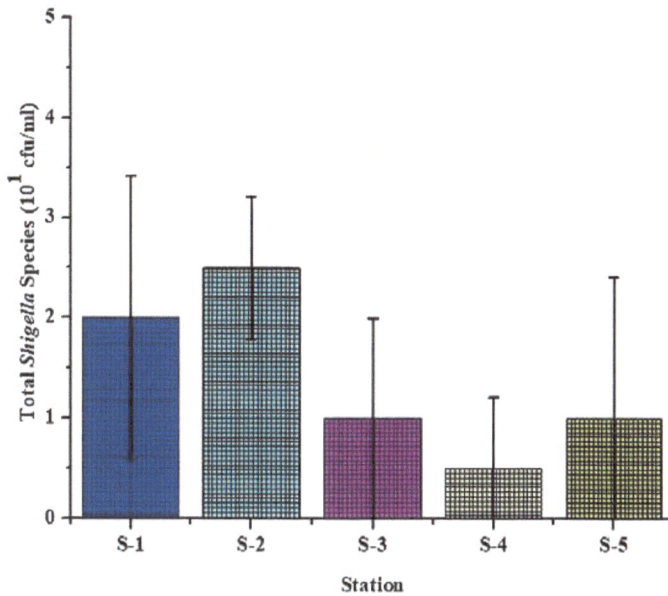

Fig. (16). The total *Shigella* species variation in the five stations.

Total Klebsiella Species

The *Klebsiella* species greatest value (39×10^1 cfu/mL) was observed at the station-II, and the greatest decline value (1×10^1 cfu/mL) was observed at the station-V. The *Klebsiella* species Pearson correlation strong positive correlation with the TSHS respectively (Fig. **17**). The *Klebsiella* species are naturally found in soil and water environment, and warm-blooded animal and human fecal matter derived a vast amount of the *Klebsiella* species in seawaters. The improper farm and poor pond management of aquaculture effluent discharges were highly *Klebsiella* species contaminated in the nearby waterbodies (Fuchs *et al.*, 1999). Pathogenic *K. pneumoniae* species contamination vigorously affected the fishes, and they are associated with the animal mortality and morbidity rates increased in the coastal marine ecosystems (Gopi *et al.*, 2016). The *Klebsiella* species cause the pyogenic liver abscess, urinary tract infections, bacteremia, and pneumonia by nearby dwellers. Particularly, *K. pneumoniae* species showed high risk for human health, which increased the infection morbidity and mortality rates from around the world (Jenney *et al.*, 2016).

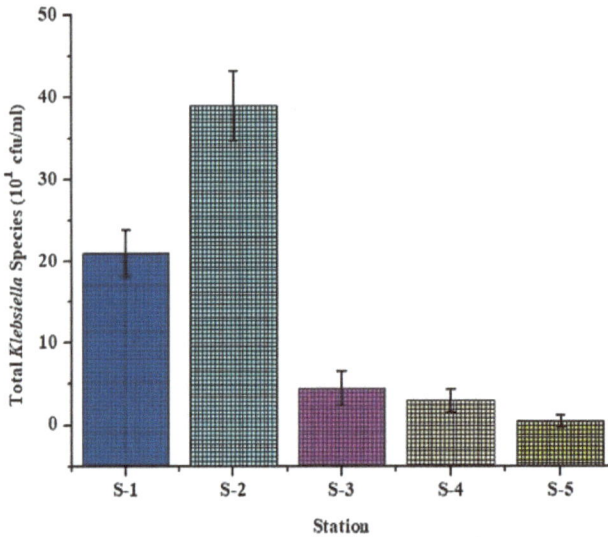

Fig. (17). The total *Klebsiella* species variation in the five stations.

With few exception, the physiochemical indexes, microbial indicators, and bacterial pathogens highlighted station-I as relatively most polluted location, which can clearly attribute to the anthropogenic activities, shrimp culturing in this case. However, the station-IV and station S-V were observed as relatively less affected sites with better water quality, as revealed by the normal values of physiochemical parameters. Overall, these finding revealed that exploitation of mangroves in an impropriate manner is causing ecological degradation in the area and increasing the ecological risks.

CONCLUSION

TThe present findings clearly revealed that the physicochemical parameters, microbiological indicator, and bacterial pathogens were strongly connected with unregulated shrimp culture effluent discharges in seawater quality, particularly high level of pollution in nearby the shrimp ponds and this pollution gradually moving the towards the offshore environment of the lagoon. This impact directly affect and decline the richest biological resources in around the mangrove ecosystems. Therefore, emergency action needs to protect the mangrove ecosystem coastal qualities, prevent the unregulated aquaculture practices, promote the sustainable aquaculture activities, and educate the fisherman communities and nearby dwellers, which will play a crucial role for protecting and enriching the diversities of biological resources in the mangrove ecosystems.

CONSENT FOR PUBLICATION

Not applicable.

CONFLICT OF INTEREST

The authors confirm that this chapter contents have no conflict of interest.

ACKNOWLEDGEMENT

Authors thank the authorities of Bharathidasan University for providing the necessary facilities and Muthupettai Fisherman.

REFERENCES

Aguirre Guzmán, G, Labreuche, Y, Ansquer, D, Espiau, B, Levy, P, Ascencio, F & Saulnier, D (2003) Proteinaceous exotoxins of shrimp-pathogenic isolates of Vibrio penaeicida and Vibrio nigripulchritudo. *Cienc Mar,* 29, 77-88.
[http://dx.doi.org/10.7773/cm.v29i1.132]

Ahmed, N & Glaser, M (2016) Coastal aquaculture, mangrove deforestation and blue carbon emissions: Is REDD+ a solution? *Mar Policy,* 66, 58-66.
[http://dx.doi.org/10.1016/j.marpol.2016.01.011]

Ahmed, N, Cheung, WW, Thompson, S & Glaser, M (2017) Solutions to blue carbon emissions: Shrimp cultivation, mangrove deforestation and climate change in coastal Bangladesh. *Mar Policy,* 82, 68-75.
[http://dx.doi.org/10.1016/j.marpol.2017.05.007]

Akoachere, JF, Oben, PM, Mbivnjo, BS, Ndip, LM, Nkwelang, G & Ndip, RN (2008) Bacterial indicators of pollution of the Douala lagoon, Cameroon: *public health implications. Afr Health Sci,* 8, 85-9.
[PMID: 19357756]

Almeida, C & Soares, F (2012) Microbiological monitoring of bivalves from the Ria Formosa Lagoon (south coast of Portugal): a 20 years of sanitary survey. *Mar Pollut Bull,* 64, 252-62.
[http://dx.doi.org/10.1016/j.marpolbul.2011.11.025] [PMID: 22197556]

Anonymous (2004) Regulation (EC) No. 854/2004 of the European Parliament and of the Council of 29 April 2004 laying down specific rules for the organisation of official controls on products of animal origin intended for human consumption. Off. J. Eur. Union L226/83, Brussels.

Anonymous (2004) Directive 2006/113/EC of the European Parliament and of the Council of 12 December 2006 on the quality required of shellfish waters. Off. J. Eur. Union L376/14, Brussels.

Babalola, OA & Agbebi, FO (2013) Physico-Chemical Characteristics and Water Quality Assessment from Kuramo Lagoon, Lagos, Nigeria. *Society for Science and Nature,* 3, 98-102.

Barbier, EB, Hacker, SD, Kennedy, C, Koch, EW, Stier, AC & Silliman, BR (2011) The value of estuarine and coastal ecosystem services. *Ecol Monogr,* 81, 169-93.
[http://dx.doi.org/10.1890/10-1510.1]

Barraza-Guardado, RH, Arreola-Lizárraga, JA, López-Torres, MA, CasillasHernández, R, Miranda-Baeza, A, Magallón-Barrajas, F & Ibarra-Gámez, C (2013) Effluents of shrimp farms and its influence on the coastal ecosystems of Bahía de Kino, Mexico. *Scientific World Journal,* 2013306370
[http://dx.doi.org/10.1155/2013/306370] [PMID: 23861653]

Barraza-Guardado, RH, Chávez-Villalba, J, Atilano-Silva, H & Hoyos-Chairez, F (2008) Seasonal variation in the condition index of Pacific oyster postlarvae (*Crassostrea gigas*) in a land-based nursery in Sonora, Mexico. *Aquacult Res,* 40, 118-28.

[http://dx.doi.org/10.1111/j.1365-2109.2008.02076.x]

Bernardino, AF, Gomes, LEO, Hadlich, HL, Andrades, R & Correa, LB (2018) Mangrove clearing impacts on macrofaunal assemblages and benthic food webs in a tropical estuary. *Mar Pollut Bull,* 126, 228-35. [http://dx.doi.org/10.1016/j.marpolbul.2017.11.008] [PMID: 29421092]

Boyd, CE & Green, BW (2002) Coastal water quality monitoring in shrimp farming areas, an example from Honduras. Report prepared under the World Bank, NACA, WWF and FAO Consortium Program on Shrimp Farming and the Environment. *Work in Progress for Public Discussion Published by the Consortium,* 29.

Bui, TD, Luong-Van, J & Austin, CM (2012) Impact of shrimp farm effluent on water quality in coastal areas of the world heritage-listed Ha Long Bay. *Am J Environ Sci,* 8, 104-16. [http://dx.doi.org/10.3844/ajessp.2012.104.116]

Bush, SR, Van Zwieten, PA, Visser, L, van Dijk, H, Bosma, R, de Boer, WF & Verdegem, M (2010) Scenarios for resilient shrimp aquaculture in tropical coastal areas. *Ecol Soc,* 15, 15. [http://dx.doi.org/10.5751/ES-03331-150215]

Cardoso-Mohedano, JG, Lima-Rego, J, Sanchez-Cabeza, JA, Ruiz-Fernández, AC, Canales-Delgadillo, J, Sánchez□Flores, EI & Páez-Osuna, F (2018) Sub-tropical coastal lagoon salinization associated to shrimp ponds effluents. *Estuar Coast Shelf Sci,* 203, 72-9. [http://dx.doi.org/10.1016/j.ecss.2018.01.022]

CPCB (1995) *Criteria for Classification and Zoning of Coastal Waters (Sea Waters SW) – A Coastal Pollution Control Series: COPOCS/6/1993-CPCB.*Central Pollution Control Board, New Delhi.

Danba, EP, David, DL, Wahedi, JA, Buba, U, Bingari, MS, Umaru, FF, Ahmed, MK, Tukur, KU, Barau, BW & Dauda, UD (2015) Microbiological Analysis of Selected Catfish Ponds in Kano Metropolis, Nigeria. *Journal of Agriculture and Veterinary Science,* 8, 74-8.

de Souza Queiroz, L, Rossi, S, Calvet-Mir, L, Ruiz-Mallén, I, García-Betorz, S, Salvà-Prat, J & de Andrade Meireles, AJ (2017) Neglected ecosystem services: Highlighting the socio-cultural perception of mangroves in decision-making processes. *Ecosyst Serv,* 26, 137-45. [http://dx.doi.org/10.1016/j.ecoser.2017.06.013]

De, N & Uchechukwu, S (2015) Physiochemical and microbiological assessment of Lagos lagoon water, Lagos, Nigeria. *Journal of Pharmacy and Biological Sciences,* 10, 78-84.

Debenay, JP, Marchand, C, Molnar, N, Aschenbroich, A & Meziane, T (2015) Foraminiferal assemblages as bioindicators to assess potential pollution in mangroves used as a natural biofilter for shrimp farm effluents (New Caledonia). *Mar Pollut Bull,* 93, 103-20. [http://dx.doi.org/10.1016/j.marpolbul.2015.02.009] [PMID: 25758645]

European Commission 2285/2015/EC, Commission Regulation No. 2285/2015 (2015) Amending Annex II to Regulation (EC) No 854/2004 of the European Parliament and of the Council laying down specific rules for the organisation of official controls on products of animal origin intended for human consumption as regards certain requirements for live bivalve molluscs, echinoderms, tunicates. and marine gastropods and Annex I to Regulation.

European Commission No 2073/2005 on microbiological criteria for foodstuffs. In E. Union (Ed.), 2285/2015 Brussels, Belgium: The European parliament and the council of the European Union.

Elsaidy, N, Abouelenien, F & Kirrella, GA (2015) Impact of using raw or fermented manure as fish feed on microbial quality of water and fish. *Egyptian Journal of Aquatic Research,* 41, 93-100. [http://dx.doi.org/10.1016/j.ejar.2015.01.002]

EU Directive 2006/7/EC (2006) The European Parliament and of the Council of 15 February 2006 concerning the management of bathing water quality. The European Parliament and the Council of the European Union: Official Journal of the European Union. 2006; L64/37.

Ferreira, NC, Bonetti, C & Seiffert, WQ (2011) Hydrological and water quality indices as management tools in marine shrimp culture. *Aquaculture,* 318, 425-33.

[http://dx.doi.org/10.1016/j.aquaculture.2011.05.045]

Food & Agriculture Organization of the United Nations (2010) The State of World Fisheries and Aquaculture 2010. *Rome* 197.

Food & Agriculture Organization (2007) *The World's Mangroves 1980–2005.* Food and Agriculture Organization of the United Nations, Rome.

Fuchs, J, Martin, JLM & An, NT (1999) Impact of tropical shrimp aquaculture on the environment in Asia and the Pacific. *European Commission Fish Bulletin,* 12, 9-13.

Gengmao, Z, Mehta, SK & Zhaopu, L (2010) Use of saline aquaculture wastewater to irrigate salt-tolerant Jerusalem artichoke and sunflower in semiarid coastal zones of China. *Agric Water Manage,* 97, 1987-93. [http://dx.doi.org/10.1016/j.agwat.2009.04.013]

Gopi, M, Kumar, TTA & Prakash, S (2016) Opportunistic pathogen Klebsiella pneumoniae isolated from Maldive's clown fish Amphiprion nigripes with hemorrhages at Agatti Island, Lakshadweep archipelago, 4,3. 464-7.

Guan, TY & Holley, RA (2003) Pathogen survival in swine manure environments and transmission of human enteric illness-a review. *J Environ Qual,* 32, 383-92. [http://dx.doi.org/10.2134/jeq2003.3830] [PMID: 12708660]

Hale, TL (1991) Genetic basis of virulence in Shigella species. *Microbiol Rev,* 55, 206-24. [PMID: 1886518]

Heenatigala, PPM & Fernando, MUL (2016) Occurrence of bacteria species responsible for vibriosis in shrimp pond culture systems in Sri Lanka and assessment of the suitable control measures. *Sri Lanka Journal of Aquatic Sciences,* 21, 1-17. [http://dx.doi.org/10.4038/sljas.v21i1.7481]

Hennani, M, Maanan, M, Robin, M, Chedad, K & Assobhei, O (2012) Temporal and Spatial Distribution of Faecal Bacteria in a Moroccan Lagoon. *Pol J Environ Stud,* 21, 1-17.

Jayanthi, M (2010) *Status of mangroves in relation to Brackishwater aquaculture development in Tamil Nadu.* 1-44.

Jenney, AW, Clements, A, Farn, JL, Wijburg, OL, McGlinchey, A, Spelman, DW, Pitt, TL, Kaufmann, ME, Liolios, L, Moloney, MB, Wesselingh, SL & Strugnell, RA (2006) Seroepidemiology of Klebsiella pneumoniae in an Australian Tertiary Hospital and its implications for vaccine development. *J Clin Microbiol,* 44, 102-7. [http://dx.doi.org/10.1128/JCM.44.1.102-107.2006] [PMID: 16390956]

Joffre, OM & Schmitt, K (2010) Community livelihood and patterns of natural resources uses in the shrimp-farm impacted Mekong Delta. Aquaculture Research, 41(12), pp.1855-1866.2073/2005/EC, R. (2005). Commission Regulation (EC) No. 2073/2005 on microbiological criteria for foodstuffs. In E. Union (Ed.), 2073/2005. Brussels, Belgium: The European parliament and the council of the European Union. [http://dx.doi.org/10.1111/j.1365-2109.2010.02588.x]

Kantharajan, G, Pandey, PK, Krishnan, P, Samuel, VD, Bharti, VS & Purvaja, R (2017) Molluscan diversity in the mangrove ecosystem of Mumbai, west coast of India. *Reg Stud Mar Sci,* 14, 102-11. [http://dx.doi.org/10.1016/j.rsma.2017.06.002]

Kauffman, JB, Heider, C, Norfolk, J & Payton, F (2014) Carbon stocks of intact mangroves and carbon emissions arising from their conversion in the Dominican Republic. *Ecol Appl,* 24, 518-27. [http://dx.doi.org/10.1890/13-0640.1] [PMID: 24834737]

Kouassi, AM, Kaba, N & Métongo, BS (1995) Land-based sources of pollution and environmental quality of the Ebrié lagoon waters. *Mar Pollut Bull,* 30, 295-300. [http://dx.doi.org/10.1016/0025-326X(94)00245-5]

Lawson, EO (2011) Physico-chemical parameters and heavy metal contents of water from the Mangrove Swamps of Lagos Lagoon, Lagos, Nigeria. *Adv Biol Res (Faisalabad),* 5, 8-21.

Mandal, RN & Naskar, KR (2008) Diversity and classification of Indian mangroves: a review. *Trop Ecol*, 49, 31-146.

Menasveta, P (2002) Improved shrimp growout systems for disease prevention and environmental sustainability in Asia. *Rev Fish Sci*, 10, 391-402.
[http://dx.doi.org/10.1080/20026491051703]

Mlejnková, H & Sovová, K (2013) Impact of fish pond manuring on microbial water quality. *Acta Univ Agric Silvic Mendel Brun*, 60, 117-24.
[http://dx.doi.org/10.11118/actaun201260030117]

Mumby, PJ, Edwards, AJ, Arias-González, JE, Lindeman, KC, Blackwell, PG, Gall, A, Gorczynska, MI, Harborne, AR, Pescod, CL, Renken, H, Wabnitz, CC & Llewellyn, G (2004) Mangroves enhance the biomass of coral reef fish communities in the Caribbean. *Nature*, 427, 533-6.
[http://dx.doi.org/10.1038/nature02286] [PMID: 14765193]

Natvig, EE, Ingham, SC, Ingham, BH, Cooperband, LR, Roper, TR & Arment, AR (2003) Salmonella enterica Serovar Typhimurium and Escherichia coli Contamination of Root and Leaf Vegetables Grown in Soils with Incorporated Bovine Manure. *Appl Environ Microbiol*, 69, 3686.
[http://dx.doi.org/10.1128/AEM.69.6.3686.2003] [PMID: 12039728]

Nellemann, C & Corcoran, E (2009) Blue carbon: the role of healthy oceans in binding carbon: a rapid response assessment. UNEP/Earthprint.Bouillon, S, Borges, AV, Castañeda☐Moya, E, Diele, K, Dittmar, T, Duke, NC, Pérez-Cataluña, A, Lucena, T, Tarazona, E, Arahal, DR, Macián, MC & Pujalte, MJ (2016) An MLSA approach for the taxonomic update of the Splendidus clade, a lineage containing several fish and shellfish pathogenic Vibrio spp. *Syst Appl Microbiol*, 39, 361-9.

Price, RG & Wildeboer, D (2017) E.coli as an Indicator of Contamination and Health Risk in Environmental Waters. *Escherichia coli-Recent Advances on Physiology, Pathogenesis and Biotechnological Applications. InTech* 125-39.

Queiroz, L, Rossi, S, Meireles, AJA & Coelho-Jr, C (2013) Shrimp aquaculture in the state of Ceará, 1970–2012: trends in mangrove forest privatization in Brazil. *Ocean Coast Manage*, 73, 54-62.
[http://dx.doi.org/10.1016/j.ocecoaman.2012.11.009]

Rajendran, V, Nirmaladevi D, S, Srinivasan, B, Rengaraj, C & Mariyaselvam, S (2018) Quality assessment of pollution indicators in marine water at critical locations of the Gulf of Mannar Biosphere Reserve, Tuticorin. *Mar Pollut Bull*, 126, 236-40.
[http://dx.doi.org/10.1016/j.marpolbul.2017.10.091] [PMID: 29421093]

Ramanathan, N, Padmavathy, P, Francis, T, Athithian, S & Selvaranjitham, N (2005) Manual on polyculture of tiger shrimp and carps in freshwater. *Tamil Nadu Veterinary and Animal Sciences University, Fisheries College and Research Institute, Thothukudi* 1.

Rogers, S & Haines, J (2005) *Detecting and mitigating the environmental impact of fecal pathogens originating from confined animal feeding operations: review EPA-600-R-06-021.*USEPA, Office of Research and Development, National Risk Management Research Laboratory, Cincinnati, OH.

Sarà, G, Lo Martire, M, Sanfilippo, M, Pulicanò, G, Cortese, G, Mazzola, A, Manganaro, A & Pusceddu, A (2011) Impacts of marine aquaculture at large spatial scales: evidences from N and P catchment loading and phytoplankton biomass. *Mar Environ Res*, 71, 317-24.
[http://dx.doi.org/10.1016/j.marenvres.2011.02.007] [PMID: 21427008]

Shrinithivihahshini, ND, Viji, R, Sheelamary, M, Chithradevi, R, Mahamuni, D & Ramesh, D (2014) An assessment of religious ceremonies and their impact on the physico-chemical and microbiological characterization of foremost seawater in Navagraha temple, Devipattinam, Tamil Nadu, India. *Global Journal of Science Frontier Research: H Environment &. Earth Sci*, 14, 71-80.

Sung, HH, Lin, SC, Chen, WL, Ting, YY & Chao, WL (2003) Influence of Timsen™ on Vibrio populations of culture pond water and hepatopancreas and on the hemocytic activity of tiger shrimp (*Penaeus monodon*). *Aquaculture*, 219, 123-33.

[http://dx.doi.org/10.1016/S0044-8486(03)00021-8]

Valiela, I, Bowen, JL & York, JK (2001) Mangrove Forests: One of the World's Threatened Major Tropical Environments: At least 35% of the area of mangrove forests has been lost in the past two decades, losses that exceed those for tropical rain forests and coral reefs, two other well-known threatened environments. *Bioscience,* 51, 807-15.
[http://dx.doi.org/10.1641/0006-3568(2001)051[0807:MFOOTW]2.0.CO;2]

Venglovsky, J, Martinez, J & Placha, I (2006) Hygienic and ecological risks connected with utilization of animal manures and biosolids in agriculture. *Livest Sci,* 102, 197-203.
[http://dx.doi.org/10.1016/j.livsci.2006.03.017]

Viji, R & Shrinithivihahshini, ND (2017) An assessment of water quality parameters and survival of indicator in pilgrimage place of Velankanni, Tamil Nadu, India. *Ocean Coast Manage,* 146, 36-42.
[http://dx.doi.org/10.1016/j.ocecoaman.2017.06.002]

Viji, R, Shrinithivihahshini, ND, Ranjeetha, R, Santhanam, P, Narayanan, PSR & Balakrishnan, S (2018) Assessment of environmental parameters with special emphasis on avifaunal breeding season in the coastal wetland of Point Calimere Wildlife Sanctuary, Southeast coast of India. *Mar Pollut Bull,* 131, 233-8.
[http://dx.doi.org/10.1016/j.marpolbul.2018.04.023] [PMID: 29886942]

Wang, J, Li, Y, Chen, J, Hua, D, Li, Y, Deng, H, Li, Y, Liang, Z & Huang, J (2018) Rapid detection of food-borne Salmonella contamination using IMBs-qPCR method based on pagC gene. *Braz J Microbiol,* 49, 320-8.
[http://dx.doi.org/10.1016/j.bjm.2017.09.001] [PMID: 29108975]

World Health Organization (2010) Safe management of shellfish and harvest waters. *IWA Publishing.* Rees, G, Pond, K, Kay, D, Bartram, J, Domingo, JS 360.

World Health Organization (1992) *Our Planet, Our Health.* WHO, Geneva.

Marine Ecology: Current and Future Developments, 2019, Vol. 1, 409-434 409

CHAPTER 16

Marine Microbial Mettle for Heavy Metal Bioremediation: A Perception

Haresh Z. Panseriya[1,2]**, Haren B. Gosai**[1,2]**, Bhumi K. Sachaniya**[1]**, Anjana K. Vala**[1,*] **and Bharti P. Dave**[1,2,*]

[1] *Department of Life Sciences, Maharaja Krishnakumarsinhji Bhavnagar University, Bhavnagar - 364001, India*

[2] *Department of Biosciences, School of Sciences, Indrashil University, Rajpur - Kadi, 382740, India*

Abstract: Marine environment gets polluted due to a range of contaminants including heavy metals. Various physicochemical methods available conventionally for heavy metal remediation suffer from one or the other limitation. Bioremediation is an encouraging solution to heavy metal pollution. Microbes are endowed with diverse potentials to combat heavy metal stress. In this chapter, major sources and effects of heavy metals, factors influencing heavy metal bioremediation, the microbial mechanism for heavy metal detoxification and transformation and involvement of marine microorganisms in heavy metal bioremediation have been discussed.

Keywords: Heavy Metal Pollution, Bioremediation, Marine Environment.

INTRODUCTION

Oceans provide food, recreation, and transportation, which sustain a substantial portion of world's economy (Ansari *et al.*, 2004). Over-population, urbanization and increased industrialization have significantly impacted this largest ecosystem as waste from any source ultimately finds its way to sea or ocean, besides this, pollution arises due to offshore drilling and related activities also (Gosai *et al.*, 2017; Gosai *et al.*, 2018a, b). Pollution of the marine environment poses ecological and economic pressures, because the marine environment holds fundamental importance owing to its biological productivity, geochemical cycling, and human utility (Dudhagara *et al.*, 2016 a, b; Gosai *et al.*, 2018 a, b). According to Article 1 (4) of the 1982 United Nations Convention on the Law of the Seas (UNCLOS), "pollution of the marine environment" is defined as "the int-

* **Corresponding authors Anjana K. Vala and Bharti P. Dave:** Department of Life Sciences, Maharaja Krishnakumarsinhji Bhavnagar University, Bhavnagar-364 001, India; Tel/Fax: +91-2782519824; E-mails: anjana_vala@yahoo.co.in; bpd8256@gmail.com

De-Sheng Pei & Muhammad Junaid (Eds.)
All rights reserved-© 2019 Bentham Science Publishers

roduction by man, directly or indirectly, of substances or energy into the marine environment, including estuaries, which results or is likely to result in such deleterious effects as harm to living resources and marine life, hazards to human health, hindrance to marine activities, including fishing and other legitimate uses of the sea, impairment of quality for use of sea water and reduction of amenities" (UNCLOS,1982).

Domestic waste, agricultural waste, and industrial wastes are among the major marine pollutants. By and large the domestic and agricultural wastes are of similar composition worldwide, however, the composition of industrial waste varies significantly depending on the type of industry. A range of pollutants including polycyclic aromatic hydrocarbons (PAHs), heavy metals, other persistent organic pollutants (POPs), radioactive wastes and plastics *etc.* are contaminating the marine environment (Gosai *et al.*, 2017; Dudhagara *et al.*, 2016a, b).

In the sediments, the heavy metals are more persistent compared to organic contaminants such as petroleum hydrocarbons and pesticides. Moreover, they are mobile in sediments subjected to change in the pH and their speciation. So a fraction of the total mass can leach to aquifer or can become bioavailable to living organisms (Alloway, 1990; Santona *et al.*, 2006). Heavy metal poisoning can result from drinking-water contamination (*e.g.* Pb pipes, industrial and consumer wastes), intake *via* the food chain or high ambient air concentrations near emission sources (Lenntech 2004). In the past decade, Love Canal tragedy in the City of Niagara, USA demonstrated the devastating effect of soil and groundwater contamination on human population (Fletcher, 2002). The diffusion phenomenon of contaminants through soil layers and the change in mobility of heavy metals in aquifers with the intrusion of organic pollutants are being studied in more details in recent years (Cuevas *et al.*, 2011).

A range of physicochemical methods have been employed conventionally for removal of pollutants. However, these methods could not be a permanent solution, as they suffer from one or the other limitations. In this case, the involvement of microorganisms has been given tremendous importance to tackle this problem and microbial bioremediation has been in focus recently (Vala and Dave, 2017; Vala, 2018).

A combined pollution due to heavy metals and PAHs is also a matter of increasing concern these days (Liu *et al.*, 2017). Hence, the discussions in the present chapter are confined to microbial remediation of pollution arising due to some heavy metals as well as combined effects of heavy metals and PAHs as pollutants.

Major Sources of Heavy Metals

Industrial Effluents

Effluents released due to increased industrial growth are one of the major sources of heavy metal pollution. The coastal and marine environments receiving heavy metal-laden effluents have been the 'hotspots' for heavy metal contamination (Naser, 2013).

Sewage

The huge amount of sewage is discharged into the coastal and marine environment. Besides high suspended solids and the heavy load of nutrients, sewage discharges may contain heavy metals also and hence, affects life (Al-Muzaini *et al.*, 1999; Shatti and Abdullah, 1999; Singh *et al.*, 2004; Naser, 2013).

Dredging and Reclamation Activities

Dredging and reclamation activities diminish biodiversity, richness, abundance, and biomass of marine biota (Smith and Rule, 2001). Further, such activities mobilize increased levels of heavy metals leading them to enter foodweb components and hence posing threats to human health (Guerra *et al.*, 2009; Hedge *et al.*, 2009; Naser, 2013).

Toes *et al.*, (2008) carried out microcosm experiments with a view to simulating the influence of dredging in heavy metal-polluted sediments and observed that transient exposure to Cu and Cd resulted in prolonged modification of the indigenous bacterial community.

Desalination Plants

To meet the need for fresh water, desalinated seawater is harnessed especially in Arabian Gulf countries due to low precipitation and high aridity (Hashim and Hajjaj, 2005; Nazer 2013). Due to the discharge of reject waters from desalination plants on a daily basis to coastal and subtidal areas, increased levels of heavy metals have been observed in the vicinities of desalination plants along the Arabian Gulf coastline (Sadiq, 2002; Naser, 2010; 2012; 2013).

Oil Pollution

Activities pertaining to oil exploration, production, and transport contribute significantly to oil pollution (MEMAC, 2003). It has been reported that major oil spill during 1991 Gulf war has led to elevated levels of heavy metals (Al-Arfaj and Alam, 1993; Naser 2013). Table **1** shows sources of various heavy metals and their effects on human health.

Marine Microbial Remediation of Heavy Metal Pollution

The term 'heavy metals' include the group of metals and metalloids with an atomic density greater than $4g/cm^3$, or 5 times or more, greater than water, all the metals in Groups 3 to 16 which are in periods 4 and greater (Hawkes, 1997; Paul, 2017). Heavy metals in the aquatic environment get absorbed in living organisms and through the food chain, and human population also is likely to accumulate such metals. Unlike most of the contaminants, heavy metals are non-degradable. Exposure to such heavy metals results in a number of health implications ranging from headache to allergy to respiratory tract disorder to various cancers depending on type and concentration of heavy metal (Kaufman, 1970; Katz and Salem, 1993; Costa, 2003; Park *et al.*, 2005; Kotas and Stasicka, 2000; Vala and Dave, 2017). Therefore, treatment of metal contaminated wastewater prior to its release into the environment is imperative and a range of treatment processes including chemical precipitation, electrodialysis, evaporation, ion exchange, liquid extraction, membrane process and reverse osmosis, *etc.* are conventionally being used for this purpose. However, these methods suffer from one or the other limitation such as inefficient removal, high-energy requirements and generation of toxic sludge and are not eco-friendly. Hence, prominence is given to bioremediation of metal contaminated sites (Eccles, 1999; Paul, 2017; Vala and Dave, 2017).

Table 1. Sources and effects of some heavy metals (Dixit *et al.*, 2015; Gupta *et al.*, 2016; Meenambigai *et al.*, 2016; IARC, 2012).

Heavy Metal	IARC Group	Source	Effects
Lead	2B	Present in petro -based materials lead acid batteries, paints, E-wastes, Smelting operations, coal-based thermal power plants, ceramics, bangle industry, and many other manufacturing amenities	Damage to the nervous system, circulatory system, blood-forming system, reproductive system, gastrointestinal tract, and kidney. The central nervous system is most insightful to the effects of lead.
Metallic chromium, chromium (III) compounds, chromium (VI) compounds	3,3 and 1, respectively	Built-up operations together with chrome plating, petroleum refining, leather, tanning, wood preserving, textile manufacturing, and pulp processing.	Irritant, sickness and nausea, low-level exposure can irritate the skin and cause ulceration. Long-term exposure can cause kidney and liver damage, and damage too circulatory and nerve tissue, carcinogenic.

(Table 1) cont.....

Heavy Metal	IARC Group	Source	Effects
Zinc	Data not available	Paint, rubber, dye, wood preservatives and ointments and electroplating industries.	Nausea and vomiting. Zinc combines with other elements to form zinc compounds; common zinc compounds found at hazardous waste sites include zinc chloride, zinc oxide, zinc sulfate, zinc phosphate, zinc cyanide, and zinc sulfide.
Nickel	1	Galvanized, paint and powder batteries processing units.	Long-term exposure can cause decreased body weight, heart and liver damage, and skin irritation.
Arsenic	1	Geogenic/natural processes, smelting operations, thermal power plants, fuel burning	Vomiting, a decrease in red and white blood cells, damage to blood vessels, abnormal heartbeat, development of skin injuries, peripheral and damage to central nervous systems, carcinogenic
Mercury and inorganic mercury compounds, methylmercury compounds	3 and 2B respectively	Industrial activities, sewage, atmospheric accumulation (coal burning, volcanoes *etc.*), domestic, medical incinerators and mishandling of E-wastes.	Autoimmune diseases, depression, drowsiness, fatigue, hair loss, insomnia, loss of memory, restlessness, disturbance of vision, tremors, temper outbursts, brain damage, and lung or kidney failure.
Cadmium	1	Mining (including mining wastes and mine waters), metallurgical industries, industrial use, municipal sewage effluents, fertilizers, domestic waste, and agricultural soils.	Carcinogenic, mutagenic, endocrine disruptor, lung damage, and fragile bones, affects calcium regulation in biological systems.

Microorganisms including bacteria and fungi play an important role in bioremediation of heavy metal contamination. Microbial remediation of heavy metal contamination is environment-friendly, efficient, cost-effective, self-reproducible and recycles bioproducts. Involvement of microbial processes epitomizes logical and long-term solution for remediation (Alvarez *et al.*, 2017). Microbial processes leading to the solubilization and immobilization of heavy metals play a vital role in bioremediation. While removal of metal contaminants from solid matrices like sediment and soils is aided by solubilization, immobilization is involved in *in situ* transformation of heavy metals and is mainly relevant to heavy metal removal from aqueous solutions (Gadd, 2004). Fig. (**1**) depicts various microbial mechanisms involved in detoxification and transformation of metals.

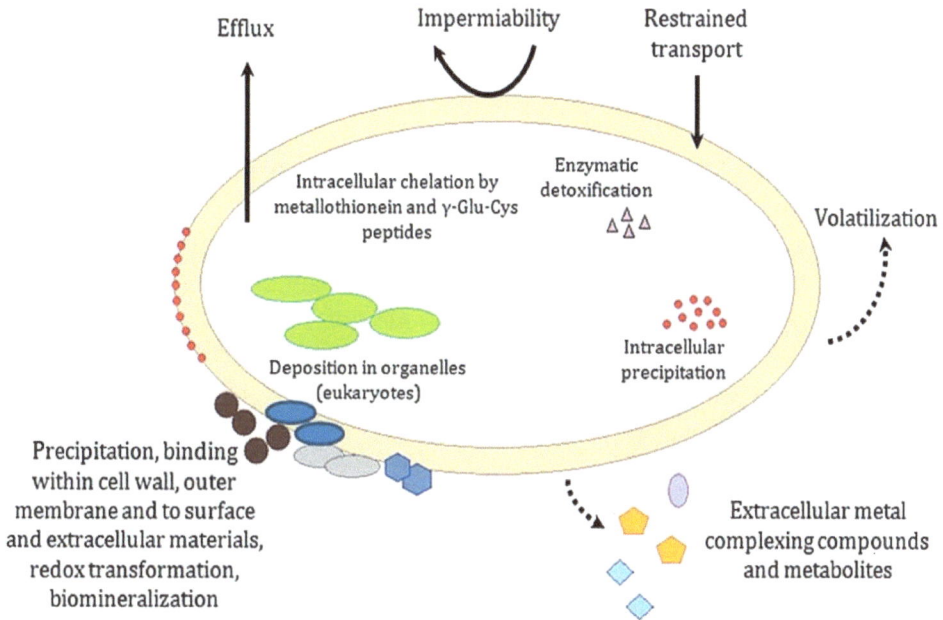

Fig. (1). Various microbial mechanisms involved in detoxification and transformation of heavy metals (Adapted from Gadd, 2010).

As depicted in Fig. (**1**), volatilization is one of the important mechanisms involved in heavy metal transformation. Vidal and Vidal, (1980) examined arsenic metabolism (As(V)) in marine yeast *Rhodotorula rubra* and based on qualitative analysis of metabolism products, reported generation of As(III), methylarsonic acid [$CH_3AsO(OH)_2$], dimethylarsinic acid $(CH_3)_2AsO(OH)$ and volatile alkylarsines. It was observed that some of the produced As(III) was transported to culture medium and the remaining got methylated. Further methylation of the dimethylarsenic acid ultimately led to the formation of volatile alkylarsine product. Vala, (2018) also reported volatilization of 15.75% supplied trivalent arsenic from the culture medium by *Aspergillus sydowii*.

In a case study, De *et al.*, (2014) explained the importance of biofilm-forming bacteria in bioremediation of heavy metals, especially mercury. The authors suggested the bacteria like *Bacillus cereus* BW-03 are capable of synthesizing EPS like polysaccharides, amyloids, biosurfactants and extracellular enzymes that make them promising candidates for remediation of inorganic mercury.

Factors Influencing Heavy Metal Bioremediation

A range of biotic and abiotic parameters affect the heavy metal toxicity to an aquatic organism.

Biotic Factors

Species and Tolerance Capacity

Response to a heavy metal *viz.* tolerance, uptake and bioremediation ability vary with each organism. Organisms use various strategies to tackle heavy metal stress. Variation in response at genus as well as species level is very likely. For instance, among different species of marine-derived fungus *Aspgerillusviz.A. candidus*, *A. flavus*, *A. niger* and *A. sydowii* tested for arsenic tolerance and removal, *A. sydowii* exhibited maximum tolerance to As (Vala, 2010; Vala *et al.*, 2010; Vala *et al.*, 2011; Vala, 2018). Divalent heavy metals (Cd, Cu, Hg, Ni, Pb, and Zn) were removed by freshwater microalgal species, *C. miniata*, *C. vulgaris*, and *C. reinhardtii*, whereas *C. vulgaris* and *S. Platensis* could remove trivalent metals (Fe and Cr), while *C. miniata* and *C. vulgaris* were observed to remove the hexavalent Cr (González *et al.*, 2011; Suresh Kumar *et al.*, 2015).

Biomass Concentration

Increased metal removal efficiency with increasing biomass concentration has been reported by several workers. More number of available binding sites could be the reason for such an observation (Monteiro *et al.*, 2012; Suresh Kumar *et al.*, 2015). However, sometimes reduction in effective surface area for biosorption that occurs due to partial aggregation of biomass, lessening the average distance between available adsorption sites and a screen effect (formation of dense outer layer of cells) causing blocking of binding sites could lead to decrease in metal removal with increasing biomass concentration (Bishnoi *et al.*, 2004; Suresh Kumar *et al.*, 2015).

Size and Volume of Biota

Smaller cells have the larger surface to volume ratios and hence, are the most effective metal sequesters. Hence, microbes and microalgae are considered as the promising tool for heavy metal bioremediation (Suresh Kumar *et al.*, 2015).

Abiotic Factors

pH

Solubility, toxicity, and speciation of heavy metals are greatly affected by pH of the solution. Metal binding to the cell surface is also affected by pH. For instance, increased binding capacity of heavy metals (Cd, Cu, and Zn) has been observed with an increase in pH from 4 to 7 (Les and Walker, 1984).

Ionic Strength

The available sites for metal uptake decrease with increasing ionic strength, hence, with the decrease in ionic strength, metal removal efficiency increases (Dwivedi, 2012; Suresh Kumar *et al.*, 2015).

Temperature

Metal speciation and solubility are greatly affected by temperature; however, the effect exerted by temperature would vary with the type of heavy metal (Suresh Kumar *et al.*, 2015).

Metal Speciation

The effect exerted by heavy metal on organisms chiefly depends on the metal ion species that in turn may be governed by pH of the solution. For instance, the hexavalent form of chromium is the most hazardous among the different valency states ranging from Cr(II) to Cr(VI) (Suresh Kumar *et al.*, 2015; Vala and Dave, 2017).

Salinity and Hardness

Salinity affects heavy metal uptake, however, the response varies from metal to metal. It has been reviewed that increase in chloride or salinity decreases metal (Ni, Zn, Sn, Cu, and Cd) toxicity (Wang, 1987).

Effect of Combined Metals

Wastewater discharge generally contains more than one metal pollutant. The combined toxic effect exerted would be different from that exerted individually by the pollutant. The combined possible effect could be (a) synergistic, where combined toxic effect is more than sum of individual toxicities (b) antagonistic – combined toxic effect is less than sum of individual toxicities and (c) Non-interactive or additive – combined effect is similar to the sum of individual effects (Suresh Kumar *et al.*, 2015).

Effect of Matrix

Bioavailability of pollutant is an important criterion while considering its bioremediation. Surfactants play an important role in restoring the adverse effects exerted by pollutants like heavy metals as well as PAHs. Surfactants could promote the transmembrane transport, enhance cell surface hydrophobicity (especially for PAHs) and can act as metal complexing agents (Li *et al.*, 2015; Liu *et al.*, 2017). Remarkable removal of Fe, Pb, and Zn has been observed by biosurfactants produced by *Candida sphaerica* (Luna *et al.*, 2016).

Fig. (**2**) depicts various influential parameters for microbial bioremediation of heavy metal as well as PAHs.

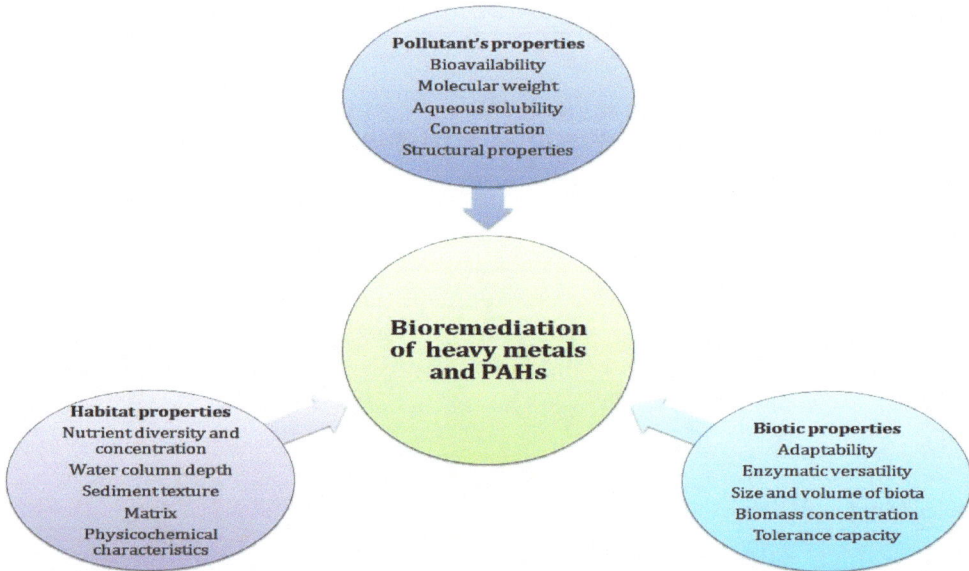

Fig. (2). Influential parameters for bioremediation of heavy metals and PAHs.

Microorganisms from the Marine Environment and Heavy Metal Bioremediation

The occurrence of metal in the natural environment and anthropogenic activities are the reasons for their encounter with microbes. Substantial variations in heavy metal concentrations are observed in the marine subsurface. Composition and concentration of heavy metals play an important role in shaping microbial community structure (Oliveira and Pampulha 2006; Ravikumar *et al.*, 2007; Pachiadaki *et al.*, 2016; Vala and Dave, 2017).

Due to relatively low oxygen concentration and higher salt concentrations in seawater, removal of metal contamination using organisms from non-marine sources could be difficult, for applying microorganisms to the marine environment they have to be able to tolerate high salt concentration and low oxygen (Kim *et al.*, 2015).

Heavy Metal Removal by Bacteria

A number of microorganisms from the marine environment have been reported to possess heavy metal tolerance and removal efficiency. Many of the reports also confirmed multimetal removal efficiency of either microbial cultures or their products. Shirdam *et al.*, (2006) isolated three marine bacteria, *Pseudomonas putida* PTCC 1664, *Bacillus cereus* PTTC 1665 and *Pseudomonas pseudoalkaligenes* PTCC 1666 from the East Anzali wetland sediments of the Caspian Sea. The isolates were resistant to cadmium (Cd), nickel (Ni) and vanadium (V) and accumulated approximately 40-50% Cd, 5-6% Ni and 10-12% V.

Immobilized cells were observed to be more efficient in biosorption of heavy metals than free cells. The uptake of metal ions was reported to be in two stages *viz.* rapid and slow stage. Metal ions adsorb onto the microbial surface during the rapid stage while in the slow stage transportation of metal ions across the cell membrane occurs (Hong and Shan-shan, 2005; Shirdam *et al.*, 2006).

Dash and Das, (2014) isolated two highly mercury resistant isolates, *Bacillus thuringiensis* PW-05 (from the marine environment) and *Bacillus* sp. SD-43 (from steel industry waste) and observed that the marine isolate exhibited more mercury volatilization ability. They suggested that marine bacteria could be a better option for enhanced bioremediation of mercury-contaminated sites.

De *et al.*, (2008) observed multimetal resistance in marine bacteria highly resistant to mercury (BHRM). The bacteria supplied with 100 ppm initial metal concentration could remove 70% Cd (72h) and 98% Pb (96h). Volatilization (for Hg), entrapment in extracellular polymeric substances (for Hg, Cd, and Pb) and/or precipitation as sulfide (for Pb) were suggested to be responsible for heavy metal detoxification.

A marine biofilm-forming mercury-resistant bacterium *B. cereus* BW-03 was reported as a suitable candidate for remediation of mercury-contaminated waste (De *et al.*, 2014). Enhanced bioremediation capability of biofilm-forming bacteria was attributed to the occurrence of EPS, extracellular enzymes and biosurfactants. The authors suggested that for mercury bioremediation priority should be given to biosorption aspect than to volatilization aspect.

Jankowska *et al.*, (2006) examined the effect of mineral ship motor oil on cadmium and lead sensitivity of bacteria isolated from marine waters and sediments of Sopot beach, Gdańsk Bay, Poland and observed increased sensitivity to heavy metals in presence of mineral ship motor oil.

Mulik and Bhadekar, (2017) screened bacterial isolates from the Antarctic oceanic region for heavy metal tolerance and metal removal efficiency. The test isolates could tolerate Cd^{2+}, Cr^{3+}, Ni^{2+}, and Pb^{2+} in the range of 300-600 ppm. Percentage metal removal was observed in the range of 6-95.39%. *Kocuria* sp. exhibited 95.4% removal of Cd^{2+} and 86.7% removal of Pb^{2+}, while *Halomonas* sp. removed 88.2% of supplied Cd^{2+}. In the case of *Kocuria* sp., 85% Cd^{2+} was observed to be mainly accumulated intracellularly while 20.29% Ni^{2+} was removed by *Halomonas* sp. through cell adsorption.

Iyer *et al.*, (2004) examined an exopolysaccharide (EPS) producing *Enterobacter cloacae* (AK-I-MB-71a) for Cr (VI) tolerance. The bacterium exhibited enhanced growth and EPS production even at 100ppm Cr(VI) concentration and could accumulate about 60–70% chromium. Further, Iyer *et al.*, (2005) reported that the exopolysaccharide produced by the same culture *Enterobacter cloacae* demonstrated excellent heavy metal chelation properties. When supplied with different concentrations of metals (cadmium, cobalt, copper, and mercury), increased chelation was observed with increasing concentrations of cadmium, cobalt and copper. Highest chelation was observed in the case of cadmium (65%) followed by copper (20%) and cobalt (8%). Mercury could not be chelated by the EPS. The cell pellet did not chelate any of the supplied metal. The authors suggested that the type of chelation of metals by bacterial EPS could be biosorption.

Das *et al.*, (2009) reported biosurfactant-mediated heavy metals removal. Biosurfactant product of a marine bacterium *Bacillus circulans* isolated from Andaman and Nicobar Islands, India was examined for metal removal potential and almost complete removal of 100 ppm lead and cadmium could be obtained using concentration 5X, its critical micelle concentration (CMC). Detailed analyses using AAS studies, Fourier transform infrared spectroscopy (FTIR) and transmission electron microscopy (TEM) equipped with energy dispersive X-ray spectroscopy (EDS) revealed metal binding by the test anionic biosurfactant even at a concentration less than the CMC.

Microbes from the marine environment have also been observed (Acharya *et al.*, 2009) to sequester uranium from aquatic system above pH 6, a condition otherwise considered less efficient due to the formation of stable carbonato complexes of uranyl ions that hampers adsorption/uptake of uranium by organisms. Acharya *et al.*, (2009) evaluated uranium sequestering properties of a marine, unicellular cyanobacterium *Synechococcus elongatus*, an organism that occurs abundantly in oceans. The organism was observed to remove 72% of supplied uranium at pH 7.8. Uranium could be removed using live/dead cells or EPS produced by the organism, hence, *Synechococcus elongatus* was suggested as a potential candidate for uranium removal from natural aquatic environments.

Interaction of uranium with a filamentous, heterocystous cyanobacterium

Anabaena torulosa was examined and it was observed that *A. torulosa* could bind uranium from supplied 100µM uranyl carbonate at pH 7.8 (Acharya *et al.*, 2012). Detailed analyses revealed the importance of viable cells in uranyl binding as heat-killed cells or EPS derived from live cells exhibited limited uranyl binding. The authors revealed the involvement of polyphosphates in uranium accumulation. Compartmentalization of heavy metal in polyphosphate bodies detoxifies metal by removing it from the active metabolic pool and delaying its toxic effects (Jensen *et al.*, 1982; Acharya *et al.*, 2012).

With high-resolution physical evidence Acharya and Apte, (2013) reported surface associated polyphosphate bodies (SAPBs) in the marine, filamentous, heterocystous microbe *Anabaena torulosa* and a novel uranium immobilization phenomenon involving SAPBs.

A novel biocarrier has been developed that consisted of zeolite as a support carrier and a sulfate-reducing bacteria (SRB) *Desulfovibrio desulfuricans* for efficient heavy metal removal (Kim *et al.*, 2015). The SRB-immobilized zeolite carriers could remove 98.2% Cu^{2+}, 90.1%, Ni^{2+}, and 99.8% Cr^{6+} from 100 ppm concentration of heavy metals. The authors envisaged SRB-zeolite carrier as an efficient tool for bioremediation of metal contaminated the marine environments.

With advancement in the field of genetics and molecular biology, genetically engineered microorganisms with toxicant bioremediation ability have been developed. However, genetic manipulation of marine bacteria for enhanced bioremediation has not been reported much (Dash *et al.*, 2013). Engineered *Synechococcus* sp. with heavy metal tolerance had been developed by Sode *et al.*, (1998) by introducing exogenic metallothionein in gene. Cheung and Gu, (2003) developed engineered sulfate-reducing bacteria (SRB) for chromate reduction. Ramanathan *et al.*, (1997) carried out a fusion of arsB gene with lux genes and developed Antimonite and arsenite sensing marine *Staphylococcus aureus*.

Heavy Metal Removal by Fungi

Among inshore microbiota, fungi are important members that come across metal ions and complexes (Millward *et al.*, 2001; Hyde *et al.*, 1998; Newell and Barlocher 1993; Vala and Dave 2017). Assessment of mycobiota for remediation of metal pollution is an exciting area of research (Vala *et al.*, 2004; Vala, 2010; Vala and Dave, 2017).

In India, Gujarat state comprises the country's 22% coastline. Mycobiota from marine coastal habitats of Gujarat state have been revealed to possess diverse ecologically significant potentials including removal of arsenic (Vaidya *et al.*, 2000; Vala *et al.*, 2000; Vala *et al.*, 2004; Khambhaty *et al.*, 2009a, b; Vala 2010;

Vala and Patel, 2011; Vala *et al.*, 2012, Vala and Dave, 2015; Vala, 2018; Vala and Dave, 2017).

Vala *et al.*, (2004) while searching for novel sources of mycobiota for heavy metal tolerance and removal tested two seaweeds associated fungi *Aspergillus flavus* and *A. Niger* for their hexavalent chromium tolerance potential. Both the test fungi were observed to exhibit notable chromium tolerance and removal potential. Removal of chromium (mg g^{-1} dry wt) was found to increase with increasing exposed Cr(VI) concentrations.

Taboski *et al.*, (2005) assessed the toxicity of Cd and Pb to two fungal species *Corollospora lacera* and *Monodictys pelagica* from the marine environment by examining their radial growth rate and biomass accumulation. Biosorption of metals was also evaluated. Radial growth of fungi was not affected by lead, however, increasing cadmium concentration reduced the radial growth of fungi, especially, *M. pelagica*. About 93% of lead sequestration by *C. lacera* was found to be extracellular. *M. Pelagica* bioaccumulated over 60mgg^{-1} Cd and over 6 mgg^{-1} Pb. Over 7mgg^{-1} Cd and up to 250 mgg^{-1}Pb was bioaccumulated by *C. lacera*.

Khambhaty *et al.*, (2008) examined dead fungal biomass of four marine *Aspergillus* species for Hg(II) biosorption and observed *Aspergillus niger* as the most efficient Hg(II) biosorbent. Dead biomass of *A. niger* exhibited 40.53 mgg^{-1} Hg(II) removal under optimized conditions. Evaluation of possible cell-metal ion interaction revealed involvement of –OH and NH2 groups present on the cell surface in Hg(II) biosorption.

Three marine-derived aspergilli *viz.Aspergillus niger*, *A. wentii* and *A. terreus* were isolated from Gujarat coast and were tested for their hexavalent chromium removal potential. Among the three, *A. niger* was observed as the most potential candidate for Cr(VI) removal (Khambhaty *et al.*, 2009a). Detailed analyses of biosorption parameters and sorption capacity revealed 117.33 mgg^{-1} Cr(VI) adsorption by *A. niger* under optimized conditions and sorption efficiency was observed to be 100%. Studies related to kinetics, equilibrium and thermodynamics on Cr (VI) biosorption by marine *A. Niger* have also been carried out (Khambhaty *et al.*, 2009b). The process of biosorption process was revealed to be endothermic. Based on FTIR analysis, amino, –CH2, hydroxyl and phosphorous groups were found to be involved in binding of Cr(VI) to fungal biomass.

Hexavalent chromium tolerant strain of *Trichoderma viride* was isolated from water samples from the Mediterranean Sea (El-Kassas and El-Taher, 2009). The fungus could remove 4.66 mgg^{-1} Cr(VI). It was observed by transmission electron microscopic examination that chromium accumulation by the fungus did not affect its mycelial and conidial structures. Upon optimization of Cr(VI) removal, parameters like biosorbent dosage, contact time, initial metal concentration and pH of the solution were revealed as influential parameters.

Two marine fungal strains of *Dendryphiella salina* were observed to absorb 80–92% Hg^{2+} from the liquid media. Strain Den32 had higher absorption efficiency than strain Den35 (Mendoza *et al.*, 2010). The study revealed the potential application of both the strains for bioremediation of mercury, especially through biosorption.

Three species of Thraustochydrids *viz. Aplanochytrium* sp., *Thraustochytrium* sp. and *Schizochytrium* sp. were examined for chromium removal efficacy (Gomathi *et al.*, 2012). Detailed studies including adsorption kinetics and optimization using *Aplanochytrium* sp. were also carried out and 69.4% chromium removal by the test fungus could be achieved.

Microbes play a substantial role in decreasing arsenic toxicity (Bełdowski *et al.*, 2016). Role of bacteria from the marine environment in remediation of arsenic, a group A and category 1 human carcinogen, have been studied by several workers, however, role of marine-derived fungi in arsenic remediation is comparatively less studied (Vidal and Vidal, 1980; Takeuchi *et al.*, 2007; Handley *et al.*, 2009; Vala, 2010; Keren *et al.*, 2015; Khambholja and Kalia, 2016; Vala and Dave 2017). Recently, Vala and Dave, (2017) have reported that in India, Bhavnagar University (now Maharaja Krishnakumarsinhji Bhavnagar University) has carried out pioneering work in arsenic bioremediation by facultative marine fungi.

Examination of arsenic tolerance and accumulation potential of *Aspergillus* sp. isolated from coastal waters of Bhavnagar, Gulf of Khambhat, West coast of India, revealed the fungus to tolerate supplied $100mgL^{-1}$ As (III) or As(V). HGAAS analysis revealed higher removal of As(V) than As(III) (Vala and Upadhyay 2008). Energy Dispersive X-ray spectroscopic (EDX) data further confirmed the presence of arsenic in fungal biomass.

Vala (2010) investigated tolerance to and removal of arsenic by *Aspergillus candidus* isolated from coastal waters of Bhavnagar, Gulf of Khambhat, West coast of India. The fungus exhibited tolerance to trivalent and pentavalent forms of arsenic (25 and $50mgL^{-1}$). Highest arsenic removal (mgg^{-1}) by the test fungus was observed on day 3. Facultative marine fungus *A. candidus* was suggested as a promising candidate for arsenic bioremediation by the author.

Facultative marine fungi *Aspergillus flavus* and *A. niger* have been reported to exhibit tolerance to and removal of arsenic by Vala *et al.*, (2010) and Vala *et al.*, (2011), respectively. Vala, (2009) has reviewed *A. niger* as potential biosorbent. This perception was corroborated by marine-derived *A. niger*. Heat-killed biomass of marine-derived *A. niger* when examined for its As(III) biosorption capability, it was observed to remove more than 90% of supplied As(III) concentrations. Maximum biosorption (108.083 mgg^{-1}) was achieved at 600 mgL^{-1} As concentration by Vala and Patel, (2011).

Vala and Sutariya, (2012) explored the degree of arsenic tolerance and removal efficiency of two facultative marine fungi *A. flavus* and *Rhizopus* sp. Upon exposure to 25 mg/L and 50 mg/L sodium arsenite (As (III)), both the fungi exhibited As tolerance and arsenic accumulation. Slightly better accumulation was observed by *Rhizopus* sp. Increase in accumulation was observed with increasing supplied As concentration indicating higher complexation rates between arsenic and arsenic complexing group on the fungal biomass.

Compared to their filamentous fungal counterparts, yeasts from the marine environment have been less explored for heavy metal removal. Strains of *Yarrowia lipolytica* have been reported to be potential hexavalent chromium remediators by several workers (Banker *et al.*, 2009; Rao *et al.*, 2013; Imandi *et al.*, 2014). Similarly, marine yeast *Rhodotorula rubra* has been harnessed for arsenic metabolism by a number of workers (Button *et al.*, 1973; Vidal and Vidal 1980; Cullen and Reimer, 1989; Maher and Butler, 1988). However, arsenic remediation by marine yeasts has not been attended much in recent past (Vala and Dave, 2017).

Abe *et al.*, (2001) isolated thirteen yeast strains from deep-sea sediment samples of Japan Trench. Among them, *Cryptococcus* sp. was observed to possess the highest tolerance to Cu^{2+}. The authors suggested the role of superoxide dismutase (SOD) to combat high Cu^{2+} stress.

Deep-sea psychrotolerant yeast isolates *Cryptococcus* sp., when grown in presence of various concentrations of heavy metal salts *viz.* CdCl$_2$, CuSO$_4$, Pb (CH$_3$COO)$_2$, and ZnSO$_4$, exhibited remarkable growth in presence of 100mgL^{-1} metal concentrations. Tolerance to these metals exhibited by the isolate was comparatively higher than other deep-sea and terrestrial yeasts. Alteration in the cell morphology was observed in presence of heavy metals. The yeast could remove 30-90% of the supplied heavy metals. The authors suggested the test *Cryptococcus* sp. as a potential candidate for bioremediation of heavy metal contaminated sites. The authors speculated the metal tolerant property of the yeast to contribute to its ecological role and adaptations in extreme environments (Singh *et al.*, 2013).

Oyetibo *et al.*, (2015) investigated mercury removal by resting and growing cells of mercury-resistant *Yarrowia* spp. isolated from estuarine sediments polluted with mercury. The resting cells of yeast strain were suggested to be applicable as a reusable bioadsorbent while the growing cells were suggested to be more suitable as efficient mercury bioreduction and volatilization agent.

Over the past few decades, many bioremediation technologies were applied all over the world to deal with contaminated habitats. Many documents and reviews on these technologies for remediating heavy metals are available (Khan *et al.*, 2004). However, there are many gaps in the understanding of microbial remediation of heavy metals especially due to the extreme complexity of soil chemistry. Thus, extensive site specific research is necessary to bring out the optimum performance from any of these technologies.

Bioremediation of Combined Pollution by Heavy Metals and PAHs

Effects exerted by combined exposure of heavy metal and PAHs on microbiota are many folds more complex than their separate exposure. For instance, the ROS generated during PAH metabolism leads to reducing the microbial activities including their ability to tackle heavy metals (Kuang *et al.*, 2013). Further, PAHs intermediates like salicylic acid can affect the heavy metal adsorption potential (Gong *et al.*, 2009; Guo *et al.*, 2015; Liu *et al.*, 2017). On the other hand, in presence of heavy metals, PAH remediation can be perplexing as heavy metals would affect ATP generation, C-mineralization, enzymatic functions and community shift (Biswas *et al.*, 2015; Liu *et al.*, 2017).

Effect of PAHs on Heavy Metal Bioremediation

Upon exposure to heavy metals, bioadsorption is the most effective and common response of microbes to combat metal stress. Presence of PAHs leads to an adverse effect on microbial membranes and brings about altered the transport of heavy metals. Biomembrane permeability is altered and allows easy penetration of heavy metal into the cell (Shen *et al.*, 2006). Altered membrane fluidity and change in electrical potential can also be imparted by PAHs resulting in inhibition of heavy metal adsorption by microorganisms (Gorria *et al.*, 2006). Ion regulation disruption can also be a consequence of the effect of PAH on membrane enzymes, decreasing the metal-ATPase activity and hence affecting heavy metal transportation (Gauthier *et al.*, 2015; Liu *et al.*, 2017).

Effect of Heavy Metal on PAHs Bioremediation

Change in the microbial surface properties and interference with microbial enzymes are the main two effects exerted by the presence of heavy metals on microbial PAH biodegradation. Adsorption areas of PAHs on microbial cells are occupied by heavy metals due to electrostatic attraction between the metal ion and microbial cell surface, which are quite stronger than the van der walls interactions between the microbial cell surface and PAHs (Zouboulis *et al.*, 2004). With increasing cationic metal concentration, the microbial cell surface eventually neutralizes, becomes less hydrophilic and promotes adsorption of PAHs (Weissenfels *et al.*, 1992; Al-Turki, 2009). Besides, the PAHs tend to be attracted to aggregation of heavy metals, facilitating the adsorption of PAHs by microbes (Zhu *et al.*, 2004; Zhang *et al.*, 2011). As PAHs impart a narcotic effect to the lipophilic compound that may exert the effect on perviousness and configuration of the microbial cell, heavy metals invade microbes easily and affect their functions (Shen *et al.*, 2005; Michalec *et al.*, 2016; Liu *et al.*, 2017). As some of the heavy metals act as cofactors for enzymes, it is very likely that such metal species with low concentration would enhance enzymatic degradation of PAHs. However, their presence in higher concentration would be deleterious (Karaca *et al.*, 2010; Guo *et al.*, 2010; Liu *et al.*, 2017).

CHALLENGES AND FUTURE DIRECTIONS

Despite the advancements (already discussed in the chapter on Microbial PAH bioremediation by our group) in bioremediation of pollutants like heavy metals and PAHs, still there is a long way to go and there is scope for future research in this line. Screening from diverse habitats may be beneficial and lead to more competent organisms. Increasing bioavailability of pollutants should also be focused besides developing new technologies aiding microbial detoxification of pollutants. Investigation of metabolic potential of the microbial community followed by cloning of potential gene using suitable expression systems could play important role in the development of large-scale profitable applications of bioremediation.

CONSENT FOR PUBLICATION

Not applicable.

CONFLICT OF INTEREST

The authors confirm that this chapter contents have no conflict of interest.

ACKNOWLEDGEMENT

Earth Science and Technology Cell (ESTC), Ministry of Earth Science (MoES), Government of India (GoI), New Delhi is gratefully acknowledged for financial support.

REFERENCES

Abe, F, Miura, T, Nagahama, T, Inoue, A, Usami, R & Horikoshi, K (2001) Isolation of a highly copper-tolerant yeast, *Cryptococcus* sp., from the Japan Trench and the induction of superoxide dismutase activity by Cu^{2+}. *Biotechnol Lett,* 23, 2027-34.
[http://dx.doi.org/10.1023/A:1013739232093]

Acharya, C & Apte, SK (2013) Novel surface associated polyphosphate bodies sequester uranium in the filamentous, marine cyanobacterium, *Anabaena torulosa. Metallomics,* 5, 1595-8.
[http://dx.doi.org/10.1039/c3mt00139c] [PMID: 23912813]

Acharya, C, Chandwadkar, P & Apte, SK (2012) Interaction of uranium with a filamentous, heterocystous, nitrogen-fixing cyanobacterium, *Anabaena torulosa. Bioresour Technol,* 116, 290-4.
[http://dx.doi.org/10.1016/j.biortech.2012.03.068] [PMID: 22522016]

Acharya, C, Joseph, D & Apte, SK (2009) Uranium sequestration by a marine cyanobacterium, *Synechococcus elongatus* strain BDU/75042. *Bioresour Technol,* 100, 2176-81.
[http://dx.doi.org/10.1016/j.biortech.2008.10.047] [PMID: 19070485]

Al-Arfaj, AA & Alam, IA (1993) Chemical characterization of sediments from the Gulf area after the 1991 oil spill. *Mar Pollut Bull,* 27, 97-101.
[http://dx.doi.org/10.1016/0025-326X(93)90013-A]

Alloway, BJ (1990) Soil Processes and the Behaviour of Metals. *Heavy Metals in Soils.* In: Alloway, B.J., (Ed.), Blackie and Son Inc., New York 7-28.

Al-Muzaini, S, Beg, M, Muslamani, K & Al-Mutairi, M (1999) The quality of marine water around a sewage outfall. *Water Sci Technol,* 40, 11-5.
[http://dx.doi.org/10.2166/wst.1999.0316]

Al-Turki, AI (2009) Microbial polycyclic aromatic hydrocarbons degradation in soil. *Research Journal of Environmental Toxicology,* 3, 1-8.
[http://dx.doi.org/10.3923/rjet.2009.1.8]

Álvarez, SP, Tapia, MAM, Duarte, BND & Vega, MEG (2017) Fungal Bioremediation as a Tool for Polluted Agricultural Soils. *Mycoremediation & Environmental Sustainability,* Springer, Cham 1-15.
[http://dx.doi.org/10.1007/978-3-319-68957-9_1]

Ansari, TM, Marr, IL & Tariq, N (2004) Heavy Metals in Marine Pollution Perspective–A Mini Review. *J Appl Sci (Faisalabad),* 4, 1-20.
[http://dx.doi.org/10.3923/jas.2004.1.20]

Bankar, AV, Kumar, AR & Zinjarde, SS (2009) Removal of chromium (VI) ions from aqueous solution by adsorption onto two marine isolates of *Yarrowia lipolytica. J Hazard Mater,* 170, 487-94.
[http://dx.doi.org/10.1016/j.jhazmat.2009.04.070] [PMID: 19467781]

Bełdowski, J, Szubska, M, Emelyanov, E, Garnaga, G, Drzewińska, A, Bełdowska, M, Vanninen, P, Östin, A & Fabisiak, J (2016) Arsenic concentrations in Baltic Sea sediments close to chemical munitions dumpsites. *Deep Sea Res Part II Top Stud Oceanogr,* 128, 114-22.
[http://dx.doi.org/10.1016/j.dsr2.2015.03.001]

Bishnoi, NR & Pant, A (2004) Biosorption of copper from aqueous solution using algal biomass. *J Sci Ind Res (India),* 63, 813-6.

Biswas, B, Sarkar, B, Mandal, A & Naidu, R (2015) Heavy metal-immobilizing organoclay facilitates polycyclic aromatic hydrocarbon biodegradation in mixed-contaminated soil. *J Hazard Mater,* 298, 129-37.
[http://dx.doi.org/10.1016/j.jhazmat.2015.05.009] [PMID: 26022853]

Button, DK, Dunker, SS & Morse, ML (1973) Continuous culture of *Rhodotorula rubra*: kinetics of phosphate-arsenate uptake, inhibition, and phosphate-limited growth. *J Bacteriol,* 113, 599-611.
[PMID: 4690960]

Cheung, KH & Gu, JD (2003) Reduction of chromate (CrO4(2-)) by an enrichment consortium and an isolate

of marine sulfate-reducing bacteria. *Chemosphere,* 52, 1523-9.
[http://dx.doi.org/10.1016/S0045-6535(03)00491-0] [PMID: 12867184]

Costa, M (2003) Potential hazards of hexavalent chromate in our drinking water. *Toxicol Appl Pharmacol,* 188, 1-5.
[http://dx.doi.org/10.1016/S0041-008X(03)00011-5] [PMID: 12668116]

Cuevas, J, Ruiz, AI, de Soto, IS, Sevilla, T, Procopio, JR, Da Silva, P, Gismera, MJ, Regadío, M, Sánchez Jiménez, N, Rodríguez Rastrero, M & Leguey, S (2011) The performance of natural clay as a barrier to the diffusion of municipal solid waste landfill leachates. *J Environ Manage,* 95, S175-81.
[http://dx.doi.org/10.1016/ j.jenvman.2011.02.014] [PMID: 21420226]

Cullen, WR & Reimer, KJ (1989) Arsenic speciation in the environment. *Chem Rev,* 89, 713-64.
[http://dx.doi.org/10.1021/cr00094a002]

Das, P, Mukherjee, S & Sen, R (2009) Biosurfactant of marine origin exhibiting heavy metal remediation properties. *Bioresour Technol,* 100, 4887-90.
[http://dx.doi.org/10.1016/j.biortech.2009.05.028] [PMID: 19505818]

Dash, HR & Das, S (2014) Bioremediation potential of mercury by *Bacillus* species isolated from marine environment and wastes of steel industry. *Bioremediat J,* 18, 204-12.
[http://dx.doi.org/10.1080/10889868.2014.899555]

Dash, HR, Mangwani, N, Chakraborty, J, Kumari, S & Das, S (2013) Marine bacteria: potential candidates for enhanced bioremediation. *Appl Microbiol Biotechnol,* 97, 561-71.
[http://dx.doi.org/10.1007/s00253-012-4584-0] [PMID: 23212672]

De, J, Dash, HR & Das, S (2014) Mercury pollution and bioremediation—a case study on biosorption by a mercury-resistant marine bacterium. *Microbial biodegradation and bioremediation,* 137-66.

De, J, Ramaiah, N & Vardanyan, L (2008) Detoxification of toxic heavy metals by marine bacteria highly resistant to mercury. *Mar Biotechnol (NY),* 10, 471-7.
[http://dx.doi.org/10.1007/s10126-008-9083-z] [PMID: 18288535]

Dixit, R, Malaviya, D, Pandiyan, K, Singh, UB, Sahu, A, Shukla, R, Singh, BP, Rai, JP, Sharma, PK, Lade, H & Paul, D (2015) Bioremediation of heavy metals from soil and aquatic environment: an overview of principles and criteria of fundamental processes. *Sustainability,* 7, 2189-212.
[http://dx.doi.org/10.3390/su7022189]

Dudhagara, DR, Rajpara, RK, Bhatt, JK, Gosai, HB & Dave, BP (2016) Bioengineering for polycyclic aromatic hydrocarbon degradation by *Mycobacterium litorale*: statistical and artificial neural network (ANN) approach. *Chemom Intell Lab Syst,* 159, 155-63. b
[http://dx.doi.org/10.1016/j.chemolab.2016.10.018]

Dudhagara, DR, Rajpara, RK, Bhatt, JK, Gosai, HB, Sachaniya, BK & Dave, BP (2016) Distribution, sources and ecological risk assessment of PAHs in historically contaminated surface sediments at Bhavnagar coast, Gujarat, India. *Environ Pollut,* 213, 338-46. a
[http://dx.doi.org/10.1016/j.envpol.2016.02.030] [PMID: 26925756]

Dwivedi, S (2012) Bioremediation of heavy metal by algae: current and future perspective. *Journal of Advanced Laboratory Research in Biology,* 3, 195-9.

Eccles, H (1999) Treatment of metal-contaminated wastes: why select a biological process? *Trends Biotechnol,* 17, 462-5.
[http://dx.doi.org/10.1016/S0167-7799(99)01381-5] [PMID: 10557157]

El-Kassas, HY & El-Taher, EM (2009) Optimization of batch process parameters by response surface methodology for mycoremediation of chrome-VI by a chromium resistant strain of marine *Trichodermaviride. Am-Eurasian J Agric Environ Sci,* 5, 676-81.

Fletcher, T (2002) Neighborhood change at Love Canal: contamination, evacuation and resettlement. *Land Use Policy,* 19, 311-23.

[http://dx.doi.org/10.1016/S0264-8377(02)00045-5]

Gadd, GM (2004) Microbial influence on metal mobility and application for bioremediation. *Geoderma,* 122, 109-19.
[http://dx.doi.org/10.1016/j.geoderma.2004.01.002]

Gadd, GM (2010) Metals, minerals and microbes: geomicrobiology and bioremediation. *Microbiology,* 156, 609-43.
[http://dx.doi.org/10.1099/mic.0.037143-0] [PMID: 20019082]

Gauthier, PT, Norwood, WP, Prepas, EE & Pyle, GG (2015) Metal–polycyclic aromatic hydrocarbon mixture toxicity in *Hyalellaazteca.* 2. Metal accumulation and oxidative stress as interactive co-toxic mechanisms. *Environ Sci Technol,* 49, 11780-8.
[http://dx.doi.org/10.1021/acs.est.5b03233] [PMID: 26308184]

Gomathi, V, Saravanakumar, K & Kathiresan, K (2012) Biosorption of chromium by mangrove-derived *Aplanochytrium* sp. *Afr J Biotechnol,* 11, 16177-86.
[http://dx.doi.org/10.5897/AJB12.1529]

Gong, JL, Wang, B, Zeng, GM, Yang, CP, Niu, CG, Niu, QY, Zhou, WJ & Liang, Y (2009) Removal of cationic dyes from aqueous solution using magnetic multi-wall carbon nanotube nanocomposite as adsorbent. *J Hazard Mater,* 164, 1517-22.
[http://dx.doi.org/10.1016/j.jhazmat.2008.09.072] [PMID: 18977077]

González, F, Romera, E, Ballester, A, Blázquez, L, Muñoz, JÁ & García-Balboa, C (2011) Algal Biosorption and Biosorbents. *Microbial Biosorption of Metals,* In: Kotrba, P., Mackova, M., Macek, T., (Eds.), Springer Dordrech, Heidelberg, London, NewYork 159-78.
[http://dx.doi.org/10.1007/978-94-007-0443-5_7]

Gorria, M, Tekpli, X, Sergent, O, Huc, L, Gaboriau, F, Rissel, M, Chevanne, M, Dimanche-Boitrel, MT & Lagadic-Gossmann, D (2006) Membrane fluidity changes are associated with benzo[a]pyrene-induced apoptosis in F258 cells: protection by exogenous cholesterol. *Ann N Y Acad Sci,* 1090, 108-12.
[http://dx.doi.org/10.1196/annals.1378.011] [PMID: 17384252]

Gosai, HB, Sachaniya, BK, Dudhagara, D, Panseriya, HZ & Dave, BP (2018) Bioengineering for multiple PAHs degradation using process centric and data centric approaches. (In communication).

Gosai, HB, Sachaniya, BK, Dudhagara, DR, Rajpara, RK & Dave, BP (2017) Concentrations, input prediction and probabilistic biological risk assessment of polycyclic aromatic hydrocarbons (PAHs) along Gujarat coastline. *Environ Geochem Health,* 1-13.
[PMID: 28801833]

Gosai, HB, Sachaniya, BK, Panseriya, HZ & Dave, BP (2018b) Functional and Phylogenetic Diversity Assessment of Microbial Communities at Gulf of Kachchh, India: An Ecological Footprint. (In communication).

Guerra, R, Pasteris, A & Ponti, M (2009) Impacts of maintenance channel dredging in a northern Adriatic coastal lagoon. I: Effects on sediment properties, contamination and toxicity. *Estuar Coast Shelf Sci,* 85, 134-42.
[http://dx.doi.org/10.1016/j.ecss.2009.05.021]

Guo, H, Luo, S, Chen, L, Xiao, X, Xi, Q, Wei, W, Zeng, G, Liu, C, Wan, Y, Chen, J & He, Y (2010) Bioremediation of heavy metals by growing hyperaccumulaor endophytic bacterium *Bacillus* sp. L14. *Bioresour Technol,* 101, 8599-605.
[http://dx.doi.org/10.1016/j.biortech.2010.06.085] [PMID: 20637605]

Guo, PY, Liu, Y, Wen, X & Chen, SF (2015) Effects of algicide on the growth of Microcystisflos-aquae and adsorption capacity to heavy metals. *Int J Environ Sci Technol,* 12, 2339-48.
[http://dx.doi.org/10.1007/s13762-014-0633-9]

Gupta, A, Joia, J, Sood, A, Sood, R, Sidhu, C & Kaur, G (2016) Microbes as potential tool for remediation of heavy metals: A review. *J Microb Biochem Technol,* 8, 364-72.

[http://dx.doi.org/10.4172/1948-5948.1000310]

Handley, KM, Héry, M & Lloyd, JR (2009) Redox cycling of arsenic by the hydrothermal marine bacterium *Marinobacter santoriniensis. Environ Microbiol,* 11, 1601-11.
[http://dx.doi.org/10.1111/j.1462-2920.2009.01890.x] [PMID: 19226300]

Hashim, A & Hajjaj, M (2005) Impact of desalination plants fluid effluents on the integrity of seawater, with the Arabian Gulf in perspective. *Desalination,* 182, 373-93.
[http://dx.doi.org/10.1016/j.desal.2005.04.020]

Hawkes, SJ (1997) What Is a" Heavy Metal"? *J Chem Educ,* 74, 1374.
[http://dx.doi.org/10.1021/ed074p1374]

Hedge, LH, Knott, NA & Johnston, EL (2009) Dredging related metal bioaccumulation in oysters. *Mar Pollut Bull,* 58, 832-40.
[http://dx.doi.org/10.1016/j.marpolbul.2009.01.020] [PMID: 19261303]

Chen, H & Pan, SS (2005) Bioremediation potential of *spirulina*: toxicity and biosorption studies of lead. *J Zhejiang Univ Sci B,* 6, 171-4.
[http://dx.doi.org/10.1631/jzus.2005.B0171] [PMID: 15682500]

Hyde, KD, Gareth, JEB, Leano, E, Pointing, SB, Poonyth, AD & Vrijmoed, LLP (1998) Role of fungi in marine ecosystems. *Biodivers Conserv,* 7, 1147-61.
[http://dx.doi.org/10.1023/A:1008823515157]

(2012) Arsenic, metals, fibres, and dusts. *IARC Monogr Eval Carcinog Risks Hum,* 100, 11-465.
[PMID: 23189751]

Imandi, SB, Chinthala, R, Saka, S, Vechalapu, RR & Nalla, KK (2014) Optimization of chromium biosorption in aqueous solution by marine yeast biomass of *Yarrowialipolytica* using Doehlert experimental design. *Afr J Biotechnol,* 13.

Iyer, A, Mody, K & Jha, B (2004) Accumulation of hexavalent chromium by an exopolysaccharide producing marine *Enterobacter cloaceae. Mar Pollut Bull,* 49, 974-7.
[http://dx.doi.org/10.1016/j.marpolbul.2004.06.023] [PMID: 15556183]

Iyer, A, Mody, K & Jha, B (2005) Biosorption of heavy metals by a marine bacterium. *Mar Pollut Bull,* 50, 340-3.
[http://dx.doi.org/10.1016/j.marpolbul.2004.11.012] [PMID: 15757698]

Jankowska, K, Olańczuk-Neyman, K & Kulbat, E (2006) The Sensitivity of Bacteria to Heavy Metals in the Presence of Mineral Ship Motor Oil in Coastal Marine Sediments and Waters. *Pol J Environ Stud,* 15

Jensen, TE, Baxter, M, Rachlin, JW & Jani, V (1982) Uptake of heavy metals by *Plectonemaboryanum (Cyanophyceae)* into cellular components, especially polyphosphate bodies: an X-ray energy dispersive study. *Environ Pollut A,* 27, 119-27.
[http://dx.doi.org/10.1016/0143-1471(82)90104-0]

Karaca, A, Cetin, SC, Turgay, OC & Kizilkaya, R (2010) Effects of heavy metals on soil enzyme activities. *Soil heavy metals,* Springer, Berlin, Heidelberg 237-62.
[http://dx.doi.org/10.1007/978-3-642-02436-8_11]

Katz, SA & Salem, H (1993) The toxicology of chromium with respect to its chemical speciation: a review. *J Appl Toxicol,* 13, 217-24.
[http://dx.doi.org/10.1002/jat.2550130314] [PMID: 8326093]

Kaufaman, DB (1970) Acute potassium dichromate poisoning in man. *Am J Dis Child,* 119, 374-81.
[http://dx.doi.org/10.1001/archpedi.1970.02100050376021]

Keren, R, Lavy, A, Mayzel, B & Ilan, M (2015) Culturable associated-bacteria of the sponge *Theonella swinhoei* show tolerance to high arsenic concentrations. *Front Microbiol,* 6, 154.
[http://dx.doi.org/10.3389/fmicb.2015.00154] [PMID: 25762993]

Khambhaty, Y, Mody, K, Basha, S & Jha, B (2008) Hg (II) removal from aqueous solution by dead fungal

biomass of marine *Aspergillus niger*: Kinetic studies. *Sep Sci Technol,* 43, 1221-38.
[http://dx.doi.org/10.1080/01496390801888235]

Khambhaty, Y, Mody, K, Basha, S & Jha, B (2009) Biosorption of Cr (VI) onto marine *Aspergillus niger*: experimental studies and pseudo-second order kinetics. *World J Microbiol Biotechnol,* 25, 1413. a
[http://dx.doi.org/10.1007/s11274-009-0028-0]

Khambhaty, Y, Mody, K, Basha, S & Jha, B (2009) Kinetics, equilibrium and thermodynamic studies on biosorption of hexavalent chromium by dead fungal biomass of marine *Aspergillus niger. Chem Eng J,* 145, 489-95. b
[http://dx.doi.org/10.1016/j.cej.2008.05.002]

Khambholja, DB & Kalia, K (2016) Seasonal variation in arsenic concentration and its bioremediation potential of marine bacteria isolated from Alang-Sosiya ship-scrapping yard, Gujarat, India. *Defence Life Science Journal,* 1, 78-84.
[http://dx.doi.org/10.14429/dlsj.1.10088]

Khan, FI, Husain, T & Hejazi, R (2004) An overview and analysis of site remediation technologies. *J Environ Manage,* 71, 95-122.
[http://dx.doi.org/10.1016/j.jenvman.2004.02.003] [PMID: 15135946]

Kim, IH, Choi, JH, Joo, JO, Kim, YK, Choi, JW & Oh, BK (2015) Development of a microbe-zeolite carrier for the effective elimination of heavy metals from seawater. *J Microbiol Biotechnol,* 25, 1542-6.
[http://dx.doi.org/10.4014/jmb.1504.04067] [PMID: 26032363]

Kotaś, J & Stasicka, Z (2000) Chromium occurrence in the environment and methods of its speciation. *Environ Pollut,* 107, 263-83.
[http://dx.doi.org/10.1016/S0269-7491(99)00168-2] [PMID: 15092973]

Kuang, D, Zhang, W, Deng, Q, Zhang, X, Huang, K, Guan, L, Hu, D, Wu, T & Guo, H (2013) Dose-response relationships of polycyclic aromatic hydrocarbons exposure and oxidative damage to DNA and lipid in coke oven workers. *Environ Sci Technol,* 47, 7446-56.
[http://dx.doi.org/10.1021/es401639x] [PMID: 23745771]

Lenntech (2004) Water Treatment. Lenntech, Rotterdamseweg, Netherlands (Lenntech Water Treatment and Air Purification).

Les, A & Walker, RW (1984) Toxicity and binding of copper, zinc, and cadmium by the blue-green alga, Chroococcusparis. *Water Air Soil Pollut,* 23, 129-39.
[http://dx.doi.org/10.1007/BF00206971]

Li, F, Zhu, L, Wang, L & Zhan, Y (2015) Gene expression of an *arthrobacter* in surfactant-enhanced biodegradation of a hydrophobic organic compound. *Environ Sci Technol,* 49, 3698-704.
[http://dx.doi.org/10.1021/es504673j] [PMID: 25680000]

Liu, SH, Zeng, GM, Niu, QY, Liu, Y, Zhou, L, Jiang, LH, Tan, XF, Xu, P, Zhang, C & Cheng, M (2017) Bioremediation mechanisms of combined pollution of PAHs and heavy metals by bacteria and fungi: A mini review. *Bioresour Technol,* 224, 25-33.
[http://dx.doi.org/10.1016/j.biortech.2016.11.095] [PMID: 27916498]

Luna, JM, Rufino, RD & Sarubbo, LA (2016) Biosurfactant from *Candida sphaerica* UCP0995 exhibiting heavy metal remediation properties. *Process Saf Environ Prot,* 102, 558-66.
[http://dx.doi.org/10.1016/j.psep.2016.05.010]

Maher, W & Butler, E (1988) Arsenic in the marine environment. ApplOrganomet. *Chemosphere,* 2, 191-4.

Marine Pollution and Controls-Need for a Comprehensive Environmental Impact Assessment Laws.
http://www.vmslaw.edu.in/marine-pollution-and-controls-need-for-a-comprehensive-environmental-impact-a ssessment-laws/

Meenambigai, P, Vijayaraghavan, R, Gowri, RS, Rajarajeswari, P & Prabhavathi, P (2016) Biodegradation of heavy metals-a Review. *Int J Curr Microbiol Appl Sci,* 5, 375-83.

[http://dx.doi.org/10.20546/ijcmas.2016.504.045]

MEMAC (Marine Emergency Mutual Aid Centre) (2003) *Oil Spill Incidents in ROPME Sea Area (1965-2002)*.MEMAC, Bahrain.

Mendoza, RAJS, Estanislao, KB, Aninipot, JFP, Dahonog, RA, De Guzman, JA, Torres, JMO & dela Cruz, TEE (2010) Biosorption of mercurBiosorption of mercury by the marine fungus y by the marine fungus *Dendryphiellasalinayphiellasalina*. *Acta Manila Ser A*, 58, 25-9.

Michalec, FG, Holzner, M, Souissi, A, Stancheva, S, Barras, A, Boukherroub, R & Souissi, S (2016) Lipid nanocapsules for behavioural testing in aquatic toxicology: Time-response of *Eurytemora affinis* to environmental concentrations of PAHs and PCB. *Aquat Toxicol,* 170, 310-22.
[http://dx.doi.org/10.1016/j.aquatox.2015.08.010] [PMID: 26362585]

Millward, RN, Carman, KR, Fleeger, JW, Gambrell, RP, Powell, RT & Rouse, MAM (2001) Linking ecological impact to metal concentrations and speciation: a microcosm experiment using a salt marsh meiofaunal community. *Environ Toxicol Chem,* 20, 2029-37.
[http://dx.doi.org/10.1002/etc.5620200923] [PMID: 11521831]

Monteiro, CM, Castro, PM & Malcata, FX (2012) Metal uptake by microalgae: underlying mechanisms and practical applications. *Biotechnol Prog,* 28, 299-311.
[http://dx.doi.org/10.1002/btpr.1504] [PMID: 22228490]

Mulik, AR & Bhadekar, RK (2017) Heavy metal removal by bacterial isolates from the antarctic oceanic region. *Int J Pharma Bio Sci,* 8, 535-43.
[http://dx.doi.org/10.22376/ijpbs.2017.8.3.b535-543]

Naser, H (2012) Metal concentrations in marine sediments influenced by anthropogenic activities in Bahrain, Arabian Gulf.*Metal contaminations: sources, detection and environmental impacts* NOVA Science Publishers, Inc., New York 157-75.

Naser, HA (2010) Testing taxonomic resolution levels for detecting environmental impacts using macrobenthic assemblages in tropical waters. *Environ Monit Assess,* 170, 435-44.
[http://dx.doi.org/10.1007/s10661-009-1244-7] [PMID: 19904622]

Naser, HA (2013) Assessment and management of heavy metal pollution in the marine environment of the Arabian Gulf: a review. *Mar Pollut Bull,* 72, 6-13.
[http://dx.doi.org/10.1016/j.marpolbul.2013.04.030] [PMID: 23711845]

Newel, SY & Bárlocher, F (1993) Removal of fungal and total organic matter from decaying cordgrasseaves by shredder snails. *J Exp Mar Biol Ecol,* 171, 39-49.
[http://dx.doi.org/10.1016/0022-0981(93)90138-E]

Oliveira, A & Pampulha, ME (2006) Effects of long-term heavy metal contamination on soil microbial characteristics. *J Biosci Bioeng,* 102, 157-61.
[http://dx.doi.org/10.1263/jbb.102.157] [PMID: 17046527]

Oyetibo, GO, Ishola, ST, Ikeda-Ohtsubo, W, Miyauchi, K, Ilori, MO & Endo, G (2015) Mercury bioremoval by Yarrowia strains isolated from sediments of mercury-polluted estuarine water. *Appl Microbiol Biotechnol,* 99, 3651-7.
[http://dx.doi.org/10.1007/s00253-014-6279-1] [PMID: 25520168]

Pachiadaki, MG, Rédou, V, Beaudoin, DJ, Burgaud, G & Edgcomb, VP (2016) Fungal and prokaryotic activities in the marine subsurface biosphere at Peru margin and Canterbury basin inferred from RNA-Based analyses and microscopy. *Front Microbiol,* 7, 846.
[http://dx.doi.org/10.3389/fmicb.2016.00846] [PMID: 27375571]

Park, D, Yun, YS & Park, JM (2005) Studies on hexavalent chromium biosorption by chemically-treated biomass of *Ecklonia* sp. *Chemosphere,* 60, 1356-64.
[http://dx.doi.org/10.1016/j.chemosphere.2005.02.020] [PMID: 16054904]

Paul, D (2017) Research on heavy metal pollution of river Ganga: a review. *Annals of Agrarian Science,* 15, 278-86.

[http://dx.doi.org/10.1016/j.aasci.2017.04.001]

Ramanathan, S, Shi, W, Rosen, BP & Daunert, S (1997) Sensing antimonite and arsenite at the subattomole level with genetically engineered bioluminescent bacteria. *Anal Chem,* 69, 3380-4.
[http://dx.doi.org/10.1021/ac970111p] [PMID: 9271073]

Rao, A, Bankar, A, Kumar, AR, Gosavi, S & Zinjarde, S (2013) Removal of hexavalent chromium ions by *Yarrowia lipolytica* cells modified with phyto-inspired Fe0/Fe3O4 nanoparticles. *J Contam Hydrol,* 146, 63-73.
[http://dx.doi.org/10.1016/j.jconhyd.2012.12.008] [PMID: 23422514]

Ravikumar, S, Williams, GP, Shanthy, S, Gracelin, NAA, Babu, S & Parimala, PS (2007) Effect of heavy metals (Hg and Zn) on the growth and phosphate solubilising activity in halophilic phosphobacteria isolated from Manakudi mangrove. *J Environ Biol,* 28, 109-14.
[PMID: 17717995]

Sadiq, M (2002) Metal contamination in sediments from a desalination plant effluent outfall area. *Sci Total Environ,* 287, 37-44.
[http://dx.doi.org/10.1016/S0048-9697(01)00994-9] [PMID: 11883759]

Santona, L, Castaldi, P & Melis, P (2006) Evaluation of the interaction mechanisms between red muds and heavy metals. *J Hazard Mater,* 136, 324-9.
[http://dx.doi.org/10.1016/j.jhazmat.2005.12.022] [PMID: 16426746]

Shatti, JA & Abdullah, TH (1999) Marine pollution due to wastewater discharge in Kuwait. *Water Sci Technol,* 40, 33-9.
[http://dx.doi.org/10.2166/wst.1999.0322]

Shen, G, Lu, Y & Hong, J (2006) Combined effect of heavy metals and polycyclic aromatic hydrocarbons on urease activity in soil. *Ecotoxicol Environ Saf,* 63, 474-80.
[http://dx.doi.org/10.1016/j.ecoenv.2005.01.009] [PMID: 16406598]

Shen, G, Lu, Y, Zhou, Q & Hong, J (2005) Interaction of polycyclic aromatic hydrocarbons and heavy metals on soil enzyme. *Chemosphere,* 61, 1175-82.
[http://dx.doi.org/10.1016/j.chemosphere.2005.02.074] [PMID: 16263387]

Shirdam, R, Khanafari, A & Tabatabaee, A (2006) Cadmium, nickel and vanadium accumulation by three strains of marine bacteria. *Iranian J Biotechnol,* 4, 180-7.

Singh, KP, Mohan, D, Sinha, S & Dalwani, R (2004) Impact assessment of treated/untreated wastewater toxicants discharged by sewage treatment plants on health, agricultural, and environmental quality in the wastewater disposal area. *Chemosphere,* 55, 227-55.
[http://dx.doi.org/10.1016/j.chemosphere.2003.10.050] [PMID: 14761695]

Singh, P, Raghukumar, C, Parvatkar, RR & Mascarenhas-Pereira, MBL (2013) Heavy metal tolerance in the *psychrotolerant Cryptococcus* sp. isolated from deep-sea sediments of the Central Indian Basin. *Yeast,* 30, 93-101.
[http://dx.doi.org/10.1002/yea.2943] [PMID: 23456725]

Smith, SD & Rule, MJ (2001) The effects of dredge-spoil dumping on a shallow water soft-sediment community in the Solitary Islands Marine Park, NSW, Australia. *Mar Pollut Bull,* 42, 1040-8.
[http://dx.doi.org/10.1016/S0025-326X(01)00059-5] [PMID: 11763214]

Sode, K, Yamamoto, Y & Hatano, N (1998) Construction of a marine cyanobacterial strain with increased heavy metal ion tolerance by introducing exogenic metallothionein gene. *J Mar Biotechnol,* 6, 174-7.
[PMID: 9701640]

Suresh Kumar, K, Dahms, HU, Won, EJ, Lee, JS & Shin, KH (2015) Microalgae - A promising tool for heavy metal remediation. *Ecotoxicol Environ Saf,* 113, 329-52.
[http://dx.doi.org/10.1016/j.ecoenv.2014.12.019] [PMID: 25528489]

Taboski, MA, Rand, TG & Piórko, A (2005) Lead and cadmium uptake in the marine fungi *Corollospora lacera* and *Monodictys pelagica. FEMS Microbiol Ecol,* 53, 445-53.

[http://dx.doi.org/10.1016/j.femsec.2005.02.009] [PMID: 16329962]

Takeuchi, M, Kawahata, H, Gupta, LP, Kita, N, Morishita, Y, Ono, Y & Komai, T (2007) Arsenic resistance and removal by marine and non-marine bacteria. *J Biotechnol,* 127, 434-42.
[http://dx.doi.org/10.1016/j.jbiotec.2006.07.018] [PMID: 16934903]

Toes, ACM, Finke, N, Kuenen, JG & Muyzer, G (2008) Effects of deposition of heavy-metal-polluted harbor mud on microbial diversity and metal resistance in sandy marine sediments. *Arch Environ Contam Toxicol,* 55, 372-85.
[http://dx.doi.org/10.1007/s00244-008-9135-4] [PMID: 18273665]

UNCLOS (1982) www.un.org/depts/los/convention_agreements/texts/unclos/part1.htm

Vaidya, SY, Vala, AK & Dube, HC (2000) Cellulase production by marine bacteria. *Indian J Geo-Mar Sci,* 29.

Vala, AK & Patel, RJ (2011) Biosorption of trivalent arsenic by facultative marine *Aspergillusniger. Bioremediation: Biotechnology, engineering and environmental management,* In: Mason, A., (Ed.),   Nova Science Publishers 459-64.

Vala, AK, Vaidya, SY & Dube, HC (2000) Siderophore production by facultative marine fungi. *Indian J Geo-Mar Sci,* 29, 339-40.

Vala, AK & Dave, BP (2015) Explorations on Marine-derived Fungi for L-Asparaginase–Enzyme with Anticancer Potentials. *Curr Chem Biol,* 9, 66-9.
[http://dx.doi.org/10.2174/2212796809666150817195019]

Vala, AK & Dave, BP (2017) Marine-derived fungi: Prospective candidates for bioremediation. *Mycoremediation & Environmental Sustainability,* Springer, Cham 17-37.
[http://dx.doi.org/10.1007/978-3-319-68957-9_2]

Vala, AK & Sutariya, V (2012) Trivalent arsenic tolerance and accumulation in two facultative marine fungi. *Jundishapur J Microbiol,* 5, 542-5.
[http://dx.doi.org/10.5812/jjm.3383]

Vala, AK & Upadhyay, RV (2008) On the tolerance and accumulation of arsenic by facultative marine *Aspergillus*sp. *Res J Biotechnol,* 366-8.

Vala, AK (2009) *Aspergillus niger* and heavy metal removal: A perception. *Res J Biotechnol,* 4, 75-9.

Vala, AK (2010) Tolerance and removal of arsenic by a facultative marine fungus *Aspergillus candidus. Bioresour Technol,* 101, 2565-7.
[http://dx.doi.org/10.1016/j.biortech.2009.11.084] [PMID: 20022490]

Vala, AK (2018) On the Extreme Tolerance and Removal of Arsenic by a Facultative Marine Fungus Aspergillus sydowii. *Metallic Contamination and Its Toxicity,* In: Gautam, A., Pathak, C., (Eds.), Daya Publishing House 37-44.

Vala, AK, Anand, N, Bhatt, PN & Joshi, HV (2004) Tolerance and accumulation of hexavalent chromium by two seaweed associated aspergilli. *Mar Pollut Bull,* 48, 983-5.
[http://dx.doi.org/10.1016/j.marpolbul.2004.02.025] [PMID: 15111047]

Vala, AK, Chudasama, B & Patel, RJ (2012) Green synthesis of silver nanoparticles using marine-derived fungus *Aspergillus niger. Micro & Nano Lett,* 7, 859-62.
[http://dx.doi.org/10.1049/mnl.2012.0403]

Vala, AK, Davariya, V & Upadhyay, RV (2010) An investigation on tolerance and accumulation of a facultative marine fungus *Aspergillus flavus* to pentavalent arsenic. *J Ocean Univ China,* 9, 65-7.
[http://dx.doi.org/10.1007/s11802-010-0065-1]

Vala, AK, Sutariya, V & Upadhyay, RV (2011) Investigations on trivalent arsenic tolerance and removal potential of a facultative marine *Aspergillus niger. Environ Prog Sustain Energy,* 30, 586-8.
[http://dx.doi.org/10.1002/ep.10511]

Vidal, FV & Vidal, VMV (1980) Arsenic metabolism in marine bacteria and yeast. *Mar Biol,* 60, 1-7.
[http://dx.doi.org/10.1007/BF00395600]

Wang, W (1987) Factors affecting metal toxicity to (and accumulation by) aquatic organisms—overview.
Environ Int, 13, 437-57.
[http://dx.doi.org/10.1016/0160-4120(87)90006-7]

Weissenfels, WD, Klewer, HJ & Langhoff, J (1992) Adsorption of polycyclic aromatic hydrocarbons (PAHs)
by soil particles: influence on biodegradability and biotoxicity. *Appl Microbiol Biotechnol,* 36, 689-96.
[http://dx.doi.org/10.1007/BF00183251] [PMID: 1368071]

Zhang, W, Zhuang, L, Yuan, Y, Tong, L & Tsang, DC (2011) Enhancement of phenanthrene adsorption on a
clayey soil and clay minerals by coexisting lead or cadmium. *Chemosphere,* 83, 302-10.
[http://dx.doi.org/10.1016/j.chemosphere.2010.12.056] [PMID: 21232783]

Zhu, D, Herbert, BE, Schlautman, MA & Carraway, ER (2004) Characterization of cation-π interactions in
aqueous solution using deuterium nuclear magnetic resonance spectroscopy. *J Environ Qual,* 33, 276-84.
[http://dx.doi.org/10.2134/jeq2004.2760] [PMID: 14964382]

Zouboulis, AI, Loukidou, MX & Matis, KA (2004) Biosorption of toxic metals from aqueous solutions by
bacteria strains isolated from metal-polluted soils. *Process Biochem,* 39, 909-16.
[http://dx.doi.org/10.1016/S0032-9592(03)00200-0]

CHAPTER 17

Polycyclic Aromatic Hydrocarbons (PAHs): Occurrence and Bioremediation in the Marine Environment

Bhumi K. Sachaniya[1], Haren B. Gosai[1,2], Haresh Z. Panseriya[1,2], Anjana K. Vala[1,*] and Bharti P. Dave[1,2,*]

[1] *Department of Life Sciences, Maharaja Krishnakumarsinhji Bhavnagar University, Bhavanagar - 364001, India*

[2] *Department of Biosciences, School of Sciences, Indrashil University, Rajpur - Kadi - 382740, India*

Abstract: Contamination by various hazardous compounds released due to sea-related activities has received great concern about the pollution of the marine ecosystem. Nowadays, polycyclic aromatic hydrocarbons (PAHs) are immerging as critical pollutant with context to the marine environment due to some distinctive properties, which makes them persistent organic pollutants (POPs) posing threat to the environment. PAHs make their way in marine environment through various natural and anthropogenic sources. Marine microorganisms have reported to be leading candidates for PAHs degradation. Recent advancements in genomics, proteomics, and metabolomics technologies have gathered significant increment in the knowledge of ecology, physiology and regulatory mechanisms of microbial communities involved in PAHs remediation. Morden technologies will be a vital approach to reveal the mechanisms involved in the bioremediation of pollutants and will offer more insights as yet more uncultivable microbial diversity attached with pollutant degradation.

Keywords: Bioremediation, Marine Environment, Polycyclic Aromatic Hydrocarbons.

INTRODUCTION

Over the preceding 60-50 years, marine environment has changed more hastily than in any other time period of the history by the human to comply with the vigorously growing demand for food, energy, fuel, and transportation. These changes have contributed to economic development and well-being but at the

* **Corresponding authors Anjana K. Vala & Bharti P. Dave:** Department of Life Sciences, Maharaja Krishnakumarsinhji Bhavnagar University, Bhavnagar-364 001, India; Tel/Fax: +91-2782519824; E-mails: anjana_vala@yahoo.co.in and bpd8256@gmail.com

De-Sheng Pei & Muhammad Junaid (Eds.)
All rights reserved-© 2019 Bentham Science Publishers

same time have made some unalterable loss to ecosystem diversity. Some common sources contributing in marine pollution are uncontrolled spew of untreated industrial wastes, various sea-based petroleum-related activities, agri-cultural and municipality run-offs, ship-breaking/recycling activities and spills/ accidents during transportation (Dudhagara *et al.*, 2016a, Gosai *et al.*, 2018a). Majority of these sources contributes organic pollutants in the marine en-vironment, amongst which polycyclic aromatic hydrocarbons (PAHs) these days are immerging as a critical pollutant of marine environment damaging the vital division of marine as well as terrestrial biota. Over the past decade, biodegradation has definitely immerged as the major acceptance for remediating PAHs contaminated environment (Dave *et al.*, 2014; Bhatt *et al.*, 2014). Potential marine organisms from the contaminated sites have proven to be the leading candidates for bioremediation of PAHs contaminated marine sites (Gosai *et al.*, 2018a, b; Sachaniya *et al.*, 2018).

Polycyclic or polynuclear aromatic hydrocarbons constitute a group of heterogeneous organic compounds comprising two or more fused benzene rings as their nuclei, which are arranged in linear, angular or cluster spatial configurations. As their name suggests, PAHs generally contain hydrogen and carbon as their atomic composition, but sometimes these atoms are substituted with oxygen, nitrogen, sulphur or sometimes a whole chemical reactive group in the benzene ring to form heterocyclic PAHs. Around 660 parent PAHs compounds solely consisting of conjoined benzene rings have been listed in the literature (Sander and Wise, 1997). These compounds can also be found occurring naturally not only on earth but also in the space, which are meant to be the indicator of possibilities of life throughout the universe.

These compounds hence have received great economic as well as the scientific concern due to their various deleterious structural and physicochemical properties. PAHs generally can be divided into two major groups. Those, having three or less than three aromatic rings are considered to be low molecular weight (LMW) PAHs and those having four or more than four aromatic rings are considered to be high molecular weight (HMW) PAHs. Diversity in the spatial configuration and size of these compounds result in a considerable discrepancy in their physi-cochemical properties. Different physical, solvation and molecular properties of some selected known PAHs are listed in Table **1**.

Generally, PAHs are lipophilic or hydrophobic in nature. Some of the LMW PAHs are partly soluble in the aqueous solvent. PAHs are highly photosensitive *i.e.*, they get decomposed when exposed to UV light as well as visible light. They are semi-volatile or have low volatility (Mackay and Callcott, 1998). Hydrophobicity or lipophilicity increases with the increase in the molecular weight as reflected by the increase in a number of aromatic rings (Ferreira, 2001). As a consequence of high hydrophobicity, HMW PAHs have higher tendency to

Table 1. Physical, solvation and molecular properties of 16 US EPA priority PAHs. Adapted and modified from [a]Larsson, 2013 and Ghosal et al., 2016.

PAHs	Structure	Molecular Formula	Molecular Weight	B. Pt. (°C)	M.Pt (°C)	V.P. (mmHg at 25°C) [a]	log Kow Value	IARC [b]	EPA [c]
Naphthalene		$C_{10}H_8$	128.17	218	80.2	8.5×10^{-2}	3.36	2B	C
Acenaphthene		$C_{12}H_{10}$	154.21	279	93.4	2.5×10^{-3}	3.98	3	D
Acenaphthylene		$C_{12}H_8$	152.20	280	91.8	6.68×10^{-3}	4.07	n.c.	D
Anthracene		$C_{14}H_{10}$	178.23	342	216.4	6.53×10^{-6}	4.45	3	D
Phenanthrene		$C_{14}H_{10}$	178.23	340	100.5	1.2×10^{-4}	4.45	3	D
Fluorene		$C_{13}H_{10}$	166.22	295	116.7	6.0×10^{-4}	4.18	3	D
Fluoranthene		$C_{16}H_{10}$	202.26	375	108.8	9.22×10^{-6}	4.90	3	D
Benzo[a]anthracene		$C_{18}H_{12}$	228.29	438	158	4.11×10^{-3}	5.61	2B	B2
Chrysene		$C_{18}H_{12}$	228.29	448	254	6.23×10^{-9}	5.16	2B	B2
Pyrene		$C_{16}H_{10}$	202.26	150.4	393	4.5×10^{-6}	4.88	3	D
Benzo(a)pyrene		$C_{20}H_{12}$	252.32	495	179	5.49×10^{-9}	6.06	1	B2

(Table 1) cont.....

PAHs	Structure	Molecular Formula	Molecular Weight	B. Pt. (°C)	M.Pt (°C)	V.P. (mmHg at 25°C) a	log Kow Value a	IARC b	EPA c
Benzo(b)fluoranthene		$C_{20}H_{12}$	252.32	481	168.3	5.0×10^{-7}	6.04	2B	B2
Dibenzo(a,h)anthracene		$C_{22}H_{14}$	278.35	524	262	9.55×10^{-10}	6.84	2A	B2
Benzo(g,h,i)perylene		$C_{22}H_{12}$	276.34	500	277	1.0×10^{-10}	6.50	3	D
Indeno(1,2,3-cd)pyrene		$C_{22}H_{12}$	276.34	536	161-3	1.25×10^{-3}	6.58	2B	B2

[[b]International Agency for Research on Cancer Classification Monographs Volume-92 (1=carcinogenic to humans; 2A=probably carcinogenic to humans; 2B=possibly carcinogenic to humans; 3=not classifiable as carcinogenic to humans; n. c.=not classified), [c]EPA carcinogenic classification: A=human carcinogenic; B1 and B2=probable human carcinogenic; C=possible human carcinogenic; D=not Classifiable as to human carcinogenicity; E=evidence of non-carcinogenicity for humans]

get sorbed to fractions of particulate organic matter in sediments and soils, rather than vaporizing or dissolving in the water like LMW PAHs (Bertilsson and Widenfalk, 2002). They are thus less bioavailable than LMW PAHs. As being semi-volatile in nature, LMW PAHs may be transported far from their original source ultimately reaching to various parts of the environment (Agarwal, 2009; Harris *et al.*, 2012; Morillo *et al.*, 2008; Stark *et al.*, 2003).

These different physical, salvation and molecular properties are the major factors driving the fate of PAHs in the environment. These properties also make PAHs persistent organic pollutants (POPs) having a higher impact on the environment. They are thus more prone to bioaccumulate posing threat to environment and living biota.

Source of PAHs in Marine Environment

PAHs make their way in marine environment through diverse sources. According to the origin, PAHs can be natural or anthropogenic (Bertilsson and Widenfalk 2002; Morillo *et al.*, 2008). Natural origins embrace volcanoes, cold seeps, meteors, oil seeps, and forest fire, *etc.* Some PAHs like Perylene is formed by biochemical transformation of natural organic matter. One of the most abundant natural sources of PAHs is lignin.

PAHs come in contact with environment anthropogenically by thermal/chemical transformation or incomplete combustion of organic materials like fossil fuels, petroleum products, biofuels, and *etc.* Accidental oil spills during transportation of oil and coal products through aquatic passages. In present days, the major source of PAHs in the marine environment is the utilization of petroleum products and its transportation through marine routes. Especially, concentration of PAHs in the coastal area increases, which are near the source of emission like urban and industrial zones that often consists of multiple point sources responsible for the release of PAHs in the marine environment. As a result, anthropogenic sources of PAHs have gone beyond the natural sources posing threat to the aquatic environment.

Sources of PAHs can be categorized into three main classes as pyrogenic, petrogenic and biogenic. Each source has its own characteristic PAHs Profile (Bertilsson and Widenfalk, 2002).

Petrogenic substances can be defined as substances, which are instigated from petroleum products like crude oil, lubricants, tar, paints *etc.* These PAHs make their way in the marine environment by accidental oil spills during transportation of oil, leakage from oil tanker, urban and municipal run-off, and shipbreaking/recycling at coastal areas. Petrogenic sources have more proportion of LMW PAHs and their alkylated derivatives and degrade at much faster rates than pyrogenic ones. Petrogenic PAHs are more bioavailable than pyrogenic ones (Baumard *et al.*, 1998) thus, are degraded much faster compared to pyrogenic PAHs (Zakaria *et al.*, 2002).

Pyrogenic materials are organic substances produced from oxygen-depleted, high-temperature combustion or pyrolysis of fossil fuels and biomass. Pyrogenic PAHs make their way in marine niches through forest fires, incineration of wastes containing PAHs, combustion of fossil fuels *etc.* Pyrogenic sources have a complex mixture of parental HMW PAHs (Zeng and Vista, 1997; Wang *et al.*, 1999; Jiang *et al.*, 2009).

PAHs produced by algae, plants, microorganisms and biological product formed by slow transformation of organic matters in marine sediments make up the biogenic sources of PAHs (Venkatesan, 1988). Perylene is the major biogenic PAH produced from organic matter, soil and marine, subtidal sediments (Guo *et al.*, 2007, Boll *et al.*, 2008). If perylene does not correlate with the total organic carbon, then it can be used as an indicator PAH of the biogenic source (Luo *et al.*, 2008).

Transportation and Fates of PAHs in Marine Environment

The marine environment comprises of the diverse ecosystem including estuaries, coastal zone, surface and deep oceans each having their own specific characteristics. Biological pump and microbial degradation, which play an important role in sequestration by numerous biological processes also control the global PAHs fluxes in the marine environment (Turner *et al.*, 2015). Just like the particulate and dissolved organic matter PAHs get integrated into the microbial loop and are subjected to mineralization (Almeda *et al.*, 2013; Binark *et al.*, 2000).

Estuaries, marinas, and harbors, the most urbanized coastal regions are said to be the hot spots for multi-contamination including PAHs, which leads to 'coastal pollution and contamination syndrome' (Newton *et al.*, 2012). Thus, this part of the marine environment is considered as 'tipping element' in earth system by Lenton *et al.*, 2008. However, the environmental risk gets reduced by dilution of pollutants in the gigantic breadth of the ocean. The concentration of PAHs decreases with increasing time and distance in the water column and sediments (Adhikari *et al.*, 2016).

Various physicochemical properties along with environmental variables, such as wind speed, wave currents, tides, and seawater surface tension, affect the fluxes of PAHs in open oceans (Gonzales-Gaya *et al.*, 2014). The affinity of PAHs to soot carbon also facilitates their deposition in seawater. Air-water exchange is enhanced due to higher primary productivity resulting in vertical fluxes of PAHs (Dachs *et al.*, 2000). From oceanic water bodies, atmospheric pollutants are transferred to depth in the water column and finally get trapped in sediments.

Accidental discharge of oil and gas from deep sea and natural oil seeps releases oil and gas hydrocarbons, which gets partially dissolved in the water column (Reddy *et al.*, 2012). Accidental oil spills during oil transportation through sea surface leads to the formation of oils slicks at the surface, which ultimately undergoes various natural processes including spreading, evaporation, dispersion, emulsification, dissolution, photo-oxidation, adsorption, sinking/sedimentation, and biodegradation. All of these contribute to the main process called weathering. Spreading, evaporation, dispersion, emulsification, and dissolution takes place during the early stage of contaminant entry while photo-oxidation, sedimentation, and biodegradation are the late stage processes, which are long-term and which decides the ultimate fate of PAHs in the marine environment. Fig. (**1**) illustrates the fates of PAHs in the marine environment in form of different weathering processes.

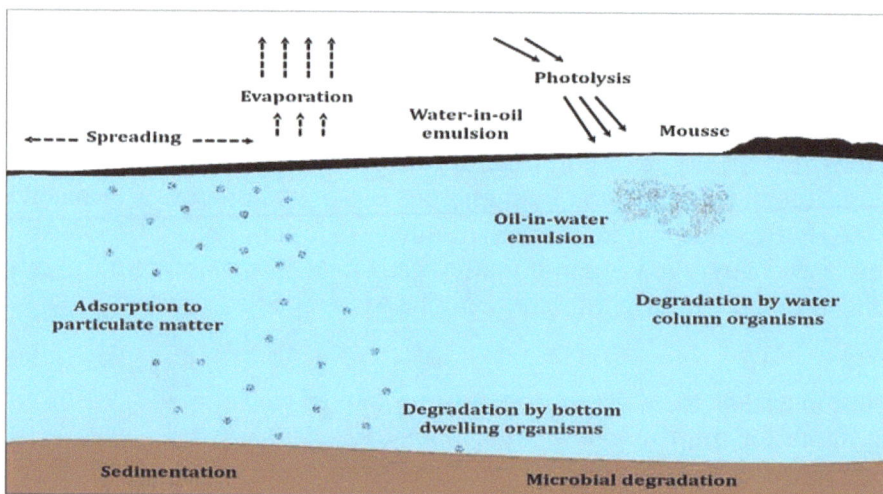

Fig. (1). Fates of PAHs in the marine environment.

Oil content gets spread onto the sea water surface from the point of entry with the help of currents and waves. Water-soluble fractions get dissolved in seawater. The low molecular weight fractions of crude oil undergo evaporation in air. Crude oil gets mixed-up into the water column by the action of waves, which disperses oil in form of small droplets. Sometimes the polar components of crude oil behave as emulsifier and forms water-in-oil emulsion, which acquires viscous consistency and sometimes called as 'chocolate mousse' by its appearance. Tar formation can also take place due to remaining heavy fractions. The portion of spilled oil on the surface comes in contact with sunlight (UV) is subjected to photo-oxidation or photochemical modification. The oxy-PAHs formed due to biological, chemical and photo-oxidation may increase the solubility of PAHs but can also lead to the formation of 'dead-end' products that are resistant to degradation and may get sunk in sediments posing threat to marine and human lives (Lundstedt *et al.*, 2007). Remobilisation takes place when sediments get disturbed by dredging, currents, tides and waves, earthquakes or bioturbation resulting in the return of colloidal or soluble forms of PAHs in the water column.

Apart from these physicochemical processes, microbial loop and microbial carbon pump are also involved in the fluxes of PAHS in the marine environment (Jiao *et al.*, 2010; Turner, 2015). PAHs are metabolized by marine organisms but some of the metabolically active products are proved to be toxic to the biota. Phytoplankton are an important lead through autotrophy and photosynthesis in PAHs uptake by adsorption and accumulation at the surface and sub-surface parts and their transport through the water column. Since PAHs are hydrophobic compounds and tend to get sorbed on the bottom, the benthic organisms are in continuous exposure of PAHs in the contaminated area. Zooplanktons through heterotrophy and respiration are involved in the cycling of PAHs at the bottom part of oceans. Microbial degradation can be observed throughout the water column and sediments but in sediments, microbes are the prime candidates playing the major role in PAHs degradation. Both aerobic (Cerniglia *et al.*, 1992; Haritash and Kaushik, 2009; Lu *et al.*, 2011) and anaerobic (Widdel and Rabus, 2001; Meckenstock and Mouttaki, 2011; Rabus *et al.*, 2016) microbial degradation processes of PAHs have been extensively studied in last few decades.

Effect of PAHs on Marine Biota

Different fates of PAHs in marine environment produces transformed products of PAHs. Some biological traits like habitat/depth of the marine organisms exposed to PAHs contamination together with the fate of PAHs in marine environment decide their effects on marine biota. PAHs and their transformation products are proven to be carcinogenic, mutagenic and teratogenic to living organisms. As previously described, when any pollutant-containing PAHs comes in contact with the marine environment, they either get sunk in sediments or remain on the seawater surface. If PAHs contamination remains on the seawater surface then wind and water currents transfer the contamination to coastal zones where it affects coastal biota. However if the PAHs are dispersed through the water column and consequently to sediments, it will pose deleterious effects on all marine biota.

PAHs are lipophilic and tend to accumulate in tissues of living organisms. PAHs enter in bodies of aquatic organisms and bioaccumulate in the food chain and are transferred to high trophic levels *via* biomagnification (Kipopoulou *et al.*, 1999).

Effect on Planktonic Organisms

Marine phytoplanktons are the most important primary producers in a marine ecosystem. Some of the phytoplanktons like *Chlamydomonas, Chlorella, Cyclotella, Navicula* can decompose and transform PAHs and other hydrocarbons (Semple *et al.*, 1999). PAHs get accumulated in phytoplankton cells resulting in suppressed photosynthesis by reducing photochemical yield and electron transport capacity (Liu *et al.*, 2009). It may also block nutrient uptake leading to decreased chlorophyll a and primary productivity (Sargian *et al.*, 2007). Researchers have also shown that PAHs may destroy the membrane system and the cell structures of microalgae (Sikkema *et al.*, 1995).

Zooplankton, an important part of the marine food web, influence the primary productivity by top-down effect (Yang *et al.*, 2006) and their population dynamics influence other marine macroorganisms by bottom-up effect (Beaugrad *et al.*,

2003) as they are important food resource for small and big fishes like baleen whales. Zooplankton like copepods, mysids, and euphausiids directly assimilate PAHs from seawater or by ingestion of contaminated foods, which may affect the fluidity of membrane lipids (Di Toro *et al.*, 2000; Barata *et al.*, 2005). Feeding, spawning, growth, and development may also get disturbed. Zooplankton having high body fat accumulate PAHs in their body.

PAHs will accumulate in the original stages of embryonic development of marine fauna like eggs and larvae through yolk of fertilized eggs having high lipid content. During the development of the PAHs contaminated embryo, PAHs will disperse in embryo resulting in deleterious effects to the physiological and biochemical processes of early embryonic development (Wassenberg and Giulio 2004), followed by organ formation and differentiation (Alvarez-Guerra *et al.*, 2008). Since the early life stages of marine organisms involve most of the vital developmental process larvae and fertilized eggs of marine organisms from the PAHs contaminated regions are most widely used for the study and toxicity test (Carls *et al.*, 1999; Wassenberg and Giulio, 2004; Bellas *et al.*, 2008; Weinstein and Garner, 2008).

Mechanisms of toxicity of PAHs in marine plankton include non-acute anaesthesia, the formation of adduct, free reactive oxygen radicals and endocrine disruption. These different mechanisms are dependent on different PAHs compounds.

Effect on Benthic Organisms

Benthic organisms like invertebrates and higher organisms, which are dwelling and filtering sediments are more susceptible to contamination than other taxa due to lack of mixed function oxygenase (MFO) system, which make them unable to metabolize PAHs and similar compounds to excretable polar metabolites. Other higher organisms like otters, walrus *etc.* which rely on bivalve molluscs are likely to get PAHs by ingestion of contaminated bivalves (Andral *et al.*, 2011). Surprisingly, marine crustaceans possessing well developed MFO have shown a high capability to readily metabolize and excrete PAHs (Lee 1981). As marine invertebrates like amphipods show quite sensitivity towards PAHs, they have studied as a model for biomarkers like catalase, acetylcholinesterase, malondialdehyde and DNA adducts, which are related to the metabolism of PAHs contaminants.

In addition to the above, corals and the organisms residing in and around the coral reefs have a high risk of exposure to PAHs and other toxic contaminants, sometimes resulting in smothering of those organisms. Deteriorated and bleaching of corals by a variety of contaminant has resulted in substantial change/ loss of coral and its related diversity around the globe.

Most of the fish do not accumulate or retain PAHs and other hydrocarbons due to well advanced hepatic mixed function oxygenase (MFO) system (Lee 1981), which facilitate the formation of less toxic metabolites during digestion resulting in a reduced transfer of contaminants in further food chain.

Marine mammals spend quite an amount of time on the surface for swimming, breathing, feeding and resting. Thus, likely to come in contact with oil slicks, tar-balls, and water-in-oil emulsions, *etc*. Mammals, *e.g*. skim-feeder baleen when obtaining their food through seawater surface in the contaminated region may lead to coat on baleen plates causing foul in feeding (Wursig *et al.*, 1985 and Neff 1988; Saadoun, 2015). Mammalian species like walrus, gray whale and some seals, which feed heavily on benthic organisms have chances of gaining contamination from their food.

Volatile PAHs, such as naphthalene and phenanthrene *etc*. can be inhaled by these organisms, which may get transferred to their bloodstream from where it may get accumulate in brain, lung, and liver, causing neurological disorder and liver malfunctioning (Geraci and Aubin, 1982; Saadoun, 2015).

Effect of Marine PAHs Contamination on Human Health

The prime itinerary of marine PAHs contamination to human is through seafood. As described earlier, benthic organisms like mussels, clams and lobsters *etc*. devour PAHs from sediments and pore-water, which in turn are consumed directly by human or by higher organisms in food chain like fishes, squids, and octopus which also used by human in their diet (Chen *et al.*, 2012). PAHs which get through human bodies accumulate in body organs, in particular, organs having adipose tissues.

All individual PAHs exert their distinct effects on a living being. PAHs are renowned for being carcinogens, mutagens, and teratogens posing a great threat to human health. Some of the PAHs are classified by International Agency for Research on Cancer (IARC monograph 2010) in various groups based on the type of carcinogenicity they exert as known carcinogen (Group 1) including Benzo(a)pyrene; possibly carcinogen (Group 2A) including naphthalene, benzo- and dibenzo- anthracenes and probably carcinogen to humans (Group 2B) including chrysene, dibenzo varieties of pyrene (IARC monogram 2010). On the bases of their toxicity and abundance in the environment, United States Environmental Protection Agency (USEPA) has listed 16 PAH compounds as priority pollutants (Kanaly and Harayama, 2000).

The health impacts of PAHs are dependent on a variety of factors like length and rout of exposure, the concentration of PAHs and their relative toxicity. Along with these, age, health history of the individual exposed to PAHs also plays a role in the severity of PAHs toxicity (ACGIH, 2005).

In countries having sea coasts, fisheries are an important source of income. Experimental studies have shown that frying, smoking, grilling and barbequing contaminated seafood may increase the concentration of PAHs and has identified 22 different PAHs in food. Out of these 22 PAHs, 11 PAHs are found to be carcinogenic in an experiment involving animals but presently, no evidence of these 11 carcinogenic PAHs producing cancer effects in human is there (Lo and Sandi, 1978; Balcioğlu 2016).

PAHs may get absorbed in the human body, metabolized in the liver and kidney and finally excreted through faeces and urine. Thus, the level of the PAHs metabolites, such as 1-hydroxypyrene, hydroxyl naphthalene, and hydroxyl phenanthrene in urine, is the most commonly used biomarkers for the estimation of exposure of PAHs in humans (Balcioğlu 2016). The intensity of carcinogenicity of a PAHs compound is generally associated with its structural complexity. The reason behind the carcinogenicity of PAHs in human and animals is because of the production of highly active metabolites, which can natively bind to DNA (Sikka and Naz, 1999). In fact, most of the immunotoxic effects reported for PAH in humans are due to their reactive epoxide metabolites rather than the parent compound (Burchiel and Luster 2001; Miyata *et al.*, 2001). The route of exposure for PAHs in human can be inhalation, ingestion or dermal contact and the effects can be acute (short term) or chronic (long term).

Symptoms, such as eye/skin irritation, nausea, vomiting, diarrhoea, and confusion, are considered as acute effects of PAHs toxicity (Unwin *et al.*, 2006) and some responsible PAHs are naphthalene, anthracene, and benzo[a]pyrene. All these three are reported to be causing skin irritation and inflammation from which anthracene and benzo[*a*]pyrene are found to be skin sensitizer producing allergic reactions not only in human but in animals also (Grover *et al.*, 1976).

Long-term or chronic effects are proved to be more dangerous. Reduced immune functionality, cataracts, kidney and liver damage, breathing problems like asthma and abnormal functioning of the lung are some results of chronic toxicity of PAHs. Breakdown of red blood cells can occur by inhalation or ingestion of PAHs specifically naphthalene (Diggs *et al.*, 2011; Olsson *et al.*, 2010; Bach *et al.*, 2005). The most lethal or life-threatening results of PAHs are carcinogenicity, mutagenicity, and teratogenicity.

Carcinogenicity

Reactive metabolites formed from PAHs, such as epoxides and dihydrodiols, bind to cellular proteins and DNA leading to cell damage and disruption of the biochemistry of cell (Armstrong *et al.*, 2004). Lab scale studies of PAHs carcinogenicity involving animals have shown that test animals have developed lung, stomach and skin cancers due to inhalation, ingestion and dermal contact with PAHs for long (Grover *et al.*, 1976). The USEPA has categorized seven PAHs namely benzo(a)anthracene, benzo(a)pyrene, benzo(b)fluoranthene, benzo(k)fluoranthene, chrysene, dibenzo(a, h)anthracene, and indeno(1,2,3-c, d)pyrene as probable carcinogens to humans.

Genotoxicity and Mutagenicity

Genotoxicity plays an imperative function in the process of carcinogenicity and also in some variety of developmental toxicity (Lewtas 2007; Arlt *et al.*, 2008). Benzo(a)pyrene and benzo(a)pyrene-7,8-diol-9,10-epoxide (BaPDE) are the main PAHs model compounds used for investigating the mechanism of mutagenicity in humans and other living organisms (Spink *et al.*, 2008; Tarantini *et al.*, 2011). Base pair substitution namely The G >T changeover has been found to be induced by BaPDE in mammalian cells *in vitro*(Knasmüller *et al.*, 2004). Other damaging results includes bulky adducts of PAH to DNA bases inducing frame-shift mutations, deletions, S-phase arrest, strand breakage and a variety of chromosomal alterations (Abdel-Shafy and Mansour, 2016).

Teratogenicity

Teratogenic effects of PAHs exposure during pregnancy may result in critical and irretrievable effects on the foetus, such as cancer, low birth weight, premature delivery, heart malformations, decreased fecundity *etc.* (Dejmek *et al.*, 2000; Perera *et al.*, 2005). After birth effects include lower Intelligence Quotient (IQ), behavioural problems, mental abnormalities, asthma. Examination of the cord blood of affected babies has shown damage of DNA, which may be linked to cancer and other life-threatening effects (Edwards *et al.*, 2010; Abdel-Shafy and Mansour, 2016).

Biodegradation of PAHs

Removal or alteration of PAHs can be carried out by conventional methods, such as excavation of contaminated soil and its incineration. There are many other techniques successfully developed by the scientific community for PAHs de-gradation like adsorption, volatilization, photolysis, and chemical degradation. Apart from being expensive, these techniques transfer the PAHs from one phase to another. Besides these techniques, bioremediation transform PAHs to less or non-hazardous forms with eco-friendly and cost-effective approaches (Providenti *et al.*, 1993; Ward *et al.*, 2003). Bioremediation transforms PAHs into less hazardous or non-hazardous metabolites or mineralizes into the inorganic minerals like H_2O, CO_2 or CH_4. Therefore, bioremediation answers the limitations associated with physicochemical processes by degrading PAHs at a reduced cost. Thus, bioremediation is a popular option for PAHs removal.

Biodegradation of PAHs and its rate depends on the environmental conditions, number, and type of the microorganisms, nature, structure and molecular weight of PAHs being degraded. Thus, to devise a bioremediation system, a number of factors are to be counted for. Both bacteria and fungi have been extensively studied for their ability to degrade PAHs.

Bacterial Degradation of PAHs

Bacteria are the nature's vital scavengers that obtain energy from almost all the organic compounds using various strategies. Due to quick adaptability, they have largely been used to remediate priority pollutants, such as PAHs. 80% of microbial biodegradation studies have been devoted to bacterial biodegradation. Historically petroleum or oil contaminated sites harbour PAHs degrading microbial community to a considerable extent (Dudhagara *et al.*, 2016b). Many reports suggest the involvement of bacteria isolated from the contaminated sediments in the degradation of PAHs (Dudhagara *et al.*, 2016b; Ghevariya *et al.*, 2011; Rajapara *et al.*, 2017; Sachaniya *et al.*, 2018). Various bacteria have been found to degrade PAHs, in which degradation of naphthalene and phenanthrene, which are low molecular weight PAHs, have been most widely studied (Cerniglia, 1992; Peng *et al.*, 2008; Seo *et al.*, 2009; Mallick *et al.*, 2011).

Degradation of PAHs depends on the presence or absence of oxygen. Apart from, being electron acceptor oxygen also contributes in oxynolytic ring cleavage and as a co-substrate for the hydroxylation during aerobic catabolism. Anaerobic catabolism involves reductive reaction mechanism to attack the aromatic ring (Foght 2008). Aerobic catabolism of PAHs has been well studied by the scientific community, whereas anaerobic catabolism still needs deeper understanding.

Generally, bacteria use oxygenase mediated metabolism involving either monooxygenase or dioxygenase enzymes. The first step of oxygenase mediated metabolism involves hydroxylation of an aromatic ring by dioxygenase, converting into *cis*-dihydrodiol. This *cis*-hydrodiol rearomatized to a diol metabolite due to the action of dehydrogenase. Diol metabolites may then convert into catechol and sequentially these intermediates enter into TCA cycle by intra or extradiol ring cleaving dioxygenase enzymes. Another pathway involved in PAHs metabolism is cytochrome P450-mediated pathway, which produces trans-dihydrodiols (Moody *et al.*, 2004; Ghosal *et al.*, 2016). Under anaerobic conditions, nitrate reduction is the main mechanism involved in PAHs metabolism (Carmona *et al.*, 2009).

Low Molecular Weight PAHs Degradation

Numbers of studies have been reported for naphthalene degrading bacteria together with enzymatic mechanisms and biochemical pathways (Peng *et al.*, 2008; Lu *et al.*, 2011; Mallick *et al.*, 2011). Naphthalene catabolic genes (*nah*), present in the plasmid of *Pseudomonas putida* G7 contain two operons: *nal* operon and *sal* operon. *nal* operon contains the enzymes involved in the conversion of naphthalene to salicylate, an upper pathway, whereas enzymes present in the *sal* operon convert salicylate to pyruvate to acetaldehyde (Simon *et al.*, 1993).A *LysR* type regulator, *NahR* regulates the operon positively by inducing the *nah* genes in the bacteria in presence of salicylate (Peng *et al.*, 2008; Ghosal *et al.*, 2016). Moreover, several reports also suggest the genes encoding for upper pathway enzymes in various *Pseudomonas* strains and the sequences are identical (Simon *et al.*, 1993; Yang *et al.*, 1994; Ghosal *et al.*, 2016).

Apart from *Pseudomonas*, many *Sphingomonads*, such as *Sphingomonas, Sphingobium* and *Novosphingobium*are, also frequently reported to be PAHs degraders. Species belonging to these genera are highly versatile with having great potential to degrade high molecular weight PAHs (Basta *et al.*, 2005; Stolz 2009). Recently, a finding suggests that *Sphingomonas* strain PNB have seven sets of ring-hydroxylatingoxygenases (RHO) with various substrate specificities involved in monoaromatic and polyaromatic compounds degradation (Khara *et al.*, 2014). Generally, the degradation of PAHs is subjected to a diversity of mobile genetic elements (MGEs), such as plasmid and transposons. These MGEs are involved in the selection of mutant bacteria, which are capable of degradation of PAHs. Besides vertical gene transfer, horizontal gene transfer also increases the catabolic traits of phylogenetically diverse bacteria (Nojiri *et al.*, 2004; Ghosal *et al.*, 2016).

Among all the PAHs degrading bacterial genus *Rhodococcus* has unique characteristics. Unlike *Pseudomonas* and other Gram-negative bacteria, *Rhodococcus* have only three naphthalene degrading genes, such as *narAa, narAb,* and *narB* (Larkin *et al.*, 2005). It has been noted that *nar* region is not clustered into a single operon, and homologous and non-homologous transcription units having direct and inverted repeated sequences have been separated in different *Rhodococcus* strain, (Larkin *et al.*, 2005). Moreover, in *Rhodococcus* strains, there is the absence of genes encoding for electron transport components reductase and ferredoxin of NDO, which are present in *Pseudomonas* (Kulakov *et al.*, 2005).

Apart from naphthalene, phenanthrene degradation has been also reported numerous times by many Gram positive and negative bacteria (Peng *et al.*, 2008, Mallick *et al.*, 2011). Mallick *et al.*, (2007) claimed the degradation of phenanthrene by *Staphylococcus* sp. Strain PN/Y dioxygenation at 1,2-position, followed by *meta*-cleavage of phenanthrene 1,2-diol, leading to the formation of leading to the formation of 2-hydroxy- 1-naphthoic acid as the metabolic intermediate. *Ortho*-cleavage yields formation of naphthalene-1, 2-dicarboxylic acid. 2-hydroxy-1-naphthoic acid was metabolized by 2-hydroxy-1-naphthoate dioxygenase enzyme acting on the *meta* position leading to the formation of *trans*-2,3-dioxo-5- (20-hydroxyphenyl)-pent-4-enoic acid. This metabolite sub-sequently degraded *via* salicylic acid and catechol (Mallick *et al.*, 2007; Ghosal *et al.*, 2016). The results of Mallick *et al.*, (2007) also supported by the study of Ghosal *et al.*, 2010, which reported the phenanthrene degradation by *Ochrobactrum* sp. Strain PWTJD, which was isolated from a municipal waste soil sample. Ghosal *et al.*, (2010) also reported the phenanthrene degradation *via* 2-hydroxy-1-naphthoic acid, salicylic acid, and catechol.

Besides, naphthalene and phenanthrene other low molecular weight PAHs, such as acenaphthene, acenaphthylene, anthracene, and fluorine, are usually found in high concentrations at PAHs contaminated sites and various bacterial species have the potential to degrade these compounds as sole carbon and energy sources (Seo *et al.*, 2009; Dudhagara *et al.*, 2016a; Gosai *et al.*, 2017). Ghosal *et al.*, (2013) reported the degradation of acenaphthene by *Acinetobacter* sp. Strain AGAT-W, isolated from a municipal waste soil sample. Acenaphtene degradation was seen *via* 1-acenaphthenol, naphthalene-1,8-dicarboxylicacid, 1-naphthoicacid, salicy-licacid, and *cis* muconic acid, subsequently resulted into the TCA cycle intermediates.

High Molecular Weight PAHs Degradation

Polycyclic aromatic hydrocarbons with more than three rings are considered as high molecular weight PAHs, such as fluoranthene, pyrene, benzo [a] pyrene, benzo [a] anthracene, and dibenzo [a,k] anthracene. These high molecular weight PAHs have potential mutagenic and carcinogenic properties due to their long persistence and high toxicity and hence they are of environmental concern (Dudhagara *et al.*, 2016; Gosai *et al.*, 2017; Kanaly and Harayama, 2000). During the last two decades, study on bacterial degradation of high molecular weight PAHs has advanced considerably and many high molecular weight PAHs degrading bacterial strains have been reported (Kanaly and Harayama 2000; Peng *et al.*, 2008; Seo *et al.*, 2009).

Chrysene degrading *Achromobacter xylosoxidans* from oil-contaminated sediments of Gulf of Kachchh was isolated at M K Bhavnagar University, Bhavnagar, India, by Dave *et al.*, (2014). They have optimized the medium components and co-substrate concentrations for the enhancement of the chrysene degradation up to 40.79% (Ghevariya *et al.*, 2011). Moreover, they have also demonstrated mixed culture experiment using *Achromobacter xylosoxidans, Pseudomonas* sp. and *Sphingomonas* sp. for the degradation of chrysene up to 66.45% (Dave *et al.*, 2014). To demonstrate the applicability of the *A. Xylosoxidans* for the field level application microcosm experiment was carried out. Microcosm system constructed in the laboratory has demonstrated the degradation of PAHs up to 6 rings using *A. Xylosoxidans* by manipulating concentrations of glucose, Triton X-100 and b-cyclodextrin (Ghevariya *et al.*, 2011).

In another study, Dudhagara *et al.*, (2016b) have reported degradation of fluoranthene by *Mycobacterium litorale* isolated from the oil-contaminated marine sediments. They have reported 51.28% fluoranthene degradation after stepwise optimization protocol including a screening of medium components.

However, there is a dearth of research in high molecular weight PAHs demand more investigation on the regulatory mechanisms of high molecular weight PAHs biodegradation and changes in the bacterial community structure during biodegradation of high molecular weight PAHs degradation. These studies will be helpful to understand microbial ecology of high molecular weight PAHs degrading bacterial community structures and the outcome will also help to aid the development of effective bioremediation strategies to restore the high molecular weight PAHs contaminated sites in near future.

Fungal Degradation of PAHs

PAHs biodegradation by fungi has been extensively studied during the last two decades and many fungal species have been accounted to degrade PAHs (Cerniglia and Sutherland, 2010; Ghosal *et al.*, 2016). Although fungi cannot use PAHs as a sole source of carbon and energy, the metabolism of PAHs has resulted in a variety of oxidized products and CO_2. Unlike bacteria and algae, fungal degradation of PAHs mostly involves monooxygenases enzymes (Cerniglia and Sutherland, 2010). However, the involvement of several enzymatic pathways in fungal PAHs degradation leads to specific conditions for different fungal species. Chiefly, two types of fungi are involved in the PAHs biodegradation: (i) ligninolytic fungi or white-rot fungi and (ii) non-lignolytic fungi. Many scientific reports accounted ligninolytic fungi, such as *Phanerochaete chrysosporium*,

Laetiporus sulphurous, Flammulina velutipes, and *Pleurotus ostreatus* for PAHs degradation (Bezalel *et al.,* 1997; Sack *et al.,* 1997; Cerniglia and Sutherland 2010). Non-ligninolytic fungus *Cunninghamella elegans* has also been reported for the degradation of PAHs (Bezalel *et al.,* 1997).

Degradation of PAHs by Non-Lignolytic Fungi

Degradation pathways of polycyclic aromatic hydrocarbons by non- ligninolytic fungi is similar those formed by mammalian enzymes. The initial oxidation of PAHs by non-lignolytic fungi is carried out by the activity of cytochrome P450 monooxygenase enzymes. The cytochrome P450 monooxygenase forms an unstable arene oxide by ring epoxidation, which is transformed into *trans*-dihydrodiol (Jerina al., 1983; Sutherland *et al.,* 1995). This pathway is carried out by *Cunninghamella elegans,* a non-lignolytic fungus and *Pleurotus ostreatus,* a lignolytic fungus to metabolite PAHs (Tortella *et al.,* 2005). In the same way, pyrene is converted into pyrene *trans-* 4,5 – dihydrodiol by *P. ostreatus.* Moreover, arene oxide reshuffle into phenol derivatives by cytochrome P450 with non-enzymatic reactions and conjugated with glucose, xylose, and sulphate (Pothuluri *et al.,* 1996; Tortella *et al.,* 2005). Many fungi oxidize PAHs to epoxides and dihydrodiols, potent carcinogens than PAHs through cytochrome P450 mediated pathway. On the other hand, peroxidize mediated oxidation of PAHs produces less toxic compounds than parent PAHs, such as quinines. This is the main reason to develop strategies for the PAHs degradation by lignolytic fungi compared to non-lignolytic fungi for the restoration of historically polluted PAHs sites.

PAHs Degradation by Lignolytic Fungi

White-rot fungi, found to be ubiquitous in nature are the basin of ligninolytic enzymes, secreted extracellularly. These enzymes are involved in the degradation of lignin present in the organic substances and woods. These enzymes are categorized into the two categories viz. (i) peroxidases and (ii) laccases. Peroxidases enzymes are of two types; lignin peroxidase and manganese peroxidase, which are involved in the oxidation of PAHs (Cerniglia and Sutherland, 2010). Apart from these enzyme machineries, laccases are also the major enzymes that take part in the utilization of PAHs (Cerniglia and Sutherland, 2010). Unlike, bacterial degrading enzymes, ligninolytic enzymes are not provoked in the presence of PAHs or by their intermediates products (Verdin *et al.,* 2004).

The main advantage of lignolytic enzymes could be that they are secreted extracellularly and easily diffuse toward the immobile PAHs compared to bacterial intracellular enzymes. Hence, these enzymes can be more useful to make an initial attack on PAHs present in the soil or sediments. Compared to bacterial PAHs degrading enzymes, lignolytic enzymes have more functional diversity and substrate specificity and therefore, they are able to utilize the wide range of substrates, even with the high potential of mutagenic properties (Tortella *et al.*, 2005; Cerniglia and Sutherland, 2010). PAHs transformation occurs by producing hydroxyl free radicals by the donation of one electron, which is involved in the oxidation of PAHs rings (Sutherland *et al.*, 1995). Thus, PAHs are converted into quinones and acids instead of dihydrodiols. There are many reports suggesting that lignolytic fungi mineralize PAHs by combination with ligninolytic enzymes, cytochrome P450 monooxygenases, and epoxide hydrolases (Bezalel *et al.*, 1997). Many reports have been accounted for the degradation of PAHs by white rot fungi (Tortella *et al.*, 2005). The metabolic pathways for the degradation of phenanthrene by ligonolytic fungus, such as *Pleurotus ostreatus,* has been revealed by Bezalel *et al.*, (1997). Moreover, PAHs degradation potential of *Pleurotus ostreatus* and *Antrodia vaillantii* is also studied through mesocosm experiment using artificially contaminated PAHs as low molecular weight PAHs *viz.* fluorine, phenanthrene and high molecular weight PAHs *viz.* pyrene and benz [a] anthracene by Anderson *et al.*, 2003. However, it has been also accounted that despite being potential degrader *P. Ostreatus* was not useful for bioremediation strategies as in this process toxic PAHs metabolites accumulated. This could be due to the inhibition of indigenous microbial population inhabiting in contaminated sites by white rot fungi resulting in incomplete mineralization of PAHs (Anderson *et al.*, 2003). In contrast to white rot fungi, brown rot fungi do not produce dead-end metabolites of PAHs, although the rate of degradation is almost similar to white rot fungi. Bogan *et al.*, (1996 a,b) have reported the mineralization of pyrene, fluorene, anthracene to quinines by lignin peroxidase, and manganese peroxidase by *Phanerochaete chrysosporium.* Even very high molecular weight and highly carcinogenic PAH B[a]P complete mineralization was also reported in two-stage pilot scale using white-rot fungus *P. Chrysosporium* by May *et al.*, (1997).

Contaminated marine sediments are a reservoir of yeasts (Leahy and Colwell, 1990; Berdicevsky *et al.*, 1993). However, there is a dearth of research in the field of PAHs degradation by yeast only a few reports are available for the degradation of PAHs by marine yeasts. Yeasts utilize PAHs by oxidation with alternative carbon sources (Romero *et al.*, 1998). Yeasts capable of degrading hydrocarbons include genera of *Yarrowia lipolytica, Candida tropicalis, Candida albicans,* and *Debaryomyces hansenii* (Gargouri *et al.*, 2015). Romero *et al.*, 1998 reported *Rhodotorula glutinis,* yeast isolated from contaminated sediments as potential degraders of phenanthrene. *R. glutinis* was able to grow on 200 mg/L phenanthrene by using phenanthrene as their sole carbon source within 3-5 days of incubation time. Moreover, the rate of degradation was also found to be almost equal to the degradation by bacteria *Pseudomonas aeruginosa*. Moreover, Hagler *et al.*, 1979 and MacGillivray and Shiaris, (1993) also reported *Candida* and *Rhodotorula* as the most frequently isolated yeasts from the phenanthrene contaminated marine sediments revealing their ability to biotransform phenanthrene. In contrast to Hagler *et al.*, (1979), MacGillivray and Shiaris (1993), and Gargouri *et al.*, (2015) reported limited applicability of yeast in the field of PAHs degradation. Gargouri *et al.*, (2015) have isolated *Candida tropicalis* and *Trichosporon asahii,* which were capable of degrading n-alkanes but not able to degrade PAHs. Thus, more studies are needed to develop PAHs remediation by yeast to restore the contaminated marine environment.

Recent Advancement in Molecular Techniques in Understanding Microbial Degradation of Pollutants

During last two decades, the scientific community is extensively interested in the development of bioremediation of hazardous contaminants like PAHs, heavy metals, *etc.* hence, the modifications and development of techniques have to lead to the future directions of biodegradation of these contaminants. Despite, potential degradation capability of microorganisms, their proficiency of degradation of contaminants might not be efficient enough for cleaning up the historically polluted marine ecosystem. Thus, more research and advancement is needed to develop applicability of microbial bioremediation process at the contaminated marine environment. To improve the catabolic potential of microorganism, bioengineering is required. Hence, recent advancements in genomics, proteomics, genetic and metabolomic technologies have gathered significant increment in the knowledge of ecology, physiology and regulatory mechanisms of microbial communities involved in pollutants remediation. Hence, practical applications of this knowledge are essential to manipulate the natural processes to develop effective bioremediation strategies for pollutants deposited at contaminated marine environment (Carmona *et al.*, 2010; Ghosal *et al.*, 2016).

Advancement in the knowledge of microbial ecology subjected to identification of conserved sequences present in microorganism could provide phylogenetic correlation of the microbial population inhabiting contaminated niches (Amann *et al.*, 1995). This approach retrieves useful information because by investigating the native microflora population in the contaminated ecosystem it may be possible to predict the possible bioremediation potential of native microflora. In context to the same, with the development of denaturing gradient gel electrophoresis (DGGE) and fluorescence *in situ* hybridization (FISH), it has now been possible to evaluate microbial community structure and dynamics of contaminated marine ecosystems (Amann *et al.*, 2001). Fig. (**2**) depicts recent advancements in molecular techniques applied for improved understanding of microbial bioremediation processes. Many reports also suggest that genes responsible for biodegradation of pollutants are present but not expressed. Thus, quantification of mRNA for important catabolic genes using RT- PCR is also important (Debruyn and Sayler, 2009). In addition to that, single-cell genomics, DNA-based stable isotope probing, and DNA microarray techniques are also promising tools for development of future bioremediation strategies (Denef *et al.*, 2006; Macaulay and Voet 2014; Mishamandani *et al.*, 2014). GeoChip, a recently developed microarray technique identifies a number of genes in a single test. Culturing of novel marine microorganisms is vital, however, over last 40 years, only small fraction of total microbial diversity has been cultured using modern microbiological and molecular biology techniques. Hence, metagenomics is the perhaps the greatest approach to study the diversity involved in the bioremediation of pollutants (Hugenholtz and Pace, 1996; Zinder, 2002). Thus, whole-genome sequencing and advancement in the next generation sequencing has changed the scenario of bioremediation studies. It is now possible to study the physiology of microorganisms involved in the pollutants removal in a more specific and comprehensive manner with the applications of metagenomics (Liang *et al.*, 2011; Kappell *et al.*, 2014; Devpura *et al.*, 2017).

Isolation of pollutant degrading microorganisms subjected to the utilization of pollutants as sole carbon and energy source, while by using metagenomic approach identification of all the genes involved in biodegradation of pollutants and their phylogenetic relationships is possible.

Thus, the inclusive knowledge of catabolic potential and phylogenetic relation reduces the bottleneck in the study of biodegradation pollutants. In addition to that, metagenome sequencing also shed light upon the dominant and rare microbial community at the contaminated marine ecosystem, which can be also helpful to understand the ecological footprints at the contaminated marine ecosystem (Gosai *et al.*, 2018; In communication).

During last decade, two more technologies have also shown promises in the field of biodegradation studies of pollutants; (i) metaproteomics, the study to recognize the proteins associated with the pollutants degradation and (ii) metabolomics exploited to identify the metabolites produces during the pollutants removal studies.

In general, in near future, functional metatranscriptomics, metaproteomics, functional metagenomics, and DNA microarrays technologies could be vital approaches to reveal the mechanisms involved in the bioremediation of pollutants and will offer more insights as yet more unculturable microbial diversity attached

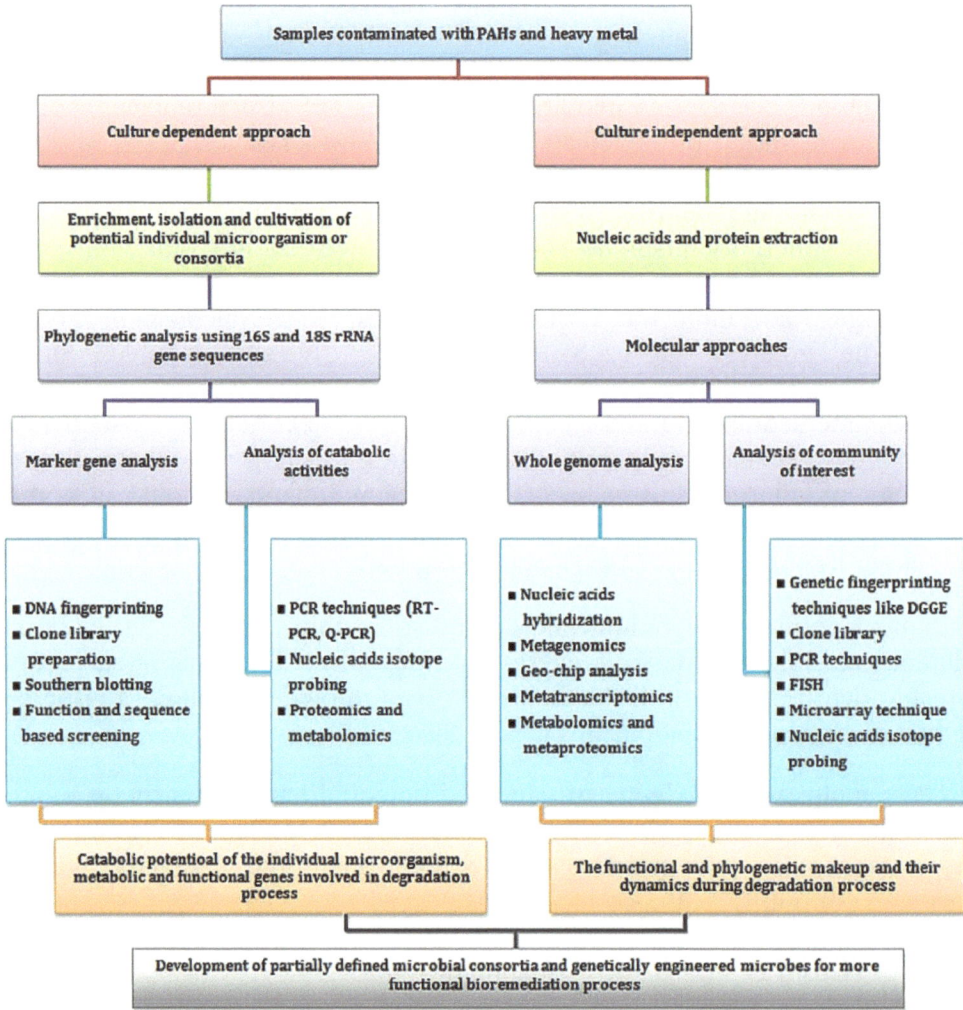

Fig. (2). Schematic diagram summarising the recent advancements in molecular techniques applied for improved understanding of microbial biodegradation processes and sustainable bioremediation of the sites contaminated with PAHs (and heavy metals) (adapted and modified from Ghosal *et al.*, 2016).

with pollutant degradation (Zhou and Thompson, 2002; Ginige *et al.*, 2004). In addition to that, *in silico* technologies are progressively used in different app-lications in bioremediation strategies (Chakraborty *et al.*, 2012). Thus, it is logical that recent advancement in molecular biology, computational biology and ana-lytical biology reduces the bottleneck involved in various aspects of biore-mediation studies and these studies should be transformed to largely empirical practice into a branch of modern science (Kweon *et al.*, 2010; Khara *et al.*, 2014).

CONCLUSION

In summary, this chapter highlighted various hazardous organic compounds through direct or indirect ways entered marine environment and caused great losses to the marine ecosystem. PAHs are emerging as critical pollutant with context to the marine environment due to some distinctive properties, which makes them POPs posing threat to the human health as well various kind of terrestrial and aquatic environment. PAHs entered to the marine environment through various natural and anthropogenic sources. Marine microorganisms have reported to be leading candidates for PAHs degradation. Recent advancements in genomics, proteomics, and metabolomics technologies have gathered significant increment in the knowledge of ecology, physiology and regulatory mechanisms of microbial communities involved in PAHs remediation. Morden technologies will be a vital approach to reveal the mechanisms involved in the bioremediation of pollutants and will offer more insights as yet more uncultivable microbial diversity attached with pollutant degradation.

CONSENT FOR PUBLICATION

Not applicable.

CONFLICT OF INTEREST

The authors confirm that this chapter contents have no conflict of interest.

ACKNOWLEDGEMENT

Earth Science and Technology Cell (ESTC), Ministry of Earth Science (MoES), Government of India (GoI), New Delhi is gratefully acknowledged for financial support.

REFERENCES

Abdel-Shafy, HI & Mansour, MS (2016) A review on polycyclic aromatic hydrocarbons: source, environmental impact, effect on human health and remediation. *Egyptian Journal of Petroleum,* 25, 107-23.
[http://dx.doi.org/10.1016/j. ejpe. 2015. 03. 011]

(2005) TLVs and BEIs Based on the Document of the Threshold Limit Values for Chemical Substances and Physical Agents & Biological Exposure Indices. *American Conference of Governmental Industrial Hygienists (ACGIH),* 8- 29.

Adhikari, PL, Maiti, K, Overton, EB, Rosenheim, BE & Marx, BD (2016) Distributions and accumulation rates of polycyclic aromatic hydrocarbons in the northern Gulf of Mexico sediments. *Environ Pollut,* 212, 413-23.
[http://dx.doi.org/10.1016/j.envpol.2016.01.064] [PMID: 26895564]

Agarwal, T (2009) Concentration level, pattern and toxic potential of PAHs in traffic soil of Delhi, India. *J Hazard Mater,* 171, 894-900.
[http://dx.doi.org/10.1016/j.jhazmat.2009.06.081] [PMID: 19615818]

Almeda, R, Wambaugh, Z, Chai, C, Wang, Z, Liu, Z & Buskey, EJ (2013) Effects of crude oil exposure on bioaccumulation of polycyclic aromatic hydrocarbons and survival of adult and larval stages of gelatinous zooplankton. *PLoS One,* 8e74476
[http://dx.doi.org/10.1371/journal.pone.0074476] [PMID: 24116004]

Alvarez-Guerra, M, González-Piñuela, C, Andrés, A, Galán, B & Viguri, JR (2008) Assessment of Self-Organizing Map artificial neural networks for the classification of sediment quality. *Environ Int,* 34, 782-90.
[http://dx.doi.org/10.1016/j.envint.2008.01.006] [PMID: 18313753]

Amann, R, Fuchs, BM & Behrens, S (2001) The identification of microorganisms by fluorescence *in situ* hybridisation. *Curr Opin Biotechnol,* 12, 231-6.
[http://dx.doi.org/10.1016/S0958-1669(00)00204-4] [PMID: 11404099]

Amann, RI, Ludwig, W & Schleifer, KH (1995) Phylogenetic identification and *in situ* detection of individual microbial cells without cultivation. *Microbiol Rev,* 59, 143-69.
[PMID: 7535888]

Andersson, BE, Lundstedt, S, Tornberg, K, Schnürer, Y, Öberg, LG & Mattiasson, B (2003) Incomplete degradation of polycyclic aromatic hydrocarbons in soil inoculated with wood-rotting fungi and their effect on the indigenous soil bacteria. *Environ Toxicol Chem,* 22, 1238-43.
[http://dx.doi.org/10.1002/ etc.56202 2060 8] [PMID: 12785579]

Andral, B, Galgani, F, Tomasino, C, Bouchoucha, M, Blottiere, C, Scarpato, A, Benedicto, J, Deudero, S, Calvo, M, Cento, A, Benbrahim, S, Boulahdid, M & Sammari, C (2011) Chemical contamination baseline in the Western basin of the Mediterranean Sea based on transplanted mussels. *Arch Environ Contam Toxicol,* 61, 261-71.
[http://dx.doi.org/10.1007/s00244-010-9599-x] [PMID: 20862467]

Arlt, VM, Stiborová, M, Henderson, CJ, Thiemann, M, Frei, E, Aimová, D, Singh, R, Gamboa da Costa, G, Schmitz, OJ, Farmer, PB, Wolf, CR & Phillips, DH (2008) Metabolic activation of benzo[a]pyrene *in vitro* by hepatic cytochrome P450 contrasts with detoxification *in vivo*: Experiments with hepatic cytochrome P450 reductase null mice. *Carcinogenesis,* 29, 656-65.
[http://dx.doi.org/10.1093/carcin/bgn002] [PMID: 18204078]

Armstrong, B, Hutchinson, E, Unwin, J & Fletcher, T (2004) Lung cancer risk after exposure to polycyclic aromatic hydrocarbons: a review and meta-analysis. *Environ Health Perspect,* 112, 970-8.
[http://dx.doi.org/10.1289 / ehp.68 9 5] [PMID: 15198916]

Bach, QD, Kim, SJ, Choi, SC & Oh, YS (2005) Enhancing the intrinsic bioremediation of PAH-contaminated anoxic estuarine sediments with biostimulating agents. *J Microbiol,* 43, 319-24.
[PMID: 16145545]

Balcıoğlu, EB (2016) Potential effects of polycyclic aromatic hydrocarbons (PAHs) in marine foods on human health: a critical review. *Toxin Rev,* 35, 98-105.
[http://dx.doi.org/10.1080/15569543.2016.1201513]

Barata, C, Calbet, A, Saiz, E, Ortiz, L & Bayona, JM (2005) Predicting single and mixture toxicity of petrogenic polycyclic aromatic hydrocarbons to the copepod Oithona davisae. *Environ Toxicol Chem,* 24, 2992-9.
[http://dx.doi.org/10.1897/ 05-18 9 R.1] [PMID: 16398138]

Basta, T, Buerger, S & Stolz, A (2005) Structural and replicative diversity of large plasmids from *sphingomonads* that degrade polycyclic aromatic compounds and xenobiotics. *Microbiology,* 151, 2025-37.
[http://dx.doi.org/10.1099/mic.0.27965-0] [PMID: 15942009]

Baumard, P, Budzinski, H, Michon, Q, Garrigues, P, Burgeot, T & Bellocq, J (1998) Origin and bioavailability of PAHs in the Mediterranean Sea from mussel and sediment records. *Estuar Coast Shelf Sci,* 47, 77-90.
[http://dx.doi.org/10.1006/ecss.1998.0337]

Beaugrand, G, Brander, KM, Alistair Lindley, J, Souissi, S & Reid, PC (2003) Plankton effect on cod

recruitment in the North Sea. *Nature,* 426, 661-4.
[http://dx.doi.org/10.1038/nature02164] [PMID: 14668864]

Bellas, J, Saco-Álvarez, L, Nieto, O & Beiras, R (2008) Ecotoxicological evaluation of polycyclic aromatic hydrocarbons using marine invertebrate embryo-larval bioassays. *Mar Pollut Bull,* 57, 493-502.
[http://dx.doi.org/10.1016/j.marpolbul.2008.02.039] [PMID: 18395228]

Berdicevsky, I, Duek, L, Merzbach, D & Yannai, S (1993) Susceptibility of different yeast species to environmental toxic metals. *Environ Pollut,* 80, 41-4.
[http://dx.doi.org/10.1016/0269-7491(93)90007-B] [PMID: 15091870]

Bertilsson, S & Widenfalk, A (2002) Photochemical degradation of PAHs in freshwaters and their impact on bacterial growth–influence of water chemistry. *Hydrobiologia,* 469, 23-32.
[http://dx.doi.org/10.1023/A:1015579628189]

Bezalel, L, Hadar, Y & Cerniglia, CE (1997) Enzymatic mechanisms involved in phenanthrene degradation by the white rot fungus *Pleurotus ostreatus. Appl Environ Microbiol,* 63, 2495-501.
[PMID: 16535634]

Bhatt, JK, Ghevariya, CM, Dudhagara, DR, Rajpara, RK & Dave, BP (2014) Application of response surface methodology for rapid chrysene biodegradation by newly isolated marine-derived fungus *Cochliobolus lunatus* strain CHR4D. *J Microbiol,* 52, 908-17.
[http://dx.doi.org/10.1007/s12275-014-4137-6] [PMID: 25359268]

Binark, N, Güven, KC, Gezgin, T & Unlü, S (2000) Oil pollution of marine algae. *Bull Environ Contam Toxicol,* 64, 866-72.
[http://dx.doi.org/10.1007/s0012800083] [PMID: 10856345]

Bogan, BW, Lamar, RT & Hammel, KE (1996) Fluorene oxidation *in vivo* by *phanerochaete chrysosporium* and *in vitro* during manganese peroxidase-dependent lipid peroxidation. *Appl Environ Microbiol,* 62, 1788-92. a
[PMID: 16535320]

Bogan, BW, Schoenike, B, Lamar, RT & Cullen, D (1996) Expression of lip genes during growth in soil and oxidation of anthracene by *Phanerochaete chrysosporium. Appl Environ Microbiol,* 62, 3697-703. b
[PMID: 8837425]

Boll, ES, Christensen, JH & Holm, PE (2008) Quantification and source identification of polycyclic aromatic hydrocarbons in sediment, soil, and water spinach from Hanoi, Vietnam. *J Environ Monit,* 10, 261-9.
[http://dx.doi.org/10.1039/B712809F] [PMID: 18246221]

Burchiel, SW & Luster, MI (2001) Signaling by environmental polycyclic aromatic hydrocarbons in human lymphocytes. *Clin Immunol,* 98, 2-10.
[http://dx.doi.org/10.1006/clim.2000.4934] [PMID: 11141320]

Carls, MG, Rice, SD & Hose, JE (1999) Sensitivity of fish embryos to weathered crude oil: Part I. Low□level exposure during incubation causes malformations, genetic damage, and mortality in larval pacific herring (*Clupea pallasi*). *Environ Toxicol Chem,* 18, 481-93.
[http://dx.doi.org/10.1002/etc.5620180317]

Carmona, M, Zamarro, MT, Blázquez, B, Durante-Rodríguez, G, Juárez, JF, Valderrama, JA, Barragán, MJ, García, JL & Díaz, E (2009) Anaerobic catabolism of aromatic compounds: a genetic and genomic view. *Microbiol Mol Biol Rev,* 73, 71-133.
[http://dx.doi.org/10.1128/MMBR.00021-08] [PMID: 19258534]

Cerniglia, CE & Sutherland, JB (2010) Degradation of polycyclic aromatic hydrocarbons by fungi.*Handbook of Hydrocarbon & lipid Microbiology* Springer Berlin Heidelberg 2079-10.
[http://dx.doi.org/10.1007/978-3-540-77587-4_151]

Cerniglia, CE (1992) Biodegradation of polycyclic aromatic hydrocarbons.*Microorganisms to Combat Pollution* Springer, Dordrecht 227-44.
[http://dx.doi.org/10.1007/978-94-011-1672-5_16]

Chakraborty, J, Ghosal, D, Dutta, A & Dutta, TK (2012) An insight into the origin and functional evolution of bacterial aromatic ring-hydroxylating oxygenases. *J Biomol Struct Dyn,* 30, 419-36.
[http://dx.doi.org/10.1080/07391102.2012.682208] [PMID: 22694139]

Chen, HY, Teng, YG & Wang, JS (2012) Source apportionment of polycyclic aromatic hydrocarbons (PAHs) in surface sediments of the Rizhao coastal area (China) using diagnostic ratios and factor analysis with nonnegative constraints. *Sci Total Environ,* 414, 293-300.
[http://dx.doi.org/10.1016/j.scitotenv.2011.10.057] [PMID: 22115615]

Dachs, J, Eisenreich, SJ & Hoff, RM (2000) Influence of eutrophication on air− water exchange, vertical fluxes, and phytoplankton concentrations of persistent organic pollutants. *Environ Sci Technol,* 34, 1095-02.
[http://dx.doi.org/10.1021/es990759e]

Dave, BP, Ghevariya, CM, Bhatt, JK, Dudhagara, DR & Rajpara, RK (2014) Enhanced biodegradation of total polycyclic aromatic hydrocarbons (TPAHs) by marine halotolerant Achromobacter xylosoxidans using Triton X-100 and β-cyclodextrin--a microcosm approach. *Mar Pollut Bull,* 79, 123-9.
[http://dx.doi.org/10.1016/j.marpolbul.2013.12.027] [PMID: 24382467]

Debruyn, JM & Sayler, GS (2009) Microbial community structure and biodegradation activity of particle-associated bacteria in a coal tar contaminated creek. *Environ Sci Technol,* 43, 3047-53.
[http://dx.doi.org/10.1021/es803373y] [PMID: 19534112]

Dejmek, J, Solanský, I, Benes, I, Lenícek, J & Srám, RJ (2000) The impact of polycyclic aromatic hydrocarbons and fine particles on pregnancy outcome. *Environ Health Perspect,* 108, 1159-64.
[http://dx.doi.org/10.1289/ehp.001081159] [PMID: 11133396]

Denef, VJ, Klappenbach, JA, Patrauchan, MA, Florizone, C, Rodrigues, JLM, Tsoi, TV, Verstraete, W, Eltis, LD & Tiedje, JM (2006) Genetic and genomic insights into the role of benzoate-catabolic pathway redundancy in *Burkholderia xenovorans* LB400. *Appl Environ Microbiol,* 72, 585-95.
[http://dx.doi.org/10.1128/AEM.72.1.585-595.2006] [PMID: 16391095]

Devpura, N, Jain, K, Patel, A, Joshi, CG & Madamwar, D (2017) Metabolic potential and taxonomic assessment of bacterial community of an environment to chronic industrial discharge. *Int Biodeterior Biodegradation,* 123, 216-27.
[http://dx.doi.org/10.1016/j.ibiod.2017.06.011]

Di Toro, DM & McGrath, JA (2000) Technical basis for narcotic chemicals and polycyclic aromatic hydrocarbon criteria. II. Mixtures and sediments. *Environ Toxicol Chem,* 19, 1971-82.
[http://dx.doi.org/10.1002/etc.5620190804]

Diggs, DL, Huderson, AC, Harris, KL, Myers, JN, Banks, LD, Rekhadevi, PV, Niaz, MS & Ramesh, A (2011) Polycyclic aromatic hydrocarbons and digestive tract cancers: a perspective. *J Environ Sci Health C Environ Carcinog Ecotoxicol Rev,* 29, 324-57.
[http://dx.doi.org/10.1080/10590501.2011.629974] [PMID: 22107166]

Dudhagara, DR, Rajpara, RK, Bhatt, JK, Gosai, HB & Dave, BP (2016) Bioengineering for polycyclic aromatic hydrocarbon degradation by *Mycobacterium litorale*: statistical and artificial neural network (ANN) approach. *Chemom Intell Lab Syst,* 159, 155-63. b
[http://dx.doi.org/10.1016/j.chemolab.2016.10.018]

Dudhagara, DR, Rajpara, RK, Bhatt, JK, Gosai, HB, Sachaniya, BK & Dave, BP (2016) Distribution, sources and ecological risk assessment of PAHs in historically contaminated surface sediments at Bhavnagar coast, Gujarat, India. *Environ Pollut,* 213, 338-46. a
[http://dx.doi.org/10.1016/j.envpol.2016.02.030] [PMID: 26925756]

Edwards, BK, Ward, E, Kohler, BA, Eheman, C, Zauber, AG, Anderson, RN, Jemal, A, Schymura, MJ, Lansdorp-Vogelaar, I, Seeff, LC, van Ballegooijen, M, Goede, SL & Ries, LA (2010) Annual report to the nation on the status of cancer, 1975-2006, featuring colorectal cancer trends and impact of interventions (risk factors, screening, and treatment) to reduce future rates. *Cancer,* 116, 544-73.
[http://dx.doi.org/10.1002/cncr.24760] [PMID: 19998273]

Ferreira, MM (2001) Polycyclic aromatic hydrocarbons: a QSPR study. Quantitative structure-property relationships. *Chemosphere,* 44, 125-46.
[http://dx.doi.org/10.1016/S0045-6535(00)00275-7] [PMID: 11444294]

Foght, J (2008) Anaerobic biodegradation of aromatic hydrocarbons: pathways and prospects. *J Mol Microbiol Biotechnol,* 15, 93-120.
[http://dx.doi.org/10.1159/000121324] [PMID: 18685265]

Gargouri, B, Mhiri, N, Karray, F, Aloui, F & Sayadi, S (2015) Isolation and characterization of hydrocarbon-degrading yeast strains from petroleum contaminated industrial wastewater. *BioMed Res Int,* 2015929424
[http://dx.doi.org/10.1155/2015/929424] [PMID: 26339653]

Geraci, JR & Aubin, DJ (1982) *Study of the Effects of Oil on Cetaceans.*Rep. U.S. Dept. of the Interior, Minerals Management Serv.Washington, DC

Ghevariya, CM, Bhatt, JK & Dave, BP (2011) Enhanced chrysene degradation by halotolerant *Achromobacter xylosoxidans* using Response Surface Methodology. *Bioresour Technol,* 102, 9668-74.
[http://dx.doi.org/10.1016/j.biortech.2011.07.069] [PMID: 21855331]

Ghosal, D, Chakraborty, J, Khara, P & Dutta, TK (2010) Degradation of phenanthrene *via* meta-cleavage of 2-hydroxy-1-naphthoic acid by *Ochrobactrum* sp. strain PWTJD. *FEMS Microbiol Lett,* 313, 103-10.
[http://dx.doi.org/10.1111/j.1574-6968.2010.02129.x] [PMID: 20964703]

Ghosal, D, Dutta, A, Chakraborty, J, Basu, S & Dutta, TK (2013) Characterization of the metabolic pathway involved in assimilation of acenaphthene in *Acinetobacter* sp. strain AGAT-W. *Res Microbiol,* 164, 155-63.
[http://dx.doi.org/10.1016/j.resmic.2012.11.003] [PMID: 23178176]

Ghosal, D, Ghosh, S, Dutta, TK & Ahn, Y (2016) Current state of knowledge in microbial degradation of polycyclic aromatic hydrocarbons (PAHs): a review. *Front Microbiol,* 7, 1369.
[http://dx.doi.org/10.3389/fmicb.2016.01369] [PMID: 27630626]

Ginige, MP, Hugenholtz, P, Daims, H, Wagner, M, Keller, J & Blackall, LL (2004) Use of stable-isotope probing, full-cycle rRNA analysis, and fluorescence *in situ* hybridization-microautoradiography to study a methanol-fed denitrifying microbial community. *Appl Environ Microbiol,* 70, 588-96.
[http://dx.doi.org/10.1128/AEM.70.1.588-596.2004] [PMID: 14711691]

González-Gaya, B, Zúñiga-Rival, J, Ojeda, MJ, Jiménez, B & Dachs, J (2014) Field measurements of the atmospheric dry deposition fluxes and velocities of polycyclic aromatic hydrocarbons to the global oceans. *Environ Sci Technol,* 48, 5583-92.
[http://dx.doi.org/10.1021/es500846p] [PMID: 24724834]

Gosai, HB, Sachaniya, BK, Dudhagara, DR, Rajpara, RK & Dave, BP (2017) Concentrations, input prediction and probabilistic biological risk assessment of polycyclic aromatic hydrocarbons (PAHs) along Gujarat coastline. *Environ Geochem Health,* •••, 1-13.
[PMID: 28801833]

Gosai, HB, Sachaniya, BK, Dudhagara, D, Panseriya, HZ & Dave, BP (2018) Bioengineering for multiple PAHs degradation using process centric and data centric approaches *(In communication).*

Gosai, HB, Sachaniya, BK, Panseriya, HZ & Dave, BP (2018) Functional and Phylogenetic Diversity Assessment of Microbial Communities at Gulf of Kachchh, India: An Ecological Footprint *In communication.*

Grover, PL, Hewer, A, Pal, K & Sims, P (1976) The involvement of a diol-epoxide in the metabolic activation of benzo(a)pyrene in human bronchial mucosa and in mouse skin. *Int J Cancer,* 18, 1-6.
[http://dx.doi.org/10.1002/ijc.2910180102] [PMID: 947857]

Guo, Z, Lin, T, Zhang, G, Zheng, M, Zhang, Z, Hao, Y & Fang, M (2007) The sedimentary fluxes of polycyclic aromatic hydrocarbons in the Yangtze River Estuary coastal sea for the past century. *Sci Total Environ,* 386, 33-41.
[http://dx.doi.org/10.1016/j.scitotenv.2007.07.019] [PMID: 17716705]

Hagler, AN, Santos, SS & Mendonca-Hagler, LC (1979) Yeasts of a polluted Brazilian estuary. *Rev Microbiol,* 10, 36-41.

Haritash, AK & Kaushik, CP (2009) Biodegradation aspects of polycyclic aromatic hydrocarbons (PAHs): a review. *J Hazard Mater,* 169, 1-15.
[http://dx.doi.org/10.1016/j.jhazmat.2009.03.137] [PMID: 19442441]

Harris, F (2012). *Global Environmental Issues.*John Wiley & Sons.
[http://dx.doi.org/10.1002/9781119950981]

Hugenholtz, P & Pace, NR (1996) Identifying microbial diversity in the natural environment: a molecular phylogenetic approach. *Trends Biotechnol,* 14, 190-7.
[http://dx.doi.org/10.1016/0167-7799(96)10025-1] [PMID: 8663938]

International Agency for Research on Cancer (2010) Some non-heterocyclic polycyclic aromatic hydrocarbons and some related exposures, IARC Press, International Agency for Research on Cancer 92

Jerina, DM (1983) The 1982 Bernard B. brodie Award Lecture. Metabolism of Aromatic hydrocarbons by the cytochrome P-450 system and epoxide hydrolase. *Drug Metab Dispos,* 11, 1-4.
[PMID: 6132787]

Jiang, YF, Wang, XT, Wang, F, Jia, Y, Wu, MH, Sheng, GY & Fu, JM (2009) Levels, composition profiles and sources of polycyclic aromatic hydrocarbons in urban soil of Shanghai, China. *Chemosphere,* 75, 1112-8.
[http://dx.doi.org/10.1016/j.chemosphere.2009.01.027] [PMID: 19201443]

Jiao, N, Herndl, GJ, Hansell, DA, Benner, R, Kattner, G, Wilhelm, SW, Kirchman, DL, Weinbauer, MG, Luo, T, Chen, F & Azam, F (2010) Microbial production of recalcitrant dissolved organic matter: long-term carbon storage in the global ocean. *Nat Rev Microbiol,* 8, 593-9.
[http://dx.doi.org/10.1038/nrmicro2386] [PMID: 20601964]

Kanaly, RA & Harayama, S (2000) Biodegradation of high-molecular-weight polycyclic aromatic hydrocarbons by bacteria. *J Bacteriol,* 182, 2059-67.
[http://dx.doi.org/10.1128/JB.182.8.2059-2067.2000] [PMID: 10735846]

Kappell, AD, Wei, Y, Newton, RJ, Van Nostrand, JD, Zhou, J, McLellan, SL & Hristova, KR (2014) The polycyclic aromatic hydrocarbon degradation potential of Gulf of Mexico native coastal microbial communities after the Deepwater Horizon oil spill. *Front Microbiol,* 5, 205.
[http://dx.doi.org/10.3389/fmicb.2014.00205] [PMID: 24847320]

Khara, P, Roy, M, Chakraborty, J, Ghosal, D & Dutta, TK (2014) Functional characterization of diverse ring-hydroxylating oxygenases and induction of complex aromatic catabolic gene clusters in *Sphingobium* sp. PNB. *FEBS Open Bio,* 4, 290-300.
[http://dx.doi.org/10.1016/j.fob.2014.03.001] [PMID: 24918041]

Kipopoulou, AM, Manoli, E & Samara, C (1999) Bioconcentration of polycyclic aromatic hydrocarbons in vegetables grown in an industrial area. *Environ Pollut,* 106, 369-80.
[http://dx.doi.org/10.1016/S0269-7491(99)00107-4] [PMID: 15093033]

Knasmüller, S, Mersch-Sundermann, V, Kevekordes, S, Darroudi, F, Huber, WW, Hoelzl, C, Bichler, J & Majer, BJ (2004) Use of human-derived liver cell lines for the detection of environmental and dietary genotoxicants; current state of knowledge. *Toxicology,* 198, 315-28.
[http://dx.doi.org/10.1016/j.tox.2004.02.008] [PMID: 15138058]

Kulakov, LA, Chen, S, Allen, CC & Larkin, MJ (2005) Web-type evolution of *rhodococcus* gene clusters associated with utilization of naphthalene. *Appl Environ Microbiol,* 71, 1754-64.
[http://dx.doi.org/10.1128/AEM.71.4.1754-1764.2005] [PMID: 15811998]

Kweon, O, Kim, SJ, Freeman, JP, Song, J, Baek, S & Cerniglia, CE (2010) Substrate specificity and structural characteristics of the novel Rieske nonheme iron aromatic ring-hydroxylating oxygenases NidAB and NidA3B3 from *Mycobacterium vanbaalenii* PYR-1. *MBio,* 1, 00135-10.
[http://dx.doi.org/10.1128/mBio.00135-10] [PMID: 20714442]

Larkin, MJ, Kulakov, LA & Allen, CC (2005) Biodegradation and *Rhodococcus*--masters of catabolic versatility. *Curr Opin Biotechnol,* 16, 282-90.
[http://dx.doi.org/10.1016/j.copbio.2005.04.007] [PMID: 15961029]

Larsson, M, Hagberg, J, Rotander, A, van Bavel, B & Engwall, M (2013) Chemical and bioanalytical characterisation of PAHs in risk assessment of remediated PAH-contaminated soils. *Environ Sci Pollut Res Int,* 20, 8511-20.
[http://dx.doi.org/10.1007/s11356-013-1787-6] [PMID: 23666634]

Leahy, JG & Colwell, RR (1990) Microbial degradation of hydrocarbons in the environment. *Microbiol Rev,* 54, 305-15.
[PMID: 2215423]

Lee, RF (1981) Mixed function oxygenases (MFO) in marine invertebrates. *Mar Biol Lett*

Lenton, TM, Held, H, Kriegler, E, Hall, JW, Lucht, W, Rahmstorf, S & Schellnhuber, HJ (2008) Tipping elements in the Earth's climate system. *Proc Natl Acad Sci USA,* 105, 1786-93.
[http://dx.doi.org/10.1073/pnas.0705414105] [PMID: 18258748]

Lewtas, J (2007) Air pollution combustion emissions: characterization of causative agents and mechanisms associated with cancer, reproductive, and cardiovascular effects. *Mutat Res,* 636, 95-133.
[http://dx.doi.org/10.1016/j.mrrev.2007.08.003] [PMID: 17951105]

Liang, Y, Van Nostrand, JD, Deng, Y, He, Z, Wu, L, Zhang, X, Li, G & Zhou, J (2011) Functional gene diversity of soil microbial communities from five oil-contaminated fields in China. *ISME J,* 5, 403-13.
[http://dx.doi.org/10.1038/ismej.2010.142] [PMID: 20861922]

Liu, X, Duan, S, Li, A, Xu, N, Cai, Z & Hu, Z (2009) Effects of organic carbon sources on growth, photosynthesis, and respiration of *Phaeodactylum tricornutum. J Appl Phycol,* 21, 239-46.
[http://dx.doi.org/10.1007/s10811-008-9355-z]

Lo, MT & Sandi, E (1978) Polycyclic aromatic hydrocarbons (polynuclears) in foods.*Residue reviews* Springer, New York, NY 35-86.
[http://dx.doi.org/10.1007/978-1-4612-6281-7_2]

Lu, XY, Zhang, T & Fang, HHP (2011) Bacteria-mediated PAH degradation in soil and sediment. *Appl Microbiol Biotechnol,* 89, 1357-71.
[http://dx.doi.org/10.1007/s00253-010-3072-7] [PMID: 21210104]

Lundstedt, S, White, PA, Lemieux, CL, Lynes, KD, Lambert, IB, Öberg, L, Haglund, P & Tysklind, M (2007) Sources, fate, and toxic hazards of oxygenated polycyclic aromatic hydrocarbons (PAHs) at PAH-contaminated sites. *Ambio,* 36, 475-85.
[http://dx.doi.org/10.1579/0044-7447(2007)36[475:SFATHO]2.0.CO;2] [PMID: 17985702]

Luo, XJ, Chen, SJ, Mai, BX, Sheng, GY, Fu, JM & Zeng, EY (2008) Distribution, source apportionment, and transport of PAHs in sediments from the Pearl River Delta and the Northern South China Sea. *Arch Environ Contam Toxicol,* 55, 11-20.
[http://dx.doi.org/10.1007/s00244-007-9105-2] [PMID: 18172566]

Macaulay, IC & Voet, T (2014) Single cell genomics: advances and future perspectives. *PLoS Genet,* 10e1004126
[http://dx.doi.org/10.1371/journal.pgen.1004126] [PMID: 24497842]

MacGillivray, AR & Shiaris, MP (1993) Biotransformation of polycyclic aromatic hydrocarbons by yeasts isolated from coastal sediments. *Appl Environ Microbiol,* 59, 1613-8.
[PMID: 8517753]

Mackay, D & Callcott, D (1998) Partitioning and physical chemical properties of PAHs.*PAHs and Related Compounds* Springer, Berlin, Heidelberg 325-45.
[http://dx.doi.org/10.1007/978-3-540-49697-7_8]

Mallick, S, Chakraborty, J & Dutta, TK (2011) Role of oxygenases in guiding diverse metabolic pathways in

the bacterial degradation of low-molecular-weight polycyclic aromatic hydrocarbons: a review. *Crit Rev Microbiol*, 37, 64-90.
[http://dx.doi.org/10.3109/1040841X.2010.512268] [PMID: 20846026]

Mallick, S, Chatterjee, S & Dutta, TK (2007) A novel degradation pathway in the assimilation of phenanthrene by *Staphylococcus* sp. strain PN/Y *via* meta-cleavage of 2-hydroxy-1-naphthoic acid: formation of trans-2,3-dioxo-5-(2'-hydroxyphenyl)-pent-4-enoic acid. *Microbiology*, 153, 2104-15.
[http://dx.doi.org/10.1099/mic.0.2006/004218-0] [PMID: 17600055]

May, R, Schröder, P & Sandermann, H (1997) *Ex-situ* process for treating PAH-contaminated soil with *Phanerochaete chrysosporium*. *Environ Sci Technol*, 31, 2626-33.
[http://dx.doi.org/10.1021/es9700414]

Meckenstock, RU & Mouttaki, H (2011) Anaerobic degradation of non-substituted aromatic hydrocarbons. *Curr Opin Biotechnol*, 22, 406-14.
[http://dx.doi.org/10.1016/j.copbio.2011.02.009] [PMID: 21398107]

Mishamandani, S, Gutierrez, T & Aitken, MD (2014) DNA-based stable isotope probing coupled with cultivation methods implicates *Methylophaga* in hydrocarbon degradation. *Front Microbiol*, 5, 76.
[http://dx.doi.org/10.3389/fmicb.2014.00076] [PMID: 24578702]

Miyata, M, Furukawa, M, Takahashi, K, Gonzalez, FJ & Yamazoe, Y (2001) Mechanism of 7,12-dimethylbenz[a]anthracene-induced immunotoxicity: role of metabolic activation at the target organ. *Jpn J Pharmacol*, 86, 302-9.
[http://dx.doi.org/10.1254/jjp.86.302] [PMID: 11488430]

Moody, JD, Freeman, JP, Fu, PP & Cerniglia, CE (2004) Degradation of benzo[a]pyrene by *Mycobacterium vanbaalenii* PYR-1. *Appl Environ Microbiol*, 70, 340-5.
[http://dx.doi.org/10.1128/AEM.70.1.340-345.2004] [PMID: 14711661]

Morillo, E, Romero, AS, Madrid, L, Villaverde, J & Maqueda, C (2008) Characterization and sources of PAHs and potentially toxic metals in urban environments of Sevilla (Southern Spain). *Water Air Soil Pollut*, 187, 41-51.
[http://dx.doi.org/10.1007/s11270-007-9495-9]

Neff, JM (1988) Composition and Fate of Petroleum and Spill-treating Agents in the Marine Environment.*Synthesis of Effect of Oil on Marine Mammals Minerals Management Service (MMS) 88-0049* Atlantic OCS Region, Canada 2-33.

Newton, A, Carruthers, TJ & Icely, J (2012) The coastal syndromes and hotspots on the coast. *Estuar Coast Shelf Sci*, 96, 39-47.
[http://dx.doi.org/10.1016/j.ecss.2011.07.012]

Nojiri, H, Shintani, M & Omori, T (2004) Divergence of mobile genetic elements involved in the distribution of xenobiotic-catabolic capacity. *Appl Microbiol Biotechnol*, 64, 154-74.
[http://dx.doi.org/10.1007/s00253-003-1509-y] [PMID: 14689248]

Olsson, AC, Fevotte, J, Fletcher, T, Cassidy, A, 't Mannetje, A, Zaridze, D, Szeszenia-Dabrowska, N, Rudnai, P, Lissowska, J, Fabianova, E, Mates, D, Bencko, V, Foretova, L, Janout, V, Brennan, P & Boffetta, P (2010) Occupational exposure to polycyclic aromatic hydrocarbons and lung cancer risk: a multicenter study in Europe. *Occup Environ Med*, 67, 98-103.
[http://dx.doi.org/10.1136/oem.2009.046680] [PMID: 19773276]

Peng, RH, Xiong, AS, Xue, Y, Fu, XY, Gao, F, Zhao, W, Tian, YS & Yao, QH (2008) Microbial biodegradation of polyaromatic hydrocarbons. *FEMS Microbiol Rev*, 32, 927-55.
[http://dx.doi.org/10.1111/j.1574-6976.2008.00127.x] [PMID: 18662317]

Perera, FP, Tang, D, Rauh, V, Lester, K, Tsai, WY, Tu, YH, Weiss, L, Hoepner, L, King, J, Del Priore, G & Lederman, SA (2005) Relationships among polycyclic aromatic hydrocarbon-DNA adducts, proximity to the World Trade Center, and effects on fetal growth. *Environ Health Perspect*, 113, 1062-7.
[http://dx.doi.org/10.1289/ehp.7908] [PMID: 16079080]

Pothuluri, JV, Evans, FE, Heinze, TM & Cerniglia, CE (1996) Formation of sulfate and glucoside conjugates of benzo [e] pyrene by *Cunninghamella elegans*. *Appl Microbiol Biotechnol,* 45, 677-83.
[http://dx.doi.org/10.1007/s002530050747]

Providenti, M, Lee, H & Trevors, J (1993) Selected factors limiting the microbial degradation of recalcitrant compounds. *J Ind Microbiol,* 12, 379-95.
[http://dx.doi.org/10.1007/BF01569669]

Rabus, R, Boll, M, Heider, J, Meckenstock, RU, Buckel, W, Einsle, O, Ermler, U, Golding, BT, Gunsalus, RP, Kroneck, PM, Krüger, M, Lueders, T, Martins, BM, Musat, F, Richnow, HH, Schink, B, Seifert, J, Szaleniec, M, Treude, T, Ullmann, GM, Vogt, C, von Bergen, M & Wilkes, H (2016) Anaerobic microbial degradation of hydrocarbons: from enzymatic reactions to the environment. *J Mol Microbiol Biotechnol,* 26, 5-28.
[http://dx.doi.org/10.1159/000443997] [PMID: 26960061]

Rajpara, RK, Dudhagara, DR, Bhatt, JK, Gosai, HB & Dave, BP (2017) Polycyclic aromatic hydrocarbons (PAHs) at the Gulf of Kutch, Gujarat, India: Occurrence, source apportionment, and toxicity of PAHs as an emerging issue. *Mar Pollut Bull,* 119, 231-8.
[http://dx.doi.org/10.1016/j.marpolbul.2017.04.039] [PMID: 28457555]

Reddy, CM, Arey, JS, Seewald, JS, Sylva, SP, Lemkau, KL, Nelson, RK, Carmichael, CA, McIntyre, CP, Fenwick, J, Ventura, GT, Van Mooy, BA & Camilli, R (2012) Composition and fate of gas and oil released to the water column during the Deepwater Horizon oil spill. *Proc Natl Acad Sci USA,* 109, 20229-34.
[http://dx.doi.org/10.1073/pnas.1101242108] [PMID: 21768331]

Romero, MC, Cazau, MC, Giorgieri, S & Arambarri, AM (1998) Phenanthrene degradation by microorganisms isolated from a contaminated stream. *Environ Pollut,* 101, 355-9.
[http://dx.doi.org/10.1016/S0269-7491(98)00056-6]

Saadoun, IMK Saadoun, IMK (2015) Impact of Oil Spills on Marine Life, Emerging Pollutants in the Environment - Current and Further Implications. In: Dr. Marcelo Larramendy (Ed.) InTech, DOI: 10.5772/60455. Available from: https://www.intechopen.com/books/emerging-pollutants- in-th--environment- current-and- further- implications/impact-of-oil-spills-on-marine-life

Sachaniya, BK, Gosai, HB, Panseriya, HZ & Dave, BP (2018) Screening of medium components for multiple PAHs degradation using marine bacterial candidates isolated from Gujarat Coast (In communication)

Sack, U, Heinze, TM, Deck, J, Cerniglia, CE, Martens, R, Zadrazil, F & Fritsche, W (1997) Comparison of phenanthrene and pyrene degradation by different wood-decaying fungi. *Appl Environ Microbiol,* 63, 3919-25.
[PMID: 9327556]

Sander, LC & Wise, SA (1997) In: Gaithersburg, MD (Ed) Polycyclic Aromatic Hydrocarbon Structure Index, US Department of Commerce, Technology Administration, National Institute of Standards and Technology

Sargian, P, Mas, S, Pelletier, E & Demers, S (2007) Multiple stressors on an Antarctic microplankton assemblage: water soluble crude oil and enhanced UVBR level at Ushuaia (Argentina). *Polar Biol,* 30, 829-41.
[http://dx.doi.org/10.1007/s00300-006-0243-1]

Semple, KT, Cain, RB & Schmidt, S (1999) Biodegradation of aromatic compounds by microalgae. *FEMS Microbiol Lett,* 170, 291-300.
[http://dx.doi.org/10.1111/j.1574-6968.1999.tb13386.x]

Seo, JS, Keum, YS & Li, QX (2009) Bacterial degradation of aromatic compounds. *Int J Environ Res Public Health,* 6, 278-309.
[http://dx.doi.org/10.3390/ijerph6010278] [PMID: 19440284]

Sikka, SC & Naz, RK (1999) *Endocrine disruptors and male infertility Endocrine disruptors Effects on male and female reproductive systems* CRC Press, Boca Raton 225-46.

Sikkema, J, de Bont, JA & Poolman, B (1995) Mechanisms of membrane toxicity of hydrocarbons. *Microbiol Rev,* 59, 201-22.
[PMID: 7603409]

Simon, MJ, Osslund, TD, Saunders, R, Ensley, BD, Suggs, S, Harcourt, A, Suen, WC, Cruden, DL, Gibson, DT & Zylstra, GJ (1993) Sequences of genes encoding naphthalene dioxygenase in *Pseudomonas putida* strains G7 and NCIB 9816-4. *Gene,* 127, 31-7.
[http://dx.doi.org/10.1016/0378-1119(93)90613-8] [PMID: 8486285]

Spink, DC, Wu, SJ, Spink, BC, Hussain, MM, Vakharia, DD, Pentecost, BT & Kaminsky, LS (2008) Induction of CYP1A1 and CYP1B1 by benzo(k)fluoranthene and benzo(a)pyrene in T-47D human breast cancer cells: roles of PAH interactions and PAH metabolites. *Toxicol Appl Pharmacol,* 226, 213-24.
[http://dx.doi.org/10.1016/j.taap.2007.08.024] [PMID: 17919675]

Stark, A, Abrajano, T, Jr, Hellou, J & Metcalf-Smith, JL (2003) Molecular and isotopic characterization of polycyclic aromatic hydrocarbon distribution and sources at the international segment of the St. Lawrence River. *Org Geochem,* 34, 225-37.
[http://dx.doi.org/10.1016/S0146-6380(02)00167-5]

Stolz, A (2009) Molecular characteristics of xenobiotic-degrading sphingomonads. *Appl Microbiol Biotechnol,* 81, 793-811.
[http://dx.doi.org/10.1007/s00253-008-1752-3] [PMID: 19002456]

Sutherland, JB, Rafii, F, Khan, AA & Cerniglia, CE (1995) Mechanisms of polycyclic aromatic hydrocarbon degradation *Microbial Transformation & Degradation of Toxic Organic Chemicals,* 269-06.

Tarantini, A, Maître, A, Lefèbvre, E, Marques, M, Rajhi, A & Douki, T (2011) Polycyclic aromatic hydrocarbons in binary mixtures modulate the efficiency of benzo[a]pyrene to form DNA adducts in human cells. *Toxicology,* 279, 36-44.
[http://dx.doi.org/10.1016/j.tox.2010.09.002] [PMID: 20849910]

Tortella, GR, Diez, MC & Durán, N (2005) Fungal diversity and use in decomposition of environmental pollutants. *Crit Rev Microbiol,* 31, 197-212.
[http://dx.doi.org/10.1080/10408410500304066] [PMID: 16417201]

Turner, JT (2015) Zooplankton faecal pellets, marine snow, phytodetritus and the ocean's biological pump. *Prog Oceanogr,* 130, 205-48.
[http://dx.doi.org/10.1016/j.pocean.2014.08.005]

Unwin, J, Cocker, J, Scobbie, E & Chambers, H (2006) An assessment of occupational exposure to polycyclic aromatic hydrocarbons in the UK. *Ann Occup Hyg,* 50, 395-403.
[PMID: 16551675]

US-EPA IRIS. United States, Environmental Protection Agency, Integrated Risk Information System http://www.epa.gov/iris/subst

Venkatesan, MI (1988) Occurrence and possible sources of perylene in marine sediments-a review. *Mar Chem,* 25, 1-27.
[http://dx.doi.org/10.1016/0304-4203(88)90011-4]

Verdin, A, Sahraoui, ALH & Durand, R (2004) Degradation of benzo [a] pyrene by mitosporic fungi and extracellular oxidative enzymes. *Int Biodeterior Biodegradation,* 53, 65-70.
[http://dx.doi.org/10.1016/j.ibiod.2003.12.001]

Wang, Z, Fingas, M & Page, DS (1999) Oil spill identification. *J Chromatogr A,* 843, 369-11.
[http://dx.doi.org/10.1016/S0021-9673(99)00120-X]

Ward, O, Singh, A & Van Hamme, J (2003) Accelerated biodegradation of petroleum hydrocarbon waste. *J Ind Microbiol Biotechnol,* 30, 260-70.
[http://dx.doi.org/10.1007/s10295-003-0042-4] [PMID: 12687495]

Wassenberg, DM & Di Giulio, RT (2004) Synergistic embryotoxicity of polycyclic aromatic hydrocarbon

aryl hydrocarbon receptor agonists with cytochrome P4501A inhibitors in *Fundulus heteroclitus. Environ Health Perspect,* 112, 1658-64.
[http://dx.doi.org/10.1289/ehp.7168] [PMID: 15579409]

Weinstein, JE & Garner, TR (2008) Piperonyl butoxide enhances the bioconcentration and photoinduced toxicity of fluoranthene and benzo[a]pyrene to larvae of the grass shrimp (Palaemonetes pugio). *Aquat Toxicol,* 87, 28-36.
[http://dx.doi.org/10.1016/j.aquatox.2008.01.002] [PMID: 18294710]

Widdel, F & Rabus, R (2001) Anaerobic biodegradation of saturated and aromatic hydrocarbons. *Curr Opin Biotechnol,* 12, 259-76.
[http://dx.doi.org/10.1016/S0958-1669(00)00209-3] [PMID: 11404104]

Würsig, B, Dorsey, EM, Fraker, MA, Payne, RS & Richardson, WJ (1985) Behavior of bowhead whales, Balaena mysticetus, summering in the Beaufort Sea: a description. *Fish Bull,* 83, 357-77.

Yang, Y, Chen, RF & Shiaris, MP (1994) Metabolism of naphthalene, fluorene, and phenanthrene: preliminary characterization of a cloned gene cluster from *Pseudomonas putida* NCIB 9816. *J Bacteriol,* 176, 2158-64.
[http://dx.doi.org/10.1128/jb.176.8.2158-2164.1994] [PMID: 8157584]

Yang, YF, Wang, Q & Chen, JF (2006) Research advance in estuarine zooplankton ecology. *Acta Ecol Sin,* 26, 576-85.

Zakaria, MP, Takada, H, Tsutsumi, S, Ohno, K, Yamada, J, Kouno, E & Kumata, H (2002) Distribution of polycyclic aromatic hydrocarbons (PAHs) in rivers and estuaries in Malaysia: a widespread input of petrogenic PAHs. *Environ Sci Technol,* 36, 1907-18.
[http://dx.doi.org/10.1021/es011278+] [PMID: 12026970]

Zeng, EY & Vista, CL (1997) Organic pollutants in the coastal environment off San Diego, California. 1. Source identification and assessment by compositional indices of polycyclic aromatic hydrocarbons. *Environ Toxicol Chem,* 16, 179-88.
[http://dx.doi.org/10.1002/etc.5620160212]

Zhou, J & Thompson, DK (2002) Challenges in applying microarrays to environmental studies. *Curr Opin Biotechnol,* 13, 204-7.
[http://dx.doi.org/10.1016/S0958-1669(02)00319-1] [PMID: 12180093]

Zinder, SH (2002) The future for culturing environmental organisms: a golden era ahead? *Environ Microbiol,* 4, 14-5.
[http://dx.doi.org/10.1046/j.1462-2920.2002.t01-3-00257.x] [PMID: 11966819]

CHAPTER 18

Tackling Marin Pollution: Final Thoughts and Concluding Remarks

De-Sheng Pei[1,*], **Muhammad Junaid**[1,2] and **Naima Hamid**[1,2]

[1] *Key Laboratory of Reservoir Aquatic Environment, Chongqing Institute of Green and Intelligent Technology, Chinese Academy of Sciences, Chongqing, 400714, China*

[2] *University of Chinese Academy of Sciences, Beijing100049, China*

Abstract: This chapter encompasses the final words and summarized the current status of marine pollution globally in terms of the distribution, potential sources, associated hazardous impacts, and possible bio-remedial measures. This chapter is divided into two sections and highlighted the viewpoints about different topics discussed in this book and the concluding remarks.

Keywords: Marine Ecosystem, Pollution Control, Public Awareness, Remedial measures, Toxicity Assessment.

SUMMARIZED VIEWPOINT

Due to the anthrophonic interventions and exploitation of resources, marine ecosystems are facing significant challenges (Lu *et al.*, 2018). The marine ecosystem is complex in nature and rich in biodiversity, where non-native species from another environment can cause significant harms in terms of its proliferation and competition for the available resources. This is one of the major problems in the marine ecosystem that needs to be addressed for protecting the innate biodiversity and to avoid economic loss. This problem has become worse due to the factors, such as the release of the blast waters by the tankers, frequent movement of the vessels, and irregular water currents. There are other natural factors, such as dissolved oxygen, temperature, pH, water circulations, and light, also affect the marine organism, but the contribution of anthropogenic pollution is most significant. Recently, the plastic pollution, including plastic debris, microplastic, and nanoplastic, is becoming a global concern that is threatening the marine ecological environment (Auta *et al.*, 2017; Rezania *et al.*, 2018).

---

[*] **Corresponding author De-Sheng Pei:** Research Center for Environment and Health, Chongqing Institute of Green and Intelligent Technology, Chinese Academy of Sciences, Chongqing 400714, China; Tel/Fax: +86-23-65935812; E-mails: peids@cigit.ac.cn and deshengpei@gmail.com

De-Sheng Pei & Muhammad Junaid (Eds.)
All rights reserved-© 2019 Bentham Science Publishers

The plastic debris may enhance the distribution and transportation of other pollutants, negatively affect the growth and survival of the organisms, and can also transfer along the food chain. Besides plastics, the other organic and inorganic pollutants that are persistent and recalcitrant in nature also induce deleterious effects on the marine ecosystem (Axel *et al.*, 2011; Lammel *et al.*, 2016; Wolska *et al.*, 2012). The intake of pollutants in the marine environment by the organism depends on different properties of the pollutants, such as partition coefficient, hydrophobicity, *etc.* Therefore, the continuous monitoring of different classes of pollutants is critically important to devise abating strategies for protecting biodiversity and reducing the burden of pollution on marine organisms. The monitoring of marine pollution can be performed through various methods, such as the use of bioindicators, biological monitoring, chemical monitoring, and isotopic analysis (Cunha *et al.*, 2017; Jr *et al.*, 2017; Wang *et al.*, 2014). However, the monitoring of real-time toxicity of pollutants to the marine life across a large spectrum is still challenging.

Recently, the integrated use of analytical, chemical, ecological, and toxicological assessment techniques is recommended for a more precise and efficient evaluation of the status and associated impacts of marine pollution. For biomonitoring and toxicology assessment, marine medaka (*O. melastigma*) has been proposed as a feasible and ideal model to investigate the estuarine and marine eco-toxicology (Kim *et al.*, 2016). *O. melastigma* possesses obvious advantages in comparison to other model animals, such as high fecundity, small size, transparent embryos, short generation cycle, the wide range of salinity adaptations, and sexual dimorphism (Bo *et al.*, 2011; Won *et al.*, 2011). Many of the previous eco-toxicological studies have employed wild-type and transgenic species of these models to elucidate the toxicogenetics endpoints and biomonitoring of organic pollutants, heavy metals, and endocrine disruptors (Huang *et al.*, 2015; Mu *et al.*, 2016). Therefore, *O. melastigma* can be used as an excellent to screen the biomarker of marine contaminants and to track the distribution of pollutants in the marine environment. In addition, the advance gene editing techniques, such as CRISPR/Cas9, can also be employed to investigate the changes in the genetic makeup of the marine organism and associated molecular pathways in response to the environmental stressors (Mu *et al.*, 2016; Xie *et al.*, 2017). This could be coupled with other genome information assessment strategies, *i.e.* transcriptome, proteome, and metabolome analysis through a feasible bioinformatics platform. The nexus of advance gene editing techniques and high throughput bioinformatic systems will provide a better understanding of molecular mechanisms and complex biological processes associated with the ecotoxicological effects of pollutants on marine species.

As far as the abatement strategies for marine pollution are concerned, bioremediation considered as the most useful and encouraging solution to control the pace of increasing marine pollution (Catania *et al.*, 2015). Microbes exhibit excellent potential to combat the stress of organic and inorganic pollutants through bioremediation (Marques, 2016; Mohanrasu *et al.*, 2018; Sakthipriya *et al.*, 2015). Despite the improvement in bioremediation efficiency for various classes of pollutants, such as PAHs, PCBs, and heavy metals, there is still room for further research in this field. Screening and isolation of microbes from diverse habitats may be useful and can assist to discover more efficient microbial strains. Apart from devising the new bioremediation techniques, there should be certain methods to enhance the bioavailability of the pollutants for assisting the microbial detoxification of the pollutants. The application of bioremediation can become commercial and profitable at large scale by genetically modifying the microbial strains for their metabolic potential.

CONCLUSION

The conservation of marine biodiversity and its biogeochemical implications are critically important for the continuation of the life cycle on planet earth. Hence, it is immensely important to devise strategies based on multidisciplinary efforts for in-depth understanding of the interaction of organic and inorganic pollutants and their different chemical species with different compartments of the marine ecosystem. Therefore, the continuous research, public awareness, outreach programs, policies making, and implementation, could prove as the building blocks for the abatement and prevention of marine pollution. In addition, the sustainable practices should be introduced regarding the treatments of agriculture, domestic, and industrial wastes to avoid the emissions of pollutants to the environment that ultimately sink in the marine resources. The industrial emissions could significantly control or reduce through making industrial and manufacturing processes much more efficient, educating the workers, and use of mitigation technologies. More international collaboration and agreements are needed in near future to combat marine pollution and preserve the endangered coastal and estuarine habitats. There is a dire need to educate the masses about the detrimental effects of marine pollution and its regional and global consequences in terms of climate change. Proactive forums also need to be established to present the novel and feasible solutions to decrease the number of pollutants entering the marine ecosystem. Lastly, scientific research must continue to elucidate the current status, toxic impacts, and available remedies for marine pollution.

CONSENT FOR PUBLICATION

Not applicable.

CONFLICT OF INTERESTS

The authors confirm that this chapter contents have no conflict of interest.

ACKNOWLEDGEMENT

The editors are grateful for the support from the CAS Team Project of the Belt and Road (to D.S.P), the Three Hundred Leading Talents in Scientific and Technological Innovation Program of Chongqing (No. CSTCCXLJRC201714 to D.S.P), the Program of China–Sri Lanka Joint Research and Demonstration Center for Water Technology and China–Sri Lanka Joint Center for Education and Research by Chinese Academy of Sciences, China(to D.S.P), and the University of Chinese Academy of Sciences (UCAS) for CAS-TWAS Scholarship (No. 2017A8018537001 to N.H).

REFERENCES

Auta, HS, Emenike, CU & Fauziah, SH (2017) Distribution and importance of microplastics in the marine environment: A review of the sources, fate, effects, and potential solutions. *Environ Int,* 102, 165-76.
[http://dx.doi.org/10.1016/j.envint.2017.02.013] [PMID: 28284818]

Möller, A, Xie, Z, Sturm, R & Ebinghaus, R (2011) Polybrominated diphenyl ethers (PBDEs) and alternative brominated flame retardants in air and seawater of the European Arctic. *Environ Pollut,* 159, 1577-83.
[http://dx.doi.org/10.1016/j.envpol.2011.02.054] [PMID: 21421283]

Bo, J, Cai, L, Xu, JH, Wang, KJ & Au, DWT (2011) The marine medaka Oryzias melastigma – A potential marine fish model for innate immune study *Marine Pollution Bulletin,* 63, 267-76.

Catania, V, Santisi, S, Signa, G, Vizzini, S, Mazzola, A, Cappello, S, Yakimov, MM & Quatrini, P (2015) Intrinsic bioremediation potential of a chronically polluted marine coastal area. *Mar Pollut Bull,* 99, 138-49.
[http://dx.doi.org/10.1016/j.marpolbul.2015.07.042] [PMID: 26248825]

Cunha, SC, Pena, A & Fernandes, JO (2017) Mussels as bioindicators of diclofenac contamination in coastal environments. *Environ Pollut,* 225, 354-60.
[http://dx.doi.org/10.1016/j.envpol.2017.02.061] [PMID: 28284552]

Huang, Q, Chen, Y, Chi, Y, Lin, Y, Zhang, H, Fang, C & Dong, S (2015) Immunotoxic effects of perfluorooctane sulfonate and di(2-ethylhexyl) phthalate on the marine fish Oryzias melastigma. *Fish Shellfish Immunol,* 44, 302-6.
[http://dx.doi.org/10.1016/j.fsi.2015.02.005] [PMID: 25687394]

Dirrigl, FJ, Jr, Badaoui, Z, Tamez, C, Vitek, CJ & Parsons, JG (2018) Use of the sea hare (Aplysia fasciata) in marine pollution biomonitoring of harbors and bays. *Mar Pollut Bull,* 129, 681-8.
[http://dx.doi.org/10.1016/j.marpolbul.2017.10.056] [PMID: 29110893]

Kim, BM, Kim, J, Choi, IY, Raisuddin, S, Au, DWT, Leung, KMY, Wu, RSS, Rhee, JS & Lee, JS (2016) Omics of the marine medaka (Oryzias melastigma) and its relevance to marine environmental research. *Mar Environ Res,* 113, 141-52.
[http://dx.doi.org/10.1016/j.marenvres.2015.12.004] [PMID: 26716363]

Lammel, G, Meixner, FX, Vrana, B, Efstathiou, CI, Kohoutek, J, Kukučka, P, Mulder, MD, Přibylová, P, Prokeš, R & Rusina, TP (2016) Bidirectional air-sea exchange and accumulation of POPs (PAHs, PCBs, OCPs and PBDEs) in the nocturnal marine boundary layer. *Atmos Chem Phys,* 9, 1-28.
[http://dx.doi.org/10.5194/acp-16-6381-2016]

Lu, Y, Yuan, J, Lu, X, Su, C, Zhang, Y, Wang, C, Cao, X, Li, Q, Su, J, Ittekkot, V, Garbutt, RA, Bush, S,

Fletcher, S, Wagey, T, Kachur, A & Sweijd, N (2018) Major threats of pollution and climate change to global coastal ecosystems and enhanced management for sustainability. *Environ Pollut,* 239, 670-80.
[http://dx.doi.org/10.1016/j.envpol.2018.04.016] [PMID: 29709838]

Marques, CR (2016) Bio-rescue of marine environments: On the track of microbially-based metal/metalloid remediation. *Sci Total Environ,* 565, 165-80.
[http://dx.doi.org/10.1016/j.scitotenv.2016.04.119] [PMID: 27161138]

Mohanrasu, K, Premnath, N, Siva Prakash, G, Sudhakar, M, Boobalan, T & Arun, A (2018) Exploring multi potential uses of marine bacteria; an integrated approach for PHB production, PAHs and polyethylene biodegradation. *J Photochem Photobiol B,* 185, 55-65.
[http://dx.doi.org/10.1016/j.jphotobiol.2018.05.014] [PMID: 29864727]

Mu, J, Jin, F, Wang, J, Wang, Y & Cong, Y (2016) The effects of CYP1A inhibition on alkyl-phenanthrene metabolism and embryotoxicity in marine medaka (Oryzias melastigma). *Environ Sci Pollut Res Int,* 23, 11289-97.
[http://dx.doi.org/10.1007/s11356-016-6098-2] [PMID: 26924701]

Rezania, S, Park, J, Md Din, MF, Mat Taib, S, Talaiekhozani, A, Kumar Yadav, K & Kamyab, H (2018) Microplastics pollution in different aquatic environments and biota: A review of recent studies. *Mar Pollut Bull,* 133, 191-208.
[http://dx.doi.org/10.1016/j.marpolbul.2018.05.022] [PMID: 30041307]

Sakthipriya, N, Doble, M & Sangwai, JS (2015) Bioremediation of Coastal and Marine Pollution due to Crude Oil Using a Microorganism Bacillus subtilis. *Procedia Eng,* 116, 213-20.
[http://dx.doi.org/10.1016/j.proeng.2015.08.284]

Wang, D, Zhao, Z & Dai, M (2014) Tracing the recently increasing anthropogenic Pb inputs into the East China Sea shelf sediments using Pb isotopic analysis. *Mar Pollut Bull,* 79, 333-7.
[http://dx.doi.org/10.1016/j.marpolbul.2013.11.032] [PMID: 24360651]

Wolska, L, Mechlińska, A, Rogowska, J & Namieśnik, J (2012) Sources and Fate of PAHs and PCBs in the Marine Environment. *Crit Rev Environ Sci Technol,* 42, 1172-89.
[http://dx.doi.org/10.1080/10643389.2011.556546]

Won, H, Yum, S & Woo, S (2011) Identification of differentially expressed genes in liver of marine medaka fish exposed to benzo[a]pyrene. *Toxicology & Environmental Health Sciences,* 3, 39-45.
[http://dx.doi.org/10.1007/s13530-011-0076-3]

Xie, SL, Junaid, M, Bian, WP, Luo, JJ, Syed, JH, Wang, C, Xiong, WX, Ma, YB, Niu, A, Yang, XJ, Zou, JX & Pei, DS (2018) Generation and application of a novel transgenic zebrafish line Tg(cyp1a:mCherry) as an *in vivo* assay to sensitively monitor PAHs and TCDD in the environment. *J Hazard Mater,* 344, 723-32.
[http://dx.doi.org/10.1016/j.jhazmat.2017.11.021] [PMID: 29154098]

SUBJECT INDEX

A

Absorption 57, 195, 239, 422

Accumulation 56, 60, 62, 68, 73, 76, 83, 85, 93, 96, 116, 120, 130, 132, 148, 175, 176, 177, 178, 179, 180, 184, 185, 186, 187, 191, 196, 197, 202, 203, 205, 206, 209, 231, 232, 234, 244, 259, 260, 265, 279, 280, 386, 397, 413, 420, 421, 422, 423, 441

Acidification 5, 12, 14, 31, 60, 64, 155, 156, 233

Acropora tenuis 101

Adaptation 16, 275, 276, 370, 385, 423, 468

Aerobic 50, 60, 119, 203, 205, 392, 442, 447, 448

Aequorea tenuis 313

Aryl hydrocarbon receptor (AHR) 6, 101, 283

Algae 13, 18, 21, 59, 60, 72, 75, 79, 104, 123, 154, 155, 156, 240, 259, 268, 270, 295, 302, 350, 352, 356, 385, 439, 450

Algal bloom 1, 23, 24, 136, 155, 156, 157, 158, 159, 259

Aluminum 3, 4, 54, 68, 69, 76, 77, 173, 185

Amphipod 296, 325, 443

Anaerobic 50, 119, 203, 442, 447, 448

Androgen 200, 282

Aquaculture 104, 105, 137, 261, 294, 344, 345, 347, 348, 359, 361, 367, 384, 385, 386, 387, 392, 395, 399, 402, 403

Arsenic (As) 3, 41, 53, 68, 71, 72, 73, 74, 75, 76, 86, 173, 185, 196, 284, 413, 414, 415, 420, 422, 423

Aspergillus sydowii 414

Assimilation 37, 40, 56, 57, 58, 60, 64, 69, 83, 85, 189, 190, 193

Atolla wyvillei 298, 311

Aurelia aurita 298, 308, 354

B

Ballast waters 294, 295, 354, 354

Balanus amphitrite 300, 327, 328

Behavior 2, 27, 83, 93, 118, 126, 127, 173, 175, 192, 193, 199, 239, 243, 265, 271, 280, 281, 386, 395

Beroe ovate 319

Benzene 263, 264, 436

Benzo*[a]*pyrene (BaP) 445

Bicarbonate 39, 40, 60

Bioaccumulation 2, 3, 4, 37, 56, 57, 68, 74, 75, 76, 82, 96, 104, 115, 117, 119, 121, 122, 127, 159, 173, 175, 178, 179, 183, 188, 189, 190, 196, 209, 245, 256, 261, 262, 270, 279, 281, 282, 283

Bioassay 98, 256, 269, 270

Bioavailability 7, 41, 42, 48, 49, 51, 52, 63, 68, 70, 76, 99, 148, 151, 174, 180, 244, 264, 276, 279, 281, 416, 425, 469

Bioindicator 93, 103, 114, 115, 123, 175, 468

Biomagnification 2, 3, 4, 56, 115, 130, 159, 177, 178, 179, 180, 184, 185, 186, 187, 442

Biomarker 1, 5, 114, 115, 127, 128, 129, 130, 131, 197, 244, 256, 269, 275, 279, 284, 285, 286, 443, 445, 468

Bioremediation 1, 6, 7, 128, 409, 410, 412, 413, 414, 415, 416, 417, 418, 420, 422, 423, 424, 425, 435, 436, 447, 450, 452, 453, 454, 455, 456, 469

Blackfordia virginica 312

Bleaching 443

Blue mussel 105, 237, 241, 360

Breeding 29, 265, 385, 400

C

Cadmium (Cd) 3, 4, 38, 41, 50, 58, 68, 69, 70, 76, 77, 78, 79, 86, 173, 186, 284, 413, 418, 419, 421

Calcium (Ca) 41, 413

Cancer 71, 85, 208, 350, 412, 438, 444, 445, 446

Carbohydrate 39, 41, 51, 179, 200, 201, 202

Carbon dioxide (CO2) 3, 40, 60, 92

Carcinus maenas 347, 348, 350, 359

De-Sheng Pei & Muhammad Junaid (Eds.)
All rights reserved-© 2019 Bentham Science Publishers

www.ingramcontent.com/pod-product-compliance
Lightning Source LLC
Chambersburg PA
CBHW050757220326
41598CB00006B/46